Coastal Research Library

Volume 28

Series Editor
Charles W. Finkl
Department of Geosciences
Florida Atlantic University
Boca Raton, FL 33431, USA

The aim of this book series is to disseminate information to the coastal research community. The Series covers all aspects of coastal research including but not limited to relevant aspects of geological sciences, biology (incl. ecology and coastal marine ecosystems), geomorphology (physical geography), climate, littoral oceanography, coastal hydraulics, environmental (resource) management, engineering, and remote sensing. Policy, coastal law, and relevant issues such as conflict resolution and risk management would also be covered by the Series. The scope of the Series is broad and with a unique cross-disciplinary nature. The Series would tend to focus on topics that are of current interest and which carry some import as opposed to traditional titles that are esoteric and non-controversial. Monographs as well as contributed volumes are welcomed.

More information about this series at http://www.springer.com/series/8795

Vanda Claudino-Sales

Coastal World Heritage Sites

Springer

Vanda Claudino-Sales
Federal University of Ceará State
Fortaleza, Ceará, Brazil

ISSN 2211-0577 ISSN 2211-0585 (electronic)
Coastal Research Library
ISBN 978-94-024-1526-1 ISBN 978-94-024-1528-5 (eBook)
https://doi.org/10.1007/978-94-024-1528-5

Library of Congress Control Number: 2018943259

Printed on acid-free paper

This Springer imprint is published by the registered company Springer Nature B.V.
The registered company address is: Van Godewijckstraat 30, 3311 GX Dordrecht, The Netherlands

To my husband Roger Morton, for his love and support
To my parents, Claudino Sales and Francy Carneiro Sales (in memoriam)

Preface

"Heritage is our legacy from the past, what we live with today, and what we pass on to future generations." With this statement, the United Nations Educational, Scientific and Cultural Organization (UNESCO) presents the world landscapes that are of outstanding value to humanity.

Supported by this understanding, the agency annually inscribes in an official list sites that – for their nature, or culture, or both – are important to human kind as a whole. The list, known as the UNESCO World Heritage List, takes into consideration the idea that World Heritage sites belong to all the peoples of the world, irrespective of the territory on which they are located.

After UNESCO, outstanding universal value is the central requirement for inscription of a site on the World Heritage List. It refers to cultural and/or natural significance, which is so exceptional as to transcend national boundaries and to be of common importance for present and future generations of all humanity.

The agency works it out using the International Convention for the Protection of the World Cultural and Natural Heritage, known as the World Heritage Convention. The World Heritage Convention was adopted by the agency in 1972 and is based on an agreement between national signatories concerning the safeguarding of the heritage sites. Up to January 2018, 193 countries signed it.

UNESCO also created a World Heritage Committee. The World Heritage Committee is responsible for the implementation of the World Heritage Convention and establishes the sites to be listed as UNESCO World Heritage sites. It also defines the use of the World Heritage Fund and allocates financial assistance upon requests from the countries that signed the convention.

The World Heritage Committee is composed of 21 state parties elected by the General Assembly of States Parties for a 4-year term. It meets annually to analyze requests of the state parties of sites to be added to the World Heritage List. The IUCN (International Union for Conservation of Nature) is the advisory body for natural heritage properties to the UNESCO World Heritage Committee. The International Council on Monuments and Sites (ICOMOS) provides the World Heritage Committee with evaluations of cultural properties. Both institutions are advisory bodies for mixed sites.

The committee also examines reports on the state of conservation of already inscribed sites and asks state parties to take action when properties are not being properly managed. The meeting also decides on the inscription or deletion of properties on the List of World Heritage in Danger.

Up to January 2018, the number of sites on the World Heritage List was 1,073. Among them, 832 are cultural sites, 206 natural sites, and 35 mixed. They are distributed in 167 countries.

This book deals with the natural and mixed properties of the UNESCO World Heritage list that are coastal. The book provides its own World Coastal Heritage list. UNESCO already has a list of marine sites, related to marine ecosystems. Up to January 2018, they are 49 in 36 countries. But there was not a coastal list associated with sites that can be terrestrial, but have segments of coastal areas in the property. This is the list prepared for this book.

The World Coastal Heritage List is composed of 84 sites, situated in 48 countries. Five among them are mixed sites; the others were inscribed in the list for their natural contents.

The country with the most coastal sites is Australia, with ten. It is followed by United States of America, with six. The United Kingdom of Great Britain and Northern Ireland and Canada have five sites each. Mexico has four sites. Brazil, Denmark, Indonesia, and Japan have three sites each. Costa Rica, Cuba, France, New Zealand, Panama, the Philippines, Russia, South Africa, and Spain have two sites. The other countries have one site each. They are Argentina, Bangladesh, Belize, Colombia, Ecuador, Finland, Germany, Honduras, Iceland, India, Italia, Kiribati, Korea, Mauritania, Namibia, the Netherlands, Norway, Palau, Portugal, Romania, Russia, Saint Lucia, Senegal, Seychelles, Solomon Islands, Sudan, Sweden, Vietnam, and Yemen. Four sites are transboundary (Bangladesh and India, Canada and United States, Denmark and the Netherlands, Finland and Sweden) and one is tripartite (Germany, the Netherlands, and Denmark).

The Coastal World Heritage sites are present in the coast of all continents with the exception of Antarctica. They also occur as islands, gulfs, and barrier reefs in all oceans. Twenty-one sites are in the Pacific Ocean and 11 in the Atlantic Ocean. The other sites are distributed in the other oceans and various seas.

The World Coastal Heritage sites present different levels of development and conservation. Three are part of the World Heritage list of sites in danger. It is worth noting that one of the motivations for the World Heritage List was the consideration that the heritage sites were threatened with destruction by traditional causes of decay and also by social and economic factors. In relation to the coastal sites, it is also necessary to consider the eminent risk of degradation and, in some cases, destruction, associated with climate change and with the rising of the ocean level related to global warming.

In light of the magnitude and gravity of the dangers to properties, UNESCO considered that it was incumbent on the international community to participate in the protection of heritage sites of outstanding universal value. This book offers the coastal community of researchers and scientists elements to deal with the knowledge of the World Coastal Heritage, with the goal of finding means to think, participate, and contribute in its preservation.

For that, we present for each of the 84 coastal sites a chapter containing an introduction with the presentation of the property with location, access, economy, population, history, and other relevant data about the geographical aspects of the site; one section with natural settings, with data about climatic conditions, geological and geomorphological aspects and biodiversity information, and data about conditions of degradation, conservation, and protection actions to preserve the sites. We hope that the scientific coastal community, as well as the people interested in the World Heritage and nature, find interest in reading it.

Fortaleza, Ceará, Brazil Vanda Claudino-Sales

Acknowledgments

This book would not have been completed without the kind help of several people, whom I thank.

First, I would like to thank my husband Roger Morton, who edited the text using his expertise as an American newspaper editor and writer. He improved the quality of the English and of the writing.

Second, I would like to thank the contribution of the authors of the photos. Their help was fundamental. They are: Judy McIlrath, Luigi Tura, Els Slots, Peter Howard, Valdir do Amaral Vaz Manso, Noam Levin, Tab Hauser, Rui Morel Carneiro, Harry Burton, Justin Watt, Robert E. Hyman, Eric Struyf, Tommy Lahti, Merete Binderup, Martin Stock, Øyvind Heen, Andrea Prida Bay, Adriana Alvim Oliveira Carvalho, Alexey Bezrukov, Roger del Moral, Peter Ryan, Harry Jans, Miguel Mir-Gual, Achmad Ariefiandy, Morten B. Andersen, Gerardo Rodriguez Quiroz, Jorge Cortés Núñez, Fernando Zapata, Peter Ryan, Christian Reepmeyer, Geoffrey Clark, Chris Roher, Andy Collins, Maria Leoni Gaffet, Roger G Dolorosa, Gary Lehman, Giordano J. Cervera, Paddy Ryan, Dave Warburton, Bram de Bruim, Maria Cristina Pulliero, Jack Shick, Vadim Semjonov, Diane Trotter, Melanie Lancaster, Sandra Mesquita, and Alex Liu.

I also thank the following institutions: Parks Australia, Kakadu National Park and Department of Enviroment, Australia; Joggins Fossil Institute and Société des Établissements de Plein Air du Québec (Sépac), Canada; United States National Park Service and National Oceanic and Atmospheric Administration (NOAA), USA; Department of Conservation Papa Atawhai, New Zealand; Historic Scotland, UK; Korea Tourism Organization, Korea; Docksta Havet, Sweden; Institut de Recherche pour le Développment (IDR), France; Dockstahavet, Sweden; Oro Verde Tropical Forest Foundation, Germany; Sistema Nacional de Areas Protegidas – SNAP, Cuba; Shiretoko Nature Foundation, Japan; and Seychelles Islands Foundation, Seychelles. I also used pictures of Shutterstock Stock Photos, Alamy, National Geographic, iStock, and Fotostock/Visual Photos. All pictures are used with permission.

Third, I would like to thank the geographer and cartographer Andrea Castro Panizza. Besides taking care of the design of the maps, she performed additional research and provided suggestions on how to produce the best illustrations possible of specific locations. I also would like to thank the geographer Lucas Lopes, who helped with the research online in the beginning of the work, he was very helpful.

Finally, I would like to thank Dr. Charles Finkl for believing in the book project since the beginning. He was also very helpful during the years while I was working on the book's script.

Fortaleza, Ceará, Brazil Vanda Claudino-Sales

Contents

Part XII World Coastal Heritage in Islands, Gulfs and Reefs: Southern Ocean

Part XIII World Coastal Heritage in Islands, Gulfs and Reefs: Caribbean Sea

Abbreviations

AD	*Anno Domini* ("in the year of our Lord Jesus Christ")
BC	Before Christ
BP	Before Present (1 January 1950)
Ca	Circa ("more or less")
IUCN	International Union for Conservation of Nature
UNESCO	United Nations Educational, Scientific and Cultural Organization
UNEP	United Nations Environmental Program
WCMC	World Conservation Monitoring Center
WWF	World Wildlife Fund
WW	World War

Part I
World Heritage on Continental Coasts: Africa

Chapter 1
Banc D'Arguin National Park, Mauritania

Abstract The Banc d'Arguin National Park, on the Atlantic coast of Mauritania, is formed of sand dunes, coastal swamps, small islands, and shallow coastal waters. It is one of the most important habitats of the western Atlantic zones and one of most important in the world for nesting birds. The park safeguards marine and coastal environments which are rich and diverse and supports important communities of fish and marine mammals. The contrast between the harsh desert environment and the biodiversity of the site has resulted in a landscape of exceptional contrast and outstanding natural significance. Due to the vast dune fields and the arid weather in the coastal area, it is considered the place where the Sahara meets the ocean.

1.1 Introduction

Bordering the Atlantic Ocean along the coast of Mauritania, the Banc d'Arguin National Park is a natural reserve established to protect natural resources and fisheries, as well as aesthetically valuable geological features. It stretches about 160 km north-south, covering more than 12,000 km^2 of surface. Its area is 50% marine and 50% terrestrial. The boundary extends a maximum of 60 km into the shallow sea and 35 km inland into the Sahara (UNESCO WHC 2016) (Fig. 1.1).

The rectangular boundaries of the park suggest that they mostly correspond to administrative limits and are not solely based on ecological criteria. The site is about 250 km from the capital, Nouakchott, in a drive northward along a paved road. Inside the park, there are only unpaved desert roads (UNEP/WCMC 2011).

The Banc d'Arguin National Park was created by the Mauritanian government in 1976. It was designated a Wetland of International Importance under the Convention of Wetlands (the Ramsar Convention) in 1982. It became a World Natural Heritage Site in 1989. The park administration is linked directly to the president of the country, and it has an important political role in the nation (UNEP/WCMC 2011).

The Banc d'Arguin National Park is inhabited by approximately 1000 tribesmen, the Imragen. The Imragen live in seven villages distributed in the property. They maintain old lifestyles, based almost exclusively on artisanal fishing with the use of traditional sailing boats (UNEP/WCMC 2011). Fishing techniques are characterized

V. Claudino-Sales, *Coastal World Heritage Sites*, Coastal Research Library 28,
https://doi.org/10.1007/978-94-024-1528-5_1

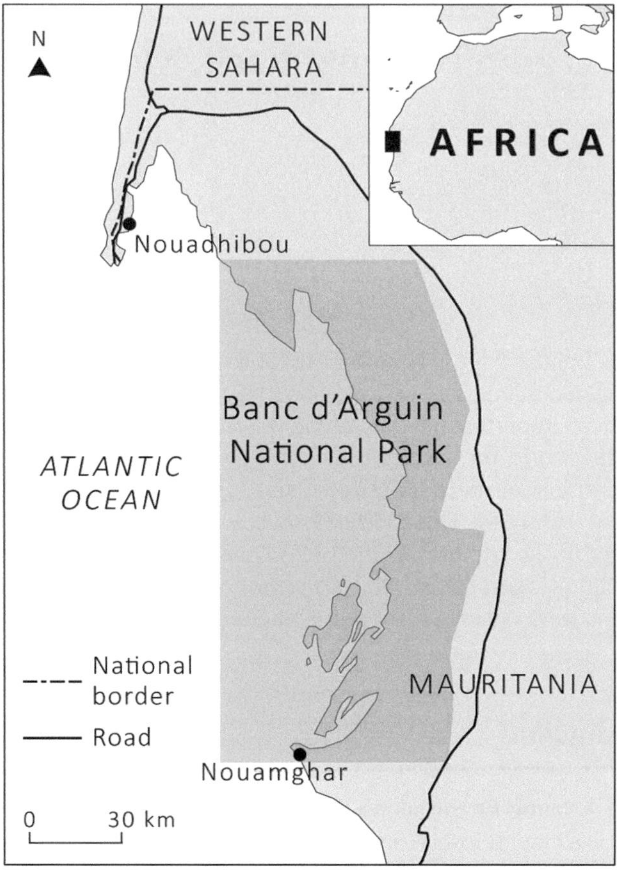

Fig. 1.1 Location of the Banc d'Arguin National Park, in Mauritania

by the symbiotic relationship with dolphins to catch schools of fish, unchanged since first recorded by fifteenth-century Portuguese explorers (Afrol 2001).

1.2 Natural Settings

1.2.1 Climatic Conditions

The park is an example of the interaction of desert and ocean. This fact, together with the latitude, controls climatic conditions.

The site lies on the boundary between temperate and tropical climates. The coastal area and the hinterland, nevertheless, are hyper-arid (Goudie and Selly 2011). The prevailing winds are the northeast trade winds. The wind speed can reach up to 29 km/h (UNEP/WCMC 2011).

The rainfall is irregular and very low, averaging 34–40 mm annually. The temperatures are seasonally stable. The cold season, from January to May, has a mean minimum of 8 °C. The hot season, from August to October, has a mean temperature in September of 32 °C (Monod 1977).

1.2.2 Geology and Geomorphology

The Banc d'Arguin National Park is located in a gulf with the presence of various islands, such as Tidra Island, at 8 km by 35 km, and 14 other smaller islands, which are up to 1 km wide and 5 km long. Four of them expose rocky outcrops (UNEP/WCMC 2011). There is also a large segment of coastal area.

The gulf is a semi-enclosed basin in a stable tectonic context, interpreted as a landlocked freshwater lake during the postglacial sea-level rise under wet climate (Aleman et al. 2014). Its sedimentary filling took place during the end of the postglacial transgression: it was flooded ca. 8700 years BP, creating a marine-estuarine environment. The aridification and the sea-level stabilization from 6500 years BP allowed the deposition of aeolian sand, with marine biogenic carbonates (Aleman et al. 2014).

The marine area is very shallow, at low tide only 5 m deep, even far offshore. The range of spring tides is 2.1 m, of neap tides 0.6 m, and of normal tides, 1.2–1.7 m (Hughes and Hughes 1992). Submarine banks extend up to 80 km from the coast, originating from the combination of aeolian transport from the desert and relict estuaries (SCBA 1993).

The coast is flat, with heights from 5 to 15 m. The shoreline exposes sandstone cliffs shaped in Miocene-Pliocene sediments of the Mauritania basin, which rise to 15 m (Lancaster et al. 2002) (Fig. 1.2). Inland, the arid landscape is characterized by linear dunes (Fig. 1.3) and dry salt lakes (sebkas) (Hughes and Hughes 1992; Lancaster et al. 2002). In the park, there are 590 km^2 of tidal swamps and marshland and 630 km^2 of mudflats (UNEP/WCMC 2011). In the north segment, the mudflat works as a functional delta derived from fossil estuaries (Diop et al. 2014).

1.2.3 Biodiversity

One hundred ninety species of plants are found in the park. The vegetation of the shallow water is characterized by vast expanses of seagrass. The seagrass anchors the mud substrate, producing oxygen and shelter to huge communities of algal epiphytes, as well as to a rich invertebrate fauna, such as mollusks and crustaceans. The shallows and islands are also the center of intense biological activity: there are 45 fish species and 11 species of shellfish. This combination provides the biggest fish feeding and spawning area in West Africa (PNBA 1986; Ramsar 2001; UNESCO WHC 2016).

Fig. 1.2 Sandstone seacliffs in the coastal area of Bank d'Arguin National Park, showing notches and coastal retreat. (Picture reproduced by permission of Peter Howard, website http://www.AfricanWorldHeritageSites.org)

Fig. 1.3 Sand dunes in the coast of Banc d'Arguin National Park, topping the sandstone seacliffs. The dunes extend inland. (Picture reproduced by permission of Peter Howard, website http://www.AfricanWorldHeritageSites.org)

The vegetation of the coastline, mudflats and islands is halophytic. There are approximately 1400 ha of white mangrove swamp on emergent mudbanks and 1700 ha in bays on the mainland. The mangrove stands on muds dating from the time when coastal wadis carried freshwater from the Sahara (Ramsar 2001). Cordgrass marsh grows behind the mangroves (Hughes and Hughes 1992).

The terrestrial vegetation is Saharan with a limited Mediterranean influence. The sand dunes are dominated by adapted psammophila species, sometimes with herbaceous strata (PNBA 1987; Ramsar 2001).

Of the estimated 7 million shorebirds which use the East Atlantic flyway, approximately 30% winter at Banc d'Arguin, coming from Northern Europe, Siberia, and Greenland (UNESCO WHC 2016). It has the world's largest concentration of wintering shorebirds and diversified communities of piscivorous birds (Hoffmann 1988).

At least 249 bird species have been recorded, from both Palaearctic (especially the migratory waders) and Afrotropical realms (IUCN/WWF 1989). The wintering shorebirds number more than 2 million, with 40,000 pairs of breeding birds (Ramsar 2001; IUCN 2002).

Mammals, terrestrial and marine, find habitat in the park. A small population of about 130–150 of one of the world's rarest animals, the monk seal *Monachus monachus*, lives at Cap Blanc, in the northern segment. Reptiles, such as marine turtles, are also present (UNESCO WHC 2016).

1.3 Conservation Challenges

The law gives the Imraguen the exclusive right to use the park's fishery, in order to allow the maintenance of their traditional style of life. Nevertheless, there is overfishing on official permits by industrial-scale fleets and mechanical shellfish harvesting by licensed foreign companies outside the park and fishing by poachers inside the park (UNEP/WCMC 2011). These activities are degrading the seabed and severely depleting fish stocks, which may cause the decline of breeding colonies of fish-eating birds (UNEP/WCMC 2011).

The overfishing is the biggest environmental problem at the Banc d'Arguin National Park. In 2013, an international agreement was signed between the European Union and the Mauritanian Fisheries. The agreement defined that the European Union would invest to protect the Banc d'Arguin fish resources that could be exploited by its fishing fleet (Thomas et al. 2013). This international agreement, the first for marine ecosystem service (e.g., a direct and indirect contribution of ecosystems to human well-being) in the world, may help to preserve the local environment (Thomas et al. 2013).

Inland, desertification and hunting are also potential concerns. Shoreline pollution by oil is also a risk, since the bank is down-current from the oil terminals on the northern coast (IUCN 2008). The possible impact of climate change has yet to be studied.

An updated report about the state of conservation of the site (Osipova et al. 2017) considered that it presents significant concerns. Its values are threatened by the mentioned factors, with significant additional conservation measures being required to preserve the values over the medium to long term.

References

Afrol.com (2001) Mauritania, the background – the Imraguen of Banc d'Arguin. African News Agency. www.afrol.com. Accessed 17 Dec 2015

Aleman N, Certain R, Barusseau JP, Courp T, Dia A (2014) Post-glacial filling of a semi-enclosed basin: the Arguin Basin (Mauritania). Mar Geol 349:126–135

Diop S, Barusseau JP, Descamps C (eds) (2014) The Land/Ocean Interactions in the Coastal Zone of West and Central Africa. Springer, Cham

Goudie A, Selly M (2011) World heritage desert landscapes: potential priorities for the recognition of desert landscapes and geomorphological sites on the world heritage list. IUCN, Gland

Hoffman L (1988) The Banc d'Arguin: nature in harmony with man. WWF News June/July:10–11

Hughes R, Hughes S (1992) Repertory of African wetlands. Report, IUCN, Gland, Switzerland

IUCN (International Union for the Conservation of Nature) (2002) Report on the state of conservation of natural and mixed sites inscribed on the world heritage list. Rapport, Gland, Switzerland

IUCN (International Union for the Conservation of Nature) (2008) State of conservation report Banc d'Arguin National Park (Mauritania). Rapport, Gland, Switzerland

IUCN/WWF (International Union for the Conservation of Nature/World Wild Life) (1989) Technical evaluation: Banc d'Arguin National Park. Rapport, Gland, Switzerland

Lancaster N, Kocurek G, Singhvi A, Pandey V, Deynoux M, Ghienne JF, Lo K (2002) Late Pleistocene and Holocene dune activity and wind regimes in the western Sahara Desert of Mauritania. Geology 30(11):991–994

Monod T (1977) Le Parc National du Banc d'Arguin et l'histoire, le cadre général, flore et végétation du PNBA. In: Association de Soutien du Parc National du Banc d'Arguin. Richesse du Parc National du Banc d'Arguin (Mauritanie). Rapport, Versailles

Osipova E, Shadie P, Zwahlen C, Osti M, Shi Y, Kormos C, Bertzky B, Murai M, Van Merm R, Badman T (2017) IUCN world heritage outlook 2: a conservation assessment of all natural world heritage sites. IUCN, Gland

PNBA (Parc National du Banc d'Arguin) (1986) Le Parc National du Banc D'Arguin. Milieu Vivant. Rapport, Nouadhibou, Mauritania

PNBA (Parc National du Banc d'Arguin) (1987) Plan de Développement Integré 1988–1992. Rapport, Nouadhibou, Mauritania

Ramsar (2001) Draft information sheet IMR001. Ramsar Convention Bureau. Rapport, Gland, Switzerland

SCBA (Scientific Council of the Banc d'Arguin) (1993) Research Master plan for the Parc National du Banc d'Arguin. Rapport, La Tour du Valat, France

Thomas B, Failler P, Chavance PN, Mayif MA (2013) First international payment for marine ecosystem services: the case of the Banc d'Arguin National Park, Mauritania. Glob Environ Chang 23(6):1434–1443

UNEP/WCMC (United Nations Environmental Programme/World Conservation Monitoring Centre) (2011) Banc d'Arguin National Park Mauritania. Rapport, Cambridge

UNESCO WHC (United Nation Educational, Scientific and Cultural Organization World Heritage Centre) (2016) Banc d'Arguin National Park. http://whc.unesco.org/en/list/506. Accessed 14 Dec 2015

Chapter 2
Namib Sand Sea

Abstract The Namib Sand Sea is located in the arid African coast of the South Atlantic, in Namibia. It is the only coastal desert in the world that includes large dune fields, among the biggest in the world, influenced by fog. Fog is the primary source of water in the site, accounting for a unique environment composed of spectacular dune accumulation of all types, with associated landforms such as inselbergs, pediplains, playas, gravel plains, coastal flats, rocky hills, coastal lagoons, and ephemeral rivers. The Namib is an extremely dry desert, but it supports a remarkable diversity of unique plants and animals that have evolved to live in this extraordinarily hostile yet beautiful environment.

2.1 Introduction

The Namib Sand Sea, in the arid coast of Namibia, covers 30,000 km^2, within Namibia's Namib-Naukluft Park, a 50–160-km-wide area on the South Atlantic (UNEP 2013) (Fig. 2.1).

The Namib Sand Sea has been under conservation management for more than 50 years with well-established management and resource allocation systems, based on regularly revised and updated management plans and long-term budgetary planning (UNESCO WHC 2016). Prior to establishment of conservation management, the area was protected for its potential as a diamond-mining area, but this was never realized (UNEP 2013). It became a World Heritage site in 2013.

There are virtually no roads or facilities of any kind within the world heritage site (African Natural Heritage 2013). Permanent visitor and management infrastructure is nonexistent within the boundaries of the property, and visitation is restricted to small, temporary point locations (UNESCO WHC 2016). There is limited use of desert plants and animals around the fringes of the property by indigenous communities and a growing tourism industry (African Natural Heritage 2013).

V. Claudino-Sales, *Coastal World Heritage Sites*, Coastal Research Library 28,
https://doi.org/10.1007/978-94-024-1528-5_2

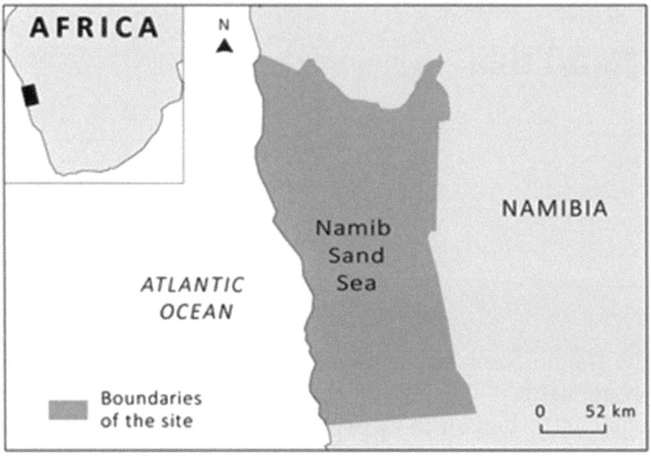

Fig. 2.1 The Namib Sand Sea, Namibia, Western Africa

2.2 Natural Settings

2.2.1 Climatic Conditions

The cold Benguela Current runs along the South West African coast and suppresses rainfall over the continent, creating one of the most arid habitats on earth (Norgaard and Dacke 2010). It is also the origin of the fog that characterizes the region, which can reach as much as 100 km inland (Pietruszka and Seely 1985). The fog brings the only water from the atmosphere to the desert of Namibia and occurs approximately 30 days per year, in the summer of the Southern Hemisphere (Lancaster et al. 1984). This climatic pattern creates a cold, foggy coastal hyper-arid desert (Goudie and Viles 2015) (Fig. 2.2).

Winds in the vicinity of the Namib Sand Sea are high-energy unimodal near the coast, becoming bimodal or complex inland (Lancaster 1985).

2.2.2 Geology and Geomorphology

Buffeted as is it by large waves, strong winds, and rapid longshore drift of sediments, the south coast of Namibia is a highly dynamic environment, with ever-changing spits and lagoons (Lancaster 1989). Landforms such as inselbergs, pediplains, playas, gravel plains, coastal flats, rocky hills, and ephemeral rivers characterize the area (Lancaster 1989). Nevertheless, the large dunefields, among the biggest in world, are the most impressive features of this coast (Goudie and Viles 2015).

Unlike most of the world's desert dunefields, which are derived from bedrock in situ, the Namib Sand Sea is derived from material transported from thousands of

Fig. 2.2 Coastal desert in Namibia – the Namib Sand Sea – in the South Atlantic coast of Africa (Skeleton Coast). The desert is hyper-arid and characterized by the presence of fog in the summer. (Picture reproduced by permission of Marzia Franceschini/Shutterstock Stock Photos)

kilometers: sand is carried to the coastal sand desert from the interior of Southern Africa by river, coastal currents, and wind (Bluck et al. 2007).

The process begins with erosion of material inland, in the headwaters of the Orange River, which is carried into the South Atlantic, where it is picked up and driven northward by strong ocean currents (UNEP 2013). It is deposited as beach sand and then mobilized and transported inland by wind, where it creates the diversified aeolian desert landforms and features that characterize the Namib Sand Sea (Bluck et al. 2007). The coastal desert sands are the response to these three process systems (river, ocean currents, and wind transport) 4785\e44A/.

The site is composed of two dune systems, an ancient semi-consolidated and fossilized one overlain by a younger active one (Goudie and Seely 2011). The dunes are mostly red, due to the presence of hematite; the redness increases inland, and it seems to be an answer to the source of the material, as well as of climatic gradient and/or abrasion of coating during transport (Walden et al. 1996).

There is an overall decrease in wind energy and sand transport rates from south to north and west to east, such that sand moves from coastal and southern source areas to accumulate in the northern and central parts of the sand sea (Livingstone et al. 2010; Lancaster 1989). Such a pattern can explain much of the variability in dune types and sizes. The most common types of dunes are linear, barchans (Fig. 2.3), and stars (Bourke and Goudie 2009; Bristow et al. 2007; Lancaster 1989).

The Namib Sand Sea is considered to be one of the oldest desert regions – over a million years – with a Tertiary-aged fossil desert underlying the modern sand sea

Fig. 2.3 Aerial view of barchan dunes in the Namib Sand Sea, Namibia. (Picture reproduced by permission of Frantisek Staud/Shutterstock Stock Photos)

(Stone 2013). The depositional ages for dune sediments vary from marine isotope stage (MIS) 5 (between 130,000 and 80,000 years ago), the Pleistocene around the Last Glacial Maximum (around 20,000 years ago), and the Holocene (Stone 2013).

The diversity of dune formations, their ever-changing form, and the range of color and texture create landscapes of outstanding natural beauty (UNESCO 2013).

Inland, the Namib Sand Sea is bordered by the Great Escarpment of Southern Africa (Goudie and Viles 2015). In effect, the Namib Sand Sea exemplifies the impact of sea floor spreading since the Cretaceous, with the emplacement of many sub-volcanic complexes and the development of the upwarped marginal escarpment (Goudie and Seely 2011; Goudie and Eckardt 1999).

2.2.3 *Biodiversity*

The fog is the only source of water for the Namib Desert organisms (Norgaard and Dacke 2010). Many species have developed ways of trapping the atmospheric water that comes ashore with the fog, so they can survive without rain (Norgaard and Dacke 2010). They have evolved special ways of living in the ever-changing dunes; small animals hide in the subsurface sand to escape the scorching heat and the risk of predation (UNEP 2013).

More than 50% of animals and plants are endemic to the sand of sea, and for arachnids it exceeds 80% (Graeme 2015). There are also endemic invertebrates, reptiles, and mammals (UNEP 2013). The outstanding combination and characteristics of the physical environment – loose sand, variable winds, and fog gradients across the property – create an ever-changing variety of microhabitats and ecological niches that is globally unique on such a scale (UNESCO WHC 2016).

2.3 Conservation Challenges

Because of its vast size, difficulty of access and current management within the protected Namib-Naukluft Park (49,768 km^2), the Namib Sand Sea is well conserved and in an excellent, undamaged state (UNESCO WHC 2016). Actually, the nature of the Namib Sand makes it sufficiently inhospitable to deter most forms of exploration or exploitation (African Natural Heritage 2013).

The most significant management challenges are regulating visitor activities in the few accessible parts of the site; controlling alien plants, especially along the seasonal rivers; and managing local community access rights (UNEP 2013). Water resources require careful management if the few seasonal rivers that penetrate the area are going to be maintained, but the key values of the site seem secure (African Natural Heritage 2013).

The Topnaar community, living in scattered villages along the northern boundary of the World Heritage site, claims ancestral rights to land and resources within the area. They maintain a limited number of livestock (mainly cattle and goats) which are grazed along the northern fringes of the site and harvest other renewable natural resources for subsistence use (African Natural Heritage 2013). There may be a risk of overexploitation of hunting, fishing, and resource use by this local community.

The effect of climate change in the site has not been analyzed, but following the most expected effects of it, it is very possible that the hyper-arid desert may suffer from an increase of aridity, with less fog created by the cold Benguela Current, putting the desert life under enormous pressure and risk of extinction.

The International Union for Conservation of Nature (IUCN) reported in its last update that the state of the conservation of the site was good, without any concern. It indicates that its values are in good condition and are likely to be maintained for the foreseeable future, provided that current conservation measures are maintained (Osipova et al. 2017).

References

African Natural Heritage (2013) http://www.africannaturalheritage.org/namib-sand-sea-namibia. Accessed 19 Feb 2016

Bluck BJ, Ward JD, Cartwright J, Swart R (2007) The Orange River, southern Africa: an extreme example of a wave dominated sediment dispersal system in the South Atlantic Ocean. J Geol Soc 164:341–351

Bourke MC, Goudie AS (2009) Varieties of barchan dunes in the Namib Desert and on Mars. Aeolian Res 1:45–54

Bristow CS, Duller GAT, Lancaster N (2007) Age and dynamics of linear dunes in the Namib Desert. Geology 35:555–558

Goudie AS, Eckardt F (1999) The evolution of the morphological framework of the Central Namib Desert, Namibia, since the Early Cretaceous. Geogr Ann 81A:443–458

Goudie A, Seely M (2011) World Heritage Desert Landscapes: potential priorities for the recognition of desert landscapes and geomorphological sites on the world heritage list. IUCN, Rapport, Gland, Switzerland

Goudie AS, Viles H (2015) Landscapes and landforms of Namibia. Springer, London

Graeme W (ed) (2015) Protected area governance and management. Australian National University, Canberra

Lancaster J, Lancaster N, Seely MK (1984) The climate of the central Namib. Modoqua 14:5–61

Lancaster N (1985) Winds and sand movements in the Namib Sand Sea. Earth Surf Process Landf 10(6):607–619

Lancaster N (1989) The Namib Sand Sea: dune forms, processes and sediments. A. A. Balkema, Rotterdam

Livingstone I, Bristow C, Bryant RG, Bullard J, White K, Wiggs GFS, Baas ACW, Bateman MD, Thomas DSG (2010) The Namib Sand Sea digital database of aeolian dunes and key forcing variables. Aeolian Res 2:93–104

Norgaard T, Dacke M (2010) Fog-basking behavior and water collection efficiency in Namib Desert darkling beetles. Front Zool 7:23–33

Osipova E, Shadie P, Zwahlen C, Osti M, Shi Y, Kormos C, Bertzky B, Murai M, Van Merm R, Badman T (2017) IUCN world heritage outlook 2: a conservation assessment of all natural world heritage sites. IUCN, Gland

Pietruszka RD, Seely MK (1985) Predictability of two moisture sources in the Namib Desert. S Afr J Sci 81:682–685

Stone AEC (2013) Age and dynamics of the Namib Sand Sea: a review of chronological evidence and possible landscape development models. J Afr Earth Sci 80:70–87

UNEP (United Nations Environment Programme) (2013) http://www.unep.org/DEWA/Africa/docs/en/aeo-2/chapters/aeo-2_ch05_COASTAL_AND_MARINE_ENVIRONMENTS.pdf. Accessed 30 Jan 2016

UNESCO WHC (United Nations Educational, Scientific and Cultural Organization World Heritage Center) (2013) Namib Sand Sea World Heritage Nomination. Report, Paris

UNESCO WHC (United Nations Educational, Scientific and Cultural Organization World Heritage Centre) (2016) Namib Sand Sea. http://whc.unesco.org/en/list/1430. Accessed 19 Feb 2016

Walden J, White K, Drake NA (1996) Controls on dune color in the Namib Sand Sea: preliminary results. J Afr Earth Sci 22:349–353

Chapter 3
Djoudj National Bird Sanctuary, Senegal

Abstract The Djoudj National Bird Sanctuary, in Senegal, is in the delta of the Senegal River, on the border with Mauritania. The river flows to the Atlantic Ocean, and its valley creates a wetland of 160 km^2, the third largest in the world. It includes seasonally flooded areas of lakes, ponds, and bayous surrounded by streams and backwaters and irregularly flooded alluvial plains. The wetland is one of the first stops after the Sahara Desert for birds migrating from Europe. The environment provides a living and fragile sanctuary for up to 1.5 million birds. The sanctuary can, during winter months, host up to 3 million migrating birds. It is characterized by a large breeding population, including white pelican, purple heron, African spoonbill, and cormorant.

3.1 Introduction

A national bird sanctuary since 1971, Djoudj lies on the southeast bank of the delta of the Senegal River, on the border of Senegal and Mauritania, northeast of the city of Saint-Louis (Fig. 3.1).

It was designated a Wetland of International Importance under the Convention of Wetlands, the Ramsar Convention, in 1977. The site became a World Heritage site in 1981, due to its beauty and importance for migratory birds (IUCN 2010). Effectively, it was established for supporting waterfowl, considering that it is one of the first freshwater sources birds reach after crossing 200 km of the Sahara Desert coming from Europe (IUCN 2010).

Human settlement in the area, mainly dependent on coastal and estuarine fishing, has been traced back 500 years. The neighboring areas are now inhabited, with the population working with agriculture in the plains of the river and in the conservation of the site (UNEP/WCMC 2006).

The sanctuary is the object of intense tourism. There are five bird observation points on the borders of the most important water bodies. Canoes can be hired (UNEP/WCMC 2006). The property is managed by the National Parks Service of the Ministry for the Protection of Nature of Senegal (UNESCO WHC 2016). The

V. Claudino-Sales, *Coastal World Heritage Sites*, Coastal Research Library 28,
https://doi.org/10.1007/978-94-024-1528-5_3

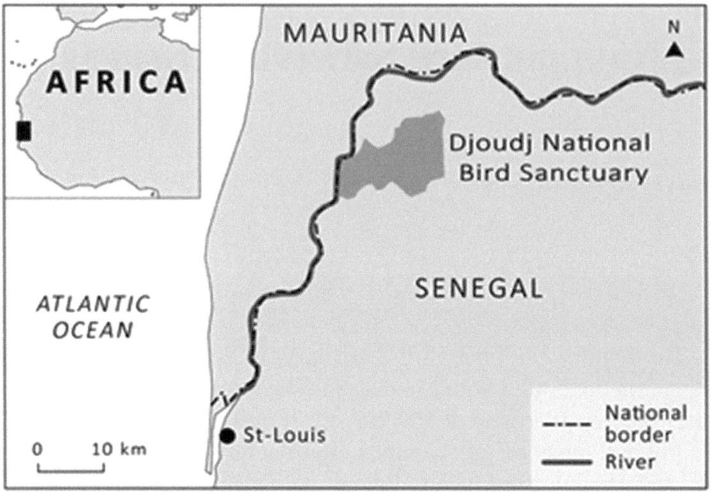

Fig. 3.1 Location of the Djoudj National Bird Sanctuary Heritage site in the delta of the Senegal River, in Senegal, on the border with Mauritania

entire area is very poorly studied, with few data about geology, sedimentology, climate change, and other natural and environmental settings.

3.2 Natural Settings

3.2.1 Climatic Conditions

The climate is a semiarid Sahelian, alternating wet and dry seasons. The weather is tempered by the ocean and sea winds. The rainfall is variable. The mean annual amount, falling mainly between July and mid-September, does not exceed 300 mm. The evaporation rate is ten times the annual rainfall. Even so, during the dry months of March to June, the sanctuary is still the most humid area in the region (IUCN 2010).

The humidity is variable, with 40% dry season and 70% wet season humidity. Flood levels are highest in October, but in recent years rainfall has been less than one-fifth of the average, and Djoudj has become much drier (Kane et al. 2014). The mean annual temperature is 27 °C, but during the cold season, between December and March, it can fall to 10 °C (UNEP/WCMC 2006).

3.2.2 Geology and Geomorphology

The property is in the Western Sahel. The altitudes go from near sea level to about 20 m. It comprises seasonally flooded areas of lakes, ponds, and bayous surrounded by streams and backwaters and irregularly flooded alluvial plains (UNESCO WHC 2016).

The sanctuary is a small part of the wide delta floodplain of the Senegal River, the second largest river in West Africa. The river marks the line between the Sahara and the Sahel and forms an important ecological barrier on the desert edge (Kane et al. 2014). The bayous, meres, and backwaters are dry for 3 months every year. The ground is covered with impermeable salt-impregnated alluvial clays forming saline flats (UNEP/WCMC 2006).

It was originally steppe, paddy fields, marshes, and lagoons subject to annual flooding, but dykes have canalized the river, and an anti-saltwater barrier was built downstream (Barusseau et al. 1995). For this reason, the tides that penetrated before for more than 50 km in the estuary are now prevented from invading the river valley. The dam induces the presence of freshwater in the basin for long periods, replacing the former balance of wet and dry, fresh- and saltwater cycles (UNEP/WCMC 2006).

3.2.3 Biodiversity

From September to April, an estimated 3 million migrant birds pass through Djoudj National Bird Sanctuary (Figs. 3.2 and 3.3). It is the only known wintering sanctuary for the aquatic warbler, Europe's most globally threatened passerine bird, and is one of the main West African sanctuaries for Palaearctic migrant birds (IUCN 2010).

The sanctuary is also a major breeding area for thousands of nesting birds: flamingoes, white pelicans, white-faced and fulvous tree ducks, bicolor spur-winged geese, purple and night heron, egrets, spoonbills, African darters, common and white-breasted cormorants, and Sudan bustards, among others (UNESCO WHC 2016). Mammals include warthogs and West African manatees. Crocodiles and gazelles have been successfully reintroduced into the area. There are also python and lizards (IUCN 2010).

The site is classified by BirdLife International as an Important Bird Area for breeding, staging, and wintering water birds. For millions of Palaearctic ducks and other aquatic birds, the sanctuary, along with the Niger swamps in Mali and Lake Chad on the edge of the Sahara, is one of the great staging and wintering areas on the west of the Sahara. Three hundred species have been recorded (UNEP/WCMC 2006).

Fig. 3.2 The Djoudj National Bird Sanctuary, where many species of migrant birds and breeding birds enjoy the wetland. (Picture reproduced by permission of Peter Howard, website http://www.AfricanWorldHeritageSites.org)

Fig. 3.3 Pelicans in the big flooded lake at the Djoudj National Bird Sanctuary, Senegal. (Picture reproduced by permission of Galen R. Frysinger)

The vegetation is a mosaic of Sahelian habitats and the product of low rainfall on halomorphic soils. The savanna is the principal vegetation type, with herbaceous level. Halophytic plants cover much of the area (UNEP/WCMC 2006).

3.3 Conservation Challenges

The Senegal River estuary has undergone significant alterations as a consequence of the hydrodynamic conditions that have prevailed there following human actions in the area. It has been, for more than 30 years, under multiple types of management and planning (Kane et al. 2014).

During the 1980s, two dams were built along the valley floor to facilitate hydroelectrical production and regular water supply for crop irrigation: they are the Diama dam, built in 1985 near the mouth of the river in Senegal, and Manantali dam, built in 1989 upriver in Mali (Niang and Kane 2014; Dumas et al. 2010). These changes are reflected in irrigation development but contribute to the general degradation of the landscape (Barusseau et al. 1995).

The Diama dam was designed to block seawater intrusion within the valley (Kane et al. 2014; Dumas et al. 2010). The dam prevented the tides that came from the Atlantic Ocean from reaching the Djoudj National Bird Sanctuary. For this reason, and also because of the arrival of the invasive water plant *Salvinia molesta* (see next paragraphs), in 1984 the site was listed in the World Heritage Sites in Danger. It was taken off from the list in 1988, when it was clear that the bird sanctuary did not suffer serious consequences because of the lack of seawater, and also considering the control of the invasive plant in the wetland by the managers (UNEP/WCMC 2006). Nevertheless, the sanctuary is permanently at risk of being colonized by *Salvinia molesta* again. This plant is one of the most invasive in the world and menaces the ecosystem and economy of the region (IUCN 2010).

An attempt to control the invasive species used biological methods by releasing plant-eating beetles provided by South Africa, but this has proved to be inadequate (UNEP/WCMC 2006). The new invasion of *Salvinia molesta* caused the heritage site to be reinscribed on the World Heritage List in Danger in 2000. The national authorities agreed to the re-list the site to facilitate their task and help them appeal for financial support from donors (UNESCO WHC 2016).

New types of biocontrol measures have enabled the managers of Djoudj to eradicate the threat of invasive plant species to the wetland, and it was excluded from the list of sites in Danger in 2006 (UNESCO WHC 2016). Now, there is continuous surveillance of the site, but weevils may be reintroduced annually (IUCN 2010).

In terms of the impacts of climate change, it is not yet clear what kind of consequences it will produce in Djoudj National Bird Sanctuary, but it is evident that the Senegal River basin is going through a decrease in rainfall (Kane et al. 2013).

Recently, the World Bank approved funding to boost water management in the river basin. The project has the goal to improve water availability for agriculture and food, upgrade water man production, support aquaculture and fisheries management, promote hydropower through feasibility studies, reduce tropical diseases, and foment pilot innovative approaches to adapt to climate change (African Research Bulletin 2014). It may be a palliative for global warming consequences.

But the last International Union for Conservation of Nature (IUCN) report about the state of conservation of the site found significant concerns. Its values are under considerable threat and in need of new management actions in order to face the problems (Osipova et al. 2017).

References

African Research Bulletin (2014) WATER: Senegal River Basin. Wiley Online Library. http://onlinelibrary.wiley.com/. Accessed 26 Dec 2014

Barusseau JP, Michel P, Richard JF, Sall MM (eds) (1995) L'après-barrages dans la vallée du Sénégal. Modifications hydro-dynamiques et sédimentologiques – Conséquences sur le milieu et les aménagements hydro-agricole. Presses Universitaires de Perpignan, Perpignan

Dumas D, Mietton M, Hamerlynck O, Pesneadu F, Kane A, Coly A, Duvail S, Baba MLO (2010) Large dams and uncertainties. The case of the Senegal River (West Africa). Soc Nat Resour 23(11):1108–1112

IUCN (International Union of the Conservation of Nature) (2010) Winter home for 3 million birds. Report, Gland, Switzerland

Kane C, Humbert J, Kane A (2013) Responding to climate variability: the opening of an artificial mouth on the Senegal River. Reg Clim Chang 13(1):125–136

Kane C, Kane A, Humbert J (2014) Management of a Tropical River: impacts on the resilience of the Senegal River Estuary. In: Diop S, Barusseau J-P, Descamps C (eds) The land/ocean interactions in the Coastal Zone of West and Central Africa. Springer, Cham

Niang A, Kane A (2014) Morphological and hydrodynamic changes in the lower estuary of the Senegal River: effects on the environment of the breach of the 'Langue De Barbarie' sand spit in 2003. In: Diop S, Barusseau J-P, Descamps C (eds) The land/ocean interactions in the Coastal Zone of West and Central Africa. Springer, Cham

Osipova E, Shadie P, Zwahlen C, Osti M, Shi Y, Kormos C, Bertzky B, Murai M, Van Merm R, Badman T (2017) IUCN world heritage outlook 2: a conservation assessment of all natural world heritage sites. IUCN, Gland

UNEP/WCMC (United Nation Environmental Programme/World Conservation Monitoring Centre) (2006) Djoudj National Bird Sanctuary. Report, Cambridge

UNESCO WHC (United Nation Educational, Scientific and Cultural Organization World Heritage Centre) (2016). Djoudj National Bird Sanctuary. http://whc.unesco.org/en/list/25. Accessed 25 Dec 2015

Chapter 4
Cape Floral Region Protected Areas, South Africa

Abstract The Cape Floral Region in South Africa is one of the richest areas for plants in the world and one of the world's great centers of terrestrial biodiversity. It represents less than 0.5% of the area of Africa but is home to nearly 20% of the continent's flora. The site displays outstanding ecological and biological processes associated with the fynbos, a shrubland vegetation adapted to Mediterranean climate and to periodic fires, which is unique to the Cape Floral Region. The outstanding diversity, density, and endemism of the flora are among the highest worldwide and are of outstanding value to science. The property is formed by parks, nature reserves, wilderness, and protected areas. The landscape is characterized by mountains, stepped surfaces, valleys, and coastal sites, such as the Cape of Good Hope. It is also the zone where the Atlantic Ocean meets the Indian Ocean.

4.1 Introduction

The Cape Floral Protected Areas heritage site is characterized by eight properties, aggregating 5575 km^2 (UNESCO WHC 2015). The flora of each area is sufficiently distinct to justify these several sites, each of which is large enough to preserve its genetic viability (UNEP/WCMC 2011) (Fig. 4.1).

The site is one of the world's 18 hotspots for biodiversity, a global center of plant diversity and an endemic bird area. Its 9000 plant species, containing 20% of Africa's flora, spring from a wide spectrum of elevations, soils, and climatic conditions. The property is considered of outstanding universal value for representing ecological and biological processes associated with the evolution of the unique fynbos biome and is one of the most intensely researched floral regions in the world (UNESCO WHC 2009).

Cape Floral Protected Areas became a World Heritage site in 2004. It doubled in size in 2015 and is now one of Africa's biggest natural preserved areas. The eight sites are administered by the Department of Environmental Affairs and Tourism (DEAT) of South Africa (UNESCO WHC 2015).

Two of the properties of Cape Floral Protected Areas are coastal – the Cape Peninsula National Park (or Table Mountain National Park) and the De Hoop Nature

V. Claudino-Sales, *Coastal World Heritage Sites*, Coastal Research Library 28,
https://doi.org/10.1007/978-94-024-1528-5_4

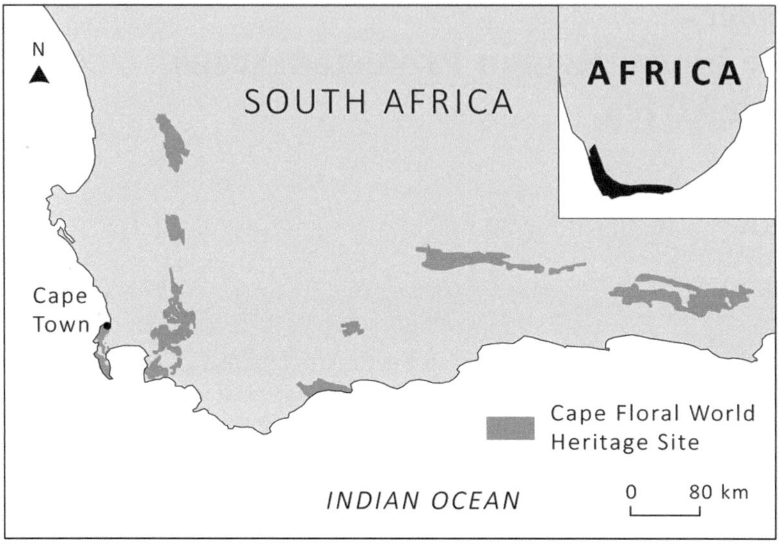

Fig. 4.1 Location of the Cape Floral Protected Areas in South Africa. From the eight properties that integrated the heritage site, two have segments of coastal areas

Reserve. These protected areas are in the Western Cape province, approximately 230 km from each other (see Fig. 4.1) (UNEP/WCMC 2011). Table Mountain, which towers over Cape Town, was crowned one of the New Seven World Wonders in 2011.

De Hoop Nature Reserve has 340 km^2 and is in the south of the Cape Province (see Fig. 4.1). The neighboring marine reserve, which extends 5 km out to sea, is one of the largest marine protected areas in Africa. It is mountainous in the continental segments and hilly in the coast (UNEP/WCMC 2011). The coastal area is bathed by the Indian Ocean.

The Cape Peninsula National Park has 170 km^2 and is in the Atlantic Ocean. It forms the southwestern tip of the Cape Floral Protected Areas, 50 km from Cape Town (see Fig. 4.1). The park has a pronounced topography, with mountains, hills, and seacliffs (Fig. 4.2). The property contains two well-known landmarks: Table Mountain and the Cape of Good Hope, the most southwestern extremity of Africa.

The Cape Floral Protected Area is a popular tourist destination, both nationally and internationally. The most visited site is the Table Mountain National Park. Mountains, capes, flowers, whales, and penguins are among the attractions. Infrastructure and reserve facilities are excellent, and effective methods are used to control visitor numbers when necessary (UNESCO WHC 2016).

Artifacts and fossils show that the region was occupied by humans at least 250,000 years ago. Stone tools from the Early Stone Age and hundreds of later shell middens have been found. In 1488, the Portuguese Bartholomew Dias named the Cape of Good Hope, and in 1652 the Dutch East India Company established a post. Settlement previously limited by the infertility of the area became possible after

Fig. 4.2 Lighthouse in a granitic seacliff in Cape Floral National Park (also known as Table Mountain National Park), at the southern tip of Africa. (Picture reproduced by permission of Peter Howard, website http://www.AfricanWorldHeritageSites.org)

European crops were introduced by the colonists, who cleared much of the lowlands for farming (UNEP/WCMC 2011).

4.2 Natural Settings

4.2.1 Climatic Conditions

The region has a semi-Mediterranean climate. Winters are cool and wet, and summers are dry and hot. Rainfall varies markedly with topography: between 300 and 500 mm in the lowlands and 1000 and 3300 mm in the mountains, where clouds and fog can persist. In the mountains, snow falls in winter. Temperatures range from below freezing to above 40 °C (UNEP/WCMC 2011).

The coastal areas near the oceans are more temperate. Coastal winds can be strong, and in winter winds occasionally blow from the interior, aggravating the natural fires, which occur at 10–20 year intervals. These facts produce a mosaic of climatic and microclimatic zones which contribute to the complexity and diversity of the flora (Harris et al. 2010). It has seasonal extremes in day length and climate: winter days are markedly shorter and colder than summer days but have higher rainfall (Van Doorn et al. 2010).

4.2.2 Geology and Geomorphology

The Cape Floral Region lies between the two oceans and mountain chains that parallel the coast. The elevations go from sea level to 2077 m. The predominant soils, derived from sandstone, are shallow, sandy, nutrient-poor, and acidic, characteristic of fynbos areas. Soils are skeletal at high elevations, and valley soils are richer clays derived from intermixed shales (UNEP/WCMC 2011).

The geology in Cape Peninsula is characterized by the Cape Supergroup, formed mainly by granites and metasediments. The metasedimentary strata originally comprised deep water marine mudrock and are mantled by relatively thin deposits composed of alluvial river terrace material (Tankard and Hobday 1977). The Cape Peninsula shows the presence of igneous activity associated with the opening of the South Atlantic during the Early Cretaceous (Backeberg et al. 2011).

Analyses of soils and paleosols showed that the Cape Peninsula has witnessed significant palaeoenvironmental and climatic oscillations after Cretaceous times (Eze and Meadows 2014). The area has experienced mild climate changes with mostly cooler and wetter conditions for at least the past 2 million years (Colette 2009).

The De Hoop Nature Reserve is lesser known and studied in terms of natural settings, but a highlight is 51 km of pristine white beaches and dunes, as well as limestone cliffs, rock pools, and wetlands (UNEP/WCMC 2011).

4.2.3 Biodiversity

The Cape Floral Region has been called the world's hottest hotspot for plant diversity and endemism. None of the eight properties has less than 1100 species. It displays the highest levels of endemism at 32%. Due to its unique floristic values, it is recognized as one of the world's six floral kingdoms (Colette 2009).

This richness is the result of a wide variety of habitats resulting from the range of elevations, soils, and climatic conditions. The flora also has concentrations of relict endemics and massive still-active speciation due to its isolation (UNEP/WCMC 2011).

The distinctive flora of the Cape Floral Region, comprising 80% of its floristic richness, is a shrubland known as fynbos (fine bush), a fine-leaved vegetation adapted to both the Mediterranean type of climate and to periodic fires. It is defined by location or by dominant species such as coastal, mountain, or grassy fynbos. It grows on the predominant coarsely sandy, acidic nutrient-poor soils, and on alkaline marine sands, as well as in slightly richer alluvial soils (UNEP/WCMC 2011) (Fig. 4.3).

Fig. 4.3 Endemic vegetation with flowers in the Cape Floral Protected Area. (Picture reproduced by permission of Peter Howard, website http://www.AfricanWorldHeritageSites.org)

The Cape Peninsula has exceptionally high plant species richness: 2285 species. A 158 are endemic, concentrated in ten families (Helme and Trinder-Smith 2006). The predominant vegetation is also fynbos: 12 broadly characterized fynbos types have been described on the peninsula (Cowling et al. 1996).

In relation to fauna, the Cape Faunal Region is a distinct zoogeographic zone that coincides roughly with the floral region. The fauna is less remarkable than the flora, except for a distinctive relict endemic invertebrate fauna which persists in upper forest streams, riverine forests, and caves, especially in the Cape Peninsula National Park (Trinder-Smith et al. 1996).

The Cape Floral Region hosted a diverse large mammal fauna until shortly after permanent European settlement (1652) (UNESCO WHC 2009). These animals survived in an exceptionally nutrient-poor environment, and it is generally believed that they restricted their movements to the more fertile shale areas (Radloff et al. 2010).

Particularly in the foothills and mountains, larger mammals such as the chacma baboon appear (Van Doorn et al. 2010). The region is also an endemic bird area with, on the coast, the jackass penguin. In the De Hoop reserve, the endemic Cape mountain zebra is an important source for reintroduction of these animals to other reserves (UNEP/WCMC 2011).

4.3 Conservation Challenges

In Cape Peninsula there are 141 species of plants that are threatened by urban development, agriculture, and the spread of invasive plants (UNESCO WHC 2009). Human occupation was limited by nutrient poverty for long time, but today some 65% of original natural habitat is either transformed by urbanization and agriculture or invaded by alien plants (UNEP/WCMC 2011).

Alien invasive plants pose the most severe threat for the continued existence of fynbos ecosystems. Large areas of the site are invaded, particularly coastal habitats (Cowling et al. 1996). But the South African government is taking action toward the control of the situation, such as the survey of native species as well as invasive ones (UNESCO WHC 2009).

Nevertheless, the most significant threat facing biodiversity in Cape Floral Protected Area over the next 50–100 years is climate change. It is believed that under global warming, warmer and dryer weather will develop, threatening the fynbos survival. At the same time, new ecosystem changes will appear, due to changes in environmental conditions. The increase of fire frequency would also be a consequence of climate change in the heritage site (Colette 2009). The national agencies are trying to prepare for harsher conditions, for example, adopting new fire management strategies and monitoring natural species of the vegetation kingdom (UNESCO WHC 2009).

For all these reasons, the last report about the state of conservation of the site made by the International Union for Conservation of Nature (IUCN) found that in general, the situation is good but with some concerns (Osipova et al. 2017).

References

Backeberg NR, Reid DL, Trumbull RB, Romer RL (2011) Petrogenesis of the False dike swarm, Cape Peninsula, South Africa; evidence for basement assimilation. S Afr J Geol 114(3–4):335–352

Colette A (2009) Case studies on climate change and world heritage. United Nations Educational, Scientific and Cultural Organization (UNESCO) World Heritage Center, Paris

Cowling R, Richadson D, Heywood D (1996) The Cape Peninsula South Africa: physiographical, biological and historical background to an extraordinary hotspot of diversity. Biodivers Conserv 5:527–550

Eze PN, Meadows ME (2014) Texture contrast profile with stonelayer in the Cape Peninsula, South Africa: autochthony and polygenesis. Catena 118:103–114

Harris C, Burgers C, Miller J, Rawoot F (2010) O- and H-isotope record of Cape Town rainfall from 1996–2008, and its application to recharge studies of Table Mountain groundwater, South Africa. S Afr J Geol 113(1):33–56

Helme NA, Trinder-Smith TH (2006) The endemic flora of Cape Peninsula, South Africa. S Afr J Bot 72(2):205–210

Osipova E, Shadie P, Zwahlen C, Osti M, Shi Y, Kormos C, Bertzky B, Murai M, Van Merm R, Badman T (2017) IUCN world heritage outlook 2: a conservation assessment of all natural world heritage sites. Report, Gland, Switzerland

Radloff F, Mucina L, Bond W, Roux P (2010) Strontium isotope analyses of large herbivore habitat use in the Cape Fynbos region of South Africa. Oecologia 164(2):567–578

Tankard AJ, Hobday DK (1977) Tide-dominated back-barrier sedimentation, early Ordovician Cape Basin, Cape Peninsula, South Africa. Sediment Geol 18(1):135–159

Trinder-Smith H, Cowling R, Linder H (1996) Profiling a besieged flora: endemic and threatened plants of the Cape Peninsula, South Africa. Biodivers Conserv 5(5):575–589

UNEP/WCMC (United National Environment Programme/World Conservation Monitoring Centre) (2011) Cape floral region protected areas Western Cape and Eastern Provinces South Africa. Report, Cambridge, England

UNESCO WHC (United Nation Educational, Scientific and Cultural Organization World Heritage Centre) (2009) State of Conservation (SOC) Cape Floral Region Protected Areas http://whc.unesco.org/en/soc/63. Accessed 24 Dec 2015

Unesco WHC (United Nations Educational, Scientific and Cultural Organization World Heritage Center) (2015) Cape Floral region protected areas. https://whc.unesco.org/en/list/1007. Accessed 22 Dec 2015

UNESCO WHC (United Nation Educational, Scientific and Cultural Organization World Heritage Centre) (2016) Cape Floral Protected Areas. http://whc.unesco.org/en/list/1007. Accessed 24 Dec 2015

Van Doorn AC, O' Riain MJ, Swedell L (2010) The effects of extreme seasonality of climate and day length on the activity budget and diet of semi-commensal chacma baboons (Papio ursinus) in the Cape Peninsula of South Africa. Am J Primatol 72(2):104–112

Chapter 5
iSimangaliso Wetland Park, South Africa

Abstract iSimangaliso Wetland Park stretches for 220 km along the eastern shore of South Africa, bathed by the Indian Ocean. It includes a wide range of pristine marine, coastal, wetland, estuarine, and terrestrial environments which are scenic and unmodified by people. The ongoing fluvial, marine, aeolian, and climatic processes in the site resulted in a variety of landforms, such as coral reefs, sandy beaches, coastal dunes, lake systems, swamps, wetlands, grasslands, forests, lakes, and savanna. The transitional geographic location between subtropical and tropical Africa has resulted in exceptional species diversity and speciation. Its vivid natural spectacles include nesting turtles and large aggregations of flamingos and other waterfowl. The site contains critical habitats for a range of species from Africa's marine, wetland, and savanna environments.

5.1 Introduction

The iSimangaliso Wetland Park stretches along 220 km in the eastern coast of South Africa, bathed by the Indian Ocean. It goes from the Cape of Saint Lucia in the south to the border with Mozambique, in the north, presenting a 1–24 km wide marine component (Fig. 5.1). The park covers an area of 230 km^2 and is Africa's largest estuarine system (UNESCO WHC 2014).

The property was declared a Wetland of International Importance under the Convention of Wetlands, the Ramsar Convention, in 1986. The convention designated four places within the park, recognizing the ecological functions of wetlands as well as their importance in terms of economic, cultural, scientific, and recreational values (UNESCO WHC 2014).

The iSimangaliso Wetland Park received its World Heritage status in 1999. The property consists of 13 separate but contiguous preservation units that include 85 km^2 of marine reserves. Its history of conservation management dates to 1985, when the first reserves were created. It is managed by the iSimangaliso Wetland Park Authority (UNESCO WHC 2014).

The first evidence of human occupation of the park dates from the Early Stone Age. It was widely settled by agriculturists in the Early and Late Iron Ages (Maggs

V. Claudino-Sales, *Coastal World Heritage Sites*, Coastal Research Library 28,
https://doi.org/10.1007/978-94-024-1528-5_5

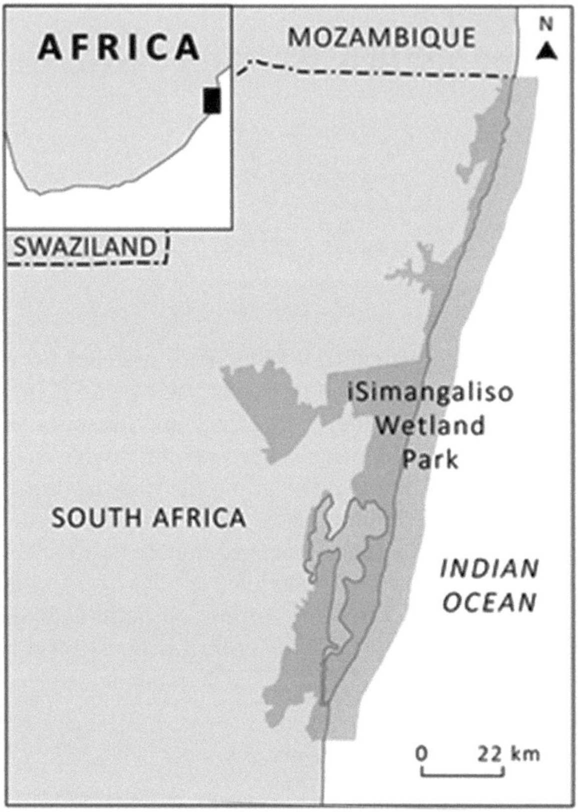

Fig. 5.1 iSimangaliso Wetland Park, in the east coast of South Africa, bathed by the Indian Ocean

1984). These early farmers probably occupied coastal sites, living and cutting fields in the forest and using black mussels for food, as shown by the shell middens found along the coast (Hall and Vogel 1980).

Portuguese navigators arrived in the area in 1576, but little is known about the nature of human settlements until the early nineteenth century. The area was occupied by the Zulu people, the predominant ethnic group in the region. Today, the park is not inhabited apart from six small townships. There are also two villages which are enclaves within the park but not part of it (UNEP/WCMC 2008).

About 100,000 people from 48 tribal groups live in villages surrounding the park. They farm, hunt, and fish, using 700-year-old fishing traditions (Hansen 2013). In that context, community conservation programs are key to minimizing conflicts and maximizing benefits (UNESCO WHC 2014).

5.2 Natural Settings

5.2.1 Climatic Conditions

The area is a transition zone between tropical and subtropical zones, in the Southern Hemisphere. It has warm, moist summers and mild, dry winters. The mean annual temperature exceeds 21 °C. The ocean water is heated by the Agulhas Current, which flows in the east coast of Africa from north to south (UNEP/WCMC 2008).

There is an east-west climatic gradient, with the coast having more rain, while inland is moderately dry. At the coast, the precipitation varies from 1200 to 1300 mm annually, with 60% of it falling in summer, between November and March. The prevailing winds parallel the coast from the south (UNEP/WCM 2008; Tyson 1987). Storms are common events in the coastal area (UNESCO WHC 2014).

5.2.2 Geology and Geomorphology

The park comprises two geomorphic units: coastal plain and continental shelf. It includes a wide range of pristine marine, coastal, wetland, estuarine, and terrestrial environments. The ongoing fluvial, marine, aeolian, and climatic processes in the site resulted in a variety of landforms, such as coral reefs, sandy beaches, coastal dunes, lake systems, swamps, wetlands, grasslands, forests, lakes, and savanna (UNESCO WHC 2012; Nustad 2011).

iSimangaliso Wetland Park is tectonically stable and has a wave-dominated coastline. It has 220 km of extension, half of it being coastal plain and half bedrock-framed, rocky coast (Salzmann and Green 2012). Stratified quaternary marine deposits related to marine transgression and regression have resulted in a series of prominent north-south oriented sandy dune ridges (Watkeys et al. 1993).

Effectively, a large semicontinuous vegetated Pleistocene dune ridge is developed along both the bedrock and coastal plain coasts (Porat and Botha 2008) (Fig. 5.2). The dunes were formed over the past 25,000 years and consist of superimposed sedimentary strata of different ages (UNESCO WHC 2012). The dunes were classified as being estuary barrier dunes, parabolic/migrating dunefield, transverse dunes, climbing dunes, headland bypass dunes, beach foredunes, and prograded beach ridge dunes. They are among the highest in the world, with altitudes of 170 m in some places (Jackson et al. 2014).

The tides in the region are semidiurnal. Mean spring tidal range is 1.72 m, and mean neap tidal range is 0.5 m. Recorded wave heights over a 2-year period ranged from 0.3 to 7.6 m with a median wave height of 1.49 m. The dominant direction of wave approach is ESE to SE. Maximum wave periods range between 17 s and 5–7 s. Longshore transport volume was calculated at 500,000 m^3 (Jackson et al. 2014).

Fig. 5.2 Sandy beach with foredunes rearward and backed by a vegetated cordon of Pleistocene dunes topping hard rock. (Picture reproduced by permission of Peter Howard, website http://www.AfricanWorldHeritageSites.org)

Along the intertidal and infratidal coast, the coastline has long sandy beaches between reefs of beach rock. On the continental shelf, the sediments being transported by the Agulhas Current are trapped by submarine canyons, allowing for remarkably clear waters for the development of coral reefs (UNESCO WHC 2012) (Fig. 5.3).

The combination of fluvial, marine, and aeolian processes initiated in the Early Pleistocene in iSimangaliso, resulting in the variety of landforms present in the site (UNESCO WHC 2014). The coastal lakes are remarkable. Two types have formed behind the coastal dunes: estuarine and freshwater, the biggest being Lake Saint Lucia. In this lake an important hydrological phenomena takes place: the shifting salinity states are linked to wet and dry climatic cycles, with the lake responding accordingly with shifts from low- to hypersaline states (UNEP/MCMC 2008).

5.2.3 *Biodiversity*

The park, lying on the interface between tropical and subtropical biota with varied geomorphic and climatic conditions, supports an exceptional ecological and biological diversity.

The distribution of the vegetation is largely determined by topography, moisture regimes, and edaphic conditions. It is a rich mosaic of savanna grasslands, wood-

Fig. 5.3 Seascape with coral in Sodwana Bay, in the north segment of iSimangaliso Wetland Park. (Picture reproduced by permission of Peter Howard, website http://www.AfricanWorldHeritageSites.org)

lands, riverine woodlands, rare swamp forests, and forested dunes. The flora is diverse, with 152 families, 734 genera, and 2173 species (UNEP/WCMC 2008).

For the fauna, the numbers are also expressive. There are 97 species of terrestrial mammals and 32 marine mammal species. The terrestrial mammals include the internationally threatened (reintroduced) black rhinoceros. There are elephants and hippopotamus. The herpetofauna is rich: 50 amphibians and 109 reptiles, 1 crocodile, 53 snakes, and 42 lizards and chameleons. In relation to invertebrates, marine and estuarine are an important group (UNESCO WHC 2014).

The ichthyofauna includes nearly 85% of the reef fish species endemic to the West Indian Ocean region, making a total of 399 species (UNESCO WHC 2012). It includes the notable coelacanth, known from Devonian fossils 370 million years old but only discovered, off South Africa, in 1938 (Venter et al. 2000).

The coral reefs contain a variety of underwater seascapes and marine flora and fauna. They are among the southernmost coral reefs in the world and almost the only reefs in South Africa (see Fig. 5.3). Among the avifauna, there are 521 species, 60% of the South African total, approximately 200 of which are water birds which the lakes attract in very large numbers (UNESCO WHC 2012).

The interaction of these environments with major floods and coastal storms in the park's transitional location has resulted in continuing speciation and exceptional species diversity. Its vivid natural spectacles include nesting turtles and large aggregations of flamingos and large breeding colonies of pelicans, storks, herons, and terns. There is an abundance of dolphins and migration of whales and whale sharks offshore (UNESCO WHC 2014).

5.3 Conservation Challenges

A major threat to the park is damage to the hydrology and salinity of the wetland system, including reduction in the water supply, by the transformation of the swamps to agriculture. Serious droughts have also raised salinity and killed the shoreline vegetation, causing bank erosion and silting of the lakes (UNESCO WHC 2014).

Other threats include damage by overuse, as tourism and overexploitation of resources such as unsustainable fishing (UNESCO WHC 2012). Infestation by alien invasive plants is a problem, although limited at present (UNEP/WCMC 2008). In addition, many species are listed as endangered or critically endangered, including rhinoceros and the coelacanth (UNESCO WHC 2014).

All human uses of iSimangaliso are subject to intensive management, research, and monitoring. They are also confined to about a third of the total area, while the remainder is free from extractive uses (UNEP/WCMC 2008).

According to UNESCO WHC (2014), the communities living in and around iSimangaliso today have a rich relationship with the land and natural environment. Cultural traditions, land use management practices, and indigenous knowledge systems continue to shape the current environment. Equally important are resources such as significant sacred spaces, oral traditions, and rituals (UNESCO WHC 2012). This brings conflicts: some consider that the park was conceived as a tool for market-based modernization and economic growth. They talk about the property being a commodity, rather than a social-ecological resource for livelihood generation (Hansen 2013).

The impacts of climate change are not clear in iSimangaliso Wetland Park, but it may have begun: a drastic decline of lake levels is observed over the past decade (Weitz and Demlie 2014).

The International Union for Conservation of Nature (IUCN) published a report at the end of 2017 about the state of conservation of the World Heritage sites and found that the iSimangaliso Wetland Park is in good condition but with situations that present some concerns. It indicates that its values are in good condition and are likely to be maintained in the long term, provided that minor additional conservation measures are put in place to address the problems (Osipova et al. 2017).

References

Hall M, Vogel JC (1980) Some recent radiocarbon dating from South Africa. J Afr Hist 21(4):431–455

Hansen M (2013) New geographies of conservation and globalization: the spatiality of development for conservation in the iSimangaliso Wetland Park, South Africa. J Contemp Afr Stud 31(3):481–502

Jackson DWT, Cooper JAG, Green AN (2014) A preliminary classification of coastal sand dunes of KwaZulu-Natal. J Coast Res 70:718–722

Maggs T (1984) The Iron Age south of Zambezi. In: Klein RG (ed) Southern Africa prehistory and palaeoenvironments. Balkem, Rotterdam

Nustad KG (2011) Performing natures and land in the iSimangaliso Wetland Park, South Africa. Ethons 76(I):88–108
Osipova E, Shadie P, Zwahlen C, Osti M, Shi Y, Kormos C, Bertzky B, Murai M, Van Merm R, Badman T (2017) IUCN world heritage outlook 2: a conservation assessment of all natural world heritage sites. IUCN, Gland
Porat N, Botha G (2008) The luminescence chronology of dune development on the Maputaland coastal plain, Southeast Africa. Quat Sci Rev 27:1024–1046
Salzmann L, Green A (2012) Boulder emplacement on a tectonically stable, wave-dominated coastline, Mission Rocks, northern KwaZulu-Natal, South Africa. Mar Geol 323–325:95–106
Tyson PD (1987) Climatic change and variability in Southern Africa. Oxford University Press, Cape Town
UNEP/WCMC (United Nation Environment Programme/World Conservation Monitoring Centre) (2008) iSimangaliso Wetland Park Kwazulu-Natal, South Africa. Report, Cambridge, England
UNESCO WHC (United Nation Educational Scientific and Cultural Organization World Heritage Centre) (2012) Assessing marine world heritage from an ecosystem perspective. The Western Indian Ocean. World Heritage Papers 32, Paris
UNESCO WHC (United Nation Educational Scientific and Cultural Organization World Heritage Centre) (2014) iSimangaliso Wetland Park. http://whc.unesco.org/en/list/914. Accessed 25 Dec 2015
Venter P, Timm P, Gunn G (2000) Discovery of a viable population of Coelacanths (Latimeria Chalumnae Smith, 1939) at Sodwana Bay, South Africa. S Afr J Sci 96:567–568
Watkeys MK, Mason TR, Goodman PS (1993) The role of geology in the development of Maputaland, South Africa. J Afr Earth Sci 16:205–221
Weitz J, Demlie M (2014) Conceptual modeling of groundwater-surface interactions in the Lake Sibayi Catchment, Eastern South Africa. J Afr Earth Sci 99:613–624

Part II
World Heritage on Continental Coasts: Asia

Chapter 6
The Sundarbans, Bangladesh and India

Abstract The Sundarbans, at the mouth of the Ganges and Brahmaputra rivers between India and Bangladesh, is the largest contiguous mangrove forest in the world. The forest is a mosaic of islands of different shapes and sizes, perennially washed by brackish water in and around endless labyrinths of water channels. The islands are of great environmental importance as a storm barrier, shore stabilizer, and nutrient and sediment trap and support a wide variety of aquatic, benthic, and terrestrial organisms. The ecosystem is an example of the ecological processes of monsoon rain flooding, delta formation, tidal influence, and plant colonization. The site supports great biodiversity in its terrestrial and marine habitats, ranging from micro to macro flora and fauna. The Sundarbans is also of important for globally endangered species. The mangrove habitat supports the largest population of tigers in the world which have adapted to an almost amphibious life.

6.1 Introduction

The Sundarbans is in the largest delta in the world, formed by the distributaries of the rivers Ganges and Brahmaputra, between India and Bangladesh (Papa et al. 2010). The Bay of Bengal forms the southern boundary of the delta, which includes the dense natural mangrove forest, the Sundarbans.

At 10,000 km^2, the Sundarbans forms the largest continuous mangrove forest in the world, with 40% of its area in India and the other 60% in Bangladesh (UNEP/WCMC 2008). There are two world heritage sites on the mangrove forest: the Sundarbans Reserve Forest in the southwest coast of Bangladesh and the Sundarbans National Park, on the east coast of India (UNESCO WHC 2016a, b) (Fig. 6.1).

The Sundarbans National Park (India) covers 1330 km^2. The area is estimated to be about 55% forest land and 45% wetlands in the form of tidal rivers, creeks, canals, and vast estuarine mouths of the Ganges River (UNESCO WHC 2016b). The Sundarbans Reserve Forest (Bangladesh) has 1395 km^2. The land area, including exposed sandbars, occupies 70% of the property, with water bodies covering 30% (UNESCO WHC 2016a). The Indian site lies about 130 km southeast of

V. Claudino-Sales, *Coastal World Heritage Sites*, Coastal Research Library 28,
https://doi.org/10.1007/978-94-024-1528-5_6

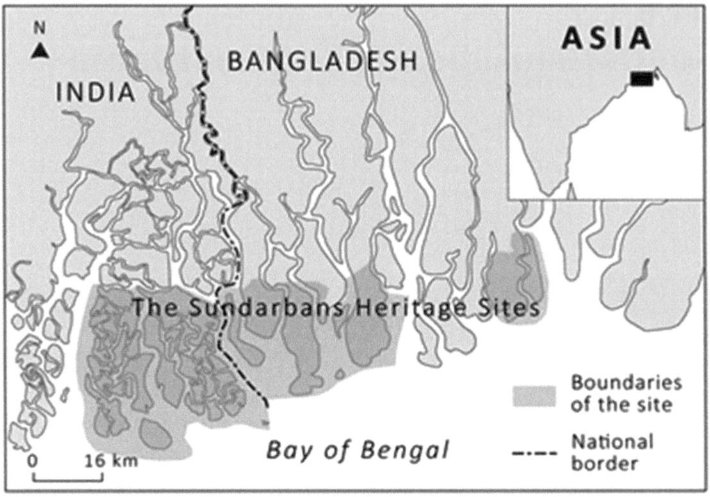

Fig. 6.1 The Sundarbans mangrove forest world heritage sites, in the Bay of Bengal, between Bangladesh and India

Calcutta, and the Bangladeshi site, about 300 km south-southwest of Dhaka. Together, the two sites cover one-fourth of the mangrove forest.

At the India side, the entire forest was made a protected reserve in 1875. Much of the area was later leased out by the government for cultivation. In 1926, the boundaries of the remaining protected forests were fixed (UNEP/WCMC 2008). In 1976, a wild sanctuary was established, and in 1984, the Sundarbans National Park was created, with a tiger reserve inside. The park was listed as a World Heritage site in 1987. It is managed by the Project Tiger program (UNESCO WHC 2016b).

On the Bangladesh side, the preservation of the mangrove forest came much later, in 1977, when the forest was divided into three wildlife sanctuaries, forming the Sundarbans Reserve Forest. It was designated a Wetland of International Importance under the Wetlands Convention (the Ramsar Convention) in 1992. In 1997, it became a World Heritage site. The site is managed by the Forest Department (UNESCO WHC 2016a, UNEP/WCMC 2008).

At the Indian site near the coast are the ruins of a city built approximately 200–300 AD. At the Bangladesh site, there is archaeological evidence of early human occupation on the deltaic islands. Nevertheless, human occupation in the Sundarbans ceased in the seventeenth century, reportedly due to pirate attacks (Christensen 1984).

Reclamation of the Sundarbans for agriculture started in 1770. Today, it provides sustainable livelihoods for millions of people who live in small villages in the vicinity of the sites. The inhabitants work as woodcutters, fishermen, honey gatherers, and leaves and grass gatherers (UNESCO WHC 2016a).

The Sundarbans support a successful tourism industry, with annual visitation increasing, especially international tourists. But 98% of the tourists are still domestic (Khanom and Buckley 2015). Tourists fly from Dhaka to Khulna, upstream of the

Fig. 6.2 Boats in a channel in the Sundarbans mangrove forest, India. The boats are the only way of transport in the area. (Picture reproduced by permission of Els Slots, website www.worldheritagesite.org)

mangroves, and then travel by boat (Fig. 6.2). Tigers are used widely in tourism advertising (Khanom and Buckley 2015).

Despite this development, tourism numbers remain relatively low due to the difficult access, transport, and lack of facilities including suitable accommodation. Effectively, there is little infrastructure within the forests (UNESCO WHC 2016a).

6.2 Natural Settings

6.2.1 Climatic Conditions

The Sundarbans has a subtropical climate moderated by the ocean, under the strong influence of the southwest or summer monsoon and weak northeast or winter monsoon. During the summer monsoon, almost half of it can be under water (Mukherjee et al. 2011).

Temperatures rise from daily minimum of 2–4 °C in winter to over 32 °C during the monsoon and a maximum around 43 °C in March. Rainfall is heavy, and the humidity averages 70–80% near the Bay of Bengal. The mean annual rainfall varies from about 1800 mm to 2800 mm on the Bangladesh coast, 80% of the rain falling during the summer monsoon between mid-June and October (UNEP/WCMC 2008).

Fig. 6.3 The macro-tidal delta-front mangrove forest of the Sundarbans, in Bangladesh. (Picture reproduced by permission of Nowak Lukasz/Shutterstock Stock Photos)

The prevailing wind is from the north and northeast from October to mid-March. However, violent southwesterlies prevail from mid-March to September. Storms, funneled up the shallow upper Bay of Bengal, are common in May and from October to November, sometimes developing into cyclones which can be accompanied by storm surges up to 7.5 m, causing enormous loss of life and damage to property and forests (UNEP/WCMC 2008). But it would be worse without the forest. The mangrove barriers absorb and reduce the direct impact of the cyclonic storms and accompanying surges moving in from the Bay of Bengal (Chatterjee et al. 2014).

6.2.2 Geology and Geomorphology

The Sundarbans, the mangrove forest on the fringes of the Ganges-Brahmaputra delta, is a macro-tidal, delta-front, and estuarine system (Fig. 6.3). Interconnecting the estuaries and forming the complex estuarine network are numerous channels, canals, and creeks (Chatterjee et al. 2014). The terrain consists of water bodies, intertidal mudflats and creeks, saline swamps, sandy shoals, and dense, impenetrable forests. The altitude goes from sea level to 3 m in Bangladesh and 10 m in India (UNEP/WCMC 2008).

The estuarine character is maintained by the semidiurnal tides at the mouths of the rivers and by the freshwater received from local runoff. The major portion of the local runoff comes annually from the summer monsoon rainfall. Considerable precipitation is also received by the frequent pre-monsoon (Attri and Tyagi 2010).

The landscape is one of low-lying forested alluvial islands, mudbanks with sandy beaches, and dunes along the coast (UNEP/WCMC 2008). The land is constantly being altered by tidal action, with erosion along estuaries and deposition along the banks, augmented by the discharge of silt from seawater (Sanyal and Bal 1986).

The area has three main hydrological zones: brackish, moderately saline, and saline, which influence the types of vegetation. As with the rest of the Bengal Plain, the alluvial deposits are geologically very recent and deep (Seidensticker and Hai 1983). The soils are saline and waterlogged, especially alkaline clay with an excess of salt, except on the coast, where sandy beaches occur (Lahiri 1973).

6.2.3 Biodiversity

The Sundarbans are classified as moist subtropical forest, comprised of a mosaic succession of four types of tidal forest communities: low mangrove forest, tree mangrove forest, saltwater Heritiera forest, and freshwater Heritiera forest (UNEP/WCMC 2008). Its exceptional biodiversity is expressed in a wide range of flora: 334 plant species belonging to 245 genera and 75 families, 165 algae, and 13 orchid species (UNESCO WHC 2016a). Mangrove swamp forest extends over half of the Sundarbans, the rest being largely brackish and salt water (Hussain and Acharya 1994), and forms 60% of the commercially useful timber (Mitra et al. 2011).

The Sundarbans is the only remaining habitat in the Bengal Basin with a great variety of fauna. Some species are now locally extinct due to agricultural reclamation and the increase in soil salinity during the twentieth century (UNEP/WCMC 2008). The rich fauna is represented by 693 species of wildlife, which includes 49 mammals, 59 reptiles, 8 amphibians, 210 white fishes, 24 shrimps, 14 crabs, and 43 mollusks species (UNEP/WCMC 2008).

On the Bangladesh side, the varied and colorful birdlife found along the waterways is one of its greatest attractions. There are 315 species of waterfowl, raptors, and forest birds including 9 species of kingfisher and the white-bellied sea eagle (UNESCO WHC 2016b).

The Sundarbans is important for globally endangered species, including the Ganges and Irrawaddy dolphins, estuarine crocodiles, and the critically endangered endemic river terrapin. The mangrove habitat also supports the single largest population in the world of the endangered Royal Bengal Tiger, which have adapted to an almost amphibious life, being capable of swimming for long distances and feeding on fish, crab, and water monitor lizards (UNESCO WHC 2016b).

6.3 Conservation Challenges

The Sundarbans mangrove forest has substantial ecological and economic importance at local, national, and global levels, but it is at risk. Because of the changes in sea level and dynamic energy system in the transition zone between land and sea due to climate change, higher rates of erosion on this ecosystem are expected (Rahman 2012).

Effectively, sea level rise, as well as alteration of water flows of the Himalayan headwaters, is among the major disturbances threatening the Sundarbans mangrove forest. Accretion rate is decreasing, and the coastline is currently in a state of net erosion (Rahman et al. 2011). From 1983 to 2003, the annual sea level rise in the area was 3.14 cm compared with the world average of 2 cm, and the outer islands began to erode (Colette 2007). Rising sea levels also increased saltwater intrusion into aquifers.

Other challenges are related to invasive species spreading significantly in the mangrove ecosystem (Biswas et al. 2007), severe degradation from oil spills, large-scale clearance and timber harvesting, overfishing by commercial fisheries, subsistence hunting, and poaching of wildlife including tiger (Aziz et al. 2013).

The International Union for Conservation of Nature (IUCN) published a report in the end of 2017 on the state of conservation of the World Heritage sites. The publication found that conservation of the Sundarbans is good but presents some concerns related to the aspects pointed out above. Nevertheless, the general conditions improved since the last report, in 2014 (Osipova et al. 2017).

References

Attri SD, Tyagi A (2010) Climate profile of India. Monograph Environment Meteorology 01/2010. India Meteorological Department, Ministry of Earth Sciences, Government of India

Aziz A, Barlos ACD, Greenwood CC, Islam A (2013) Prioritizing threats to improve conservation strategy for the tiger Panthera tigris in the Sundarbans Reserve Forest of Bangladesh. Oryx 47:510–518

Biswas SR, Choudhury JK, Nishat A, Rahman MM (2007) Do invasive plants threaten the Sundarbans mangrove forest of Bangladesh? For Ecol Manag 245(1):1–9

Chatterjee M, Shankar D, Sen G, Sanyal P, Sundar D, Michael G, Chatterjee A, Amol P, Mukherjee D, Suprit K, Mukherjee A, Vijith V, Chaterjee S, Basu A, das M, Chakraborti S, Kalla A, Misra S, Mukhopadhyay S, Mandal G, Sarkar K (2014) Tidal variations in the Sundarbans estuarine system, India. J Earth Syst Sci 122(4):899–933

Christensen B (1984) Ecological aspects of the Sundarbans. Food and Agricultural Organization (FAO), Rome

Colette A (2007) Case studies of climate change and world heritage. UNESCO World Heritage Centre, Paris

Hussain Z, Acharya G (1994) Mangroves of Sundarbans, vol 2. IUCN, Gland, Switzerland

Khanom S, Buckley R (2015) TIger tourism in the Bangladesh Sandarbans. Ann Tour Res 55(C):178–180

Lahiri R (1973) Management plan of Tiger Reserve in Sundarbans, West Bengal, India. Department of Forests, Kolkata

Mitra A, Sengupta K, Banerjee K (2011) Standing biomass and carbon storage of the above-ground structures in dominant mangrove trees in the Sundarbans. For Ecol Manag 216(7):1325–1335

Mukherjee I, Chakraborty N, Mandal T (2011) Water, air, & soil pollution 215(1):477–486

Osipova E, Shadie P, Zwahlen C, Osti M, Shi Y, Kormos C, Bertzky B, Murai M, Van Merm R, Badman T (2017) IUCN world heritage outlook 2: a conservation assessment of all natural world heritage sites. IUCN, Gland

Papa F, Durand F, Rossow WB, Rahman A, Bala SK (2010) Satellite altimeter-derived monthly discharge of the Ganga–Brahmaputra River and its seasonal to interannual variations from 1993 to 2008. J Geophys Res 115:1–19

Rahman MM (2012) Time-series analysis of the coastal erosion in the Sundarbans mangrove. Int Arch Photogramm Remote Sens Spat Inf Sci XXXIX-B8:425–429

Rahman AF, Dragoni D, El-Masri B (2011) Response of the Sundarbans coastline to sea level rise and decreased sediment flow: a remote sensing assessment. Remote Sens Environ 115(12):3121–3128

Sanyal P, Bal A (1986) Some observations on abnormal adaptations of mangrove in Indian Sundarbans. Indian Soc CoastAgric Res 4:9–15

Seidensticker J, Hai M (1983) The Sundarbans wildlife management plan: conservation in the Bangladesh coastal zone. International Union for the Conservations of Nature (IUCN), Gland

UNEP/WCMC (United Nations Environmental Programme/World Conservation Monitoring Centre) (2008) Sundarbans National Park India The Sundarbans Bangladesh. Rapport, Cambridge, UK

UNESCO WHC (United Nations Educational Scientific and Cultural Organization World Heritage Centre) (2016a) The Sundarbans. http://whc.unesco.org/en/list/798. Accessed 10 Jan 2016

UNESCO WHC (United Nations Educational Scientific and Cultural Organization World Heritage Centre) (2016b) Sundarbans National Park. http://whc.unesco.org/en/list/452. Accessed 10 Jan 2016

Chapter 7
Puerto Princesa Subterranean River National Park, The Philippines

Abstract Puerto Princesa Subterranean River National Park, in the Philippines, contains one of the world's most impressive cave systems with spectacular limestone karst landscapes. It includes an underground river that flows directly to the sea. The lower half of the river is brackish and subject to ocean tide, which makes this an important natural phenomenon. The river's cavern exhibits dramatic speleothems and several large chambers. Its accessibility and navigability inland allows it to be experienced by the general public, who can view the rock formations on spectacular river cruises. The park contains significant habitat for biodiversity conservation. It includes a full mountain-to-sea ecosystem, protecting the largest limestone forest in Asia. The park was voted one of the New Seven Wonders of the World in 2012.

7.1 Introduction

Puerto Princesa Subterranean River National Park is in the southwestern part of the Philippine archipelago on the coast of the Palawan Island district, 360 km southwest of Manila. The park, in the Saint Paul Mountain Range, has an area of approximately 222 km^2, containing an 8.2 km long underground river and a marine zone of 2,9 km^2 (UNESCO WHC 2016) (Fig. 7.1).

The area was declared a national park (then named Saint Paul Subterranean River National Park) by the government of the Philippines in 1971. It was added to the World Heritage List in 1999. The park is administered by the National Park Management Board (PAMB) of the City of Puerto Princesa, 70 km from the park (UNEP/WCMC 2005).

The park territory and surroundings are the ancestral lands of the negrito Batak people, of whom only 200 to 250 survive, and of the Tagbanua communities who live around the boundaries, including the coast (UNEP/WCMC 2005). Inside the park, the only residents are its staff members.

Tourism is very well developed in the park. After 2012, when the site was voted as one of the New Seven Wonders of the World, tourist activities increased by 300% (Rubite et al. 2015; Fitzgerald 2012). Now, the activities are subjected to control by the park administration for conservation purposes (UNESCO WHC 2016).

V. Claudino-Sales, *Coastal World Heritage Sites*, Coastal Research Library 28,
https://doi.org/10.1007/978-94-024-1528-5_7

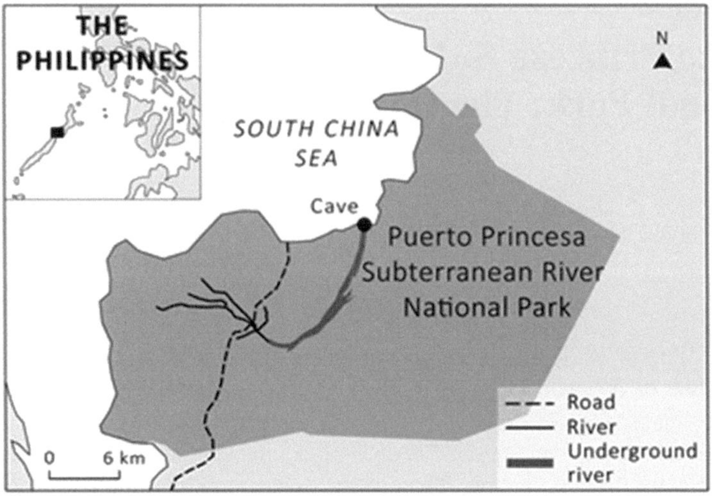

Fig. 7.1 Puerto Princesa Subterranean River National Park, in the district of Palawan, archipelago of the Philippines

7.2 Natural Settings

7.2.1 Climatic Conditions

The climate in Palawan is tropical. The mean annual rainfall averages between 2000 and 3000 mm. The wet season is from May to October and the dry season, from November to April. The average temperature is 27 °C (UNEP/WCMC 2005).

The archipelago is susceptible and vulnerable to weather-related hazards. Extreme weather events, which include tropical cyclones, monsoon rains, and dry spells, have happened and triggered hazards such as floods and landslides (Yumul et al. 2011).

Effectively, the area is exposed to outermost storms, such as the Typhoon Haiyan that devasted the country in 2013 (Gocotano et al. 2015). It was the most extreme event of its type registered in history (PAGASA 2013).

7.2.2 Geology and Geomorphology

The Subterranean River National Park includes various landforms from flat plains and rolling hinterlands to hills, cliffs, mountain peaks, and beaches (Fig. 7.2). The heights go from sea level to 1028 m at Mount Saint Paul. It centers on the Cabayugan River which runs underground for 8.2 km through a massive cave, flowing directly into the sea at St. Paul's Bay (UNEP/WCMC 2005).

Fig. 7.2 Limestone seacliffs in the small marine area that is part of the Puerto Princesa Subterranean River National Park, in the Philippines. (Photo reproduced by permission of Justin Watt, website http://justinsomnia.org/2011/01/a-week-in-palawan/)

The underground river is the site of a huge mixing phenomenon between fresh river water and salty sea water. The lower half of the river is brackish and subject to ocean tides. The river's outflow is about 200 m from the coastline, and tides push their influence as much as 6 km inside the cave (Vivo et al. 2009). The system may reasonably be considered the world's most classical example of an underground estuary (Forti et al. 1993).

The park has one of the world's most impressive cave systems, featuring spectacular limestone karst landscapes. The most striking of the features is the geologically young karst limestone mountain landscape of the Saint Paul Range, which is part of a series of rounded limestone peaks aligned north-south along the western coast of Palawan. More than 90% of the park is made of sharp karst limestone ridges around Mount St. Paul (UNEP/WCMC 2005; see Fig. 7.2).

The karst covers an area of approximately 35 km and is made up of a massive to roughly stratified, micritic, light to dark gray colored, lower Miocene limestone showing levels rich with fossils. The cave, more than 24 km long, is composed of an active level and huge fossil branches (Vivo et al. 2009; Hashimoto et al. 1973).

The karst features extremely steep shapes known as "towers" (Balazs 1973). The cathedral-like cave includes major formations of stalactites and stalagmites, and there are several large chambers up to 120 m wide and 60 m high (UNEP/WCMC 2005). Its accessibility and navigability up to 4.5 km inland allows it to be experienced by the general public, which can view the rock formations on a river cruise (UNESCO WHC 2016) (Fig. 7.3).

Fig. 7.3 Boat entering the cathedral-like cave developed under Saint Paul Mount, in the island of Palawan, the Philippines. The boats can go 4.5 km inland. (Picture reproduced by permission of Justin Watt, website http://justinsomnia.org/2011/01/a-week-in-palawan/)

7.2.3 Biodiversity

The rich flora of Palawan Island is radically different from that of the rest of the Philippines, preserving relics of the period when the island was connected by a land bridge with Borneo: Palawan Island is a remnant of the land bridges that formed during glacial periods and consequently supports a biologically rich flora closer to Borneo's than to the rest of the Philippines (UNEP/WCMC 2005).

The park contains a full mountain-to-sea ecosystem which provides significant habitat for biodiversity conservation and protects the most intact and noteworthy forests within the Palawan biogeographic province (UNESCO WHC 2016).

Eight different forest formations are present: forest on ultramafic soil, forest on limestone soil, montane forest, freshwater swamp forest, lowland evergreen tropical rainforest, riverine forest, beach forest, and mangrove forest. There are 800 recorded plant species, 295 of them trees (Rubite et al. 2015).

The fauna is moderately rich, especially in invertebrates. There are 30 mammal, 18 reptile, 10 amphibian, 41 butterfly, 62 reef fish, and 91 bird species (UNEP/WCMC 2005).

7.3 Conservation Challenges

Puerto Princesa Subterranean National Park holds the distinction of being the first national park devolved and successfully managed by a local government in the Philippines. The park's effective management system is a symbol of commitment by the Filipino people to the protection and conservation of their natural heritage. The property, including the karst mountain landscapes and the underground river, is in excellent condition (UNESCO WHC 2016).

Tourism, identified as a potential threat, is being addressed through management. The biggest threats to the property are from adverse activities in adjacent catchment areas, primarily forest clearing and agricultural activities (UNESCO WHC 2016; Mallari et al. 2013).

Households and communities surrounding the park used to gather rattan and almaciga (a giant coniferous tree which is the source of a lacquer resin) or practice swidden agriculture within the buffer areas, but it was stopped by the government to protect the park (UNEP/WCMC 2005). Some consider these preservation actions risk marginalization and loss of livelihood (Wolfram 2014).

Another environmental problem that potentially put the park at risk is global warming. The studies related to the site in particular are not available, but some researchers consider that the Philippines have recently experienced distinct changes in the weather patterns with disastrous results. These changes included too much precipitation, dry spell and drought conditions, and abnormalities and variance in weather patterns (Yumul et al. 2011). The situation may be part of climate change (Yumul et al. 2013).

The International Union for Conservation of Nature (IUCN) reported, in its last outlook on the conservation of the site, that the state of conservation of the park is good, but the situation indicated above creates some concerns that need to be dealt with by the managers (Osipova et al. 2017).

References

Balazs D (1973) Karst types in the Philippines: proceedings of the 6th International Congress of Speleology 2:13–38, Praha

De Vivo A, Piccini L, Mecchia M (2009) Recent explorations in the Saint Paul Karst (Palawan, the Philippines) Proceedings of the 15th International Congress of Speleology 3:1786–1792, Kerville, USA

Fitzgerald E (2012) PPUR officially confirmed as one of the New 7 Wonders of Nature. http://news.n7w.com/2012/01/28/3743/?utm-source-rss&utm_medium. Accessed 4 Jan 2016

Forti P, Piccini L, Rossi G, Zorzin R (1993) Note preliminari sull'idrodinamica del sistema carsico di St. Paul (Palawan, Filippine). Bulletìn Societé Geographìque de Liege 29:37–44

Gocotano AE, Dico FD, Calungsod NR, Hall JL, Counahan ML (2015) Exposure to cold weather during a mass gathering in the Philippines. Bull World Health Organ 93(11):810–814

Hashimoto P, Sato A, Tadashi R (1973) Geologic structure of North Palawan and its bearing on the geological history of the Philippines. Geol Palaeontol SE Asia 13:145–161

Mallari NAD, Collar NJ, McGowan PJK, Marsden SJ (2013) Science-driven Management of Protected Areas: a Philippine case study. Environ Manag 51:1236–1246

Osipova E, Shadie P, Zwahlen C, Osti M, Shi Y, Kormos C, Bertzky B, Murai M, Van Merm R, Badman T (2017) IUCN world heritage outlook 2: a conservation assessment of all natural world heritage sites. IUCN, Gland

PAGASA (Philippine Atmospheric, Geophysical and Astronomical Services Administration) Tropical Cyclone Warning: Typhoon "Yolanda" (Haiyan) Severe Weather Bulletin Number One (Report) November 6, 2013. Accessed 3 Jan 2016

Rubite RR, Hughes M, Blanc P, Chung KF, Yang HA, Kono Y, Alejandro GJD, Layola LBD, Virata AGN, Peng CI (2015) Three new species of Begonia endemic to the Puerto Princesa Subterranean River National Park, Palawan. Bot Stud 56:19

UNEP/WCMC (United Nations Environmental Programme/World Conservation Monitoring Centre) (2005) Puerto Princesa Subterranean River National Park the Philippines. Report, Cambridge, England

UNESCO WHC (United Nations Educational, Scientific and Cultural Organization World Heritage Centre) (2016) Puerto Princesa Subterranean River National Park. http://whc.unesco.org/en/list/652. Accessed 3 Jan 2016

Wolfram D (2014) Green governmentality and Swidden decline on Palawan Island. Trans Inst Br Geogr 39(2):250–264

Yumul GP, Cruz NA, Servando NT, Dimalanta CB (2011) Extreme weather events and related disasters in the Philippines, 2004–08: a sign of what climate change will mean? Disasters 35(2):362–382

Yumul G, Dimalanta C, Servando N, Cruz N (2013) Abnormal weather events in 2009, increased precipitation and disastrous impacts in the Philippines. Clim Chang 118(3–4):715–727

Part III
World Heritage on Continental Coasts: Central America

Chapter 8
Área De Conservación Guanacaste, Costa Rica

Abstract The Área de Conservación Guanacaste contains land and sea in the Northwest of Costa Rica. The property extends from the shore of the Pacific to the lowland rainforests in the hinterland. The many forest types include a large tract of tropical dry forest, oak forests, and savannas, as well as extensive wetlands and numerous watercourses. The largely intact coastal-marine area has estuaries, sea-cliffs, dunes, and sandy and cobble beaches surrounding the mostly uninhabited islands and islets. It contains important natural habitats for the conservation of biological diversity, and expressive ecological processes in both its terrestrial and marine-coastal environments continue.

8.1 Introduction

The Área de Conservación Guanacaste (Guanacaste Conservation Area), in Costa Rica, comprises a single biogeographical block of uninterrupted wildlife-protected areas of 1630 km^2. The property encompasses roughly 100 km from the shore of the Pacific to the rainforests of the hinterland: the site extends from the marine area of the Gulf of Papagayo, at the Archipelago Murciélago, through the Plateau Santa Rosa to the top of the Orosi, Cacao, and Rincon de la Vieja volcanoes, and continuing through the lowlands of the Caribbean side of Costa Rica (UNESCO WHC 2016).

The coastal segment is 430 km long. The marine sector extends for 22 km offshore, following the shoreline (SINAC 2012) (Fig. 8.1). The property is formed by four protected areas: Santa Rosa National Park, Rincon de la Vieja National Park, Guanacaste National Park, and Bahia Junquillal Wildlife Refuge. Two of these protected areas have coastal and marine segments: Santa Rosa National Park and Bahia Junquillal Wildlife Refuge.

The Área de Conservación Guanacaste was declared a protected zone by the Costa Rican government in 1986. It became a World Heritage site in 1999 (UNESCO WHC 2016). It is a fully functional decentralized governmental unit with a management endowment, professional Costa Rican resident staff members, and programs in education, research, tourism, reforestation, biodiversity development, fire control, and police protection (UNEP/WCMC 2003).

V. Claudino-Sales, *Coastal World Heritage Sites*, Coastal Research Library 28,
https://doi.org/10.1007/978-94-024-1528-5_8

Fig. 8.1 Location of Área de Conservación Guanacaste, Costa Rica

The Área de Conservación Guanacaste is an important productive element of Costa Rica economy, due to domestic and international tourism. The area has numerous facilities for tourists (SINAC 2012).

The coastal segments of Área de Conservación Guanacaste are pristine and uninhabited (UNESCO WHC 2016). The biological aspects of the forest ecosystems and volcanoes of the protected area are the object of some research, but general studies concerning the coastal sectors are practically nonexistent.

8.2 Natural Settings

8.2.1 Climatic Conditions

The climate is tropical with a dry season and a wet season. The dry season is November through May. It is characterized by high ambient temperatures regularly exceeding 37 °C, low relative humidity, and the near absence of precipitation (Campos and Fedigan 2009).

In the coast, the ocean moderates the temperatures, which are stable and average 28 °C (Costa Rica Tourism Board 2016). During the dry season, the area is also affected by strong trade winds. The wet season, from December to April, is characterized by rain coming from the Pacific Ocean, with an average of 1400 mm per year (SINAC 2012).

8.2.2 *Geology and Geomorphology*

The pristine coastal-marine interface features estuaries, seacliffs, rocks, dunes, and sandy and cobble beaches rimming the marine area with its various, mostly uninhabited nearshore islands and islets (UNESCO WHC 2016) (Figs. 8.2 and 8.3). The coast is formed by volcanic rocks, especially basalts, of Tertiary age. The basalts support the islands and islets (UNEP/WCMC 2003). There are also marine sedimentary rocks from the Jurassic-Triassic periods and Precambrian rocks of the mantle. Recent deposits include alluviums, sandbars, and dunes. Pleistocene dunes occur, with ages estimated in 17000 years (SINAC 2012).

The area's elevation reaches 1900 m inland, in Rincon de la Vieja Volcano. The volcanoes that characterize the continental sectors are active and include hot spring zones, fumarole lagoons, mud pots, and steam emanations (Costa Rica Tourism Board 2016).

8.2.3 *Biodiversity*

The many forest types include a large tract of tropical dry forest as well as oak forests and savannas (UNESCO WHC 2016). It is formed by a continuous block of marine-coastal ecosystems, dry tropical forest, tropical rainforest, and nebular forest. This is the only transect of this type in the Americas (UNESCO WHC 2016).

Fig. 8.2 Sandy beaches and Tertiary marine rocks in seacliffs in the coast of the Gulf of Papagayo, in Santa Rosa National Park, Guanacaste Conservation Area, Costa Rica. (Picture reproduced by permission of Ruth Peterkin/Shutterstock Stock Photos)

Fig. 8.3 Pocket beaches in the basaltic seacliffs in the coast of Santa Rosa National Park, Gulf of Papagayo, Guanacaste Conservation Area, Costa Rica. (Picture reproduced by permission of Eqroy/Shutterstock Stock Photos)

The area contains approximately 235000 species (65% of the species of Costa Rica), representing 2.6% of the world biodiversity (SINAC 2012).

Santa Rosa National Park has one of the largest protected dry forests in Central America. There are nine different habitats, such as deciduous forests, forests of altitude, evergreen forests, dry forests, mangrove, swamp, coastal vegetation, savanna, and pasture (Costa Rica Tourism Board 2016). The beaches are important for turtle nesting. In the marine area, there are coral reefs, some containing species that are unique in the Pacific Oriental (SINAC 2012; Servicio de Parques Nacionales 1995).

The site is considered an altitudinal biological corridor that enables the seasonal migration of a large number of species including insects, birds, and mammals (Janzen 1996). They come down to the coast, at Santa Rosa National Park, during the dry season, and ascend to the high areas including the Orosi Volcano and the Cacao Volcano in the wet season (Costa Rica Tourism Board 2016).

It is believed that 960 species of vertebrates, more than 7000 species of plants, and more than 8000 species of butterflies and moths live in the protected area (UNEP/WCMC 2003). The large vertebrates, such as tapirs, jaguars, and mountain lions, require large land areas to maintain a healthy breeding population, which was one reason for the creation of the protected area (Janzen 1988).

8.3 Conservation Challenges

The largely intact coastal-marine interface is remarkable in a region where coasts have disproportionally suffered from human pressure (UNESCO WHC 2016). Nevertheless, these protected areas are threatened by agricultural chemicals, solid waste, and heavy metals, as well as by construction, fishing, and tourism without planning (SINAC 2012).

Fire produced by humans is also a problem in the Área de Conservación Guanacaste. The tropical dry forest is suffering from agricultural activities near the protected area (UNEP/WCMC 2003). Invasive plants and animals also require attention (SINAC 2012).

Climate change is threatening the heritage site. It is changing the altitudinal migration between the dry forest and the humid mountains of some species of birds, insects, and mammals (Colette 2009). The migration is expected to increase in coming years. Climate change has also reduced biodiversity in recent years (Colette 2009).

The last report of the International Union for Conservation of Nature (IUCN) about the state of conservation of the World Heritage sites indicated that the Área de Conservación Guanacaste presents significant concerns for the maintenance of the values that justify its listing as a World Heritage site. The situation needs urgent attention from managers (Osipova et al. 2017).

References

Campos FA, Fedigan LM (2009) Behavioral adaptions to heat stress and water scarcity in white-faced capuchins (Cebus Capucinus) in Santa Rosa National Park, Costa Rica. Am J Phys Anthropol 138(1):101–111

Colette A (2009) Case studies on climate changes and world heritage. United Nations Educational Scientific and Cultural Organization (UNESCO) World Heritage Center, Paris

Costa Rica Tourism Board (2016) http://www.visitcostarica.com. Accessed 16 Jan 2016

Janzen DH (1988) Guanacaste National Park: tropical ecological and biocultural restoration. In: Cairns JJ (ed) Rehabilitating damaged ecosystems, vol II. CRC Press, Boca Raton

Janzen DH (1996) On the importance of systematic biology in biodiversity development. ASC Newsletter 24:23–28

Osipova E, Shadie P, Zwahlen C, Osti M, Shi Y, Kormos C, Bertzky B, Murai M, Van Merm R, Badman T (2017) IUCN world heritage outlook 2: a conservation assessment of all natural world heritage sites. IUCN, Gland

Servicio de Parques Nacionales (1995) Parque Nacional Santa Rosa http://www.nacion.co.cr/netinc/costarica/parques/santa.rosa.html. Accessed 17 Jan 2016

SINAC (Sistema Nacional de Áreas de Conservación) (2012) Área de Conservación Guanacaste. http://wwwsinacgocr/AC/ACG/. Accessed 17 Jan 2016

UNEP/WCMC (United Nations Environment Programme/World Conservation Monitoring Centre) (2003) Area de Conservación Guanacaste, Costa Rica. Report, Cambridge, England

UNESCO WHC (United Nations Educational, Scientific and Cultural Organization World Heritage Centre) (2016) Area de Conservación Guanacaste. http://whcunescoorg/en/list/928. Accessed 17 Jan 2016

Chapter 9
Río Plátano Biosphere Reserve, Honduras

Abstract Río Plátano Biosphere Reserve is the largest surviving area of undisturbed tropical rainforest in Honduras and is one of the few remaining in Central America. The Biosphere Reserve covers virtually the entire watershed of the Plátano River. The preserved area includes steep mountains which form 75% of the site and a flat to undulating coastal plain. Its main ecotypes are estuarine and marine, mangrove swamps, coastal pine savanna, broadleaf gallery forest, humid subtropical forest, very humid tropical forest, and, on the ridge tops, elfin forest. The forested valley and its coastal plain contain a wide range of habitats with abundant plant and wildlife. There, more than 2000 indigenous people have preserved their traditional way of life. The reserve also contains the site where Christopher Columbus discovered America in 1492.

9.1 Introduction

Río Plátano Biosphere Reserve is located on the watershed of the Plátano River, in Honduras, Central America. The site occupies an area of 3500 km^2 and is the biggest preserved area in the country (UNESCO WHC 2016a, b). It is in the northeastern region of Honduras, in the hinterland of the Mosquitia Coast, extending inland southwest approximately 150 km by 25 km (UNEP/WCMC 2008) (Fig. 9.1).

The area was designated a biosphere reserve under the UNESCO Man and Biosphere Program in 1980. In 1982, it was added to the World Heritage List (UNEP/WCMC 2008). In 1996, it was listed in the World Heritage List as in danger, where it remained until 2007. In 2011, it was returned to list, where it remains (UNESCO WHC 2016a).

The region has long been inhabited by humans. There are more than 200 archaeological sites. One is the Ciudad Blanca (White City), one of the most important archaeological sites of the Mayan civilization. For that reason, the biosphere is also an archaeological park (Herrera-MacBryde 2004).

V. Claudino-Sales, *Coastal World Heritage Sites*, Coastal Research Library 28,
https://doi.org/10.1007/978-94-024-1528-5_9

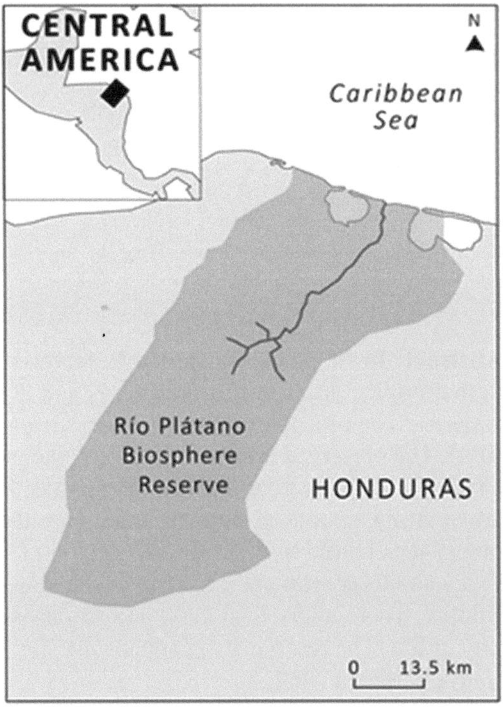

Fig. 9.1 Río Plátano Biosphere Reserve, in the hinterland of the coastal plain in the Caribbean Sea, Honduras

The reserve also contains the site where Christopher Columbus discovered the Americas in 1492 (UNEP/WCMC 2008). In its mountainous landscape sloping down to the Caribbean Coast, more than 2000 indigenous people – the Miskitos – have preserved their traditional way of life (UNESCO WHC 2016b). It is also an Amerindian reserve (Herrera-MacBryde 2004).

The area is very poorly studied, especially its natural settings. Tourism is an increasing activity in the site and is done mainly by boat (UNEP/WCMC 2008).

9.2 Natural Settings

9.2.1 Climatic Conditions

The climate is hot and humid all year, especially from May to November. The mean annual temperature is 26.6 °C. Annual precipitation varies locally from less than 2850 mm to 4000 mm during the rainiest months of November to January (UNEP/WCMC 2008).

The dry season from February to May is more pronounced since the deforestation of the last 20 years (Richards 1998). In an average decade, the region is hit by four tropical storms and two hurricanes (UNEP/WCMC 2008).

Fig. 9.2 Plátano River in one of its rocky valley segments, in the Río Plátano Biosphere Reserve. It has his mouth at the Mosquitia Coast, in the Caribbean Sea. (Picture reproduced by permission of Robert E. Hyman)

Along the coast, the climate is dictated by easterly trade winds which cause a rainy season for approximately 8 months of the year (Harbone et al. 2001). Rainfall is greater than 2000 mm per year, and air temperatures range from 25 °C to 29 °C (D'Croz et al. 1998).

9.2.2 Geology and Geomorphology

The reserve comprises two main geomorphological areas: steep mountains, which form three quarters of the site, and a flat to undulating coastal plain. The Biosphere Reserve covers virtually the entire 3500 km^2 watershed of the 115 km-long Plátano River which bisects the reserve (Fig. 9.2) (UNEP/WCMC 2008).

The mountainous segment is characterized by many steep ridges, with altitudes of 1500 m. Remarkable geological formations in the rugged upland region include the exposed granitic pinnacle El Viejo or Pico de Dama (Cruz 1986). Cataracts and cascading waterfalls are found. In one cataract the Plátano River almost disappears among massive boulders in a gorge flanked by forested escarpments 100 m high (Cruz 1986).

The undulating to flat segment of Mosquitia Coast extends from a few to over 40 km inland and rises gradually from sea level to almost 100 m in altitude, where the foothills begin abruptly (Herrera-MacBryde 2004). The Plátano River meanders

Fig. 9.3 Mosquitia Coast, showing a barrier island with sandy beaches isolating a lagoon (Brus Lagoon). There are also mangroves and reefs in the coastal and marine areas, respectively. (Picture reproduced by permission of Nature Picture Library/Alamy)

for 45 km through this area, forming oxbow lakes, backwater swamps, and natural levees which are used for agricultural plots (Herrera-MacBryde 2004).

The Mosquitia Coast (Fig. 9.3) has a complex environment of sandy beaches, reefs, a lagoon, a lake, wetlands, and barrier beaches in an expansive savanna which plays a key role in fisheries health (Sealey and Bustamante 1999). The savanna formation occurs on deeply weathered quartz sandy gravels of Pleistocene age in a strip averaging 45 km wide (to 180 km maximum) (Herrera-MacBryde 1995).

9.2.3 Biodiversity

Río Plátano Biosphere Reserve is part of the largest surviving area of undisturbed tropical rainforest in Honduras and is one of the few remaining in Central America (UNESCO WHC 2016b). The upland portion of site is covered by moist to wet forests which are poorly known. Probably over 2000 vascular plant species occur in the area

(Nelson 1978). Its topographic variety results in a range of 25 ecosystems (DGIS-WWF 2001). Its main ecotypes are estuarine and marine, mangrove swamps, coastal pine savanna, broadleaf gallery forest, humid subtropical forest, very humid tropical forest (80% of the area), and, on the ridge tops, elfin forest (UNEP/WCMC 2008).

The Mosquitia may be the least known region of Honduras, where species new to the country, phytogeographically distinctive, and/or new to science are found whenever a research trip can be carried out (Froehlich and Schwerin 1983). Mangrove vegetation is common. The most extensive mangrove ecosystems fringe the large coastal lagoon of Brus (brackish) (see Fig. 9.3) and the lake Ibans (freshwater). Although some mangroves have been cut, the area still retains much of the original formation, with *Rhizophora mangle* as the characteristic species (Herrera-MacBryde 1995).

The reserve is said to have 70% of the fish, 57% of the birds, and 57% of the reptiles of Honduras. Thirty-nine species of mammals, 377 species of birds, and 200 species of reptiles and amphibians have been recorded (UNEP/WCMC 2008). The park is within one of the world's Endemic Bird Areas (Stattersfield 1998).

9.3 Conservation Challenges

The Mosquitia region's natural and cultural richness can rarely be seen elsewhere in the world (UNESCO WHC 2016a, b). But the future of this exceptional richness, in a country torn by conflict, poverty, and adverse socioeconomic conditions, strongly depends on the readiness of its authorities to consolidate efforts to manage and conserve it (IUCN 2011).

The reserve was placed on the World Heritage List in Danger in 1996 and in 2011, because of encroachment by cattle ranchers, loggers, illegal hunters, wildlife traders, and burn colonists and also because of the inadequate management of its natural and cultural resources by the government (UNESCO WHC 2016b).

Illegal livestock grazing and agricultural intrusions, invasion by exotic species, and deficiencies in management continue to threaten the reserve (UNESCO WHC 2016a, b). Besides these problems, global warming and climatic changes probably will reduce the humidity in the uplands of the humid forest, as is the case in other humid forests in the world. Coastal erosion will probably also be a problem in the Mosquitia Coast with climate changes.

The International Union for Conservation of Nature (IUCN) in its last report about the state of conservation of World Heritage sites considered that Río Plátano Biosphere Reserve has a critical conservation outlook. Sites with a critical conservation outlook are severely threatened and require urgent, additional, and large-scale conservation measures, or the values that justify their presence in the World Heritage List may be lost (Osipova et al. 2017). These sites face a range of threats, and in many cases, managers have low capacity to address them, as seem to be the case in relation to Río Plátano Biosphere Reserve.

References

Cruz GA (1986) Areas silvestres de Honduras: guía de los parques nacionales, refugios de vida silvestre, reservas biológicas y monumentos naturales de Honduras. Asociación Hondureña de Ecología, Tegucigalpa

D'Croz RD, Jackson JB, Best MMR (1998) Siliciclastic-carbonate transition along the shelf transects through the Cayos Conchinos archipelago, Honduras. Revista de Biologia Tropical 45(4):57–66

DGIS-WWF (Directorate-General for International Cooperation-World Wild Fund) (2001) Tropical Forest portfolio quarrels in a crowded reserve. The Rio Platano project in Honduras. World Wildlife Found (WWF) International, Gland

Froehlich JW, Schwerin KH (eds) (1983) Conservation and indigenous human land use in the Río Plátano watershed, Northeast Honduras, Research Paper Series No. 12. Latin American Institute, University of New Mexico, Albuquerque

Harbone AR, Afzal DC, Andrews MJ (2001) Honduras: Carribean coast. Mar Pollut Bull 42(12):1221–1235

Herrera-MacBryde O (1995) Northeast Honduras and Rio Platano biosphere reserve. Department of Botany, Smithsonian Institution, Washington, DC

Herrera-MacBryde O (2004) North-east Honduras and Rio Platano biosphere reserve. https://www.nmh.si.edu/botany/projects/cpd/ma/ma15.htm. Acessed 28 Dec 2016

IUCN (International Union for the Conservation of Nature) (2011) A promising future for Rio Platano. http://www.iucn.org/?7579. Accessed 24 Feb 2016

Nelson C (1978) Contribuciones a la flora de La Mosquitia, Honduras. Ceiba 22(1):41–64

Osipova E, Shadie P, Zwahlen C, Osti M, Shi Y, Kormos C, Bertzky B, Murai M, Van Merm R, Badman T (2017) IUCN world heritage outlook 2: a conservation assessment of all natural world heritage sites. IUCN, Gland

Richards M (1998) Protected areas, people and incentives: the search for sustainable forest conservation in Honduras. Overseas Development Institute, Regent's College, London

Sealey SK, Bustamante G (1999) Setting geographic priorities for marine conservation in Latin America and in the Caribbean. The Nature Conservancy, Arlington/Virginia

Stattersfield A (1998) Endemic bird areas of the world: priorities for their conservation. BirdLife International, Cambridge

UNEP/WCMC (United Nations Environmental Programme/World Conservation Monitorign Centre) (2008) Rio Platano biosphere reserve Honduras. Report, Cambridge

UNESCO WHC (United Nations Educational, Scientific and Cultural Organization World Heritage Centre) (2016a) World heritage list. http://whc.unesco.org/en/list/. Accessed 26 Feb 2016

UNESCO WHC (United Nations Educational, Scientific and Cultural Organization World Heritage Centre) (2016b) Río Plátano biosphere reserve. http://whc.unesco.org/en/list/196. Accessed 26 Feb 2016

Chapter 10
Darien National Park, Panama

Abstract The Darien National Park forms the physiographic link between Central and South America. The park extends along about 80% of the Colombian border and includes part of the Pacific coast. It forms a transfrontier site with Los Katios National Park, in Colombia, also a World Heritage site. It is a hot, humid area typified by tropical rainforests, mangrove swamps, sandy beaches, rocky coasts, freshwater marshes, palm forest swamps, and low mountain ranges with cloud forest vegetation. It is the largest Panamanian park and one of the largest protected areas in Central America. It conserves the largest tropical rainforest on the Pacific slope. The property is also culturally and ethnically diverse, as evidenced by major archaeological findings, as well as Afro-descendants and indigenous peoples of the Embera, Wounaan, and Kuna living within the property.

10.1 Introduction

The Darien National Park, in Panama, extends along the Colombian border, eastern Darien Province, and the Pacific coast. It has 5790 km^2 (UNESCO WHC 2016), making it the largest Panamanian park and one of the largest protected areas in Central America, which conserves the largest tropical rainforest on the Pacific slope (Herrera-MacBryde and Ancon 1992) (Fig. 10.1).

The area has been under protection since 1972, with the establishment of Alto Darien Protection Forest. In 1980 it was declared a national park by presidential decree. UNESCO added the park as a World Heritage site in 1981. In 1983, it was recognized as a biosphere reserve under the UNESCO Man and Biosphere Programme (UNEP/WCMC 2006).

The property is also culturally and ethnically diverse. There have been major archaeological findings, as well as Afro-descendants and indigenous peoples (UNESCO WHC 2016). The national park represents the ancestral lands of the Embera, Wounaan, and Kuna peoples (Korten 1995).

The area was visited by Spanish conquistadors and the coast was explored by Christopher Columbus in 1502. As a result of the Spanish presence, many of the

V. Claudino-Sales, *Coastal World Heritage Sites*, Coastal Research Library 28,
https://doi.org/10.1007/978-94-024-1528-5_10

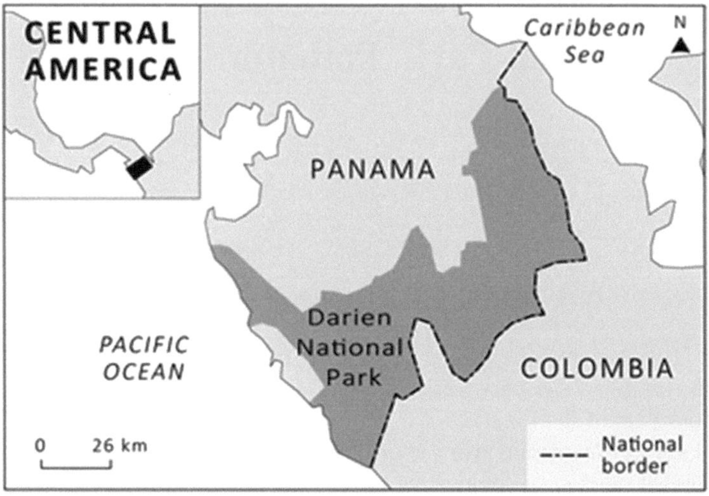

Fig. 10.1 Location of the Darien National Park, in Panama, along the Colombia border and the Pacific Ocean

Indians moved away. Today, conservation of the Indian culture is considered important (Encyclopedia of the Earth 2004).

The state owns over 90% of the land. Almost 10% is in the hands of the tribes. The park is under the authority of the National Environmental Authority (ANAM) (UNESCO WHC 2016).

The site is in a unique geographical position, forming a land bridge between Central and South America. The access to the park is difficult and is done by boat or heavy truck (Encyclopeadia Britannica 2016).

Only a few studies have been published about the Darien National Park and the Darien Province. Those available are old. Newer information comes mostly from Internet websites.

10.2 Natural Settings

10.2.1 Climatic Conditions

The climate is tropical, hot, and humid (UNEP/WCMC 2006). The Pacific coast of the Darien Province receives about 4000 mm of rainfall each year and has no persistent dry season. The mean annual temperature is 26.8 °C, the minimum being 16 °C in average (McDonald 1969).

Summer lasts from December to April. In May, the rains start gradually with frequent showers in the afternoon, mostly lasting no more than 2 h. By July or August, daily downpours resulting from thunderstorms take place, especially along the mountains. Often these thunderstorms result in flash flooding along the rivers. But Panama is completely outside the hurricane belt and experiences few climate disasters. The winds that characterize the area are the northeast trade winds (Weather and Climate 2017).

10.2.2 Geology and Geomorphology

The Darien Province had a late Miocene-Pliocene collision with northwestern South America (Coates et al. 2004). Before this collision, the area existed as a large stable island, intermittently connected by sea-level variations to western Panama (McDonalds 1969).

The mountains are largely of volcanic origin, while the inland valleys are of Cenozoic sediments (UNEP/WCMC 2006). The middle Eocene to Oligocene Darien Formation is up to ~1500 m thick. It consists mostly of fine and medium tuff, agglomerate, radiolarian chert, and basalt in its lower part and of calcareous and siliceous mudstone, micritic calcarenite, and volcanic clastics in its upper part (Coates et al. 2004).

The soils in Darien are (1) alluvial, subject to flooding and valuable for agriculture; (2) latosolic, occupying hilly and mountains, limited to use as pasture or forest; and (3) residual but not latosolic, occurring in mountain valleys. These are of recent origin but are in degradation (Martini 1960).

Along the Pacific Ocean, the southwestern edge of the province is a somewhat undulating coastline with rocky shores and sandy beaches (Herrera-MacBryde and Ancon 1992) (Figs. 10.2 and 10.3). The altitude in the park goes from sea-level to 1875 m (Encyclopedia of the Earth 2004).

10.2.3 Biodiversity

As a land bridge between Central and South America, the Darien Province is a place where species from these continents intermingle (Korten 1995).

The Darien National Park is one of the most diverse regions in Central America, with vegetation from the littoral strand and coastal dry forests to mangroves, brackish and freshwater swamps, and various lowland to lower montane rainforest life zones (Encyclopedia of the Earth 2004). There are 2440 species recorded, and the most widespread type is semi-deciduous tropical moist forest, which occupies non-flooded lowlands (Herrera-MacBryde and Ancon 1992). The forests are relatively undisturbed (UNEP/WCMC 2006).

The area protects a highly diverse assortment of fauna. There are approximately 770 vertebrate species, and over 60% of Panama's mammal species occur in the province. Endemic reptile and amphibian species include snakes, lizard, salamander, and frogs (Herrera-MacBryde and Ancon 1992). Among the wildlife are capy-

Fig. 10.2 Mountainous and forested coast at the Pacific Ocean in the littoral of the Darien National Park. (Photo reproduced by permission of Jess Kraft/Shutterstock Stock Photos)

Fig. 10.3 Dry forest and coconuts in the hilly coast of the Darien Province. (Picture reproduced by permission of Jeff Kraft/Shutterstock Stock Photos)

baras, jaguars, ocelots, giant anteaters, howler monkeys, harpy eagles, caimans, American crocodiles, and the endemic gray-headed chachalaca (Encyclopaedia Britannica 2016). Three major migration routes for intercontinental birds converge at this spot (Korten 1995).

10.3 Conservation Challenges

Despite the remote location and contiguity with adjacent conservation areas in neighboring Colombia, the property has not escaped human pressure. The expanding agricultural frontier and related colonization near the property have resulted in major deforestation and timber extraction (UNESCO WHC 2016).

Of particular conservation concern is the possible completion of the Pan-American Highway. The influx of people due to the construction of the highway could lead to uncontrolled forestry, mining, agriculture, and hunting resulting in deforestation, soil erosion, disruption of the fragile ecological equilibrium, and dislocation of traditional indigenous practices (Encyclopaedia Britannica 2016).

Climate changes and global warming are also threats to the forests. The area can experience increasing lower levels of precipitation or, the opposite, more precipitation. These two scenarios impose high risks to the environment, causing in one case in loss of biodiversity and, in the second case, in growing of the landslides and degradation of the soils that already characterizes the area.

The last report of the International Union for Conservation of Nature (IUCN) about the state of conservation of World Heritage sites found that the Darien National Park presents significant concerns. The extraordinary nature conservation values, and in particular the biodiversity values, have been suffering from a mix of pressures caused by a growing population, an advancing agricultural frontier and uncontrolled resource use (IUCN 2017).

While collection of wild biodiversity, hunting, and poaching are fundamental parts of local livelihood systems, its uncontrolled and excessive use threatens many conservation values while also casting shadows over the future productivity and use options. The combination of permanent pressure from conversion of forests and overuse of wild biodiversity is advancing (IUCN 2017; Osipova et al. 2017).

References

Coates AG, Collins LS, Aubry MP, Berggren WA (2004) The geology of the Darien, Panama, and the late Miocene-Pliocene collision of the Panama arc with northwestern South America. Bull Geol Soc Am 116(11–12):1327–1344

Encyclopaedia Britannica (2016) Darién. http://www.britannica.com/place/Darien-region-Panama. Accessed 2 May 2016

Encyclopedia of the Earth (2004) Darién National Park. http://www.eoearth.org/view/article/151645/. Accessed 2 May 2016

Herrera-MacBryde O, Asociación Nacional para la Conservación de la Naturaleza (ANCON) (1992) Darien province and Darien National Park. Smithsonian Institution, Rapport, Washington DC
IUCN (International Union for Conservation of Nature) (2017) Darien National Park. http://www.worldheritageoutlook.iucn.org/explore-sites/wdpaid/2554. Accessed 16 Dec 2017
Korten A (1995) Paving the Pan-American gap. Multinational Monitor 16:11–19
Martini JA (1960) Forest soils of Darien Province, Panama. Tropical Woods 112:28–39
McDonald MC (1969) Geologic evaluation of radar imagery from Darien Province, Panama. CRES Technical Report 133–6, Panama City
Osipova E, Shadie P, Zwahlen C, Osti M, Shi Y, Kormos C, Bertzky B, Murai M, Van Merm R, Badman T (2017) IUCN world heritage outlook 2: a conservation assessment of all natural world heritage sites. IUCN, Gland
UNEP/WCMC (United Nations Environmental Programme/World Conservation Monitoring Centre) (2006) Darien National Park, Panama. Report, Cambridge
UNESCO WHC (United Nations Educational, Scientific and Cultural Organization World Heritage Centre) (2016) Darien National Park. http://whc.unesco.org/en/list/159. Accessed 2 May 2016
Weather and Climate (2017) Climate and average weather in Panama. https://weather-and-climate.com/average-monthly-Rainfall-Temperature-Sunshine-in-Panama. Accessed 2 May 2016

Part IV
World Heritage on Continental Coasts: Europe

Chapter 11
Stevns Klint, Denmark

Abstract Stevns Klint is an exceptional testimony to the impact of meteorite in Earth. The site provides evidence of the Chicxulub meteorite impact that took place at the end of the Cretaceous Period and caused the end of the Age of the Dinosaurs. Comparative analysis indicates this is the most significant and readily accessible site to see the sedimentary record of the ash cloud formed by the meteorite impact, the actual site of the impact being deep underwater offshore the Yucatan Peninsula. The site is very significant in terms of its past, present, and future contribution to science, especially pertaining to the definition of and explanation of the Cretaceous-Tertiary (K/T) boundary. The outstanding fossil record at Stevns Klint provides a succession of three biotic assemblages including the most diverse end-Cretaceous marine ecosystem known.

11.1 Introduction

Stevns Klint is a 15-km-long and up to 41-m-high rugged and scenic beautifully exposed coastal cliff south of Copenhagen in eastern Denmark. It is a classical area for the study of the Cretaceous-Tertiary geological period boundary (Lykke-Andersen and Surlyk 2004) (Fig. 11.1).

The site has played a significant role in the international study of the causes of mass extinction and the effect of extraterrestrial impact on life on Earth. It was among the original areas that first prompted scientists to consider an asteroid impact as a cause for mass extinction (Damholt and Surlyk 2012). It was added to the World Heritage List in 2014 (UNESCO WHC 2016).

The area around Stevns Klint offers a great scenic experience which annually attracts thousands of tourists. It also has architectural interests, with buildings of the thirteenth century (Copenhagen Portal 2016).

The site is governed and managed through a steering group with representation from state, regional governments, and landowners including private and public agents (UNESCO WHC 2016).

V. Claudino-Sales, *Coastal World Heritage Sites*, Coastal Research Library 28,
https://doi.org/10.1007/978-94-024-1528-5_11

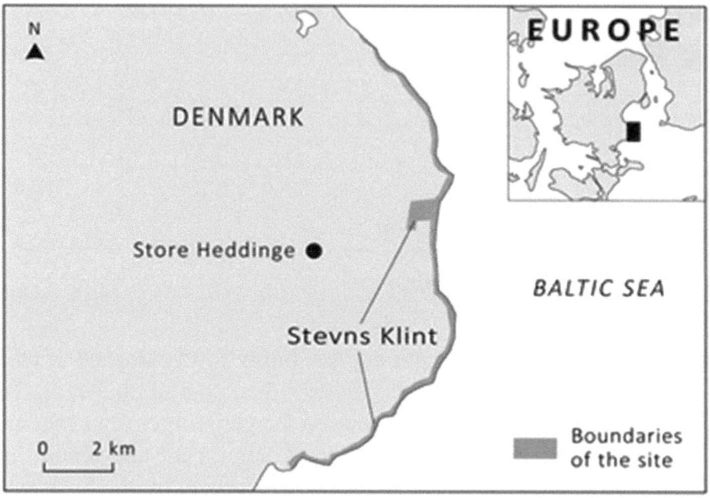

Fig. 11.1 Location of Stevns Klint, Denmark

11.2 Natural Settings

11.2.1 Climatic Conditions

The average temperature for Stevns Klint remains relatively cool throughout the year with values ranging from 0° to 17° C. The site has a moderate coastal climate and receives a great deal of precipitation, mostly rain and occasional snow showers. The months of November and December have the most rainfall. The winds are predominantly from the west, but other directions also occur (Miami University 2015).

11.2.2 Geology and Geomorphology

The Cretaceous-Tertiary boundary marks a severe ecological crisis and mass extinction that put an end to the Mesozoic Era 65.5 million years ago. This event has been intensely studied, partly because it includes the extinction of the spectacular non-avian dinosaurs and partly because of its significant role in the discussion of a possible extraterrestrial impact (Damholt and Surlyk 2012).

The Cretaceous-Tertiary boundary shows a complex development, which was not understood until 1920 (Lykke-Andersen and Surlyk 2004). Since then, the studies have improved and increased after the discovery that the Cretaceous-Tertiary boundary records a mass life extinction event caused by a bolide impact – the Chicxulub meteorite – which impacted the Yucatan Peninsula, in the Gulf of Mexico, by Alvarez et al. (1980) (Damholt and Surlyk 2012).

Fig. 11.2 The seacliff of Stevns, Denmark, showing Cretaceous chalk on the base, overtopped by Tertiary limestone. Between them, a dark layer of clay, which shows the ecological changes to Earth by the impact of a meteorite at the end of the Mesozoic Era, causing a large mass life extinction event. (Picture reproduced by permission of Merete Binderup)

Stevns Klint became a site of great interest after this discovery, because the cliff provides evidence of the Chicxulub meteorite impact (Lykke-Andersen and Surlyk 2004). The geological site has a record of the cloud of ash formed by this impact (UNESCO WHC 2016).

The fall of the asteroid caused the fall of finely dispersed metallic particles of extraterrestrial nature related to the asteroid fragments or to micrometeorites following the asteroid or to the intense supply of cosmic dust (Korchagin and Tsel'movich 2011). Comparative analysis indicates that Stevns Klint is the most significant and readily accessible site to see the sedimentary record of the ash cloud formed by the meteorite impact (UNESCO WHC 2016).

The Cretaceous-Tertiary boundary succession at Stevns Klint was developed on an underlying seafloor topographic relief of about 40 m (Lykke-Andersen and Surlyk 2004). The Cretaceous is represented by a soft chalk constituting a characteristic example of deposition in the epicontinental Chalk Sea that covered most of Northwestern Europe in Late Cretaceous times; on the top of the chalk overlie Tertiary limestones (Damholt and Surlyk 2012). The Cretaceous-Tertiary boundary separates the soft Cretaceous chalk from the harder overlying Tertiary limestone, representing a layer of clays rich in iridium (Schack and Damholt 2012) (Fig. 11.2).

Due to the continuous erosion from the sea, the profile of the cliff is constantly changing and kept fresh and well exposed (Schack and Damholt 2012). The boundary of the site accommodates the natural processes of coastal erosion, and as the cliff face migrates landward, so too do the property boundaries (Damholt and Surlyk 2012).

Fig. 11.3 Cliff collapse at Stevns Klint, with rock fall. (Picture reproduced by permission of Merete Binderup)

The erosion due to the sea, as well as by chemical weathering, is high and has an average rate of 15 cm/year. It frequently causes rock falls (Schack and Damholt 2012) (Fig. 11.3).

11.2.3 Biodiverstiy

The outstanding fossil record at Stevns Klint provides a succession of three biotic assemblages, including the most diverse end-Cretaceous marine ecosystem known (UNESCO WHC 2016).

The chalk deposits are rich in macrofossils, representing a highly diverse marine fauna dominated by bryozoans, bivalves, brachiopods, echinoids, serpulids, solitary corals, and sponges. The exposed Upper Cretaceous chalk at Stevns Klint, in total, contains more than 450 species of macrofossils and hundreds of species of nano- and microfossils (Surlyk et al. 2006).

At the boundary of the periods, more than half of these living Cretaceous species became extinct, including land-living dinosaurs and large marine reptiles. After the mass extinction, a full ecosystem gradually evolved, including new species (Hart et al. 2016).

Stevns Klint is located within the temperate broadleaf and mixed forest biome and is a part of the Baltic mixed forest ecoregion. This region has a very diverse set of plant and animal species. The Stevns Klint's region supports mostly lowland to submontane beech and mixed beech forests (Miami University 2015).

11.3 Conservation Challenges

There are some threats to the property that require continued attention. There are notable visitation and projections that this will increase. This has the potential to negatively impact the fossil heritage through uncontrolled or poorly managed fossil collecting (UNESCO WHC 2016).

Climate change with rising of sea level can also be a threat, considering that Stevns Klint is near a great body of water, the Baltic Sea, already presenting heavy cliff erosion and retreat of beaches in some places (Schack and Damholt 2012).

Nevertheless, the site benefits from a strong legislative framework, effective collaboration of local organizations and a very strong support and engagement of the local community and private landowners. The current threats to the site's values are very low and well regulated. Only the anticipated increase in visitation will need to be properly addressed and will require additional resources (IUCN 2017).

The last conservation outlook of the International Union for Conservation of Nature (IUCN) of the World Heritage sites found that the Stevns Klint presents a good state of conservation (Osipova et al. 2017). It indicates that its values are likely to be maintained for the foreseeable future, provided that current conservation measures are maintained.

References

Alvarez LW, Alvarez W, Asaro F, Michel HV (1980) Extraterrestrial cause for the Cretaceous–Palaeogene extinction. Science 208:1095–1108

Conpenhagen Portal. http://copenhagenet.dk/. Accessed 28 Mar 2016

Damholt F, Surlyk F (2012) Nomination of Stevns Klint for inclusion in the world heritage list. Østsjællands Museum, St. Heddinge

Hart MB, Fitzpatrick MEJ, Smart CW (2016) The Cretaceous/Paleocene boundary: foraminifera, sea grasses, sea level change and sequence stratigraphy. Palaeogeogr Palaeoclimatol Palaeoecol 441:420–430

IUCN (International Union for Conservation of Nature) (2017) Stevns Klint conservation outlook. http://www.worldheritageoutlook.iucn.org/explore-sites/wdpaid/555577556. Accessed 17 Dec 2017

Korchagin O, Tsel'movich V (2011) Cosmic particles (micrometeorites) and nanospheres from the Cretaceous-Paleogene (K/T) boundary clay layer at the Stevns Klint Section, Denmark. Dokl Earth Sci 437(2):449–454

Lykke-Andersen H, Surlyk F (2004) The Cretaceous–Palaeogene boundary at Stevns Klint, Denmark: inversion tectonics or sea-floor topography? J Geol Soc 161:343–352

Miami University (2015) Geo 121 Wiki Spring 2015. https://sites.google.com/a/miamioh.edu/geo-121-wiki-spring-2015/home/europe-stevns-klint-team6. Accessed 28 Mar 2016

Osipova E, Shadie P, Zwahlen C, Osti M, Shi Y, Kormos C, Bertzky B, Murai M, Van Merm R, Badman T (2017) IUCN world heritage outlook 2: a conservation assessment of all natural world heritage sites. IUCN, Gland

Schack SA, Damholt T (2012) Cliff collapse at Stevns Klint, Southeast Denmark. Geol Surv Den Greenl Bull 26:33–36

Surlyk F, Damholt T, Bjerager M (2006) Stevns Klint, Denmark: uppermost Maastrichtian chalk, cretaceous-tertiary boundary, and lower Danian bryozoan mound complex. Bull Geol Soc Den 54:1–48

UNESCO WHC (United Nations Educational, Scientific and Cultural Organization World Heritage Centre) (2016) Stevns Klint. http://whc.unesco.org/en/list/1416. Accessed 28 Mar 2016

Chapter 12
High Coast and Kvarken Archipelago, Finland and Sweden

Abstract The High Coast in Sweden and the Kvarken Archipelago in Finland are on opposite sides of the Gulf of Bothnia, in the northern part of the Baltic Sea. The area is continuously rising from the sea in a process of rapid glacio-isostatic uplift, whereby the land, previously weighed down under the weight of a glacier, lifts at rates that are among the highest in the world. This serial transboundary property serves as an outstanding example of dynamic ongoing geological processes forming the land and seascape, including interesting interactions with biological processes and ecosystem development. It is also an area for studying moraines.

12.1 Introduction

The Kvarken Archipelago (Finland) and the High Coast (Sweden) are on opposite sides of the Gulf of Bothnia, a northern extension of the Baltic Sea. The property has 3464 km^2, of which about 1007 km^2 is terrestrial (UNESCO WHC 2016) (Fig. 12.1).

The sites are about 150 km from each other. In High Coast, 43.9% is land, and 55.1% is sea. At Kvarken, 15.1% is land, and 84.9% is shallow sea. In both sites, the area is owned by the state and by private owners (UNEP/WCMC 2008).

High Coast was defined as a World Heritage site in 2000. In 2006, the Kvarken Archipelago was designated as a World Heritage site, as an extension of the 2000 site (UNEP/WCMC 2008).

There is a record along the High Coast of human presence in the late Stone Age from dwellings about 4000 years old. There are once-coastal grave mounds dating from about 3500 to 2500 years ago and remains of longhouses dating back to around 100–600 AD. More recent remains of fishing camps with log cabins date back to the sixteenth century (UNEP/WCMC 2008). Some areas were used as mountain pasture in the nineteenth century. The historical succession of settlement followed the changing level of the coastline: the oldest remains now lie some distance up the hillsides (IUCN 2006).

Settlement on the Kvarken islands is traced back to 1800 to 1600 years BC, mainly by fishermen and seal hunters. Permanent fishing villages grew up in the

V. Claudino-Sales, *Coastal World Heritage Sites*, Coastal Research Library 28,
https://doi.org/10.1007/978-94-024-1528-5_12

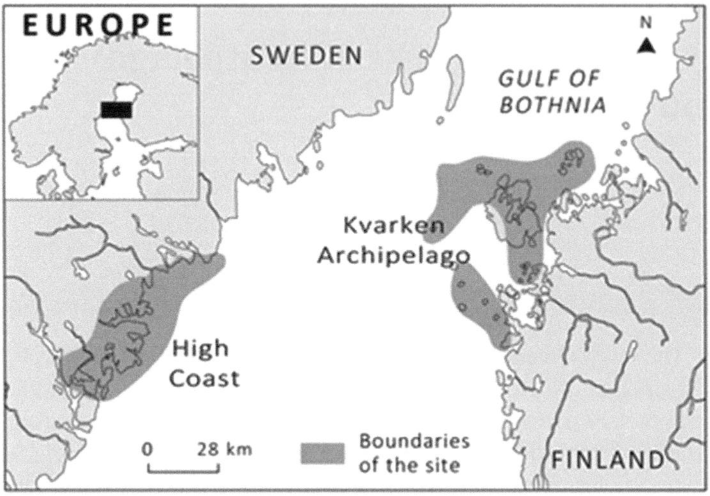

Fig. 12.1 The High Coast, Sweden, and Kvarken Archipelago, Finland, that are World Heritage by UNESCO

fourteenth century, although some islands were uninhabited till the nineteenth century. Farming and summer pasturing on the seashore meadows were less important (UNEP/WCMC 2008).

Because of its dynamic nature, the Kvarken Archipelago is an important scientific resource for understanding the processes of glaciated land uplift which originated with the theory of isostatic rebound first advanced by the nineteenth-century geologist G. De Geer, after whom the morainal ridges are named (UNEP/WCMC 2008).

In both Sweden and Finland, World Heritage management issues are dealt with at the regional level, by established bodies with representatives from authorities, municipalities, and local stakeholders. They established a transnational consultative body, to ensure that this serial transnational site has a joint management strategy (UNESCO WHC 2016).

12.2 Natural Settings

12.2.1 Climatic Conditions

The climate is southern boreal, influenced by the sea. Snow and ice cover lasts 140–150 days a year, and rainfall is 400 mm (IUCN 2006). The sea freezes annually, and the mean temperature is 3.4° C (Kotilainen et al. 2012).

The sea is brackish, being reinforced by the vigorous freshwater flow from the many rivers entering the Bay of Bothnia to the north. Tides are very low, but winds and air pressure can create sea level changes of 250 cm, the lowest levels occurring in spring and the highest in fall (UNEP/WCMC 2008).

Fig. 12.2 High Coast, in Sweden, in the Gulf of Bothnia, northern Baltic Sea, showing hilly and vegetated coastal landscape. (Picture reproduced by permission of www.dockstahavet.se)

12.2.2 Geology and Geomorphology

The High Coast and the Kvarken Archipelago are in the European glaciated epicontinental basin of the Baltic Sea, one of the largest brackish water bodies in the world (Kotilainen et al. 2012). The area is known for its rapid uplift. Today, about 7 km^2 of new land is rising from the sea every year (Poutanen and Steffen 2014).

The source of land uplift is found in the past, in times where Northern Europe was covered with a huge ice sheet. The Last Glacial Maximum occurred about 20000 years ago. The thickness of the ice was about 2500–3000 m, and its weight pressed the crust down by 500–900 m (Peltier 1994). The melting period started 18,000 years ago, and about 10,500 years ago, the Kvarken Archipelago/High Coast was ice-free (IUCN 2017).

The load of the ice sheet followed by the melting and release of the load caused a rapid elastic rebound of the crust (Berglund 2012). The land has elevated a total of 800 m, which is the highest uplift in the world after the last Ice Age (UNESCO WHC 2016). Glacio-isostatic land uplift is 8.0–8.5 mm year (Kotilainen et al. 2012).

The properties have different geomorphological characteristics. High Coast, in Sweden, is a steep landscape consisting of hilly scenery with high islands, steep shores, smooth cliffs, and deep inlets (Fig. 12.2), while Kvarken, in the other site, consists of almost flat planes with thousands of low-lying islands, shallow bays, moraine ridges, and massive boulder fields (Poutanen and Steffen 2014).

Fig. 12.3 Ridged washboard moraines in a shallow glacial lake-type landscape in Kvarken Archipelago, at Korsholm municipality, Finland. (Picture reproduced by permission of Tommy Lahti)

As a consequence of the advancing shoreline, islands appear and unite, peninsulas expand, and lakes evolve from bays and develop into marshes and peat fens, resulting in an unusual variety of environmental gradients, both topographic and hydrographic (IUCN 2006, 2017).

For the geology, the crystalline bedrock belongs to Precambrian schist belt, consistent mainly of gneiss and granites. Quaternary deposits are made up of young, mainly glacial sediments approximately 13,000 to10,000 years BP. Typical glacial formations are the moraines at the Kvarken Archipelago (Breilin et al. 2004) (Fig. 12.3).

12.2.3 Biodiversity

Because the area is a transition zone of salinity level at critical levels (salinity varies between 3 parts per thousand and 6 parts per thousand) to both marine and limnic species, the diversity of marine life is poor (Kotilainen et al. 2012).

For terrestrial biodiversity, the situation is different, especially at the High Coast side. The vegetation there displays marked altitudinal zonation and great spatial variability, with high floristic diversity, due to the complex pattern of soils and substrate on an uplifted, high-relief land surface. The vegetation is typical of West Eurasian Taiga, dominated by southern boreal forest (IUCN 2006).

The Kvarken Archipelago is a dynamic landscape, most obvious in flat and shallow areas where uplift is supplemented by sedimentation. The continually emerging shores are colonized by pioneer species which are gradually replaced by a succession of plant communities (IUCN 2006). The flat landscape provides very good conditions for wetland development. The succession from shallow bays to wetlands occurs by overgrowth to swamps or swamp forests, by the formation of peatland in a herb-rich marsh, or by peat forming directly on the newly exposed shore (UNEP/WCMC 2008).

Terrestrial mammals in the High Coast region include lynx, brown bear, deer, and elk, as well as numerous rodents. The terrestrial birdlife is rich, reflecting the diversity of habitats. The coastal birdlife is typical for the region, but offshore islands are of some importance to seabirds (UNEP/WCMC 2008). The Kvarken Archipelago is on an important migratory route and offers excellent breeding habitats for birds (IUCN 2006).

12.3 Conservation Challenges

Increasing visitors, due to the tourist attractions, put great pressure in the environment. Oil or chemical spills in the sea are potential threats. Global warming and rising sea levels would influence the visible effects of land uplift in the coastal landscape, by reducing the area of new land emerging from the sea each year (UNESCO WHC 2016).

The marine environment within the site is under serious pressure and is suffering a multitude of threats, mostly from the conditions facing the Baltic Sea as a whole. Effectively, the marine environment is already showing signs of deterioration (eutrophication, oxygen-free bottoms, regime shifts) (IUCN 2017). Nevertheless, the site itself is in a good state of conservation, as attested by the last outlook about the state of conservation of the World Heritage sites (Osipova et al. 2017).

References

Berglund M (2012) The highest postglacial shore levels and glacio-isostatic uplift pattern in northern Sweden. Geogr Ann Ser B 94:321–337

Breilin O, Kotilainen AT, Nenonen K, Virransalo P, Ojalainen J, Stén CG (2004) Geology of the Kvarken Archipelago. Geological Survey of Finland, Oslo

IUCN (International Union for the Conservation of Nature) (2006) World heritage nomination – IUCN high Coast and Kvarken Archipelado, Finland and Sweden. Technical Evaluation. Gland, Switzerland

IUCN (International Union for the Conservation of Nature) (2017) High Coast/Kvarken Archipelago Conservation Outlook. http://www.worldheritageoutlook.iucn.org/explore-sites/wdpaid/478642. Accessed 17 Dec 2017

Kotilainen AT, Kaskela AM, Bäck S, Leinikki J (2012) Submarine De Geer moraines in the Kvarken Archipelago, The Baltic Sea. In: Harris PT, Baker EK (eds) Seafloor geomorphology as benthic habitat. Elsevier, London

Osipova E, Shadie P, Zwahlen C, Osti M, Shi Y, Kormos C, Bertzky B, Murai M, Van Merm R, Badman T (2017) IUCN world heritage outlook 2: a conservation assessment of all natural world heritage sites. IUCN, Gland

Peltier WR (1994) Ice age paleotopography. Science 265:195–201

Poutanen M, Steffen H (2014) Land uplift at Kvarken Archipelago/high coast UNESCO world heritage area. Geophysica 50(2):49–64

UNEP/WCMC (United Nations Environmental Programme/World Conservation Monitoring Centre) (2008) High coast/Kvarken Archipelago Sweden/Finland. Report, Cambridge, England

UNESCO WHC (United Nations Educational, Scientific and Cultural Organization World Heritage Centre) (2016) High coast/Kvarken Archipelago. http://whc.unesco.org/en/list/898. Accessed 26 Apr 2016

Chapter 13
West Norwegian Fjords: Geirangerfjord and Nærøyfjord, Norway

Abstract Geirangerfjord and Nærøyfjord, among the world's longest and deepest, are archetypical fjord landscapes and among the most scenic. Their dramatic landscapes are exceptional in scale and grandeur. Their natural beauty is derived from their narrow and steep-sided crystalline rock walls that rise up to 1400 m from the Norwegian Sea and extend 500 m below sea level and vary in breadth from just 250 m to 2.5 km wide. Waterfalls, rivers, coniferous woodlands and forests, glacial lakes, glaciers, and rugged mountains combine to make Geirangerfjord and Nærøyfjord among the most scenic landscapes in the world. There is a wide range of supporting natural phenomena, such as submarine moraines and marine mammals.

13.1 Introduction

Geirangerfjord and Nærøyfjord, set 120 km from one another in southwestern Norway, are part of the west Norwegian fjord landscape. Their narrow, steep-sided valleys, gouged by glaciers, stretch from 500 m below the surface of the Norwegian Sea to 1400 m above it (UNESCO WHC 2016).

The property is composed of eight protected areas forming a serial World Heritage site. Geirangerfjord has an area of 544 km^2 and Nærøyfjord 683 km^2. The total area is 1227 km^2, of which 1123 km^2 is land and 107 km^2 is sea (IUCN 2010) (Fig. 13.1).

The fjords were listed as a natural World Heritage site in 2005. Nevertheless, remnants of old and mostly abandoned transhumant farms add a cultural aspect to the natural landscape that complements and adds human interest to the properties (UNESCO WHC 2016). Many carefully recorded relics are present in the area, such as Neolithic tools, burial mounds, hunting hides, and reindeer pitfall traps in the mountains, some of which were still in use 400 years ago. Viking houses a thousand years old are present, and there are more than 350 registered old buildings (UNEP/WCMC 2005).

The site has been of interest for science for a long time. There has been continuous investigation of aspects of the natural settings since the mid-eighteenth century, now

V. Claudino-Sales, *Coastal World Heritage Sites*, Coastal Research Library 28,
https://doi.org/10.1007/978-94-024-1528-5_13

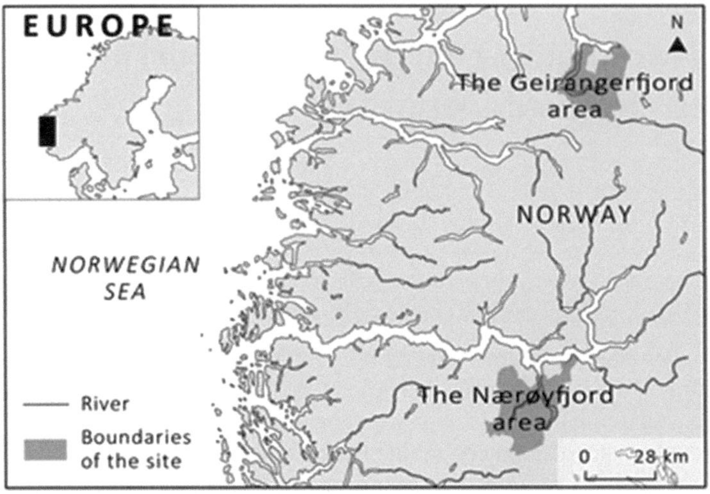

Fig. 13.1 Location of Geirangerfjord and Nærøyfjord, 120 km from each other, in southwest Norway

recorded on the national database of Norway. The site was also a place of inspiration to artists (UNEP/WCMC 2005).

The fjords are one of the most scenic draws of tourism in Europe, and tourism is a very important economic activity in the area (UNESCO WHC 2016). There are many hotels, guesthouses, holiday flats, cabins, and campgrounds in both areas, ready to host thousands of visitors every year (UNEP/WCMC 2005).

13.2 Natural Settings

13.2.1 Climatic Conditions

The climate is transitional between oceanic and continental but varies markedly with aspect and altitude. The average annual temperature for the Geirangerfjord area is 6.6 °C, ranging from 0.5 °C in January to 13.8 °C in July with an average rainfall of 1180 mm (UNEP/WCMC 2005). Nærøyfjord is around 1.5° to 3 °C colder, and rainfall averages 1000 mm.

Snow persists from October to late May on the mountains and from late November to March in the valleys of the northern site, but only 2–3 weeks at Nærøyfjord. Winter periodically ices over the fjord heads for 1–3 weeks, though in Nærøyfjord, it lasts longer (UNEP/WCMC 2005).

The narrow valleys can generate strong down-valley winds, and avalanche air blasts can be very dangerous (Jaedicke et al. 2009).

Fig. 13.2 Geirangerfjord, in Norway, showing a narrow glacial canyon on the Precambrian gneisses. (Picture reproduced by permission of Øyvind Heen, www.fjords.com)

13.2.2 Geology and Geomorphology

Fjords represent deep incisions into land masses, thus defining important repositories between continents and the open ocean beyond (Hjelstuen et al. 2009). The coastal zone along Western Norway is heavily dissected by submarine inlets that repeatedly have been covered and excavated by ice sheets during glacial stages (Sejrup et al. 2005; Sætre and Ljøen 1972).

The Geirangerfjord and the Nærøyfjord fjords are pristine, well-developed examples of an active glacial landscape (Sejrup et al. 2005) (Figs. 13.2 and 13.3).

Geirangerfjord and Nærøyfjord are classic examples of the upper sections of two of the world's longest and deepest fjords. Their sheer rock walls, hung with waterfalls, rise 1400 m directly from sea level to a quite different landscape of glaciated mountains (UNEP/WCMC 2005). Nærøyfjord is the world's narrowest fjord, at one point just 180 m across, with mountains which tower on either side up to 1400 m (Amodeo 2005).

Each landscape is at the upper end of a major fjord system that developed along faults and fracture zones at right angles, giving them a characteristic zigzag form. Both fjords are submarine hanging valleys (UNEP/WCMC 2005).

Relatively recently, in geologic terms, the products of glacial weathering were removed, leaving ice- and wave-polished surfaces on the steep fjord sides which provide superbly exposed and continuous three-dimensional sections through the bedrock (Aarseth et al. 2014).

Fig. 13.3 Nærøyfjord encloses a small bay. The seacliffs show marks of landslides. (Picture reproduced by permission of Jess Kraft/Shutterstock Stock Photos)

In Nærøyfjord, these are Precambrian gneisses of the West Gneiss Region, a world-class example of deeply subducted continental crust and of well-preserved high-pressure rocks. In Geirangerfjord, there are outcrops of peridotite and serpentinite in the predominant gneiss bedrock (Aarseth et al. 2014).

Both fjord areas are important for geomorphological research, because of the rockslides and related tsunamis that happen there. The continuing active mass movements on the steep slopes create opportunities for useful research into geohazards (IUCN 2010). Large avalanches, of both powdery snow and wet slabs, regularly occur (Jaedicke et al. 2009).

13.2.3 Biodiversity

There are 76 marine species in Nærøyfjord, where the water is unpolluted, including the Atlantic salmon, Norway lobsters, and sea pens, a kind of aquatic invertebrate such as medusa and sea anemone. A common seal colony of 15–30 individuals has a birth area in Nærøyfjord, and the common porpoise is common in both fjords. White-beaked dolphins, sperm whales, minke whales, and killer whales occasionally occur (IUCN 2010).

Around 100 bird species breed in the area, ranging from coastal birds to high mountain species. The alpine and subalpine mires attract many waders, and the cliffs host several raptors. The invertebrate fauna is not well known, but most of the habitats may have a rich range of species (IUCN 2010).

The vegetation ranges with altitude between hemiboreal in the valleys and taiga on the plateaus. Around Nærøyfjord, woodlands vary with the soil, moisture, exposure, and altitude above sea level and use. The Alpine flora of the Geirangerfjord area is dependent on snow cover and ranges from dwarf birch and polar willow to herb-rich grassland (IUCN 2010).

13.3 Conservation Challenges

Mining and underground quarrying are a concern, considering that they occur in the vicinity of the protected areas. The risks include the export of the mined material across the area. Geohazards such as avalanches and mass movements are a concern for inhabited areas and existing infrastructure (UNESCO WHC 2016).

Climate change and global warming have a great potential to make the geohazards even bigger, especially with the melting of the glaciers. Fauna and flora of the cold environment are also going to suffer differentiation from higher temperatures.

The last report on the state of conservation of the World Heritage sites made by the International Union for Conservation of Nature (Osipova et al. 2017) found that some concerns exist, particularly related to cruise ships. Nevertheless, the conservation outlook was determined to be good, thanks to the robustness of its values, limited threats, and effective management (IUCN 2017).

References

Aarseth I, Nesje A, Fredin O (2014) West Norwegian Fjords. UNESCO World Heritage. Guide to geological excursion to Geirangerfjord and Nærøyfjord. Geological Society of Norway Geological Guides 3, Oslo

Amodeo C (2005) Eight new world heritage sites designated. Geographical 77(10):7

Hjelstuen BO, Haflidason H, Sejrup HP, Lyså A (2009) Sedimentary processes and depositional environments in glaciated fjord systems – evidence from Nordfjord, Norway. Mar Geol 258:88–99

IUCN (International Union for Conservation of Nature) (2010) The world's most scenic fjords http://iucn.org/about/work/programmes/forest/?5819. Accessed 10 Apr 2016

IUCN (International Union for Conservation of Nature) (2017) West Norwegian Fjords – Geirangerfjord and Nærøyfjord Conservation Outlook. http://www.worldheritageoutlook.iucn.org/explore-sites/wdpaid/902489. Accessed 17 Dec 2017

Jaedicke C, Karstein L, Kalle Norges K (2009) Integrated database for rapid mass movement in Norway. Nat Hazards Earth Syst Sci 9(2):469–479

Osipova E, Shadie P, Zwahlen C, Osti M, Shi Y, Kormos C, Bertzky B, Murai M, Van Merm R, Badman T (2017) IUCN world heritage outlook 2: a conservation assessment of all natural world heritage sites. IUCN, Gland

Sætre R, Ljøen R (1972) The Norwegian coastal current. Technical University of Norway, Bergen

Sejrup HP, Hjelstuen BO, Dahlgren KIT, Haflidason H, Kuijpers A, Nygård A, Praeg D, Stoker MS, Vorren TO (2005) Pleistocene glacial history of the NW European continental margin. Mar Pet Geol 22:1111–1129

UNEP/WCMC (United Nations Environment Programme/World Conservation Monitoring Centre) (2005) West Norwegian Fjords Geirangerfjord and Nærøyfjord. Report, Cambridge, England

UNESCO WHC (United Nations Educational, Scientific and Cultural Organization World Heritage Centre) (2016) West Norwegian Fjords Geirangerfjord and Nærøyfjord http://whc.unesco.org/en/list/1195. Accessed 11 Apr 2016

Chapter 14
Danube Delta, Romania

Abstract The Danube Delta is the second largest and best preserved of Europe's deltas. It flows to the Black Sea and is covered with numerous lakes, ponds, and marshes which attract more than 300 species of birds and 90 species of fish. Formed over a period of more than 5000 years, it has the third largest biodiversity in the world (over 5500 flora and fauna species), exceeded only by the Great Barrier Reef in Australia and the Galapagos archipelago in Ecuador. It also is home to the world's largest reed bed expanse, a tall, slender-leaved plant of the grass family that grows on marshy ground. Some 15,000 people inhabit the delta area, living in 28 villages and 1 city (Sulina). In 1991, it became a natural World Heritage site and also a Wetland of International Importance under the Ramsar Wetland Convention.

14.1 Introduction

The site is on the coast of the Black Sea at the mouth of the Danube River, in Romania. It has 3124 km^2 of area (UNESCO WHC 2016) (Fig. 14.1).

In 1979, the delta was designated a biosphere reserve under the UNESCO Man and Biosphere Programme. In 1991, it became a natural World Heritage site and also a Wetland of International Importance under the Ramsar Wetland Convention (UNEP/WCMC 2009).

The state, in the form of the Danube Delta Biosphere Reserve Authority of the Ministry of Water, Forestry and Environment, owns more than 90% of the reserve; the rest is in private hands. The authority administers the reserve and its institutional, agency, and enforcement staff (UNEP/WCMC 2009).

The very long history of trading along the Danube is evident from the remains of Greek and Roman settlements, including a lighthouse. Some 15,000 people inhabit the delta area, living in 28 villages and 1 city (Sulina) (Romania Tourism 2016).

The local population has depended on small-scale, low-intensity use of the natural resources fairly well integrated with the natural heritage: fishing, hunting, cattle grazing, subsistence farming and beekeeping, and viticulture supplemented by outside incomes (UNEP/WCMC 2009). Tourism is also a very important economic activity (Romania Tourism 2016).

V. Claudino-Sales, *Coastal World Heritage Sites*, Coastal Research Library 28,
https://doi.org/10.1007/978-94-024-1528-5_14

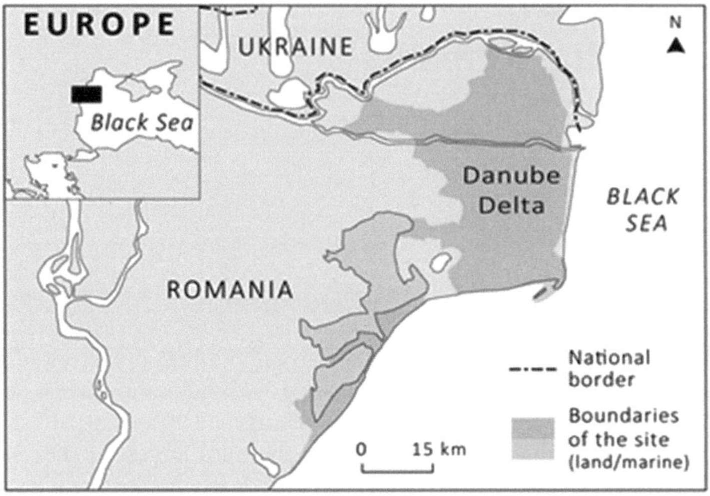

Fig. 14.1 Location of Danube Delta, in the coast of the Black Sea, Romania

14.2 Natural Settings

14.2.1 Climatic Conditions

The prevailing climate is continental with only 450 mm of rain annually, falling mainly between March and May. This is supplemented by floodwaters from the Danube's headwaters snow melts. The weather is influenced by proximity to the sea, and the high humidity rises from inland lakes and small waterways. The average annual temperature ranges between 0° C in December and 26 ° C in high summer (UNEP/WCMC 2009).

The area is subjected to storms, especially in winter. On average, this coast experiences 30 storms per year, some of which are severe (categories III–IV). The extreme hurricane storms of category V also arrive and most endanger the coastal settlements and the back-beach ecosystems as sand dunes, wetlands, and lagoons (Zăinescu et al. 2017).

14.2.2 Geology and Geomorphology

The Danube River flows 2877 km from its springs in Germany's Black Forest to the Black Sea. Just before reaching the sea, it forms the second largest and best preserved of Europe's deltas: 5697 km^2 of rivers, canals, marshes, tree-fringes, and lakes. It is the largest continuous marshland in Europe (Romania Tourism 2016) (Fig. 14.2).

Fig. 14.2 Coastal lake in the Danube Delta, Romania. (Picture reproduced by permission of Els Slots, webpage www.worldheritagesite.org)

The Holocene Danube Delta started to build out of a Black Sea embayment ~5200 years ago (Giosan et al. 2012; Giosan et al. 2006). The delta is formed around the three main channels of the Danube, named after their respective ports: Chilia (in the north), Sulina (in the middle), and Sfantu Gheorghe (in the south) (Romania Tourism 2016).

At the coast, there is a strong longshore sediment transport. Tidal processes are not important because the mouth of the Danube is a microtidal environment (Giosan et al. 1999). Flooding is a major natural risk: the coastal wetlands have been seriously impacted by floods in 2002, 2005, 2006, and 2010 (Niculescu et al. 2015).

The northern part of the delta is slowly sinking, resulting in a measurable increase of water flow in the northernmost arm. The altitude is from sea level to 15 m, and only 9% of the area is permanently above water (UNEP/WCMC 2009).

The delta has three distinct zones: the fluvial zone in the upper part, characterized by sandy levees, and small, densely vegetated lakes; the transitional zone with less sedimentation and numerous larger lakes; and the marine zone dominated by sand-dune barrier beach complexes (Ramsar 1994).

14.2.3 Biodiversity

The delta is the meeting point of Palaearctic and Mediterranean biogeographic zones and represents a unique dynamic wetland ecosystem containing a rich biodiversity of wetland habitats (Romania Tourism 2016).

Fig. 14.3 White pelicans in the Danube Delta waters. (Picture reproduced by permission of Els Slots, webpage www.worldheritagesite.org)

Different groups of plants related to different habitats, such as water, swamps, pastures, sands, and salty areas, are present (Ciocârlan 2011). It is home to the world's largest reed bed extension. The reed bed is a tall, slender-leaved plant of the grass family that grows on marshy ground (UNEP/WCMC 2009). Forests, pastures, and coastal vegetation in sand dunes are also present (Romania Tourism 2016).

The complex Danube-Black Sea geoecosystem, created by a unique combination of integrated biotopes and biocoenoses, forms a rich "ichthyosystem" (Bănăduc et al. 2016). Effectively, 90 species of fish inhabit the area (UNESCO WHC 2016).

Some 300 species of birds make Danube Delta their home (Fig. 14.3), including cormorants, white-tailed eagles, and glossy ibises. The bird-watching season lasts from early spring to late summer. There is also a rich community of animals, from wildcats, foxes, and wolves to an occasional boar or deer. Altogether, 3450 animal species can be seen here, as well as 1700 plant species (Romania Tourism 2016).

14.3 Conservation Challenges

Erosion in the coastal area of the Black Sea is threatening the ecosystem of wetlands and coastal lakes in the Danube Delta. This situation results (Giosan et al. 1999) of a general decrease in the river sediment discharge due to damming, dredging, and agricultural practices in the Danube drainage basin, as well as because of sea level rising related to global warming and climate changes. Effectively, the Danube Delta has become mainly inactive over the last few decades, due to the interventions of man-made origin, which interrupted the feeding on sediments and reinforced the erosion in the littoral segment (Panin and Jipa 2002).

The last conservation outlook of the International Union for Conservation of Nature (Osipova et al. 2017) found that the site's unique features remain well-preserved overall. The state of conservation was found to be good with some concerns, related to climate changes, the preservation of avifauna, and future financing (IUCN 2017).

References

Bănăduc D, Rey S, Trichkova T, Lenhardt M, Curtean-Bănăduc A (2016) The lower Danube River-Danube Delta-North West Black Sea: a pivotal area of major interest for the past, present and future of its fish fauna – a short review. Sci Total Environ 545–546:137–151

Ciocârlan V (2011) Vascular Flora of the Danube Delta. Scientific Annals of Alexandru Ioan Cuza University of Iasi. Vegetal Biol 57(1):41–64

Giosan L, Bokuniewicz H, Panin N, Postolache I (1999) Longshore sediment transport along the Romanian Danube Delta coast. J Coast Res 15(4):859–871

Giosan L, Donelly JP, Constantinescu S, Filip F, Ovejanu I, Vespremeanu-Stroe A, Vespremeanu E, Duller GAT (2006) Young Danube delta documents stable Black Sea level since the middle Holocene: Morphodynamic, paleogeographic, and archaeological implications. Geology 34(9):757–760

Giosan L, Donnelly JP, Vespremeanu E, Bhattacharya JP, Olariu C, Buonaiuto FS (2012) River delta morphodynamics: examples from the Danube Delta. Society for Sedimentary Geology, special issue

IUCN (International Union for Conservation of Nature) (2017) Danube Delta Conservation Outlook. http://www.worldheritageoutlook.iucn.org/explore-sites/wdpaid/67728. Accessed 17 Dec 2017

Niculescu S, Lardeux C, Hanganu J, MercierG DL (2015) Change detection in floodable areas of the Danube delta using radar images. Nat Hazards 78(3):1899–1916

Osipova E, Shadie P, Zwahlen C, Osti M, Shi Y, Kormos C, Bertzky B, Murai M, Van Merm R, Badman T (2017) IUCN world heritage outlook 2: a conservation assessment of all natural world heritage sites. IUCN, Gland

Panin N, Jipa D (2002) Danuber River sediment input and its interaction with the north-western Black Sea. Estuar Coast Shelf Sci 54(3):551–562

Ramsar (1994) A directory of wetlands of international importance. Brocure, Romania

Romania Tourism (2016) http://romaniatourism.com/index.html. Accessed 27 Apr 2016

UNEP/WCMC (United Nations Environmental Programme/World Conservation Monitoring Centre) (2009) Danube Delta, Romania. Report, Cambridge

UNESCO WHC (United Nations Educational, Scientific and Cultural Organization World Heritage Centre) (2016) Danube Delta. http://whc.unesco.org/en/list/588. Accessed 27 Apr 2016

Zăinescu F, Tatui F, Valche N, Vespremeanu-Stroe A (2017) Storm climate on the Danube Delta coast: evidence of recent storminess change and links with large-scale teleconnection patterns. Nat Hazards. https://doi.org/10.1007/s11069-017-2783-9

Chapter 15
Doñana National Park, Spain

Abstract Doñana National Park, in Andalusia, Spain, is a vast coastal marshland, productive, well-preserved, and inaccessible. There, the Guadalquivir river meets the Atlantic Ocean, creating a large estuary. It is notable for the great diversity of its biotopes, especially lagoons, marshlands, fixed and mobile dunes, scrub woodland, and maquis. Doñana boasts 38 km of pristine beaches on one of the few large stretches of undeveloped coastline remaining in Spain. Its evolution is related to Pleistocene and especially Holocene climatic and eustatic events. The marshes harbor five threatened bird species, one of the biggest heronries in the Mediterranean, more than 500,000 wintering waterfowl, and millions of migrant birds. It is probably the most important national park in the European community.

15.1 Introduction

Doñana National Park is in the southwestern Spain, in Andalusia. It is the place where the Guadalquivir river meets the Atlantic Ocean, creating a small estuary (UNESCO WHC 2016) (Fig. 15.1).

In 1965, Doñana received legal protection as a biological reserve. In 1980, the site was designated a UNESCO biosphere reserve and a Ramsar Wetland site. In 1994, it became a World Heritage site. An extension to the original area was approved by UNESCO in 2005. Today, the site has 542 km^2 (UNESCO WHC 2016).

The park is administered by the Institute for the Conservation of Nature (Instituto Nacionale para la Conservación de la Naturaleza, ICONA) under the Ministry of Agriculture, Fisheries and Food (UNEP/WCMC 2005).

The Doñana National Park is probably the most important national park in the European community (Carretero et al. 2011). It has a known history of over 700 years. It was the favorite hunting reserve of Spanish kings. The palace of Doñana remains as a testimony to this exalted past (IUCN 2005). Tourism is a very important activity in the park.

V. Claudino-Sales, *Coastal World Heritage Sites*, Coastal Research Library 28,
https://doi.org/10.1007/978-94-024-1528-5_15

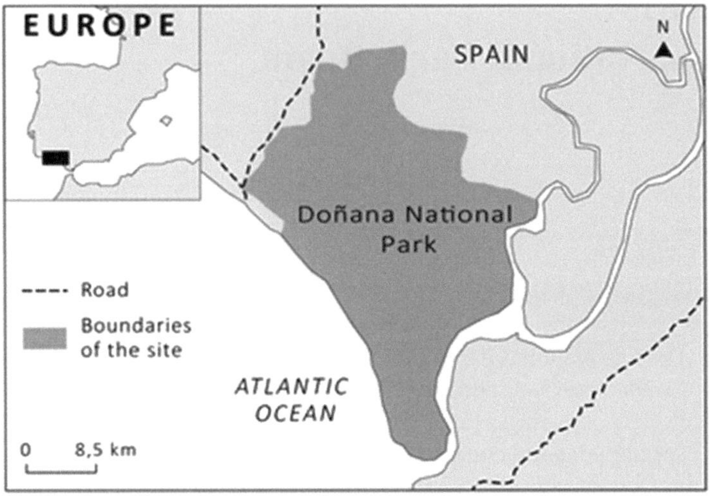

Fig. 15.1 Location of Doñana National Park, in southwestern Spain, on the Atlantic Ocean

15.2 Natural Settings

15.2.1 Climatic Conditions

Doñana Park has a mild, Mediterranean climate. The main characteristics of the Mediterranean climate are its relatively wet winters and dry summers. The rainiest seasons are early spring and late fall. In winter, however, thermal anticyclones may occur locally, bringing dry weather. Temperatures are mild throughout the year with an annual average of about 15 °C (UNEP/WCMC 2005).

The most significant characteristic of this area's climate is the three to five dry months during the summer, when the weather is under the effects of the subtropical anticyclone. Precipitation is about 400 mm/year (Parques Naturales Andalucia 2010).

15.2.2 Geology and Geomorphology

Marshes, lagoons, scrub woodlands, and mobile and fixed dunes characterize Doñana National Park (Figs. 15.2 and 15.3), along the Guadalquivir river at its Atlantic Ocean estuary (UNEP/WCMC 2005). Doñana National Park has around 40 km of pristine beaches on one of the few large stretches of undeveloped coastline remaining in Spain (National Geographic 2016).

The altitude goes from sea-level to 40 m in the dunes. The dunes, each about 2 km long by 200 m across, parallel the coast for 25 km in four rows 3–5 kilometers wide (Parques Naturales Andalucia 2010). The seaward dunes are mobile, moving

Fig. 15.2 Grassy mobile sand dunes in Doñana National Park, Andalusia, Spain. (Picture reproduced by permission of Martijn Smeets/Shutterstock Stock Photos)

Fig. 15.3 Birds in the marshes of Doñana National Park, Spain. (Picture reproduced by permission of Eric Struyf)

46 m a year before the southwesterly winds, burying pine woods as they go. The inland dunes are stabilized by scrub vegetation and in the hollows between them are lagoons and marshes (UNEP/WCMC 2005).

It is actually a vast undeveloped coastal marshland, cut off from the Atlantic by extensive dunes and subject to seasonal variations in water level and salinity. During the summer it is a dry plain which becomes a wide shallow lake in winter. It overlies quaternary deposits mainly of sand sheets (UNEP/WCMC 2005).

The evolution of Doñana National Park dates from the end of the Pleistocene and also takes the Holocene (Carretero et al. 2011). It is characterized by a first phase, which goes from 12,000 to 6500 year BP, when the area was occupied by freshwater/brackish marshes with alternation of dry periods and humid intervals. A second phase comes after that, when the marshes were inundated during the postglacial transgression (Carretero et al. 2011). In the sequence, the progressive infilling of a brackish lagoon with marine connection and permanent freshwater inputs (6500–3500 years BP) took place, partially enclosing the Doñana spit (Pozo et al. 2010).

About 2600 year BP, progradation gave way to a period of intense erosion. The resulting morphology of littoral strands and erosional surfaces permitted the formation of a marine environment. From 2300 years BP, coastal progradation has prevailed, with an erosional interruption at 1000 years BP (Rodriguez-Ramírez et al. 1996). The present-day outlet of the Guadalquivir is an estuarine delta of inactive marshland (the Doñana National Park), and the dominant environment is fluvial (Rodriguez-Ramírez et al. 1996).

Nevertheless, part of the Doñana National Park is in the marshland and part in the recharge area where the aquifer is phreatic. The functioning of the site ecosystems is closely related to the geohydrology of the area (Suso and Llamas 1993). The aquifer system consists essentially of the unconsolidated Plio-Quaternary materials. Under the central marshy area, the aquifer system is confined below low-permeability estuary deposits. Around most of the marshland, the aquifer crops out and is recharged by rainfall (Suso and Llamas 1993).

High-energy events are relatively common in Doñana National Park area. In the last 7000 years, there happened at least 20 tsunamis (Ruiz et al. 2005). Evidences of these high-energy events are washover fans, bioclastic coarse-sized layers within salt marsh deposits, or the presence of giant boulders and pebbles (Ruiz et al. 2005).

15.2.3 Biodiversity

More than half a million birds winter in the park each year, and perhaps half of Europe's bird species can be spotted here at one time or another (National Geographic 2016). It is one of the largest heronries in the Mediterranean region and is the wintering site for more than 500,000 waterfowl each year (UNESCO WHC 2016).

It contains breeding populations of several globally threatened animal (marbled teal, white-headed duck, Adalbert's eagle, Spanish lynx) and plant species. It is an important wintering site for wildfowl, receiving hundreds of thousands of migratory duck and geese every year (IUCN 2005).

15.3 Conservation Challenges

Unlike many large European estuaries, this area remains largely unmarred by development or agriculture. But there are threats, as agriculture in the borders of the park, problem of freshwater supply within the park, and increasing number of visitors with flourishing tourism, that can damage the environment (National Geographic 2016; IUCN 2005). Climate change can also be a concern, indicating a tendency of aridization, that may produce changes in the hydrological budget of the lagoons and marshes.

The International Union for Conservation of Nature (IUCN) considered, in its last outlook about the state of conservation of World Heritage sites, that Doñana National Park presents significant concern in relation to its environmental qualities (Osipova et al. 2017). Effectively, the rapport states that Doñana National Park is threatened by a significant number of issues, the most serious being the long-term and continuing reduction in water quantity and quality within the property (IUCN 2017).

References

Carretero MI, Pozo M, Ruiz F, Rodríguez VJ, Cáceres L, Abad M, Muñoz J, Gómez F, Campos J, González-Regalado M, Olías M (2011) Trace elements in Holocene sediments of the southern Doñana National Park (SW Spain): historical pollution and applications. Environ Earth Sci 64(5):1215–1223

IUCN (International Union for Conservation of Nature) (2005) World Heritage Nomination – IUCN Summary. Doñana National Park, Spain. Rapport, Gland

IUCN (International Union for Conservation of Nature) (2017) Doñana National Park Conservation Outlook. http://www.worldheritageoutlook.iucn.org/explore-sites/wdpaid/61611. Accessed 17 Dec 2017

National Geographic (2016) Doñana National Park, Spain. http://travel.nationalgeographic.com/travel/parks/donana-spain/. Accessed 27 Apr 2016

Osipova E, Shadie P, Zwahlen C, Osti M, Shi Y, Kormos C, Bertzky B, Murai M, Van Merm R, Badman T (2017) IUCN world heritage outlook 2: a conservation assessment of all natural world heritage sites. IUCN, Gland

Parques Naturales Andalucia (2010) http://parquesnaturalesandalucia.com/eparque-donana/clima-parque-donana.html. Accessed 28 Apr 2016

Pozo M, Ruiz F, Carretero MI, Vidal JR, Cáceres LM, Abad M, González-Regalado ML (2010) Mineralogical assemblages, geochemistry and fossil associations of Pleistocene-Holocene complex siliciclastic deposits from the Southwestern Doñana National Park (SW Spain): a palaeoenvironmental approach. Sediment Geol 225:1):1–1)18

Rodríguez-Ramírez A, Rodríguez-Vidal J, Cáceres L, Clemente L, Belluomini G, Manfra L, Improta S, Andrés JR (1996) Recent coastal evolution of the Doñana National Park (SW Spain). Quat Sci Rev 15(8–9):803–880

Ruiz F, Rodríguez-Ramírez A, Cáceres LM, Rodríguez Vidal J, Carretero MI, Abad M, Olías M, Pozo M (2005) Evidence of high-energy events in the geological record: mid-holocene evolution of the southwestern Doñana National Park (SW Spain). Palaeogeogr Palaeoclimatol Palaeoecol 229(3):212–229

Suso J, Llamas MR (1993) Influence of groundwater development on the Doñana National Park ecosystems (Spain). J Hydrol 141(1):239–269

UNEP/WCMC (United Nations Environmental Program/World Conservation Monitoring Center) (2005) Doñana National Park, Spain. Report, Cambridge

UNESCO WHC (United Nations Educational, Scientific and Cultural Organization World Heritage Centre) (2016). Doñana National Park. http://whc.unesco.org/en/list/685. Accessed 28 Apr 2016

Part V
World Heritage on Continental Coasts: North America

Chapter 16
Joggins Fossil Cliffs, Canada

Abstract The Joggins Fossil Cliffs along the coast of Nova Scotia, Eastern Canada, have been described as the "Coal Age forests" and the "Coal Age Galapagos" due to their richness of fossils from the Carboniferous period (354 to 290 million years ago). The cliffs reveal one of the most significant events in the history of life. They contain the world's most complete fossil record of terrestrial life in the "Coal Age," including the first vertebrates able to reproduce on land, as well as the forests in which they lived. With almost 15 km of seacliffs, low bluffs, rock platforms, and beach, the property presents remains of three ecosystems: estuarine bay, floodplain rainforest, and fire-prone forested alluvial plain with freshwater pools. These spectacular seacliffs fringing the Bay of Fundy have been carved out by the world's highest tides. Their complete and accessible fossil-bearing rock exposures provide the best evidence known of the iconic features of Carboniferous period of Earth history.

16.1 Introduction

The Joggins Fossil Cliffs is located along the coast of Nova Scotia, Eastern Canada, on the shores of Chignecto Bay, the most northerly arm of the Bay of Fundy. It extends for 14.7 km and has 6.9 km^2 of area (UNESCO WHC 2016) (Fig. 16.1).

The preservation of the fossil seacliffs started in 1970, by means of Canadian provincial laws. In the following decades, the protection was strongly reinforced. In 1990, the property was closed to mineral exploration. In 2008, it was added to the World Heritage list. The province of Nova Scotia owns approximately 95% of the area (UNEP/WCMC 2008).

The fossil seacliffs of Joggins (Fig. 16.2) have been well studied for over 150 years. It was, for example, an important place of research for Sir Charles Lyell, the founder of modern geology. The impact of this place on Lyell, and its subsequent incorporation into Darwin's *On the Origin of Species*, has led to Joggins being described as the "Coal Age Galapagos" (Falcon-Lang 2006).

The site is well resourced and is managed by the Joggins Fossil Institute. The Institute provides safety and interpretive services, conducts studies, carries out research, and promotes educational programs to help maintain environmental integ-

V. Claudino-Sales, *Coastal World Heritage Sites*, Coastal Research Library 28,
https://doi.org/10.1007/978-94-024-1528-5_16

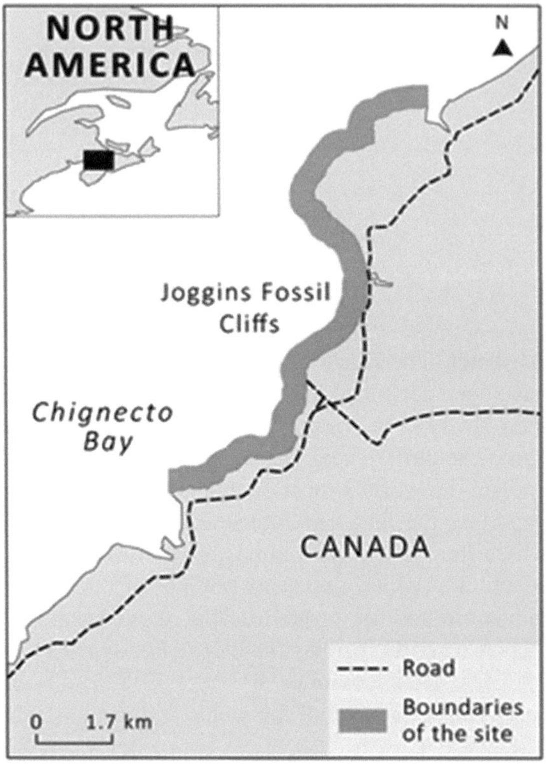

Fig. 16.1 Location of Joggins Fossil Cliff, in the Chignecto Bay, the most northerly arm of the Bay of Fundy, Canada

rity (Joggins Fossil Institute 2016). The involvement of the community and partnerships between scientists, museums, and economic interests are strong (UNESCO WHC 2016).

Tourism is an increasing activity in the area, especially scientific tourism. The visitors from schools and universities are invited to discover in the property the Coal Age ecosystem and rare fossils of plants and animals preserved where they lived 300 million years ago (Joggins Fossil Institute 2016).

16.2 Natural Settings

16.2.1 Climatic Conditions

The present climate character of the site is maritime, having cool wet winters with storms from the northeast and very warm humid summers. The mean temperature ranges from 8 °C to 36 °C, and the annual rainfall is 1144 mm (UNEP/WCMC 2008).

Fig. 16.2 Fossil cliffs showing the red sandstones of the Joggins Formation, in the Joggins Fossil Cliffs World Heritage site, Canada. (Picture reproduced by permission of Joggins Fossil Institute, website www.jogginsfossilcliffs.net)

During the Carboniferous (from about 359 to 299 million years ago during the late Paleozoic Era), the region was located within palaeoequatorial latitudes (Dullo and Devey 2016), so the climate was hot and humid with a seasonal precipitation flux, subject to storms resulting in flash floods that carried sediment loads heavy enough to half-bury forests (UNEP/WCMC 2008).

16.2.2 Geology and Geomorphology

The Joggins section is the finest natural exposure of Carboniferous period anywhere in the world (Falcon-Lang 2006). The cliffs provide insight of the most important interval in Earth's past for the formation of coal (Grey and Finkel 2011). The swamp forests produced massive quantities of organic matter that, over millions of years, creating the coal deposits for which this period of history is named (Joggins Fossil Cliff Net 2016).

The spectacular seacliffs of the World Heritage site fringe the Bay of Fundy in eastern Canada and have been carved out by the world's highest tides (Falcon-Lang and Calder 2004). The elevations go from sea level to 300 m (UNEP/WCMC 2008).

The coal beds at Joggins have been described as having been deposited mostly on small floodplains (Kaplan et al. 1985). More than that, with its low bluffs, rock platforms, and beaches, the site groups remains of three ecosystems: estuarine bay,

Fig. 16.3 Joggins Fossil Cliffs, showing fossilized trees with vestiges of animals, with a narrow beach in the foot of the cliffs. (Picture reproduced by permission of V J Matthew/Shutterstock Stock Photos)

floodplain rainforest, and fire-prone forested alluvial plain with freshwater pools (UNESCO 2008).

At the site, the Joggins Formation dips around 20° to the south and can be studied over a stratigraphic thickness of more than 600 m (Dullo and Devey 2016). The Joggins Formation represents the "classic section," where nearly all of the important paleontological discoveries have been made (Grey and Finkel 2011).

The formation is divided into 14 cycles, most of which commence with major transgressions. Higher in the cycles, the readvance of coastal and alluvial systems yielded poorly and well-drained facies assemblages, respectively (Davies et al. 2005; Falcon-Lang 2003).

Joggins Formation shows fossils of the first reptiles in Earth history, which are the earliest representatives of the amniotes, a group of animals that includes, besides the reptiles, dinosaurs, birds, and mammals (UNEP/WCMC 2008). Fossils include the world's earliest-known reptile, the world's earliest-known land snail and the tracks of the largest-known land invertebrate (Grey and Finkel 2011).

Fossil forests also occur throughout much of the Joggins section. The trees are preserved in situ and commonly contain remains of other organisms. The fossils are the buried relics of a wetland forest found still standing upright (Davies et al. 2005).

The cliffs disclose one of the most significant events in the history of life. Effectively, the Coal Age trees stand where they grew, "the footprints of creatures are frozen where they once walked, the dens of amphibians are preserved with remnants of their last meal, and the earliest reptiles remain entombed within once hollow trees" (Joggins Fossil Institute 2016) (Fig. 16.3).

16.2.3 *Biodiversity*

The fossil record at Joggins contains 195 species (Joggins Fossil Institute 2016). Evidence of the entire food chain of a terrestrial ecosystem is present, from the primary producer plants to invertebrate detritivores and tetrapod carnivores (UNESCO WHC 2016).

Presently, the vegetation is boreal forest and wooded cliff-edge farmland, included in the taiga biogeographical domain (UNEP/WCMC 2008). The fauna is represented by marine and coastal species of cold weather.

16.3 Conservation Challenges

The property has effective legal protection and has the strong support of all levels of government (UNESCO WHC 2016). The high tides of the Bay of Fundy inhibit strong development, which is an indirect way of protection of the site (UNEP/WCMC 2008). The biggest threat is global warming and climate change.

The rising of the sea level is a potential threat, considering that it produces erosion of the cliffs and destruction of the fossil elements. The Bay of Fundy water level has risen 40 cm during the past century (UNEP/WCMC 2008). Recession of the cliffs is expected in such a context.

Nevertheless, Joggins Fossil Cliffs is a model of high-quality performance within the World Heritage Convention, protecting geological values of global significance while providing benefits to local people. Threats are low, and there is highly effective and creative management that has built and maintains community support for the conservation and presentation of this site (IUCN 2017). The state of conservation of the site is good, as pointed out by the International Union for Conservation of Nature in its last conservation outlook of World Heritage sites (Osipova et al. 2017).

References

Davies SJ, Gibling MR, Rygel MC, Calder JH, Skilliter DM (2005) The Pennsylvanian Joggins Formation of Nova Scotia: sedimentological log and stratigraphic framework of the historic fossil cliffs. Atl Geol 41:115–142

Dullo WC, Devey C (2016) Joggins Cliff on Nova Scotia: the most spectacular view into upper carboniferous time. Int J Earth Sci 105(2):581–582

Falcon-Lang H (2003) Response of late carboniferous tropical vegetation to transgressive-regressive rhythms at Joggins, Nova Scotia. J Geol Soc Lond 160:643–647

Falcon-Lang H (2006) A history of research at the Joggins Fossil Cliffs of Nova Scotia, Canada, the world's finest Pennsylvanian section. Proc Geol Assoc 117:377–392

Falcon-Lang H, Calder JH (2004) Feature UNESCO world heritage and the Joggins Cliffs of Nova Scotia. Geol Today 20(4):139–143

Grey M, Finkel ZV (2011) The Joggins Fossil Cliffs UNESCO world heritage site: a review of recent research. Atl Geol 47:185–200

IUCN (International Union for Conservation of Nature) (2017) Joggins fossil cliffs conservation outlook. http://www.worldheritageoutlook.iucn.org/explore-sites/wdpaid/903133. Accessed 18 Dec 2017

Joggins Fossil Institute (2016) http://jogginsfossilcliffs.net/. Accessed 9 May 2016

Kaplan SS, Donahue J, Carr JD, Kelter PB (1985) Analysis of the trace-element content of coals from the Carboniferous Cumberland Group, near Joggins, Nova Scotia, Canada. Can J Earth Sci 22:626–629

Osipova E, Shadie P, Zwahlen C, Osti M, Shi Y, Kormos C, Bertzky B, Murai M, Van Merm R, Badman T (2017) IUCN world heritage outlook 2: a conservation assessment of all natural world heritage sites. IUCN, Gland

UNEP/WCMC (United Nations Environmental Programme/World Conservation Monitoring Centre) (2008) The Joggins Fossil Cliffs, Nova Scotia, Canada. Report, Cambridge

UNESCO WHC (United Nations Educational, Scientific and Cultural Organization World Heritage Centre) (2016) Joggins Fossil cliffs. http://whc.unesco.orgen/list/1285. Accessed 9 May 2016

Chapter 17
Miguasha National Park, Canada

Abstract Miguasha National Park, in Quebec, Canada, is considered to be the world's most perfect example of the Devonian period known as the "Age of Fishes." The Upper Devonian rocks contain fossils of fish from the period, such as the fishes that gave rise to the first four-legged, air-breathing terrestrial vertebrates, the tetrapods. Besides the fishes, there are also invertebrate and plant fossils displaying life as it existed on Earth 380 million years ago. The park's fossil collection has 18000 specimens. A large number of these fossils are exceptionally well preserved, with complete specimens, 3D specimens, and fossilized soft tissues.

17.1 Introduction

Miguasha National Park is in Quebec, Canada, on the Gaspe Peninsula, lying along the north shore of the Restigouche River estuary. The property is formed by seacliffs and has an area of 0.87 km^2 (UNESCO WHC 2016) (Fig. 17.1).

The seacliffs were first discovered by the geologist Abraham Gesner in 1842. In 1885, it was declared a provincial park by national law (UNEP/WCMC 2009). In 1990, the provincial mining act decree was issued to prohibit oil, gas, and mineral exploration in the periphery of the park. It became a World Heritage site in 1999, in recognition of its paleontological significance (UNESCO WHC 2016).

The Miguasha fossil site was among the first major paleontological localities to have been discovered and excavated in North America (Cloutier 2013). The area has been subject to scientific studies since the 1880s. Thousands of fossil specimens were collected and shipped to museums and universities around the world, helping to confirm the site's scientific importance (Parks Canada 2015).

The area of the Gaspe Peninsula was part of the hunting and fishing territory of the indigenous nation Micmac. The Micmac people called the site Megouasag, meaning "red cliffs." Over time, the name changed to Miguasha (Miguasha National Park 2007).

The park belongs to the Government of Quebec (UNEP/WCMC 2009). It is managed by the governmental agency that takes care of the outdoor facilities in the

V. Claudino-Sales, *Coastal World Heritage Sites*, Coastal Research Library 28,
https://doi.org/10.1007/978-94-024-1528-5_17

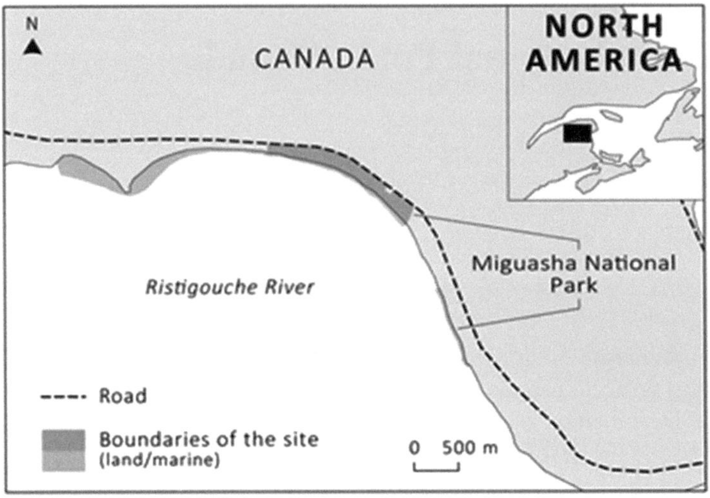

Fig. 17.1 Location of Miguasha National Park, in Quebec, Canada, on the Gulf of Gaspe

province, the Société des établissements de plein air du Québec – Sépaq (Parc National de Miguasha 2016).

17.2 Natural Settings

17.2.1 Climatic Conditions

The present climate is cool maritime, influenced by the relatively warm Chaleur Bay, with cold wet winters and humid moderate summers. The average temperatures are −6 °C to −16 °C in winter and 14 °C to 26 °C in summer with an average rainfall of 1024 mm, with wet season between August and November (UNEP/WCMC 2009). The Devonian climate was tropical (Cloutier 2013).

17.2.2 Geology and Geomorphology

At Miguasha National Park, the fossil fish and plants tell the stories about the life that existed on our planet 380 million years ago, in the Devonian period, at a time when species were making the transition from sea to land (Parc National de Miguasha 2016). The fossils are exposed in a large coastal cliff which extends along Miguasha point and the north shore of the Ristigouche River. The altitudes go from sea level to 220 m, averaging 40 m (UNEP/WCMC 2009) (Fig. 17.2).

Fig. 17.2 The Escuminac Formation, exposed in a fossil coastal cliff which extends along Miguasha point and the north shore of the Restigouche River. (Picture reproduced by permission of the Société des Établissements de Plein Air du Québec – Sépaq)

The most important sector in the park is the Escuminac Formation (119 m thick), which characterizes an area of tropical estuary inhabited by primitive forms of life (Chevrinais et al. 2016). It is composed of alternating gray layers of thick sandstone, mudstone, and calcareous schists, with the presence of 20 fish species (Chevrinais et al. 2016).

The rocks are contained within a large-folded geological structure that extends beyond Quebec into the USA. The folded rock structure is known as the Gaspe synclinorium. It is also commonly referred to as the Gaspe basin because it represents great accumulations of ancient sea deposits (Miguasha National Park 2007).

The paleoenvironmental context and diagenetic history of the Escuminac Formation indicated the presence of seawater (Matton et al. 2012), as well as of depositional setting, and its fauna is typical of transitional environments associated with land-sea interactions (Cloutier et al. 2011). Effectively, the Devonian is the period when fish first diversified, took to freshwater, and gave rise to the first terrestrial vertebrates.

Besides the Escuminac Formation (Fig. 17.2), there are additional satellite outcrops, located between 3 and 40 km from the main cliff, discovered over the past decade (Cloutier 2013).

Fig. 17.3 Fossil of a *Eusthenopteron foordi*, a Devonian lobe-finned fish that gave origin to tetrapods, found in the Escuminac Formation. (Picture reproduced by permission of Johanne Kerr/ Société des Établissements de Plein Air du Québec – Sépaq)

17.2.3 Biodiversity

The diversity of the Devonian Miguasha biota includes an allochthonous continental component, composed of plants and invertebrates, and an autochthonous aquatic component, composed of invertebrates and lower vertebrates (Cloutier 2013).

The fossil assemblage is not only fairly diverse, but it is abundant. More than 18000 specimens of fish have been found in the Escuminac Formation since its initial discovery (Cloutier et al. 2011). A large number of these fossils are exceptionally well preserved, with complete specimens, 3D specimens, and fossilized soft tissues (UNESCO WHC 2016).

In the warm tidal waters, the fish were spiny or armor-plated. Others had lungs and lobe-like fins that enabled them to crawl across mudflats and enact one major step in evolution, as fish evolved into four-limbed animals (Fig. 17.3). In the land, primitive trees, scorpions, and spiders were present (Parks Canada 2015).

There are some 60 such Devonian period fossil sites around the world. But none matches Miguasha National Park in abundance of specimens, quality of fossil preservation, and representation of evolutionary events for vertebrates. It is the only Devonian site on the World Heritage List (Parks Canada 2015).

17.3 Conservation Challenges

Protection of the integrity of the site is guaranteed by Miguasha's conservation park status. In 2004, the Government of Quebec transformed its buffer zone to a state reserve, which means that no development can be undertaken there (UNESCO WHC 2016), increasing the preservation of the entire area.

But there are still challenges to the site. Coastal erosion is a problem. To control the process, a major reforestation project was launched in 2012 on the top of the cliff, to help reduce the effects of runoff, infiltration, and congelifraction (Parc National de Miguasha 2016).

In such an erosional context, global warming and climate change are a major threat to the integrity of the site. The sea level rising tends to grow the erosion on the foot of the cliff, and it is possible that the cold weather shifts to more hot conditions, increasing the rainfall and accelerating the erosion on the top of the feature.

The International Union for Conservation of Nature (IUCN), in a conservation outlook for the World Heritage sites published in November 2017 (IUCN 2017), found that the Miguasha National Park conservation presents no immediate cause for concern. The current state of the geological values and attributes of the property is good, and the trend is stable (Osipova et al. 2017). New fossil discoveries are being made as a result of continuing field investigation, and an active research program is yielding new insights into the evolution of fish and tetrapods in Devonian times (IUCN 2017).

References

Chevrinais M, Balan E, Cloutier R (2016) New insights in the ontogeny and taphonomy of the Devonian Acanthodian Triazeugacanthus affinis from the Miguasha Fossil-Lagerstaette, Eastern Canada. Fortschr Mineral 6(1):1–16

Cloutier R (2013) Great Canadian Lagerstätten 4 the Devonian Miguasha biota (Québec): UNESCO world heritage site and a time capsule in the early history of vertebrates. Geosci Can 40:149–163

Cloutier R, Proust JN, Tessier B (2011) The Miguasha fossil-fish-Lagerstätte: a consequence of the Devonian land sea interactions. Palaeobio Palaeoenviron 91:293–323

IUCN (International Union for Conservation of Nature) (2017) Miguasha National Park Conservation Outlook. http://www.worldheritageoutlook.iucn.org/explore-sites/wdpaid/198294. Accessed 18 Dec 2017

Matton O, Cloutier R, Stevenson R (2012) Apatite for destruction; isotopic and geochemical analyses of bioapatites and sediments from the Upper Devonian Escuminac Formation (Miguasha, Quebec). Palaeogeogr Palaeoclimatol Palaeoecol 361–362:73–83

Miguasha National Park (2007) http://miguasha.ca/. Accessed 13 May 2016

Osipova E, Shadie P, Zwahlen C, Osti M, Shi Y, Kormos C, Bertzky B, Murai M, Van Merm R, Badman T (2017) IUCN world heritage outlook 2: a conservation assessment of all natural world heritage sites. IUCN, Gland

Parc National de Miguasha (2016) https://www.sepaq.com/pq/mig/index.dot?language_id=1. Accessed 13 May 2016

Parks Canada (2015) World Heritage in Canada. http://www.pc.gc.ca/eng/progs/spm-whs/sec02/sec02m.aspx. Accessed 13 May 2016

UNEP/WCMC (United Nations Environmental Programme/World Conservation Monitoring Centre) (2009) Miguasha National Park, Quebec, Canada. Report, Cambridge

UNESCO WHC (United Nations Educaitional, Scientific and Cultural Organization World Heritage Centre (2016) Miguasha National Park. http://whc.unesco.org/en/list/686. Accessed 14 May 2016

Chapter 18
Mistaken Point, Canada

Abstract Mistaken Point is a fossil site located on the island of Newfoundland, in Eastern Canada. It consists of rugged coastal cliffs of deep marine origin, representing the oldest known assemblages of large fossils in the world. Placed in the Ediacaran period, dated to the geological time interval of 580–560 million years, the fossils illustrate the first appearance of biologically complex organisms. Fossil evidence shows that life originated nearly 3 billion years before those of Mistaken Point, but most Precambrian life was generally single-celled and microscopic. The coastal cliffs of Mistaken Point represent the moment in biological evolution "when life got big." The fossils there range from a few centimeters to nearly 2 m in length and are outstanding examples of major stages of Earth's record of life.

18.1 Introduction

Mistaken Point, at the southeastern tip of Newfoundland Island, is on the Avalon Peninsula, in the Newfoundland and Labrador Province, Eastern Canada. It consists of a narrow and long (17 km) rugged coastal cliff, named for the navigational hazard it poses at the often-foggy tip of Newfoundland (Government of Newfoundland Labrador 2016) (Fig. 18.1).

The assemblage of fossils is preserved in the 5.7 km^2 Mistaken Point Ecological Reserve, which was established in 1984 (Government of Newfoundland Labrador 2016). It is a property of the Government of Canada and is managed by the Government of Newfoundland and Labrador for the benefit of the public (Mistaken Point Nomination Dossier 2016). The site has 1.46 km^2 of area and was added to the World Heritage List in 2016 (UNESCO WHC 2016).

V. Claudino-Sales, *Coastal World Heritage Sites*, Coastal Research Library 28,
https://doi.org/10.1007/978-94-024-1528-5_18

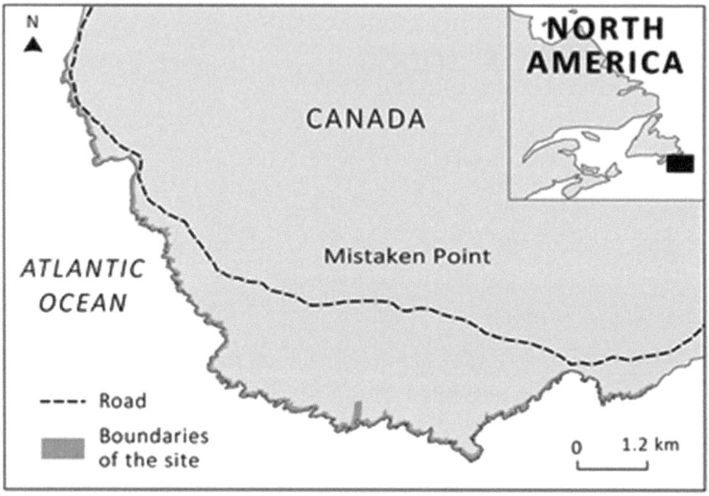

Fig. 18.1 Location of Mistaken Point, on the Newfoundland Island, Eastern Canada

18.2 Natural Settings

18.2.1 Climatic Conditions

The climate is strongly influenced by the proximity of the ocean, with cool, foggy, and wet weather (Government of Newfoundland and Labrador 2016).

The fog is most common in spring to early summer and in some places can occur for 200 days a year (Darack 2006). Foggy seas typically lie calm, but here they are often accompanied by high winds and rough conditions – creating a lethal combination for ships (Darack 2006). Fog is formed when warm air carried north with the Gulf Stream mixes with cold air brought south with the Labrador Current (Government of Newfoundland and Labrador 2016).

Despite having some of the world's wildest weather, the region's overall temperature profile is tame: January's average is −3.8 °C, with 15 ° C in July (Darack 2006). Annual rainfall ranges between 1200 and 1600 mm and annual snowfall between 2–2.5 m (Government of Newfoundland and Labrador 2016).

18.2.2 Geology and Geomorphology

The fossiliferous layers on Mistaken Point consist of 4.2 km of sedimentary rocks, tuffs, and rare basaltic lavas, deposed on a basement of granite and volcanic rocks. They accumulated within a forearc basin formed on continental crust, uplifted as part of an ancient subduction complex or accreted terrane (Retallack 2014).

Fig. 18.2 Fossiliferous seacliffs in Mistaken Point, modeled in deep ocean, originate Precambrian rocks. (Picture reproduced by permission of Dr. Alex Liu, University of Cambridge)

In general terms, the geology of the area is characterized by the presence of Late Precambrian and early Paleozoic sedimentary, volcanic, and plutonic rocks, representing the most easterly flank of the Appalachian folded belt (McCartney 1969). In the coastal area, ocean-washed fossiliferous rugged seacliffs on Precambrian rocks are present, showing evidences of glaciation, with intense coastal erosion (Shaw and Potter 2015) (Fig. 18.2).

18.2.3 Biodiversity

Known to scientists as the Ediacaran biota, the fossils of Mistaken Point represent creatures that lived 575–542 million years ago, in the Proterozoic, when all life was in the sea (Mistaken Point Nomination Dossier 2016). They were preserved by being blanketed with layers of fine volcanic ash: usually only bones, shells, and other hard parts of animals are fossilized, but the ashes allowed the imprints of the soft bodies of the Ediacaran biota to be preserved (Government of Newfoundland and Labrador 2016).

The coastal cliffs of Mistaken Point represent the moment in biological evolution "when life got big." The fossils there range from a few centimeters to nearly 2 m in length and are outstanding examples of major stages of Earth's record of life (UNESCO WHC 2016).

The fossils represent creatures that lived on the bottom of a deep ocean, considerably below the depths that waves or light could reach. But there are also fossils in

Fig. 18.3 Ediacaran life fossilized in Mistaken Point, representing creatures that lived in a deep bottom sea and that are extinct in modern life. (Picture reproduced by permission of Dr. Alex Liu, University of Cambridge)

horizons with shallow-marine sediments (Liu et al. 2010). Many are small, stemmed frond fossils (Mason and Narbonne 2016).

Surface locomotory trace fossils suggest that organisms capable of controlling locomotion and possessing muscular tissue may have existed among the Ediacaran macrofossil assemblages at Mistaken Point (Liu et al. 2014) (see Fig. 18.2).

Many generations of Ediacaran organisms are found in the site (Antcliffe et al. 2015). More than 30 species of fossil are present, many representing extinct groups unknown in the modern world (Mistaken Point Nomination Dossier 2016). (Fig. 18.3). Fossils of these multicellular organisms demonstrate the advent of mobility, heterotrophy by multicellular animals, skeletonization, sexual reproduction, and the assembly of complex ecosystems (Droser and Gehling 2014).

There is palaeoecological evidence suggesting that evolutionary innovation, ecosystem engineering, and biological interactions may have caused the first mass extinction of complex organisms, ending life as it is seen at Mistaken Point (Darroch et al. 2015).

Today, the area is in the "Eastern Hyper-Oceanic Barrens" ecoregion, which is almost completely devoid of tree cover. Coastal barren vegetation is the typical feature of the landscape. The ecoregion also has immense value for its bird life (Government of Newfoundland and Labrador 2016).

18.3 Conservation Challenges

The key coastal exposures are protected by various national and provincial laws. Access to the area is by permit only, and visits to the fossiliferous seacliffs need the presence of a guide. Undertaking activities such as scientific research at Mistaken Point requires a permit issued by the managing agency. Development is prohibited within the ecological reserve designated as the World Heritage site (Mistaken Point Nomination Dossier 2016).

The most significant threat to the integrity of the area is from coastal erosion. With climate change and global warming, this problem will increase, putting the fossils and visitations in danger.

Generally, the values of the site are robust and stable. Mistaken Point's relative isolation and exposed windswept coastal location have protected it from past development. Current threats are very low. The area is well managed, with highly effective legal and management systems in place (IUCN 2017). The state of conservation of the site is considered good (Osipova et al. 2017).

References

Antcliffe JB, Hancy AD, Brasier MD (2015) A new ecological model for the ~565 Ma Ediacaran biota of Mistaken Point, Newfoundland. Precambrian Res 268:227–242

Darack E (2006) Avalon Peninsula, Canada. Weatherwise 59:14–15

Darroch SAF, Sperling EA, Boag TH, Racicot RA, Mason SJ, Morgan AS, Tweedt S, Myrow P, Johnston DT, Erwin DH, Laflamme M (2015) Biotic replacement and mass extinction of the Ediacara biota. Proc R Soc B Biol Sci 282:129–138

Droser ML, Gehling JG (2014) The advent of animals: the view from the Ediacaran. Proc Natl Acad Sci U S A 116:4865–4870

Government of Newfoundland and Labrador (2016) Department of Environment and Conservation. Mistaken Point Ecological Reserve. http://www.env.gov.nl.ca/env/parks/wer/r_mpe/. Accessed 28 Oct 2016

IUCN (International Union for Conservation of Nature) (2017) Mistaken Point Conservation Outlook. http://www.worldheritageoutlook.iucn.org/explore-sites/wdpaid/555622046. Accessed 18 Dec 2017

Liu AG, McIlroy D, Brasier MD (2010) First evidence for locomotion in the Ediacara biota from the 565 Ma Mistaken Point Formation, Newfoundland. Geology 38(2):123–126

Liu AG, McIlroy D, Cilroy D, Matthews JJ, Brasier MD (2014) Confirming the metazoan character of a 565 Ma trace-fossil assemblage from Mistaken Point, Newfoundland. PALAIOS 29:420–430

Mason S, Narbonne GM (2016) Two new Ediacaran small fronds from Mistaken Point, Newfoundland. J Paleontol 90:183–194

McCartney WD (1969) Geology of Avalon Peninsula, Southeast Newfoundland: Chapter 8: Southeastern border of the orogenic belt. Am Assoc Pet Geol A106:115–129

Mistaken Point Nomination Dossier (2016). When life get big. Report, Government of Newfoundland Labrador, Canada

Osipova E, Shadie P, Zwahlen C, Osti M, Shi Y, Kormos C, Bertzky B, Murai M, Van Merm R, Badman T (2017) IUCN world heritage outlook 2: a conservation assessment of all natural world heritage sites. IUCN, Gland

Retallack GJ (2014) Volcanosedimentary paleoenvironments of Ediacaran fossils in Newfoundland. Geol Soc Am Bull 126:619–638
Shaw J, Potter DP (2015) Surficial geology, coastal waters, Island of Newfoundland and Labrador. Geological Survey of Canada Bulletin 605. Otawa, Canada
UNESCO WHC (United Nations Educational, Scientific and Cultural Organization World Heritage Centre) (2016) Mistaken Point. http://whc.unesco.org/en/list/1497. Accessed 1 Nov 2016

Chapter 19
Kluane, Wrangell-St. Elias, Glacier Bay, and Tatshenshini-Alsek, Canada and USA

Abstract The Kluane/Wrangell-St. Elias/Glacier Bay/Tatshenshini-Alsek national parks along the boundary of the USA and Canada are the largest nonpolar ice field in the world and present examples of some of the world's longest and outstanding glaciers – more than 380 – as well as the longest valley glacier and the largest piedmont glacier on the North American continent. It is the largest transboundary protected area in the world, presenting coastal and taiga forests, alpine meadows, and arctic and glacial wilderness. The four parks are remote, of immense size and relatively undisturbed. The parks are a refuge for large self-regulating populations of grizzly bear, Dall sheep, mountain goats and many other species.

19.1 Introduction

The Kluane/Wrangell-St. Elias/Glacier Bay/Tatshenshini-Alsek national parks are on the boundary of the state of Alaska, in the USA, and the province British Colombia and the federal territory of Yukon, Canada (UNEP/WCMC 2007). The composite site represents the largest transboundary protected area in the world. It has an area of 98390 km^2 (UNESCO WHC 2016) (Fig. 19.1).

Glacier Bay, in the USA, was declared a national monument in 1925. In 1939, its area was doubled, and it was expanded again in 1978 to its present size. In 1980, it was proclaimed a national park. The Kluane, in Canada, was made a national park reserve in 1976. Wrangell-St. Elias, in the USA, is the largest unit in the US National Park System (Winkler 2000). It was proclaimed a national monument in 1978. Tatshenshini-Alsek, in Canada, was established as a provincial park in 1993 (UNEP/WCMC 2007).

In 1979, Kluane and Wrangell-St. Elias were designated a World Heritage site. In 1992, Glacier Bay was added to the World Heritage List as an extension of the site. In 1994, Tatshenshini-Alsek was entered the World Heritage List as another extension to the previous site (UNESCO WHS 2016).

Kluane National Park is co-administered by Parks Canada Agency and the Champagne and Aishihik First Nations. Tatshenshini-Alsek is co-administered by British Columbia Parks and the Champagne and Aishihik First Nations. Wrangell-St.

V. Claudino-Sales, *Coastal World Heritage Sites*, Coastal Research Library 28,
https://doi.org/10.1007/978-94-024-1528-5_19

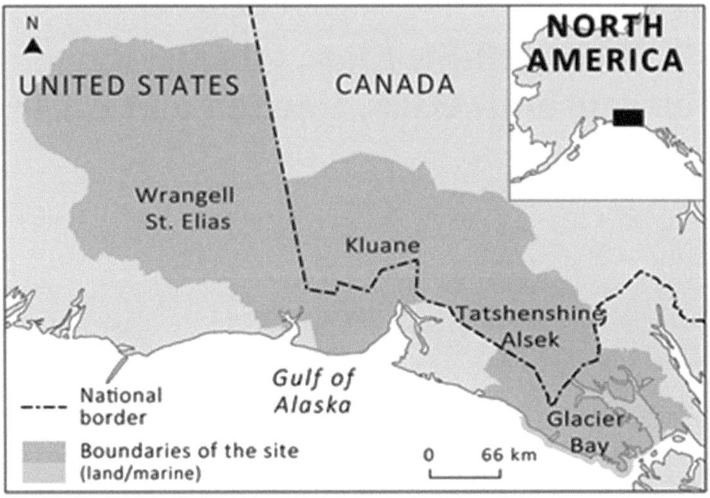

Fig. 19.1 Location of Kluane/Wrangell-St. Elias/Glacier Bay/Tatshenshini-Alsek national parks along the boundary of USA and Canada

Elias and Glacier Bay are administered by the US National Park Service (UNEP/WCMC 2007; USGS (United Station Geological Survey) 2016a, b).

With the exception of Glacier Bay, the other parks are on traditional lands of the First Nations in Canada and of the Native American population in the USA. In Kluane and Wrangell-St. Elias, they have the legal right to use protected areas for sustenance hunting, fishing, gathering, and ceremonial and spiritual practices (UNEP/WCMC 2007).

19.2 Natural Settings

19.2.1 Climatic Conditions

The climate lies at the boundary between two major climate systems, the cold, dry arctic air masses and that of the warm Pacific air masses (Krebs and Boonstra 2001).

Cyclones from the Gulf of Alaska result in very high coastal precipitation with cool, mild temperatures but with intense wind storms over the coastal mountains (UNEP/WCMC 2007). About 40–50% of the precipitation falls as snow. The rain shadow of the St. Elias Mountains causes the Kluane area to have less than 300 mm of annual precipitation and makes the region semiarid (Krebs and Boonstra 2001). Summer days are long and light, winter days, short and dark. Frost may occur all year-round (UNEP/WCMC 2007).

Fig. 19.2 Mountains with glaciers in the coastal area of Glacier Bay National Park, Alaska. (Picture reproduced by permission of Luigi Tura, webpage www.wnhs-experience.org)

19.2.2 Geomorphology and Geology

The site is characterized by high mountains, ice fields, and glaciers (Fig. 19.2). The parks have some of the best examples of glaciation and modification of landscape by glacial action in a region tectonically active (UNESCO WHC 2016). There are coastal and marine environments, snow-capped mountains, calving glaciers, deep river canyons, and fjord-like inlets (UNEP/WCMC 2007).

Over 380 glaciers form some of the world's largest and longest glaciers, several of which stretch to the sea. There is a large range of glacial processes, including classic examples of moraines and hanging valleys (UNESCO WHC 2016) (Fig. 19.3).

St. Elias Mountains are Canada's highest and most massive mountains, and Mount St. Elias, one of its peak, is the second highest peak in the USA (Parks Canada 2016). Tatshenshini-Alsek is considered to be one of the most magnificent river systems on earth (British Columbia Parks 2016). The Wrangell Mountains are a large young volcanic massif (USGS 2009). Terranes aggregated in the collision of the Pacific plate with the North American plate form the Glacier Bay region (NPS 2016).

The area is among the most geologically active in the world. The ranges were created by the successive accretions of Pacific plates against and subducted under the North American plate and overlies a subduction zone, with numerous volcanoes (UNEP/WCMC 2007).

Fig. 19.3 Glacial erosion in igneous rocks in Kluane National Park and Reserve, Canada. (Picture reproduced by permission of Sieradzki/Shutterstock Stock Photos)

The southern Alaskan continental margin has more than doubled in size in approximately the past 200 million years due to underthrusting by the Pacific oceanic plate, which prompted rapid uplift, voluminous volcanism, and powerful earthquakes. The uplift of the coastal mountains has engendered large and scenic systems of alpine, piedmont, and tidewater glaciers (Winkler 2000).

Yukon Territory (Canada) is located in the northern part of the North American Cordillera, the mountainous backbone of the western part of the continent. It is composed of rocks that record more than a billion years of Earth history, documenting the evolution of the western margin of ancestral North America (Laurentia) and of the various displaced terranes that were accreted to it since late Paleozoic (Yukon Government 2015).

In British Columbia, the bedrock geology shows a wide variety of depositional, intrusive, and tectonic features ranging from early Late Proterozoic to Tertiary. Thrust belts, volcanoes, and granite canyons, scarps, plateaus, and cordilleras characterize the area (BCGS 2016).

Several faults run through the parks and earthquakes occur regularly (UNEP/WCMC 2007).

19.2.3 Biodiversity

Wildlife species common to Alaska and Northwestern Canada are well represented, some in numbers exceeded nowhere else. The marine components support a great variety of fauna, including marine mammals and anadromous fish. Populations of bears, wolves, caribou, Dall sheep, and mountain goats that are endangered elsewhere are self-regulating there (UNESCO WHC 2016).

The site transitions from the interior to coastal zones, resulting in high biodiversity with plant and animal communities ranging from marine, coastal forest, montane, sub-alpine, and alpine tundra. The Tatshenshini and Alsek river valleys are pivotal because they allow ice-free linkages from coast to interior for plant and animal migration (UNESCO WHC 2016).

The overlapping of Pacific and arctic air masses above Kluane Park results in the greatest diversity of flora north of the 60th parallel, although only about 18% of the park is vegetated and only 7% is forested. Tatshenshini-Alsek Provincial Park has three biogeoclimatic zones: moist ocean-influenced lower reaches with spruce-willow-birch forest and scrub and both wet and drier boreal forests of white and black spruce and alpine tundra. Wrangell-St. Elias National Park has a great diversity of sub-arctic plant communities. Glacier Bay National Park has four major terrestrial biomes: coastal forests, wet tundra, alpine tundra above 2500 m, and glaciers with ice fields (UNEP/WCMC 2007; USGS (Unites States Geological Survey) 2016a, b).

In Kluane National Park, 29 mammals and 118 breeding species of birds have been recorded. The fish include several species of salmon. In Tatshenshini-Alsek Provincial Park, the marginal flora harbors a parallel fauna where the conditions encourage adaptive evolution. Historically, animal populations have been scarce in the mountains of Wrangell-St. Elias National Park, probably due to the recentness of their deglaciation. Twenty-eight terrestrial mammals, 230 bird species, and more than 230 species of fish have been found in Glacier Bay National Park (UNEP/WCMC 2007).

19.3 Conservation Challenges

This site is one of the few places remaining in the world where ecological processes are governed by natural stresses and evolutionary changes in a glacial and ecological continuum (UNESCO WHS 2016).

The great threat comes from global warming and climate change, already happening. Warmer weather has caused a retreat of the ice covers, increasing landslides and coastal and continental erosion by water, changing the ecology and the whole environment. Populations of several species are strongly affected by the declining sea ice environment (USGS 2009).

The International Union for Conservation of Nature, in its last conservation outlook about World Heritage sites, found that the site's ecological integrity is triggered by climate change-related reductions in glaciers, the potential negative effects of ocean acidification, and also by increasing visitation (IUCN 2017). Other aspects of ecological integrity are good and the site represents a pristine system with minimal impacts. The general conservation outlook of the parks can be considered good, with some concerns (Osipova et al. 2017).

References

British Columbia Geological Survey (2016) Introduction to the Geology of British Columbia. http://www.empr.gov.bc.ca/Mining/Geoscience/Pages/GeologyofBC.aspx. Accessed 17 June 2016

British Columbia Park (2016) Tatshenshini-Alsek Provincial Park. http://www.env.gov.bc.ca/bcparks/explore/parkpgs/tatshens/. Accessed 17 June 2016

IUCN (International Union for Conservation of Nature) (2017) The Kluane/Wrangell-St. Elias/Glacier Bay/Tatshenshini-Alsek Conservation Outlook. Gland, Switzerland

Krebs CJ, Boonstra R (2001) The Kluane Region. In: Krebs CJ, Boutin S, Boonstra E (eds) Ecosystem dynamics of the Boreal Forest, the Kluane Project. Oxford University Press, New York

National Park Service (NPS) (2016) Glacier Bay National Park and Preserve, Alaska. https://www.nps.gov/glba/learn/nature/. Accessed 17 Jan 2016

Osipova E, Shadie P, Zwahlen C, Osti M, Shi Y, Kormos C, Bertzky B, Murai M, Van Merm R, Badman T (2017) IUCN world heritage outlook 2: a conservation assessment of all natural world heritage sites. IUCN, Gland

Parks Canada (2016) Kluane National Park and Reserve. http://www.pc.gc.ca/eng/pn-np/yt/kluane/index.aspx. Accessed 17 June 2016

UNEP/WCMC (United Nations Environmental Programme/World Conservation Monitoring Centre) (2007) Kluane/Wrangell-St Elias/Glacier Bay/Tatshenshini-Alsek. Report, Cambridge, England

UNESCO WHC (United Nations Educational, Scientific and Cultural Organization World Heritage Centre) (2016) Periodic Report – Second Cycle. http://whc.unesco.org/archive/periodicreporting/EUR/cycle02/section2/groupa/72.pdf. Accessed 15 June 2016

UNESCO WHS (United Nations Educational, Scientific and Cultural Organization World Heritage Sites) (2016). Kluane/Wrangell-St. Elias/Glacier Bay/Tatshenshini-Alsek. http://whc.unesco.org/en/list/72. Accessed 17 June 2016

USGS (Unites States Geological Survey) (2016a). U.S. Geological Survey (USGS) Western region: Alaska Coastal and Ocean Science. Report, Anchorage, USA

USGS (United States Geological Survey) (2016b) Wrangell Mountains. http://pubs.usgs.gov/pp/p1386k/pdf/09_1386K_wrangell.pdf. Accessed 17 June 2017

Winkler GR (2000) A geologic guide to Wrangell–Saint Elias National Park and preserve, Alaska. USGS, Denver

Yukon Government (2015) Overview of Yukon Geology. http://www.geology.gov.yk.ca/overview_bedrock_geology.html. Accessed 17 June 2016

Chapter 20
Sian Ka'an, Mexico

Abstract Located on the east coast of the Yucatán Peninsula, Mexico, the Sian Ka'an area is a national biosphere reserve. It is an almost undisturbed complex hydrological system containing tropical forests, wetlands, mangroves, swamps, lagoons, beaches, dunes, sinkholes, and marshes, as well as a large marine section intersected by a barrier reef. The diversity of life in Sian Ka'an is exceptional. It is one of Mexico's largest protected areas, established to manage marine, coastal, and terrestrial ecosystems. The tropical forests are home to a great number of mammals. The site also provides habitat for a large number of resident and migratory bird species, and there is also a great diversity of marine life.

20.1 Introduction

Along its roughly 120 km of coastline, Sian Ka'an, a biosphere reserve in the Yucatán Peninsula, Mexico, covers more than 4000 km^2, with 1200 km^2 as marine (UNESCO WHC 2016) (Fig. 20.1).

The site was declared a reserve by the local government in 1982. In 1986, it was designated a national biosphere reserve by the federal government and an international biosphere reserve by UNESCO (UNEP/WCMC 2008). The reefs of the protected area were designated a national biosphere reserve in 1998. It became a wetland of importance by the Wetlands Convention (the Ramsar Convention) in 2002. The biosphere reserve became a World Heritage site in 1987 (UNESCO WHC 2016).

Ninety nine percent of Sian Ka'an is federally owned; the other 1% is in the hands of private owners. The reserve is managed by the National Commission for Natural Protected Areas (Comision Nacional de Areas Naturales Protegidas – CONANP) since 2000, under the National Secretariat for the Environment, Natural Resources and Fish (Secretaría de Medio Ambiente, Recursos Naturales y Pesca-SEMARNAP) (UNEP/WCMC 2008).

The Caribbean spiny lobster fishery is the main economic activity within the Sian Ka'an Biosphere Reserve, Mexico, where restricted access fishing is allowed (Ley-Cooper et al. 2013). The reserve also supports an important tourism industry (Mazzotti et al. 2005).

V. Claudino-Sales, *Coastal World Heritage Sites*, Coastal Research Library 28,
https://doi.org/10.1007/978-94-024-1528-5_20

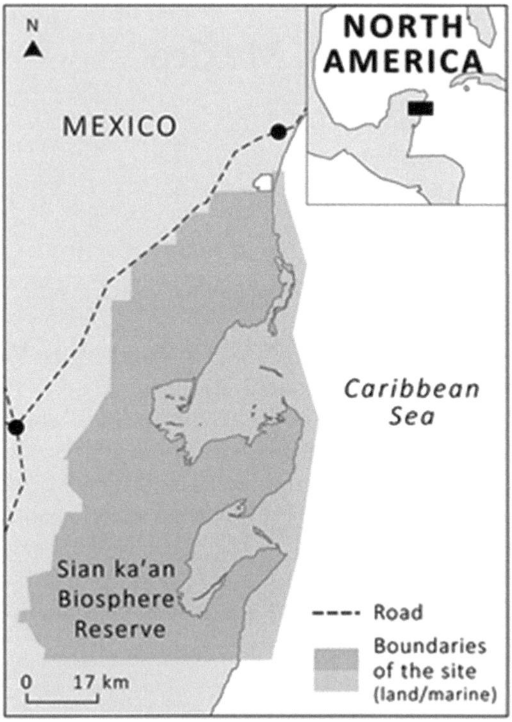

Fig. 20.1 Location of Sian Ka'an Biosphere Reserve, Yucatán Peninsula, Mexico

The land was historically occupied by the indigenous group Maya. The Mayan civilization, one of the most advanced Amerindian cultures of the ancient Americas, began in the Yucatán Peninsula near 2500 years BC, and they still inhabit the vicinity of the biosphere reserve (Encyclopedia of the Nations 2016).

Twenty-three Mayan sites are registered in the reserve (UNEP/WCMC 2008). The community historically depended on working the forest by shifting agriculture, hunting, and household gardening. But these activities are not enough to provide all family needs, and so there is a push for alternative sources of livelihoods and subsistence. Tourism is an important one (Martinez-Reyes 2014).

20.2 Natural Settings

20.2.1 Climatic Conditions

The warm waters of the Caribbean Sea contribute to the climate of the Yucatán Peninsula, helping to create tropical warm and humid weather. It has an average temperature of 25–27 °C, rarely dropping below 16 °C or rising above 49 °C. The heaviest rainfall occurs in the summer months. Average annual rainfall in this area is 1150 mm (Encyclopedia of the Nations 2016).

The area is cloudy for 200 days a year and the mean relative humidity is around 80%. Hurricanes strike between June and October. Strong marine tornados also occur between May and October, although their action is limited and brief (CIQRO 1983). During the last century, there has been an average of one hurricane every 8 years, which causes severe damage to structures and vegetation (UNEP/WCMC 2008).

20.2.2 Geology and Geomorphology

The Yucatán Peninsula sits on a horizontal bed of limestone, ranging from sea level to 10 m of altitude. The peninsula is a low-lying carbonate terrane with no geomorphic expressions of stream channels (Hanshaw and Back 1980). Mesozoic and Cenozoic limestone, dolomite, and anhydrite overlie deep buried Paleozoic crystalline and sedimentary rocks; they are covered by Pleistocene and Holocene carbonate rocks and sediments along the coast (Perry et al. 2003).

The peninsula is one of the world's largest karstic aquifer systems (Gondwe et al. 2011). Sian Ka'an in particular lies on a partially emerged coastal limestone plain which forms part of the extensive Pleistocene barrier reef system along the eastern coast of Central America (UNEP/WCMC 2008). This is the second largest barrier reef in the world, the growth of which is partly due to the lack of erosion inland and consequent silt-free water (Lopez 2003).

The biosphere reserve is an almost undisturbed complex hydrological system containing, besides the reefs, tropical forests, wetlands, mangroves, swamps, lagoons, beaches, dunes, sinkholes, and marshes (UNESCO WHC 2016) (Figs. 20.2 and 20.3).

Three geological faults cross the reserve from southwest to northeast under the bays, which have influenced the topography and hydrology. There is a large number of sinkholes characteristic of the karst landscape of Yucátan (UNEP/WCMC 2008).

The wetlands of Sian Ka'an Biosphere are fed by groundwater coming from the karst aquifer of the Yucátan Peninsula. Therefore, the wetlands are groundwater dependent. The main water input areas are associated with water-filled faults that transport the groundwater from the catchment to the wetlands (Gondwe et al. 2010).

Mixing of fresh groundwater with subterranean Caribbean Sea water generates a highly reactive geochemical zone, creating a major geomorphic process in developing caves, coves, and crescent-shaped beaches along the Yucátan coast, including Sian Ka'an (Back et al. 1986).

20.2.3 Biodiversity

The diversity of life in Sian Ka'an is exceptional. It is one of Mexico's largest protected areas, established to manage marine, coastal, and terrestrial ecosystems (UNESCO WHC 2016).

Fig. 20.2 Channel in the wetlands of Sian Ka'an Biosphere Reserve, colonized by mangroves. (Picture reproduced by permission of FCSCAFEINE/Shutterstock Stock Photos)

Fig. 20.3 Pebble beach and sandy beach in the coast of Sian Ka'an Biosphere Reserve with coastal vegetation in the shore. (Picture reproduced by permission of Andrea Prida Bay)

Plants cover 35–40% of the substrate, with red mangrove *Rhizophora mangle* up to 2 m high dominating. A total of 859 vascular plants was recorded in the reserve (Kramer and Kramer 2002). Main vegetation types are semi-deciduous forest, flooded forests, mangroves, coastal swamps, and dune vegetation (UNEP/WCMC 2008; Hooghiemstra 1999). Grass communities cover large areas and among mangroves and inland forests, though not in areas of high salinity (UNEP/WCMC 2008).

The flood lowland forest in particular is a typical ecosystem of the Sian Ka'an Biosphere Reserve (Trueba et al. 1999). It is subdivided into low forest with closed and, in lower wetter areas, open canopy tree communities (UNEP/WCMC 2008). Coastal dunes stretch along 68 km of the coast. The cultivation of introduced coconut has replaced about 60% of the natural vegetation of the dunes (Espejel 1983).

There is also a great diversity of fauna. There are a total of 44 species of fish (Camargo-Guerra et al. 2013), 103 species of mammals, 42 species of reptile and amphibians, and 339 species of bird, of which 219 breed in the reserve (UNEP/WCMC 2008). Since the diversity of aquatic habitats is very high, marine and wading birds are well represented, and this is the second largest waterbird community in Mexico (Kramer and Kramer 2002).

The 110 km of barrier reef which parallels much of the coast is one of the richest coral reefs in Mexico, with 83 recorded species of reef-building coral, representing 63% of Caribbean coral species (UNEP/WCMC 2008).

20.3 Conservation Challenges

Both fisheries and tourism industries are threatened by growing human populations and associated development, as well as unsustainable uses, agricultural and urban development, and increased extraction of natural resources (Mazzotti et al. 2005).

There is a clear decline of fisheries in Sian Ka'an and continuing deterioration of the reefs (Velez et al. 2014). In the northern part of the reserve, an elevated road built to facilitate the traffic in the Caribbean Coast disrupts the flow of fresh water through the estuarine zone, creating hypersaline conditions and a mangrove dead zone (Lagomasino et al. 2015).

The integrity of the ecosystem in Sian Ka'an is also threatened by increasing groundwater abstractions and pollution of the freshwater resource (Gondwe et al. 2011). These stressors are driven by local and national societal needs (Mazzotti et al. 2005).

Global warming is a threat to the barrier reef, which can suffer with higher sea level and higher temperatures of the seawater. The barrier reef protects the coast from erosion, and therefore the damage imposed on it by climate changes may signify a large modification in all ecosystems.

Out of the large number of current threats, coastal development, mostly related to tourism, stands out as a major concern with multiple direct and indirect impacts on land and in the sea, including pressure on species, disturbance, depletion of freshwater, and sewage and waste. Generally, the conditions of conservation are deteriorating (IUCN 2017). But the last conservation outlook on the reserve considered it good with some concerns (Osipova et al. 2017).

References

Back W, Hanshaw BB, Van Driel JN (1986) Differential dissolution of a Pleistocene reef in the ground-water mixing zone of coastal Yucatan, Mexico. Geology 142:137–140

Camargo-Guerra T, Escalera-Vázquez LH, Zambrano L (2013) Fish community structure dynamics in cenotes of the Biosphere Reserve of Sian Ka'an, Yucatán Peninsula, Mexico. Rev Mex Biodiversidad 84(3):901–911

CIQRO (Centro de Investigaciones de Quintana Roo) (1983) Sian Ka'an: Estudios Preliminares de una Zona en Quintana Roo Propuesta como Reserva de la Biosfera. Centro de Investigaciones de Quintana Roo, Puerto Morelos

Encyclopedia of the Nations (2016) Yucatan. http://www.nationsencyclopedia.com/mexico/Michoac-n-Zacatecas/Yucatan.html. Accessed 26 May 2016

Espejel I (1983) Vegetation de las dunas costeras de la Peninsula de Yucatan. 1. Reserva de la Biosfera de Sian Ka'an, Quintana Roo, Mexico. Biotica 9(2):182–210

Gondwe BRN, Bauer-Gottwein P, Hong SH, Wdowinski S (2010) Hydrologic dynamics of the ground-water-dependent Sian Ka'an wetlands, Mexico, derived from InSAR and SAR data. Wetlands 30(1):1–13

Gondwe BRN, Merediz-Alonso G, Bauer-Gottwein P (2011) The influence of conceptual model uncertainty on management decisions for a groundwater-dependent ecosystem in karst. J Hydrol 400(1–2):24–40

Hanshaw BB, Back W (1980) Chemical mass-wasting of the northern Yucatan Peninsula by groundwater dissolution. Geology 85:222–224

Hooghicmstra H (1999) Palynology of the Sian Ka'an biosphere reserve, Quintana Roo, Mexico; book review. Rev Palaeobot Palynol 105(1–2):119–120

IUCN (International Union for Conservation of Nature) (2017) Sian Ka'an Conservation Outlook. http://www.worldheritageoutlook.iucn.org/explore-sites/wdpaid/20062. Accessed 18 Dec 2017

Kramer P, Kramer P (2002) Ecoregional conservation planning for the Meso-american Caribbean Reef. World Wildlife Fund, Washington, DC

Lagomasino D, Price R, Herrera-Silveira J, Miralles-Wilhelm F, Merediz-Alonso G, Gomez-Hernandez Y (2015) Connecting groundwater and surface water sources in groundwater dependent coastal wetlands and estuaries: Sian Ka'an Biosphere Reserve, Quintana Roo, Mexico. Estuar Coasts 38(5):1744–1763

Ley-Cooper K, Lestang S, Phillips B, Lozano-Alvarez E (2013) Estimates of exploitation rates of the spiny lobster fishery for Panulirus argus from tagging within the Bahia Espiritu Santo to Sian Ka'an Biosphere Reserve, Mexican Caribbean. Mar Biol Res 9(1):88–96

Lopez A (2003) Sian Ka'an Coastal Biospheren Reserve and surrounding Forests. Report, Ramsar Convention Bureau, Paris, France

Martinez-Reyes J (2014) Beyond nature appropriation: towards post-development conservation in the Maya Forest. Conserv Soc 12(2):162–167

Mazzotti F, He F, Merediz G, Lazcano M, Lasch C, Barnes T (2005) Conceptual ecological model of the Sian Ka'an Biosphere Reserve, Quintana Roo, Mexico. Wetlands 25(4):980–997

Osipova E, Shadie P, Zwahlen C, Badman T (2017) IUCN World Heritage Outlook 2: A conservation assessment of all natural World Heritage sites. IUCN, Gland, Switzerland

Perry E, Velasquez-Oliman G, Socki RA (2003) Hydrogeology of Yucatan Peninsula. In: Gomez-Pompa A, Allen MF, Fedick SL, Jimenez-Osornio JJ (eds) The lowland Maya area: three millennia at the human-wildland interface. Haworth Press, London

Trueba DP, González MMV, Aragonés CR (1999) Comunidades de la mesofauna edáfica en una selva baja inundable de la Reserva de la Biósfera de Sian Kaan, Quintana Roo, México. Rev Biol Trop 47(3):489–492

UNEP/WCMC (United Nations Environmental Programme/World Conservation Monitoring Centre) (2008) Sian Ka'an Yucatan Mexico. Report, Cambridge, England

UNESCO WHC (United Nations Educational, Scientific and Cultural Organization World Heritage Centre) (2016) Sian Ka'an. http://whc.unesco.org/en/list/410. Accessed 26 May 2016

Velez M, Adlerstein S, Wondolleck JV (2014) Fishers' perceptions, facilitating factors and challenges of community-based no-take zones in the Sian Ka'an Biosphere Reserve, Quintana Roo, Mexico. Mar Policy 45:171–181

Chapter 21
Whale Sanctuary of El Vizcaino, Mexico

Abstract Whale Sanctuary of El Vizcaino is located in Baja California, Mexico, facing the Pacific Ocean. It is a site formed by two coastal lagoons and a complex mosaic of wetlands, marshes, halophytes, mangroves, dunes, and desert habitats. The lagoons are the world's most important place for the reproduction of the once endangered North Pacific gray whale. The lagoons are home to numerous other marine mammals. Four marine turtle species live in the shallow waters which are also habitat for a large number of fish, crustaceans, and migratory bird species. The sanctuary conserves both marine and terrestrial ecosystems and their interface. The surrounding desert displays highly diverse flora and fauna.

21.1 Introduction

The Whale Sanctuary of El Vizcaino is in the Pacific Coast of the central part of Mexico's Baja California Peninsula. The site comprises the Laguna Ojo de Liebre lagoon with the adjoining Guerrero Negro and Manuela and Laguna San Ignacio lagoons and their surroundings (UNESCO WHC 2016). The lagoons are around 100 km from each other (UNEP/WCMC 2008) (Fig. 21.1).

The Laguna Ojo de Liebre is 9 km wide by 48 km long and 5–12 m deep. Laguna San Ignacio is 6 km wide, 35 km long, and 2–4 m deep (SEDUE 1989). The total extent of the two components of the sanctuary is 3709 km^2, inserted in the much larger El Vizcaino Biosphere Reserve, Mexico's largest protected area (UNESCO WHC 2016) (see Fig. 21.1).

In 1993, the serial site was designated as a Biosphere Reserve under the UNESCO Man and the Biosphere Programme. In 1993, it was added to the World Heritage List. The lagoons were defined as Wetlands of International Importance under the Ramsar Convention of Wetlands in 2004 (UNEP/WCMC 2008).

The lagoons are subject to fishing, ecotourism (Fig. 21.2), salt extraction, and oyster farms, having a high ecological and economical relevance (Aguero 2000).

The area where they are inserted – the El Vizcaino Biosphere Reserve – has one of the lowest population densities of the region (Ortega et al. 2015). But it has been occupied for many centuries. There are a number of prehistoric sites, as well as

V. Claudino-Sales, *Coastal World Heritage Sites*, Coastal Research Library 28,
https://doi.org/10.1007/978-94-024-1528-5_21

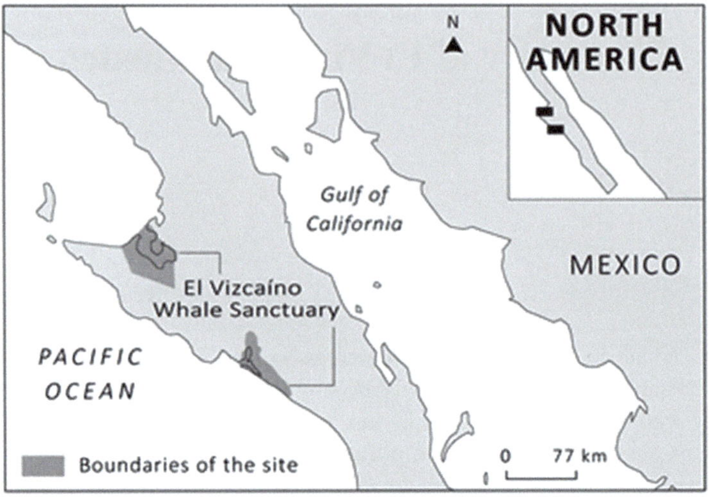

Fig. 21.1 Location of the two properties (Ojo de Liebre Lagoon up and Saint Ignacio Lagoon below) that form the World Heritage site Whale Sanctuary of El Vizcaino, in Mexico

Fig. 21.2 Tourists play with a gray whale in the Ojo de Liebre Lagoon, in El Vizcaino Whale Sanctuary. (Photo reproduced by permission of Diane Trotter, website www.bahiaasuncion.org)

petroglyphs and ancient ruined structures, together with evidence of the early European colonization (IUCN 1990).

The ownership at Laguna Ojo de Liebre is 40% national, 50% communal, and 10% private. In the Laguna Santo Ignacio, the ownership is 80% national and 20% communal. The biosphere reserve is managed by the Instituto Nacional de Ecología

(INE) for the Secretariat of Environment, Natural Resources and Fisheries (SEMARNAP) (UNEP/WCMC 2008).

21.2 Natural Settings

21.2.1 Climatic Conditions

The climate of the area is warm temperate and very dry. The mean annual temperature ranges between 18 and 22 °C during the day in winter (UNEP/WCMC 2008). Temperatures reach 40–45 °C in summer (IUCN 1990). The annual evaporation rate is 98%, and coastal fog is common. The annual rainfall averages 80 mm and fluctuates between 0 and 200 mm (UNEP/WCMC 2008).

The wind is not strong, averaging 2 m/s in winter. The humidity of the air is very low, in general around 46%. The dew point is at 1 °C. The visibility of the air close to the coast is of the order of 27 km (Time and Date 2017).

21.2.2 Geology and Geomorphology

Geologically, the area is submitted to the tension field that characterizes the Gulf of California. The Gulf of California is a proto-ocean that is subjected to a permanent stress of extension, in the contact of oblique divergent tectonic plates. The area is permanently under seismic hazard (Ortega et al. 2015).

Sedimentary rocks form the substratum, with intrusions of rhyolitic lavas, andesites, and pyroclasts. A series of shallow, sandy bays, inlets, and saltwater lagoons is found in Laguna Ojo de Liebre and Laguna San Ignacio. The majority of soils are calcareous regosols derived from volcanic material (IUCN 1990). Coastal dunes are found around Laguna Ojo de Liebre (UNEP/WCMC 2008).

The area has limited hydrologic resources because there are no permanent surface streams. Ephemeral streams appear only during wetter seasons, and the inhabitants need to explore the underground aquifers (Almendarez-Hernández et al. 2013).

21.2.3 Biodiversity

In the lagoons of the western coast of Baja California, the fauna is characterized by a mix of species of temperate-cold and temperate-warm and tropical (Aguero 2000).

Fig. 21.3 Gray whale wintering in Ojo de Liebre Lagoon, with a bird swimming nearby. Birds and whales are characteristic of the waters of this World Heritage site. (Photo reproduced by permission of Diane Trotter, website www.bahiaasuncion.org)

For Santo Ignacio Lagoon, the inventory of the ichthyofauna indicates the presence of 107 species, included in 54 families. From this total, 40 families are also present in the Ojo de Liebre Lagoon (Aguero 2000).

The lagoons are important as a refuge for wintering wildfowl: 10% of the wildfowl wintering on the west coast of Mexico winter in the reserve. This includes brent geese, 63% of Mexico's winter population (IUCN 1990). They host more than 80% of the birds which winter along the Peninsula's Pacific coast (Carmona and Danemann 1998).

The vegetation is representative of arid or hyperarid environments. Mangrove is typical of the lagoons (IUCN 1990), and it is the northernmost mangrove forests on the continent. The plants scattered around the lagoons are tolerant of the saline and alkaline soils, constant wind, and high solar radiation (UNEP/WCMC 2008).

El Vizcaino protects a large number of animal and plant species of which a considerable part is endemic and threatened (UNESCO WHC 2016). The coastal lagoons of Ojo de Liebre and San Ignacio are a site for reproduction and wintering for gray whales.

The area is recognized as the world's most important place for the reproduction of this mammal (UNESCO WHC 2016) (Fig. 21.3). The protection of the site has been fundamental in the recovery of this species after near extinction as a result of commercial whaling, including in these very lagoons (UNESCO WHC 2016).

Other mammals also live in the lagoons, such as the bottlenose dolphin, the harbor seal, the California sea lion, the northern elephant seal, and the blue whale. The site also protects four species of endangered marine turtle (IUCN 1990).

This exceptional sanctuary conserves both marine and terrestrial ecosystems and their delicate interface. The surrounding desert, biogeographically part of the Sonoran Desert, boasts highly diverse flora and fauna (UNESCO WHC 2016).

21.3 Conservation Challenges

Although the migratory whale population has recovered well since protection from commercial whaling in 1938, it is still threatened by industrial and economic development and an increase in coastal shipping (UNEP/WCMC 2008).

Despite the protected status of the Whale Sanctuary of El Vizcaino, the area is susceptible to the impacts of economic activities, in particular benthic and pelagic fisheries, large-scale salt extraction, and tourism (UNESCO WHC 2016).

The area is very dry and surrounded by desert environments. Climatic change and global warming can reduce the humidity and may pose a threat to the survival of the ecological relationships (salinity included) that support fauna and flora in the site.

All indicators of the current status of biodiversity values in the site show stable conditions, but the expected future impacts associated with the depletion of freshwater aquifers and climate change in all its dimensions can be significant (IUCN 2017). For these reasons, the state of conservation of the site is considered to be good, with some concerns (Osipova et al. 2017).

References

Aguero JC (2000) Origen y distribución de la ictiofauna de la Laguna de San Ignacio, Baja California Sur, México. Ciencia Ergo Sum 7(2):157–163

Almendarez-Hernández MA, Jaramillo-Mosqueira LA, Avilés-Polanco G, Beltrán-Morales LF, Hernández-Trejo V, Ortega-Rubio A (2013) Economic valuation of water in a natural protected area of an emerging economy: recommendations for el Vizcaino Biosphere Reserve, Mexico. Interciencia 38(4):245–252

Carmona R, Danemann G (1998) Distribución espaciotemporal de aves en la salina de Guerrero Negro, Baja California Sur, México. Cienc Mar 24(4):389–408

IUCN (International Union for Conservation of Nature) (1990) World heritage nomination – summary. Report, Gland, Switzerland

IUCN (International Unior for Conservation of Nature) (2017) Whale sanctuary of El Vizcaino conservation outlook. http://www.worldheritageoutlook.iucn.org/explore-sites/wdpaid/68918. Accessed 18 Dec 2017

Ortega R, Carciumaru D, Quintanar L, Rubio R (2015) Estimación del peligro sísmico en el centro de Baja California: Un modelo consistente entre las regiones del Golfo de California y la Peninsular. Bol Soc Geol Mex 67(1):87–102

Osipova E, Shadie P, Zwahlen C, Osti M, Shi Y, Kormos C, Bertzky B, Murai M, Van Merm R, Badman T (2017) IUCN world heritage outlook 2: a conservation assessment of all natural world heritage sites. IUCN, Gland

SEDUE (1989) Propuesta del Programa de Manejo de la Reserva de la Biosfera El Vizcaino. Secretaria de Desarrollo Urbano y Ecologia, La Paz, Mexico
Time and Date (2017) Annual weather averages near Mexicali. https://www.timeanddate.com/weather/mexico/mexicali/climate. Accessed 18 Dec 2017
UNEP/WCMC (United Nations Environmental Programme/World Conservation Monitoring Centre) (2008) Whale sanctuary of El Vizcaino Mexico. Report, Cambridge, England
UNESCO WHC (United Nations Educational, Scientific and Cultural Organization World Heritage Centre) (2016) Whale sanctuary of El Viszcaino. http://whc.unesco.org/en/list/554. Accessed 31 May 2016

Chapter 22
Everglades National Park, USA

Abstract Everglades National Park, located in South Florida, USA, protects an unparalleled landscape. It is at the interface of temperate and subtropical climates, fresh and brackish water, shallow bays, and deeper coastal waters, which creates a complex of habitats supporting a high diversity of flora and fauna. It is characterized by the presence of wetlands, sawgrass prairies, dry tropical hardwood hammocks, pinelands, and mangrove swamps, in addition to Florida Bay and Gulf Coast estuaries. It provides important habitat for numerous rare and endangered species, like the manatee, the American crocodile, and the endangered Florida panther. This site has been called "a river of grass flowing imperceptibly from the hinterland into the sea."

22.1 Introduction

Located at the southern tip of Florida, Everglades National Park is a wetland that has a segment of coast on the Gulf of Mexico (Fig. 22.1). It has 5670 km^2 of area (UNESCO WHC 2016a).

The Everglades National Park is a very large subtropical coastal wetland within a large watershed that extends 160 km from the Kissimmee River bed basin through Lake Okeechobee to the Florida Keys in southeast Florida (Harvey and McCormick 2009).

The Park is bordered by the Big Cypress National Preserve to the north and Key Largo National Marine Sanctuary to the southeast, and nearby are Biscayne National Park, ten national wildlife refuges, and the Florida Keys National Marine Sanctuary (UNEP/WCMC 2007).

The site was established in 1934 to conserve the natural landscape and prevent further degradation of a portion of the Greater Everglades (NPS 2016; UNEP/WCMC 2007). In 1976, it was recognized as a Biosphere Reserve under the UNESCO Man and the Biosphere Programme. It became a World Heritage site in 1979. In 1987 it was designated as a Wetland of International Importance under the Ramsar International Convention of Wetlands (UNEP/WCMC 2007).

Paleo- and Archaic Indian cultures have been traced from 10000 years ago. In historic times, 5 tribes numbering about 20000 were known before contact with

V. Claudino-Sales, *Coastal World Heritage Sites*, Coastal Research Library 28,
https://doi.org/10.1007/978-94-024-1528-5_22

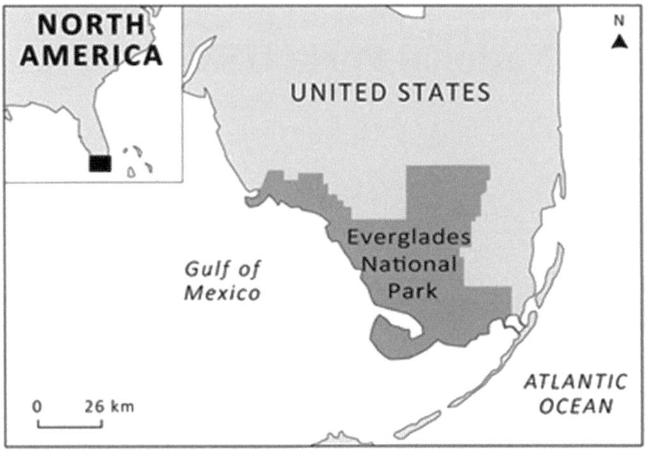

Fig. 22.1 Location of Everglades National Park, facing the Gulf of Mexico in southern Florida, USA

Westerners reduced them. A rich record of settlements, farming, and fishing remains in some 200 known archaeological sites (UNEP/WCMC 2007).

The Everglades National Park was included in the list of World Heritage sites in danger between 1993 and 2007, due to major degradation of the water regime (UNEP/WHCM 2007). It was reinscribed in the list in 2010, where it remains (UNESCO WHC 2016a).

The park is visited by tourists especially in winter, considering that it is very buggy in wet season (summer). Tourism services offer trails for hiking, campgrounds, aquatic sports such as kayak and canoeing, and bicycle and car tours (NPS 2016).

22.2 Natural Settings

22.2.1 Climatic Conditions

The subtropical climate in Everglades has two seasons: a wet season during the warm months, which goes from May to October, and a cooler dry season, from November to April (McCormick et al. 1998).

Summers are hot and humid, with temperatures between 23 and 34 °C and over 90% of humidity. Annual rainfall averages 1520 mm. Afternoon thunderstorms are common, and the area is subjected to hurricanes from June to November (UNEP/WCMC 2007).

Fig. 22.2 Estuarine environment in the Everglades National Park, with the presence of mangrove and pelican. (Picture reproduced by permission of iStock)

22.2.2 Geology and Geomorphology

The Everglades National Park is the largest subtropical wilderness in the USA (NPS 2016). It is a unique wetland ecosystem due to its carbonate sedimentary environment (Sutula et al. 2001).

This natural context is fundamental to the hydrology of the ecosystem. Groundwater flows are thought to play a major role in the hydrologic budget of Everglades' wetlands because of the porous nature of the limestone bedrock (Sutula et al. 2001).

The landscape is characterized by the presence of terrestrial and freshwater wetlands and estuarine (Fig. 22.2) and saltwater wetlands. The terrestrial and freshwater wetlands support pinelands, sawgrass marshes, wet prairies, and sloughs. The estuarine and saltwater wetlands support mangrove forests (Fig. 22.3), beaches, salt marshes, and oscillating salinity systems (Schomer and Drew 1982). There are also the Florida Bay and Gulf Coast estuaries (UNESCO WHC 2016a). The altitude goes from sea level to 2.5 m (UNEP/WCMC 2007).

Fig. 22.3 Mangroves in the Everglades National Park. Birds winter in these mangrove forests. (Picture reproduced by permission of iStock)

22.2.3 Biodiversity

Located at the confluence of temperate North America and the tropical Caribbean, the subtropical Everglades National Park is home to representative flora and fauna from both climates. The park serves as an important habitat for a number of endemic and legally protected species (NPS 2016).

It is one of the world's largest subtropical peatlands with biological communities adapted to waters low in total dissolved solids and nutrients (Harvey and McCormick 2009). The park protects 800 species of land and water vertebrates, including 14 nationally threatened species (NPS, 2016). There are 40 mammal species, 25 being native (UNEP/WCMC 2007).

The relatively warm winters attract the largest variety of wading birds and their predators: 366 bird species, many of limited distribution in the USA, were recorded (NPS 2016). More than 275 species of fish are known from the Everglades, most inhabiting the marine and estuarine waters (UNEP/WCMC 2007). The park provides important habitat for numerous rare species, like the manatee, the American crocodile, and the Florida panther (UNESCO WHC 2016a).

The great floral variety of the Everglades is one of the key resources of the park. Despite the challenges of human alterations, abundant endemic plant species remain. From photosynthetic periphyton, which is the tiniest building block of the Everglades food chain, to the largest remaining stand of pine rockland forest in south Florida, the flora is very rich (NPS 2016).

Among the plants are bromeliads and orchids. Thirty-nine native orchid species occur, in addition to 750 other kinds of native seed-bearing plants. A total of 164 plant species have been listed by the State of Florida, including 47 as threatened, 113 as endangered, and 4 as commercially exploited (NPS 2016).

This site has been called a river of grass flowing imperceptibly from the hinterland into the sea (UNESCO WHC 2016a).

22.3 Conservation Challenges

Much of the historical Everglades has been degraded as a result of human activities (McCormick et al. 2009). Effectively, the wetlands of this ecosystem were drained to create arable and developable land suited for agriculture and human habitation and growth (NPS 2016). These activities have resulted in wetlands loss, increasing nutrients inputs, and diversion of fresh water to the Atlantic coasts by canals (Sutula et al. 2001).

Many of these activities continue near the park, threatening the environment, which justified the reinclusion of the national park in the list of World Heritage sites in danger in 2010, at the request of USA.

The activities that were considered to reinscribe the park on the list of the World Heritage in Danger are alterations of the hydrological regime, adjacent urban and agricultural growth, increased nutrient pollution from upstream agricultural activities, and management of Florida Bay resulting in significant reduction of both marine and estuarine biodiversity (UNESCO WHC 2016b).

In 2016, UNESCO WHC list as threats to the park, groundwater and surface water pollution, housing, invasion of alien freshwater species (with native plant species gradually disappearing and being replaced by exotic plants, some of which were formerly cultivated and used for landscaping or in the nursery trade), water infrastructure, and mercury contamination of fish and wildlife (NPS 2016; UNESCO WHC 2016b).

Global warming and climate changes are also a threat to the Everglades, because it potentially increases the number of storms and hurricanes, and storms are already a problem for the conservation of the park and its ecosystem. Additionally, any resulting fast rise in sea level would be disastrous for most of the mangroves, seagrasses, turtle beaches, tree islands, the fisheries, and coastal freshwaters of Florida Bay and the park (NPS 2016; South Florida Natural Resources Center, 2005).

Unless more restoration projects are enacted outside the area, the essential qualities and habitat of the site will continue to be lost. The threats will lead to further reductions in populations of wading birds, which are key indicators of the health of the ecosystem and perhaps more threatened and endangered species within the site (IUCN 2017).

The state of conservation outlook published by the International Union for Conservation of Nature (IUCN) at the end of 2017 (Osipova et al. 2017) considered the degree of conservation and the general situation in Everglades National Park to be critical and presenting signs of increasing deterioration. Major issues with water quantity, quality, distribution and timing, invasive species, and climate change are creating overriding impacts to the system, and the continued deteriorating trend of so many values puts the park's World Heritage values in a critical situation (IUCN 2017).

References

Harvey JW, McCormick PV (2009) Groundwater's significance to changing hydrology, water chemistry, and biological communities of a floodplain ecosystem, Everglades, South Florida, USA. Hydrogeol J 17(1):185–201

International Union for Conservation of Nature (IUCN) (2017) Everglades National Park conservation outlook. http://www.worldheritageoutlook.iucn.org/explore-sites/wdpaid/2012. Accessed 18 Dec 2017

McCormick PV, Shufford RBE III, Backus JG, Kennedy WC (1998) Spatial and seasonal patterns of periphyton biomass and productivity in the northern Everglades U.S.A. Hydrobiologia 362:185–202

McCormick PV, Newman S, Vilchek L (2009) Landscape responses to wetland eutrophication: loss of slough habitat in the Florida Everglades, USA. Hydrobiologia 621(1):105–114

NPS (National Park Service) (2016) Everglades National Park. https://www.nps.gov/ever/index.htm. Accessed 31 May 2016

Osipova E, Shadie P, Zwahlen C, Osti M, Shi Y, Kormos C, Bertzky B, Murai M, Van Merm R, Badman T (2017) IUCN world heritage outlook 2: a conservation assessment of all natural world heritage sites. IUCN, Gland

Schomer NS, Drew RD (1982) Ecological characterization of the lower Everglades, Florida Bay, and the Florida Keys. Florida State Dept. of Environmental Regulation, Technical Report. Tallahassee, USA

South Florida Natural Resources Center (2005) South Florida Natural Resources Center (n.d.). Restoration of Everglades National Park. National Park Service. Report, Miami, USA

Sutula M, Day JW, Cable J, Rudnick D (2001) Hydrological and nutrients budgets of freshwater and estuarine wetlands of Taylor slough in southern Everglades, U.S.A. Biogeochemistry 56:287–310

UNEP/WCMC (United Nations Environmental Programme/World Conservation Monitoring Centre) (2007) Everglades National Park, Florida, United States of America. Report, Cambridge, England

UNESCO WHC (United Nations Educational, Scientific and Cultural Organization World Heritage Centre) (2016a) Everglades National Park. http://whc.unesco.org/en/list/76/. Accessed 31 May 2016

UNESCO WHC (United Nations Educational, Scientific and Cultural Organization World Heritage Centre) (2016b) Everglades National Park State of conservation 2016. http://whc.unesco.org/en/soc/3377. Accessed 31 May 2016

Chapter 23
Olympic National Park, United States of America

Abstract Located in northwestern Washington State, USA, Olympic National Park is an area of impressive natural beauty that combines a stormy coastline with glacier-capped mountains and alpine parklands rising above an extensive mountain-slope rainforest. The forest is the best example of intact and protected temperate rainforest in the world having some of the world's largest conifers. The 100 km of wilderness coastline represents the longest undeveloped coast in the contiguous USA and is rich in native and endemic animal and plant species. It features spectacular coastline. The rocky islets along the coast are remnants of a continuously receding and changing coastline. The biological evolution, ecological variety, and splendor of Olympic National Park make it a very special place.

23.1 Introduction

Olympic National Park is located in the northwest segment of Washington State, USA. It combines a stormy coastline with glacier-capped mountains, alpine parklands, and mountain-slope rainforest. It has 3523 km^2 in the Olympic Mountains and 173 km^2 in the separated Pacific coastal area (UNEP/WCMC 2007). The protected coastline extends for around 100 km (Fig. 23.1).

In 1897, the Olympic Forest Reserve was established by presidential decree. In 1938 the area was declared a National Park by Act of Congress. In 1953 the Pacific coastal segment was added, with an expansion in 1978. It was designated as a Biosphere Reserve under the UNESCO Man and the Biosphere Programme in 1976. The park was added to the World Heritage list in 1981 (UNEP/WCMC 2007).

The earliest prehistoric people occupied the area between 12000 and 6000 years BP, hunting large land animals as a means of subsistence. By about 3000 years BP, as the human population increased, inhabitants shifted their focus to lowland rivers and lakes, forming the foundation of rich and complex maritime cultures (NPS 2016).

The land is 99% property of the Federal Government and 1% privately owned. It is surrounded by the Olympic National Forest and bordered by six smaller wilderness

V. Claudino-Sales, *Coastal World Heritage Sites*, Coastal Research Library 28,
https://doi.org/10.1007/978-94-024-1528-5_23

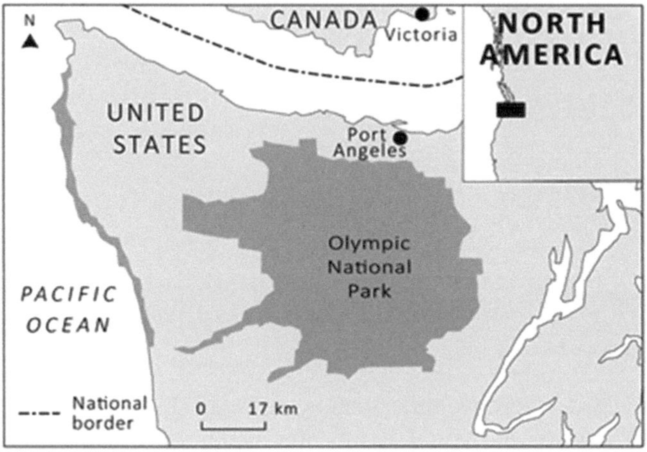

Fig. 23.1 Location of Olympic National Park with its coastal segment, Washington State, USA

areas. The park is managed by the National Park Service (NPS) of the Department of the Interior. (UNEP/WCMC 2007).

Because Olympic National Park encompasses three distinctly different ecosystems – glacier-capped mountains, wild Pacific coast, and old-growth temperate rainforest – it gets a large number of visitors, especially in summer. Tourists can hike, camp, kayak, canoe, ski, fish, bathe in the hot springs, or just appreciate the outstanding landscape.

23.2 Natural Settings

23.2.1 Climatic Conditions

Weather in Olympic National Park is variable and unpredictable, no matter the time of year. It is common for different weather conditions to exist within the park at the same time. Even in the summer, temperatures may vary from 10 to 20 °C, especially along the coast and higher elevations (NPS 2016).

But the climate is moderate and temperatures rarely drop below −7 °C or rise above 27 °C. The mean annual temperatures are 10 °C at lower areas. The precipitation, influenced by the warm air from the Pacific, is 4000 mm a year in the western valleys and 5000 mm on Mount Olympus, where most of it falls as snow in winter. Summers are dry (UNEP/WCMC 2007).

Fig. 23.2 Sea stacks creating a headland on a gravel beach in the Olympic National Park, with seacliff covered by temperate vegetation. (Picture reproduced by permission of United States (US) National Park Service Photos)

23.2.2 Geology and Geomorphology

The coastal segment of the Olympic National Park is characterized by being a storm-wave environment (Davies 1980). It is a high-energy coast, with mean annual significant wave height of 2.8 m (Corson et al. 1987). The highest waves occur in December with a mean wave height of 4.3 m, and the lowest in August, with 1.5 m (Bourke 1971). Tides are semidiurnal (US Army Corps of Engineers 1974).

The geology is characterized by unconsolidated glacial Pleistocene deposits, with outcrops forming seacliffs and sea stacks in the intertidal area and along the shore (Mckay and Terich 1992).

Fig. 23.3 Sunset at Olympic National Park seashore, highlighting sea stacks the intertidal area as well as rocky islets offshore. (Picture reproduced by permission of United States (US) National Park Service Photos)

The glacial Pleistocene deposits are very susceptible to erosion, which provides coarse-grained sediments to build the many gravel barriers that are found along the coast of Washington State, including the segment of the Olympic National Park (Mckay an Terich 1992).

Sea erosion attacking the glacial Pleistocene outcrops is also responsible for the formation of the sea stacks seen all along the shoreline (Figs. 23.2 and 23.3). The sea stacks and rocky islets are remnants of a continuously receding and changing coastline. It features spectacular coastline (UNESCO WHC 2016).

The Olympic Mountains are the highest mountains of the Pacific coastal range and form the core of the park. Geologically, their ruggedness results from the collision of continental plates, e.g., the subduction of the Gorda plate under the North American plate in Tertiary times (NPS 2016; Tabor 1975).

In the center of the park, a dome 95 km across (Mount Olympus), of contorted beds of shale, slate, and sandstone with interspersed lavas is present. Ten major rivers and many glaciers, 60 of which remain, have carved the dome into a vast landscape of deep canyons and jagged peaks. The altitude goes from sea level to 2428 m (UNEP/WCMC 2007; Tabor 1975).

23.2.3 Biodiversity

The park's varied topography, affected by high rainfall, has produced complex and varied vegetation zones, providing habitats of unmatched diversity on the Pacific coast (UNESCO WHC 2016).

Over 1450 types of vascular plants grow on the park, in addition to hundreds of species of nonvascular mosses, liverworts, and hornworts (NPS 2016). The cool, perpetually damp environment is the home range of 70 m old-growth conifers and 12-m-wide moss-covered maples with a dense undergrowth of fallen trees, shrubs, ferns, liverworts, epiphytes, fungi, lichens, and flowers (UNEP/WCMC 2007).

The native fauna is intact, except for the local subspecies of wolf, which was exterminated by man before the park was established. Fifty-six species of terrestrial mammals occur, at least 16 being endemic. Twenty-four marine mammals are seen in the park from time to time. There are 300 species of birds, 13 amphibian, 4 reptile species, and some 37 native species of fish (UNEP/WCMC 2007).

23.3 Conservation Challenges

The core of the Olympic Mountains is still largely undisturbed. No timber harvesting is permitted in the park, but there is some illegal felling at the boundaries.

The park has suffered two large oil spills from offshore shipping since 1988 and remains vulnerable to such accidents. Air and water quality in coastal areas is also threatened by large-scale applications of herbicides in timber-producing areas adjacent to the park. Tourists and other visitors have an adverse effect on the environment (UNEP/WCMC 2007).

The presence of invasive plants is also a permanent worry in the area (NPS 2016). Global warming and climate change are also threats to the park. In the coastal area, sea erosion tends to be higher with sea level rising, and the unconsolidated Pleistocene deposits, sea stacks, and rocky islets are susceptible to destruction. The amount of precipitation may also change, endangering the rainforest.

Rising temperatures are already causing an increase in wildland fire scale and intensity, as well as an increase in insect infestations for native trees, with increased mortality from both causes, well-beyond levels that have occurred historically. Particularly notable will be the loss of subalpine habitat, glaciers and more frequent flooding, and other weather-related events (IUCN 2017).

Internally the park is well managed and protected despite these threats that will need continued attention. Within the parameters of normal fluctuation, most park resources are viewed as stable or improving (IUCN 2017). For all these reasons, the conservation level of the site is considered to be good with some concerns (Osipova et al. 2017).

References

Bourke RH (1971) Waves. In: Oceanography of the nearshore coastal waters of the Pacific Northwest relating to possible pollution. Oregon State University, Corvallis, USA

Corson WD, Abel CE, Brooks RM, Farrar PD, Groves BJ, Payne JB, McAnemy DS, Tracy BA (1987) Pacific coast hindcast phase II wave information: Coastal Engineering Research Center, WIS report 16, US Army Corps of Engineering. Vicksburg, USA

Davies JL (1980) Geographical variation in coastal development. Longman Group, London

IUCN (International Union for Conservation of Nature) (2017) Olympic National Park conservation outlook. http://www.worldheritageoutlook.iucn.org/explore-sites/wdpaid/2579. Accessed 18 Dec 2017

McKay PJ, Terich Thomas A (1992) Gravel barrier morphology: Olympic National Park, Washington State, U.S.A. J Coast Res 8(4):813–829

NPS (National Park Service) (2016) Olympic National Park, Washington. https://www.nps.gov/olym/. Accessed 5 June 2016

Osipova E, Shadie P, Zwahlen C, Osti M, Shi Y, Kormos C, Bertzky B, Murai M, Van Merm R, Badman T (2017) IUCN world heritage outlook 2: a conservation assessment of all natural world heritage sites. IUCN, Gland

Tabor RW (1975) Guide to the geology of the Olympic National Park. University of Washington Press, Seattle

UNEP/WCMC (United Nations Environmental Programme/Word Conservation Monitoring Centre) (2007) Olympic National Park, Washington State, USA. Report, Cambridge, England

UNESCO WHC (United Nations Educational, Scientific and Cultural Organization World Heritage Centre) (2016) Olympic National Park. http://whc.unesco.org/en/list/151. Accessed 5 June 2016

US Army Corps of Engineering (1974) Draft EIS, Quillayute Harbor Project, Vicksburg, USA

Chapter 24
Redwood National and State Parks, United States of America

Abstract Redwood National and State Parks, in northern California, USA, is a long, thin strip of coastal mountains and coastal plain bordering the Pacific Ocean. It is forested with groves of redwoods, the tallest living things on earth. The redwood is a surviving remnant of the group of trees that has existed for 160 million years, now confined to the wet regions of the west coast of North America. Effectively, Redwood National Park preserves the largest remaining contiguous ancient coast redwood forest in the world in its original forest and streamside settings. The marine and land life are remarkable, in particular the sea lions, the bald eagle, and the endangered California black pelican. The area of the parks and the adjacent offshore area are the most seismically active region in the USA, capable of producing magnitude 9 earthquakes.

24.1 Introduction

The Redwood National and State Park is located in northwestern California, 450 km north of San Francisco, bordering the Pacific Ocean. The composite site, having 568 km^2 of area, parallels the coast for 85 km, varying in width between 0.3 and 14 km. The boundary along the coast extends 0.4 km into the sea (UNEP/WCMC 2007) (Fig. 24.1).

The site is formed by four parks: Redwood National Park, Prairie Creek Redwoods State Park, Jedediah Smith Redwoods State Park, and Del Norte Coast Redwoods State Park (NPS 2016).

The movement to preserve the redwood forest is not recent. In 1918, the "Save the Redwoods League" was founded, with the goal to save the redwood forests from heavy logging. In the 1920s, the Prairie Creek, Jedediah Smith, and Del Norte Coast State Parks were established. The three existing state parks joined into Redwood National and State Parks in 1968 (UNEP/WCMC 2007). The ensemble became a World Heritage site in 1980 (UNESCO WHC 2016).

Prehistoric sites 4500 years old have been uncovered, and there are sites of later settlement and subsistence systems. Some villages, cemeteries, and sacred or ceremonial sites remain places of importance to contemporary Native Americans

V. Claudino-Sales, *Coastal World Heritage Sites*, Coastal Research Library 28,
https://doi.org/10.1007/978-94-024-1528-5_24

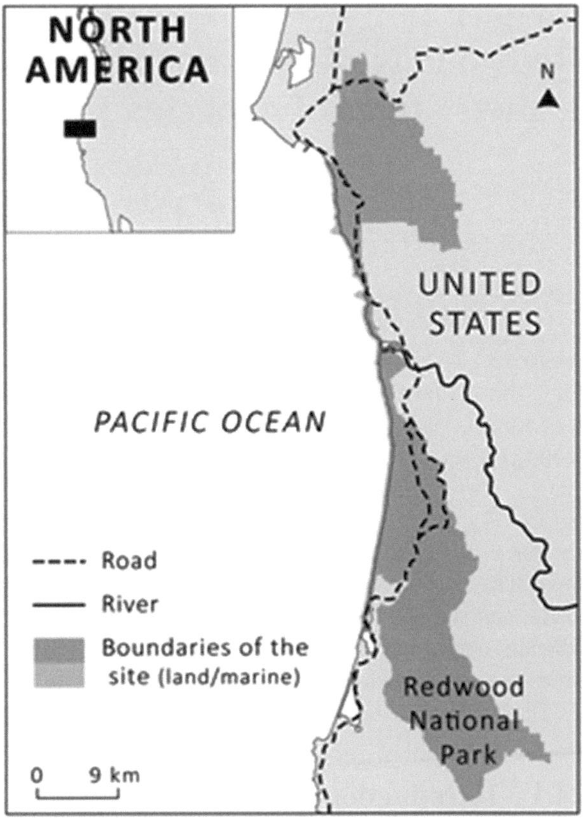

Fig. 24.1 Location of Redwood National and State Parks, in northern California

(UNEP/WCMC 2007). The parks are of importance for their prehistoric and historic cultural remains with several sites listed on the National Register of Historic Places. American Indians in and around the park form almost 10% of the region's population (UNEP/WCMC 2007).

24.2 Natural Settings

24.2.1 Climatic Conditions

The climate is temperate maritime with cool wet winters (temperatures between 4 and 13 °C) and mild summers (temperatures between 11 and 19 °C). The annual precipitation is 2540 mm, occurring mainly during winter. Coastal fog is very frequent during the dry months of the summer. Fog drip around the trees almost doubles the amount of precipitation from rainfall (UNEP/WCMC 2007).

The existence of the redwood forest is closely associated with the presence of the summer marine fog along the Pacific coast of California (Johnstone and Dawson 2010).

24.2.2 Geology and Geomorphology

The park covers two distinct environments, a narrow coastal plain and the mountains of the Coast Range of northern California (UNEP/WCMC 2007). The elevation goes from sea level to 900 m (California State Parks 2005).

The Coast Range geology is represented by remnants of Jurassic ocean crust (Hagstrum 1992). Tectonic models explain the origins of the Coast Range ophiolite as having a formation at a mid-ocean ridge, in a marginal basin above a subduction zone (Robertson 1989). It can be inferred that these rocks were formed together along the ancient California margin (Hagstrum 1992). The recent margin and emergence of the Coast Range is strongly affected by the evolution of the San Andreas transform fault system (DeCourten 2004).

Despite the age of the rocks, the mountainous landscape is not too old: the Coast Ranges rose to their present elevations in the past 3 million years to 4 million years, probably because of compression between fault slices in the San Andreas system and the northward migration of the Mendocino triple junction (DeCourten 2004).

The area of the park and the adjacent offshore area are the most seismically active region in the USA (NPS 2016), seismically capable of producing magnitude 9 earthquakes (California State Parks 2005). As a result of frequent earthquakes, rapid uplift rates have led to landslides and rapid coastal erosion. The reason for all this activity is the geologic setting of the North Coast region. Three tectonic plates contact each other offshore near Cape Mendocino, which is about 160 km southwest of the site (NPS 2016). Tsunamis are an occasional hazard (UNEP/WCMC 2007).

The bedrock is formed by the highly deformed Cretaceous marine sandstones, siltstones, and shales of the Franciscan assemblage. There are chert, volcanic greenstone, and metamorphic rocks, present as blocks, in belts mixed with the Franciscan sedimentary rocks, some isolated as offshore stacks (UNEP/WCMC 2007). The Franciscan sedimentary rocks are covered in some parts by a thin veneer of young Plio-Pleistocene shallow marine to fluvial sandstones, mudstones, and conglomerates (UNEP/WCMC 2007).

The rocks of the Franciscan assemblage extend to the coastal area. The coastal strip, 55 km long, is characterized by steep rocky cliffs broken by rolling slopes, as well as by stacks and large sandy beaches with occasional lagoons and marshes (UNEP/WCMC 2007) (Figs. 24.2 and 24.3). Udic soils are present on the hills, the beaches, and the dunes (USDA/NRCS 2008).

Fig. 24.2 A rocky segment of the coast of Redwood State and National Parks, with seacliffs and stacks sustained by Cretaceous rocks. (Picture reproduced by permission of National Park Service, USA)

24.2.3 Biodiversity

The redwood is a surviving remnant of the group of trees that has existed for 160 million years, now confined to the wet regions of the west coast of North America (UNESCO WHC 2016). The flora numbers 856 species, 699 of which are native (UNEP/WCMC 2007).

The Redwood National and State Parks preserve around 45% of the remaining redwood forest in USA (Fritschle 2009). The redwood tree found there is the sole surviving species of the genus Sequoia (Noss 2000). California's redwoods may grow to a height of 112 m and a width of 7 m at the base. They can live for 2000 years (NPS 2016). California and Oregon are the only environment in the world capable of accommodating such an ecosystem, resulting from a combination of longitude, climate, and elevation. The cool, moist air created by the Pacific Ocean keeps the trees continually damp, even during summer droughts (NPS 2016).

For the fauna, particularly important are the mammals, birds, and fishes. There are 75 mammal species, including Roosevelt elk, black bear, mountain lion, bobcat, marten, fisher, and river otter. The region's avifauna exceeds 400 species, with 398 native and 35 nonnative (UNEP/WCMC 2007).

Fig. 24.3 Crescent beach with sandy shoreline in the coastal area of Redwood State and National Parks, surrounded by vegetated seacliffs. (Picture reproduced by permission of National Park Service, USA)

Freshwater marshes, ponds, and streams provide valuable nesting and feeding areas for several migratory waterfowl species. Birds include peregrine falcon, osprey, and a variety shorebirds and waterfowl along the coastal part of the region. Streams and rivers are used by anadromous fish (UNEP/WCMC 2007; California State Parks 2005).

The marine and land life are also remarkable, in particular the sea lions, the bald eagle, and the endangered California black pelican (UNESCO WHC 2016).

24.3 Conservation Challenges

Extensive logging destroyed a large part of the coniferous forest in the area of the Redwood National Park prior to the establishment of the park (Fritschle 2009). The park is now protecting the most impressive segments of the redwood forest.

Global warming and climate change are threats to the ecosystem where the redwood is adapted. The redwood may be increasingly drought stressed under a summer climate of reduced fog frequency and greater evaporative demand (Johnstone and Dawson 2010). For the coastal area, sea-level rise will increase the risks associated with coastal hazards of flooding and erosion (Revell et al. 2011).

Most of the park values are stable and adequately protected. While the established coast redwood trees are likely to persist, many associated species in the park are threatened. The known and anticipated threats to the park's resources such as invasive species, previous land use, changed fire regimen, and future environmental changes require immediate attention to prevent resource loss. Declining funding for park management is the main barrier preventing the park from addressing the issues, including restoration and monitoring (IUCN 2017).

The last state of conservation outlook from the International Union for Conservation of Nature (IUCN) related to the World Heritage sites attested that the Redwood National and State Parks present good conditions with some concerns, associated with the abovementioned threats (Osipova et al. 2017).

References

California States Parks (2005) Natural Parks report. Natural Resources Division. Sacramento, USA

DeCourten F (2004) Geology of Northern California. USGS, Sacramento

Fritschle JA (2009) Pre-EuroAmerican settlements forests in Redwood National Park, California: a reconstruction using line summaries in historic land surveys. Landsc Ecol 24:833–847

Hagstrum JT (1992) Remagnetization of the Coast Range ophiolite at Stanley Mountain, California, during accretion near 10°N paleolatitude. Geology 20:503–506

International Union for Conservation of Nature (IUCN) (2017) Redwood National and State Parks conservation outlook. http://www.worldheritageoutlook.iucn.org/explore-sites/wdpaid/4325. Accessed 18 Dec 2017

Johnstone JA, Dawson TE (2010) Climatic context and ecological implications of summer fog in the coast redwood region. Proc Natl Acad Sci USA 107(10):4533–4538

Noss RF (ed) (2000) The redwood forest, history, ecology and conservation of the coast redwood. Island Press, Washington, DC

NPS (National Park Service) (2016) Redwood National and State Parks, California. https://www.nps.gov/redw/index.htm. Accessed 14 June 2016

Osipova E, Shadie P, Zwahlen C, Osti M, Shi Y, Kormos C, Bertzky B, Murai M, Van Merm R, Badman T (2017) IUCN world heritage outlook 2: a conservation assessment of all natural world heritage sites. IUCN, Gland

Revell D, Battalio R, Spear B, Ruggiero P, Vandever J (2011) A methodology for predicting future coastal hazards due to sea-level rise on the California coast. Clim Chang 109:251–276

Robertson AHF (1989) Paleoceanography and tectonic setting of the Jurassic Coast Range ophiolite, Central California: evidence from the extrusive rocks and the volcaniclastic sediment cover. Mar Pet Geol 6:194–220

UNEP/WCMC (United Nations Environmental Programme/World Conservation Monitoring Centre) (2007) Redwood National & State Parks, California, United States. Report, Cambridge, England

UNESCO WHC (United Nations Educational, Scientific and Cultural Organization World Heritage Centre) (2016) Redwood National and State Parks. http://whc.unesco.org/en/list/134. Accessed 14 June 2016

USDA/NRCS (United States Department of Agriculture/Natural Resources Conservation Service) (2008) Soil survey of Redwood National and State Parks, California. http://soils.usda.gov/survey/printed_surveys/. Accessed 14 June 2016

Part VI
World Heritage on Continental Coasts: Oceania

Chapter 25
Kakadu National Park, Australia

Abstract Kakadu National Park in northern Australia is a mixed World Heritage site, inscribed in the list of sites of humankind outstanding values by its cultural and natural characteristics. It is the largest national park in Australia and one of the largest in the world's tropics. It is characterized by the presence of extensive areas of savanna woodlands, open forest, floodplains, coastal areas, islands, mangroves, tidal mudflats, and monsoon forests. It has been home to Aboriginal people for more than 40,000 years, providing a window into human civilization in the days before the last ice age. The hunting-and-gathering tradition illustrated in the art and archaeological record continues today. The park also has a huge diversity of flora and is one of the least impacted areas of Australia. It is a unique archaeological and ethnological reserve covering almost the entire catchment of a major tropical monsoonal river system.

25.1 Introduction

Kakadu National Park, in northern Australia, has a coastal section bathed by the Pacific Ocean (Fig. 25.1). With 1980 km^2 of area, it is the largest national park in Australia (UNESCO WHC 2016).

The area was proclaimed a national park in 1979. In 1980 it was declared a Wetland of Importance under the Ramsar Convention of Wetlands. In 1989 the area of the wetland defined by the Ramsar Convention was extended. It was added to the World Heritage list as a mixed site (cultural and natural) in 1981, with extensions approved in 1987 and 1992 (UNEP/WCMC 2005).

Aboriginal people have occupied northern Australia for at least 40,000 years, developing a rich culture of law, ceremony, oral history, and detailed ecological knowledge (Schooler et al. 2011). There is a large range of cave paintings, rock carvings, and archaeological sites. The art records the skills and way of life of the region's inhabitants, from the prehistoric times to the present, providing an outstanding record of human interaction with the environment over tens of thousands of years (UNEP/WCMC 2005).

V. Claudino-Sales, *Coastal World Heritage Sites*, Coastal Research Library 28,
https://doi.org/10.1007/978-94-024-1528-5_25

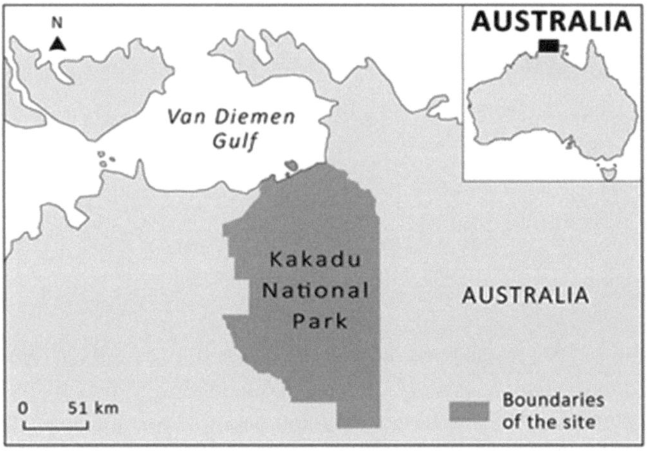

Fig. 25.1 Location of Kakadu National Park, in northern Australia

Kakadu's rock art offers a window into human civilization in the days before the last ice age. And today, the hunting-and-gathering tradition illustrated in the art and archaeological record continues (UNESCO WHC 2016). With paintings up to 20,000 years old, it is one of the longest historical records of any group of people in the world (Parks of Australia 2016). The region has had relatively little impact from European settlement (UNESCO WHC 2016).

The park is managed by Aboriginal people and the Parks of Australia, under the Department of the Environment, Sports and Territories (UNEP/WCMC 2005). Tourism is a very important economic activity in the park. Nevertheless, many areas of Kakadu National Park are inaccessible to people other than the indigenous traditional owners and park managers, so cultural sites are subject to little interference (UNESCO WHC 2016; McGregor et al. 2010).

25.2 Natural Settings

25.2.1 Climatic Conditions

In northern Australia's tropical area, hot summer conditions go from November to March, which also marks the wet season, represented by the presence of monsoonal rains, with storms (Parks of Australia 2016).

The mean annual rainfall decreases from the coast toward the interior, from 1560 mm to about 1100–1200 mm in the south. The heavy monsoonal rains drain off the plateau and inundate Kakadu's floodplains (UNEP/WCMC 2005).

The dry season goes from April to October. The temperatures vary from 17 to 37 °C (Parks of Australia 2016).

Fig. 25.2 Mangrove swamp in a tide channel in the coastal area of Kakadu National Park. (Picture reproduced by permission of Parks Australia)

25.2.2 Geology and Geomorphology

The area is tectonically stable, formed by Precambrian and Mesozoic rocks (UNEP/WCMC 2005). The three major landscape types in the park are plateau, lowlands, and floodplain (Gill et al. 2000). The lowlands have extensive areas of savanna woodlands, open forest, wetlands, monsoon forests, coastal areas with estuaries, mangroves, tidal mudflats (UNESCO WHC 2016) (Fig. 25.2), and oxbow lakes (Schooler et al. 2011). Monsoon rainforest is a key habitat in sustaining the natural and cultural values in Kakadu National Park (Banfai and Bowman 2006).

The floodplains inundate seasonally, and swamp forests form a fringe around this area (Bowman et al. 2010b) (Fig. 25.3). The floodplains develop in the watershed of the South Alligator River and parts of the West Alligator, Wildman, and East Alligator rivers (UNEP/WCMC 2005).

The plateau is formed by sandstones that create the "Arnhem Land Plateau." The escarpment ranges in height from about 30 to 330 m over some 500 km (UNEP/WCMC 2005). It consists of vertical and stepped cliff faces that extend in a jagged and unbroken line (UNESCO WHC 2016). It is relatively dry but contains many waterfalls. At the south segment of the park, there are also highlands and hills, created by volcanic rocks (Parks Australia 2006).

Fig. 25.3 Vestiges of swamp forest in an inundated fluvial plain in Kakadu coastal area. The inundation happens seasonally, during the summer of the Southern Hemisphere. (Picture reproduced by permission of Parks Australia)

Kakadu National Park contains an impressive contrast between the wetlands and the spectacular rocky escarpment and its outliers. It is also contrasting in terms of geology: Kakadu's escarpment spans more than 2 billion years of geological history, whereas the floodplains are recent and illustrate the ecological and geomorphological effects of Holocene climate change and sea level rise (UNESCO WHC 2016).

25.2.3 Biodiversity

The park has a relatively unmodified natural vegetation and largely intact faunal composition, providing a unique opportunity to investigate large-scale evolutionary processes in a relatively intact landscape (UNESCO WHC 2016).

There are numerous microhabitats. The biota of the plateau is ecologically very diverse, containing a distinctive assemblage of species, many of restricted range. Some species are relict and many are endemic. The park is the most floristically diverse part of monsoonal northern Australia (UNEP/WCMC 2005).

It has more than 2000 plant species, and many of the plants have been used by local Aboriginal people for generations as bush foods, medicines, and weaving materials (Parks Australia 2006).

The 64 native mammal species known from the park comprise around a quarter of the total number of terrestrial mammal species in Australia. Reptile species totalize 128, including two crocodile species (UNEP/WCMC 2005).

Fifty-five species of freshwater fish make the park Australia's richest in this matter. The extremely rich avifauna of 274 species includes 33% of those found in Australia, and in the fall up to one million water birds of 60 species gather there (UNEP/WCMC 2005).

25.3 Conservation Challenges

The park suffers with the frequent advance of the invasive plant *Salvinia molesta* (Schooler et al. 2011; Cowie and Werner 1993). Fire can also be a threat: the wet/dry tropics of the Australian savannas are particularly prone to fire due to the highly seasonal rainfall and accumulation of grassy fuels (Townsend and Douglas 2004).

High densities of feral buffalo between 1960 and 1980 changed the hydrology of the floodplains, allowing increased penetration of saltwater and causing extensive death of forests (Bowman et al. 2010a). The forests continue to decline because of salinization driven by sea level rise associated with climate change (Bowman et al. 2010a). Climate change and global warming can also cause erosion in the tidal flats and swamps of mangroves, due the rising sea level. The humidity also may be altered, with possibilities of expansion of the dry season and/or less volume of rain in the wet season.

The last conservation outlook of World Heritage sites published in November of 2017 by the International Union for Conservation of Nature (IUCN) found that the state of conservation of the park presents significant concern. Its values are considered to be threatened, with significant additional conservation measures being required to preserve the site over the medium to long term (Osipova et al. 2017).

References

Banfai DS, Bowman DMJS (2006) Forty years of lowland monsoon rainforest expansion in Kakadu National Park, Northern Australia. Biol Conserv 131(4):553–565

Bowman DMJS, Brown JK, Braby MF, Brown JR, Cook LG, Crisp MD (2010a) Biogeography of the Australian moonson tropics. J Biogeogr 37:201–216

Bowman DMJS, Prior LD, De Little SC (2010b) Retreating Melaleuca swamp forests in Kakadu National Park: Evidence of synergistic effects of climate change and past feral buffalo impacts. Austral Ecol 35(8):898–905

Cowie ID, Werner PA (1993) Alien plant species invasive in Kakadu National Park, tropical Northern Australia. Biol Conserv 63(2):127–135

Gill AM, Ryan PG, Moore PHR, Gibson M (2000) Fire regimes of World Heritage Kakadu National Park, Australia. Austral Ecol 25(6):616–625

McGregor S, Lawson V, Christophersen P, Kennett R, Boyden J, Bayliss P, Liedloff A, McKaige B, Andersen AN (2010) Indigenous wetland burning: conserving natural and cultural resources in Australia world heritage-listed Kakadu National Park. Hum Ecol 38(6):721–729

Osipova E, Shadie P, Zwahlen C, Osti M, Shi Y, Kormos C, Bertzky B, Murai M, Van Merm R, Badman T (2017) IUCN world heritage outlook 2: a conservation assessment of all natural world heritage sites. IUCN, Gland

Parks Australia (2006) Kakadu National Park. https://parksautralia.gov.au/kakadu? Accessed 20 June 2016

Schooler SS, Salau B, Julien MH, Ives AR (2011) Alternative stable states explain unpredictable biological control of Salvinia molesta in Kakadu National Park. Nature 470:86–90

Townsend SA, Douglas MM (2004) The effect of a wildfire on stream water quality and catchment water yield in a tropical savanna excluded from fire for 10 years (Kakadu National Park, North Australia). Water Res 38(13):3051–3058

UNEP/WCMC (United Nations Environmental Programme/World Conservation Monitoring Centre) (2005) Kakadu National Park Northern Territory, Australia. Report, Cambridge, England

UNESCO WHC (United Nations Educational, Scientific and Cultural Organization World Heritage Centre) (2016) http://whc.unesco.org/en/list/147. Accessed 20 June 2016

Chapter 26
Ningaloo Coast, Australia

Abstract Ningaloo Coast is in Western Australia, on the Indian Ocean. The site tells an extraordinary story of biogeography, climate change, assembling of continents, opening of oceans, and environmental conservation. Its dramatic setting and rich biology reveal a long record of continental change, isolation, and evolution. Annual gatherings of whale sharks occur, making it one of the best places in the world to interact with the biggest fish in the world. The coastal segments present subterranean water bodies with a substantial network of caves, conduits, and groundwater streams. The property contains exceptional underwater scenery, including coral reef structures, contrasting with the vivid colors of the arid terrestrial landscape.

26.1 Introduction

Located in the western coast of Australia, the Ningaloo Coast World Heritage site, bathed by the Indian Ocean, is an area of 7050 km^2 (Fig. 26.1). It is composed of various state parks, preserves, and local subdivisions, among them the Ningaloo Marine Park (Department of Environment Australia 2010).

The Western Australian Government declared the state waters of the Ningaloo Coast as a marine park in 1987 (Department of Environment Australia 2010). The ensemble of the area, including all the parks, preserve, and subdivisions, became a National Heritage site in 1999 and was added to the World Heritage List in 2011 (UNESCO WHC 2016).

The area has a remarkable history of human occupation, stretching back at least 35000 years. Archaeological evidence suggests that indigenous people episodically occupied rock shelters on the coast on a seasonal basis. The region also provides the earliest evidence of human ornamentation found in Australia (Department of Environment of Australia 2010).

The Ningaloo Coast is in the Gascoyne region, one of the nine nonmetropolitan areas of Australia. The economy was based in pastoral industry, but nowadays, the

V. Claudino-Sales, *Coastal World Heritage Sites*, Coastal Research Library 28,
https://doi.org/10.1007/978-94-024-1528-5_26

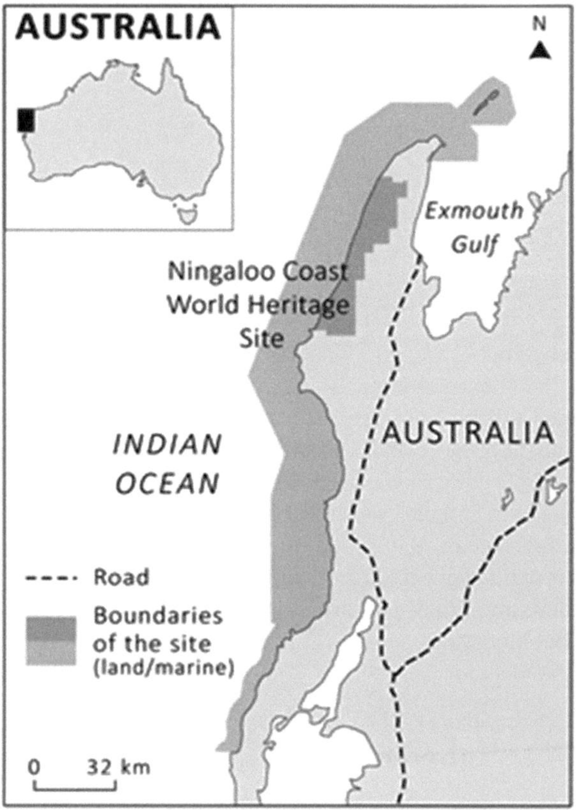

Fig. 26.1 Location of Ningaloo Coast Heritage site in Western Australia

industry of tourism is dominant, followed by fishing and horticulture (Wood et al. 2012). The management of the Ningaloo Coast demonstrates some of the world's best practice in tropical marine, terrestrial, and karst management (Department of Environment Australia 2010).

26.2 Natural Settings

26.2.1 Climatic Conditions

The climate is transitional between temperate and tropical. The coast is affected by a variety of weather systems, commonly including anticyclonic high-pressure systems, extratropical cyclones, strong sea breezes, and occasional tropical cyclones (Eliot et al. 2012).

The area is characterized by an arid to semiarid climate, with variable summer and winter rainfall. Annual regional rainfall of between 200 and 300 mm along the coast is far exceeded by evaporation rates of around 2000–3000 mm (Department of Environment Australia 2010).

Cyclones move across the Ningaloo Coast once every 3–5 years, bringing intense rain. They play a critical role in recharging underground aquifers and revitalizing cave life. Cyclonic winds may be severe, exceeding speeds of 150 km per hour (Department of Environment Australia 2010).

The name Ningaloo refers to the episodic occurrence of anomalously warm ocean conditions along the subtropical coast of Western Australia. The Ningaloo Niño typically develops in austral spring with peaks in summer, decaying in fall (Feng et al. 2015). The anomalous advection of warm water is caused by changes in the poleward flowing Leeuwin Current along the western coast of Australia due to both local and remote wind forcing (Benthuysen et al. 2014).

26.2.2 Geology and Geomorphology

The outstanding value of the Ningaloo Coast derives from its integrated reef and karst system lying along an arid coastline (Department of Environment Australia 2010) (Fig. 26.2). The Ningaloo coastal waters host a nearshore reef system (Fig. 26.3) and a directly adjacent limestone karst landform in the coastal area, with dolinas and caves (UNESCO WHC 2016).

On the coast, there are also sandy beaches backed by dunes (Cassata and Collins 2008). Intertidal systems such as rocky shores, estuaries, and mangroves are also found (UNESCO WHC 2016). There are also shoreline salients, cuspate forelands, and tombolos (Sanderson and Eliot 1996).

Much of the regional geomorphology is Late Cenozoic in age, being dominated by geological structures which reflect tectonic instability of this time. Uplift has provoked a strong and distinctive erosional and depositional geomorphological response. This is especially true at Cape Range, in the northern tip of Ningaloo Coast World Heritage site (Wyrwoll et al. 1993).

Ningaloo Reef is Australia's largest fringing reef system (Fig. 26.3), extending for over 200 km. Its narrow lagoon is backed by a coastal plain, composed of an emergent last Interglacial reef on the flank of folded Tertiary limestones (Collins et al. 2003). Climatic aridity, cyclones, tsunamis, and the poleward flowing Leeuwin Current all influence the reef system. The reefs developed in two periods: Holocene and last interglacial (Collins et al. 2003).

The continental margin of the Ningaloo Coast and Western Australia dates from the middle Mesozoic era (around 150 million years ago), when tectonic forces broke the eastern segment of the supercontinent Pangaea apart (Powell et al. 1988). The coast formed part of the Tethys Ocean when Australia was the southeastern peninsula of the supercontinent Pangaea (Department of Environment Australia 2010).

Fig. 26.2 Aerial view of the arid coast of Ningaloo, integrated with a fringing reef in the coastal waters. (Picture reproduced by permission Department of Environment, Australia)

26.2.3 Biodiversity

The remarkable biota of the Ningaloo Coast is a record of the global biogeographical, oceanographic, and plate tectonic changes that have occurred in the area. As Pangaea broke, ocean currents shifted and climates were transformed (Department of Environment Australia 2010).

Fig. 26.3 Fringing reefs at Ningaloo Coast, with shoreline sustained by Tertiary folded limestones. (Picture reproduced by permission of Environment Department, Australia)

The area gives examples of key biogeographic processes, such as isolation and adaptation, leading to change within both populations and species (Department of Environment Australia 2010).

A wide range of animals and plants have adapted to the hot, arid conditions of the surface. The coast shelters a large number of species and subspecies that are endemic to the region, including an unusual richness of reptiles (UNESCO WHC 2016).

Ningaloo coastal waters is one of the best places to encounter the remarkable whale shark, the world's largest fish, which aggregates here in higher numbers than have been recorded anywhere else. Eight cetacean species regularly visit the Ningaloo Coast. There are also turtles and a spectacular number of fish species. Crustaceans are the richest group of invertebrates (Department of Environment 2010).

There is a part of the marine fauna that is relict of the Tethys Ocean – the stygal communities – surviving initially as opportunists and specializing in their unusual setting. They give a unique contribution to the understanding of global biogeography and plate tectonics from the end of the Mesozoic era (Department of Environment Australia 2010).

There are 630 taxa of vascular plants recorded. The flora is comprised of communities that have adapted to the wide range of generally arid or semiarid landscapes found across the coast (Department of Environment Australia 2010).

26.3 Conservation Challenges

Tourism in the Ningaloo Coast is on the increase, leading to threats such as damage to vegetation, sewage and waste disposal, and disturbance to wildlife. Illegal fishing is also a concern (UNESCO WHC 2016).

Learmonth Air Weapons Range Facility, within the property, was one of Australia's most active bombing ranges until around 1990, and future bombing activities may pose a threat (UNESCO WHC 2016).

Global warming and the Ningaloo Niño started influencing the coastal ocean off Western Australia since the 1990s, with warmer ocean driving rainfall variability regionally (Doi et al. 2015). Sea level rise and warm waters due to climate change also menace the integrity of the reef ecosystem, of the coastal caves, and of the coastal landscape.

The International Union for Conservation of Nature (IUCN) published a report about the state of conservation of the World Heritage sites in the end of 2017 (Osipova et al. 2017) finding that despite these threats, the general state of conservation of Ningaloo Coast is good.

References

Benthuysen J, Feng M, Zhong L (2014) Spatial patterns of warming off Western Australia during the 2011 Ningaloo Niño: quantifying impacts of remote and local forcing. Cont Shelf Res 91:232–246

Cassata L, Collins LB (2008) Coral reef communities, habitats, and substrates in and near sanctuary zones of Ningaloo Marine Park. J Coast Res 24(1):139–151

Collins LB, Zhu ZR, Wyrwoll KH, Eisenhauer A (2003) Late Quaternary structure and development of the northern Ningaloo Reef, Australia. Sediment Geol 159(1):81–94

Department of Environment Australia (2010) Ningaloo coast: world heritage nomination. Department of the Environment, Water, Heritage and the Arts, Canberra

Doi T, Behera S, Yamagata TD (2015) An interdecadal regime shift in rainfall predictability related to the Ningaloo Nino in the late 1990s. J Geophys Res Oceans 120(2):1388–1396

Eliot I, Gozzard JR, Eliot M, Stul T, McCormack G (2012) The coast of the shires of Shark Bay to Exmouth, Gascoyne, Western Australia: geology, geomorphology & vulnerability. Geological Survey of Western Australia, Perth

Feng M, Hendon HH, Xie SP, Marshall AG, Schiller A, Kosaka Y, Caputi N, Pearce A (2015) Decadal increase in Ningaloo Niño since the late 1990s. Geophys Res Lett 42(1):104–112

Osipova E, Shadie P, Zwahlen C, Osti M, Shi Y, Kormos C, Bertzky B, Murai M, Van Merm R, Badman T (2017) IUCN world heritage outlook 2: a conservation assessment of all natural world heritage sites. IUCN, Gland

Powell CM, Roots SR, Veevers JJ (1988) Pre-breakup continental extension in East Gondwanaland and the early opening if the eastern Indian Ocean. Tectonophysics 155:261–283

Sanderson PG, Eliot I (1996) Shoreline salients, cuspate forelands and tombolos on the coast of Western Australia. J Coast Res 12(3):761–773

UNESCO WHC (United Nations Educational, Scientific and Cultural Organization World Heritage Centre (2016) Ningaloo Coast. http://whc.unesco.org/en/list/1369. Accessed 21 June 2016

Wood D, De la Mare B, Fulton B, Jones T (2012) Participatory tourism planning at Ningaloo reef, Western Australia. In: Kenshigton R, Stocker L, Wood D (eds) Sustainable coastal Mangement and climate adaptation – global lessons from regional approaches in Australia. CSIRO Publishing, Collingwood

Wyrwoll KH, Kendrick GW, Long JA (1993) The geomorphology and late Cenozoic geomorphological evolution of cape range – Exmouth gulf region. In: Humprheys WF (ed) The biogeography of cape range, Western Australia, vol 45. Records of the Western Australia Museum, Perth, pp 1–23

Chapter 27
Wet Tropics of Queensland, Australia

Abstract The Wet Tropics of Queensland follow the northeast coast of Australia for 450 km. The site supports tropical rainforests, subject to a dry season and to frequent cyclonic events. The area presents a record of evolutionary processes that shaped the flora and fauna of Australia, containing the relicts of the Gondwanan forest that covered the country 50–100 million years ago. All of Australia's marsupials originated in the rainforest, and their closest surviving relatives occur in the Wet Tropics. It also provides important information for the interpretation of fossils found elsewhere in Australia. In addition to its complex pattern of species and life forms, the Wet Tropics is also recognized as possessing outstanding scenic features, natural beauty, and magnificent landscapes.

27.1 Introduction

The Wet Tropics of Queensland stretches along the northeast coast of Australia for 450 km, covering an area of 8944 km^2 of tropical rainforest (UNESCO WHC 2016) (Fig. 27.1). The area borders the Great Barrier Reef World Heritage site. The property includes within its 3125 km area almost 730 separate parcels of land, including freehold property, 19 national parks, 31 state forests, 5 reserves, and 1 Aboriginal and Islander reserve (Department of Environment Australia 2003). The ensemble became a World Heritage site in 1988 (UNESCO WHC 2016).

The Wet Tropics of Queensland preserves the only recognized extant Aboriginal rainforest culture – with 18 distinct rainforest Aboriginal peoples – which has a long continuous history in the World Heritage area for at least 100,000 years (Marrie and Marrie 2014). Today, about 300,000 people live in or within 50 km of the Wet Tropics World Heritage Area. Its benefits to the community range from the environmental to the economic, cultural, spiritual, educational, and medicinal topics (Wet Tropics Management Authority 20,016).

The Wet Tropics Management Authority manages the site. Indigenous people, who were marginalized in the initial World Heritage listing processes (Marrie and Marrie 2014), are now present on its board. The tourism industry is recognized by

V. Claudino-Sales, *Coastal World Heritage Sites*, Coastal Research Library 28,
https://doi.org/10.1007/978-94-024-1528-5_27

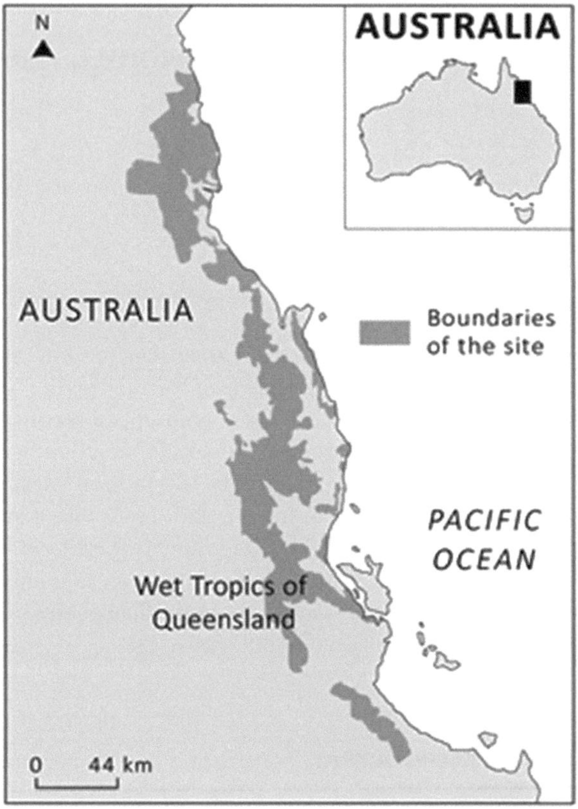

Fig. 27.1 Location of Wet Tropics of Queensland, northeast Australia

them as an ambassador for the property and a major presenter of its attributes (Department of Environment Australia 2016).

The Wet Tropics attract scientists, conservationists, and all sort of tourists from across the globe. First-class recreational facilities for bushwalking, water sports, and camping are all within easy reach. There are 600 km of scenic roads, including 40 scenic routes and over 100 scenic places, to visit (Wet Tropics Management Authority 2016).

27.2 Natural Settings

27.2.1 Climatic Conditions

The tropical climate of northeast Queensland has a high rainfall season, concentrated within a 4-month period. Mean annual rainfall ranges from 4000 mm near the coast to 1200 mm in the hinterland, with more than 60% falling in summer (UNEP/WCMC 1995).

Fig. 27.2 Rocky coast with rainforest cover in the Wet Tropics of Queensland. The coastal erosion on the slopes produces boulders and large sediments that lie on the pocket beach. (Picture reproduced by permission of ChameleonsEye/ Shuttertstock Stock Photo)

Compared with tropical rainforests elsewhere in the world, the wetter parts of the northeast Queensland region lie at the "wet" to "extreme wet" end of the hydrological spectrum. Despite the high rate of precipitation, the site is subject to a dry season, as well as to frequent cyclonic events (UNESCO WHC 2016). Widespread overland run-off is a common phenomenon, uncharacteristic of other wet tropical rainforests in the world. It results from the high rainfall intensity combined with saturated soil profiles (Walsh 1980; Bonell et al. 1983).

Mean temperatures range from 31 °C to 23 °C on the coast and are about 5 °C lower in winter. The tablelands and uplands are cooler, with mean daily temperatures ranging from 28 °C to 17 °C in summer and from 22 °C to 9 °C in winter (DASETT 1986).

27.2.2 Geology and Geomorphology

The Wet Tropics of Queensland is a region of spectacular scenery and rugged topography with rivers, gorges, waterfalls, and mountains and one of the largest rainforests in Australia. Also, the combination of fringe coral reefs and rainforest coastline is rare in the world (Department of Environment of Australia 2016) (Fig. 27.2).

The area is represented by three major geomorphic regions: the tablelands of the Great Divide, the lower coastal belt, and the intermediate Great Escarpment (UNEP/WCMC 1995).

The undulating tablelands are remnants of an elevated and warped landscape. The Great Escarpment marks the limit of headward erosion of these tablelands from the coastal plain (UNEP/WCMC 1995). These features are recognized as possessing outstanding scenic features, natural beauty, and magnificent landscapes (UNESCO WHC 2016).

The rocks identified in the site are basalts, sandstones, granites, and rhyolites. Many of the granites were probably exposed by the end of the Paleozoic and beginning of the Mesozoic. Uplift of the tablelands shed drainage northwest during the Jurassic to the Tertiary period (Department of Environment of Australia 2003).

The volcanism occurred in the tablelands. They are characterized by scoria cones, lava cones, and maars. Lakes occupy the youngest maars. Effectively, the tablelands and some segment of the coast were greatly disturbed by basalt flows throughout the Pliocene-Pleistocene (UNEP/WCMC 1995).

The deposits, present mostly in the coastal area, are alluvium and colluvium, aeolian sediments (dunes), clay plains, and sandy plains (Wet Tropics Management Authority 2016). Extensive paleo-alluvial (fan and debris) flow deposits occur along the base of the escarpment of the highlands in the property (Nott et al. 2001).

The high nutrient status of the developing basalt soils may have favored the rainforest in resisting stresses during the fluctuating climatic conditions associated with the Pleistocene glacial cycles (Department of Environment of Australia 2003).

27.2.3 Biodiversity

The site presents a record of the evolutionary processes that shaped the flora and fauna of Australia over the past 415 million years, when first it was part of the Pangaean landmass, then the ancient continent Gondwana, and for the past 50 million years an island continent. During this time, speciation, extinction, and adaptation have been in place (Department of Environment of Australia 2003). It also provides important information for the interpretation of fossils found elsewhere in Australia (UNESCO WHC 2016).

The main vegetation type is wet tropical rainforests. The rainforest had a great expansion in the Late Quaternary (Nott et al. 2001). This development appears to be at least partly controlled by regional climate, though more local environmental variables and human activity are also significant factors (Haberle 2005).

Besides the rainforests, there are also sclerophyll forests (dominated by plants that have hard leaves adapted to drought and are fire tolerant), woodlands, swamps, and mangrove ecosystem (UNEP/WCMC 1995) (Fig. 27.3).

Although the Wet Tropics of Queensland represents only about 0.1% of the land surface of the continent, it has a large and diverse fauna. Within the property occur 35% of Australia's mammal species, including the monotremes (mammals that lay eggs) (Department of Environment of Australia 2003).

All of Australia's marsupials originated in the rainforest and their closest surviving relatives occur in the Wet Tropics (UNESCO WHC 2016). More than 370 spe-

Fig. 27.3 Mangrove forest at the coastal area of northeast Queensland. (Picture reproduced by permission of Niar/Shutterstock Stock Photos)

cies of birds have been recorded, and there are 131 species of reptiles. The site has a greater diversity of freshwater fish than any other in Australia (UNEP/WCMC 1995).

27.3 Conservation Challenges

There is a wide range of threats to the biodiversity and scenery of the Wet Tropics of Queensland. The main direct threats are fragmentation, climate change, weeds, feral animals, altered fire regimes, altered water flows, water quality, and cyclones (Wet Tropics of Queensland Authority 2016).

The indirect treats are regional population growth, urban development and pollution, demand for community infrastructure (water, roads, and electricity), farming (agriculture, grazing, and aquaculture), and pressure from tourism and recreation (Wet Tropics of Queensland Authority 2016).

The Australian Wet Tropics World Heritage site contains a number of highland vertebrates predicted to face extinction due to a warming climate (Rohan et al. 2007). The global warming also may reduce the amount of precipitation of the area, putting stress on the rainforest. At the same time, the coastal area is under threat of erosion with the sea level rising, due to climate change.

These threats create a situation of significant concern to the preservation of the values of the site, as pointed out by the International Union for Conservation of Nature (IUCN) in its last conservation outlook, published in the end of 2017 (Osipova et al. 2017).

References

Bonell M, Gilmour DA, Cassells DS (1983) Runoff generation in tropical rainforests of northeast Queensland, Australia, and the implications for land use management. Proceedings of the Hamburg Symposium, International Association of Hydrological Sciences 140

DASETT (Department of Arts, Sports, the Environment, Tourism and Territories of Australia) (1986). Nomination of wet tropical rainforests of Northeast Australia by the government of Australia for inclusion in the world heritage list. Report, Canberra, Australia

Department of Environment of Australia (2003) Wet tropics of Queensland periodic report, Brisbane, Australia

Department of Environment of Australia (2016) The Wet Tropics of Queensland Overview. https://www.environment.gov.au/heritage/places/world/wet-tropics. Accessed 26 June 2016

Haberle SG (2005) A 23,000-year pollen record from Lake Euramoo, wet tropics of NE Queensland, Australia. Quat Res 64(3):343–356

Marrie H, Marrie A (2014) Rainforest aboriginal peoples and the wet tropics of Queensland World Heritage Area: the role of indigenous activism in achieving effective involvement in management and recognition of the cultural values. In: Disko S, Tugendhat H (eds) World Heritage sites and indigenous peoples' rights. International World Group for Indigenous Affairs (WGIA), Document 129, Copenhagen

Nott JF, Thomas MF, Price DM (2001) Alluvial fans, landslides and Late Quaternary climatic change in the wet tropics of northeast Queensland. Aust J Earth Sci 48(6):875–882

Osipova E, Shadie P, Zwahlen C, Osti M, Shi Y, Kormos C, Bertzky B, Murai M, Van Merm R, Badman T (2017) IUCN world heritage outlook 2: a conservation assessment of all natural world heritage sites. IUCN, Gland

Rohan DW, John WHT, Stephen EW, Dk Y (2007) Altitudinally restricted communities of Schizophoran flies in Queenland's wet tropics: vulnerability to climage change. Biodivers Conserv 16:3163–3177

UNEP/WCMC (United Nations Environmental Programme/World Conservation Monitoring Centre) (1995) The wet tropics world heritage rainforest tropical North Queensland, Australia. Report, Cambridge, England

UNESCO WHC (United Nations Educational, Scientific and Cultural Organization World Heritage Centre) (2016) Wet Tropics of Queensland. http://whc.unesco.org/en/list/486. Accessed 26 June 2016

Walsh R (1980) Runoff processes and models in humid tropics. Zeitschrift fur Geomorphologie N.F. Suppl 36:176–202

Wet Tropics of Queesnland Authority (2016) Wet Tropics World Heritage Site. http://www.wettropics.gov.au/wet-tropics-management-authority. Accessed 26 June 2016

Part VII
World Heritage on Continental Coasts: South America

Chapter 28
Península Valdés, Argentina

Abstract Península Valdés is in Patagonia, southern Argentina. It is a flat-topped promontory protruding like an island 120 km into the Atlantic Ocean, linked to the continent by a narrow isthmus. The peninsula has an arid climate, with only 232 mm/year of precipitation. The coastal area is characterized by the presence of rocky seacliffs, a series of gulfs and shallow bays, beaches, dunes, marshes, mudflats, and salt pans. The site is of global significance for the conservation of marine mammals. It has a large concentration of southern right whales, an endangered species. It is also important because of the breeding populations of southern elephant seals and southern sea lions. The orcas in this area have developed a unique hunting strategy to adapt to local coastal conditions.

28.1 Introduction

Península Valdés, in Patagonia, southern Argentina, is a flat-topped promontory protruding like an island 120 km into the Atlantic Ocean, linked to the continent by a narrow and short isthmus (25 km long and 11 km wide on average) (UNEP/WCMC 2009) (Fig. 28.1). Lying between the Gulf of San José to the west, the Nuevo Gulf to the south, and the Atlantic to the east, this World Heritage site has an area of 3600 km^2 (UNESCO WHC 2016).

Península Valdés World Heritage site is formed by a series of reserves and parks. The first of them was created in 1969. The ensemble was included in the World Heritage list in 1999 (UNEP/WCMC 2009).

Neolithic man left evidence of obsidian tools, and historically the land was part of the range of the nomadic Tehuelches, which lived off the land and sea prior to colonization (Secretaria de Turismo de Argentina 1998). Today, most of the land is privately owned in large farms. The reserves are under the administration of the Provincial Tourism Organization (UNEP/WCMC 2009).

Tourism is an important activity in the area, especially linked to whale watching and other forms of wildlife viewing on land and water (UNESCO WHC 2014).

V. Claudino-Sales, *Coastal World Heritage Sites*, Coastal Research Library 28,
https://doi.org/10.1007/978-94-024-1528-5_28

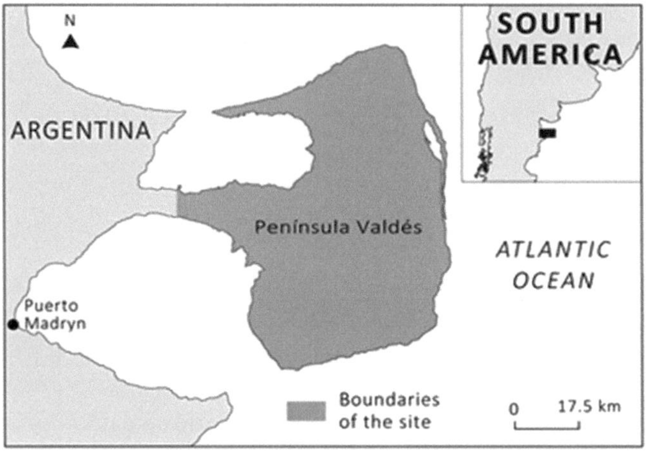

Fig. 28.1 Location of Península Valdés, Argentina

28.2 Natural Settings

28.2.1 Climatic Conditions

Península Valdés is a cool arid coastal area with an average rainfall of 232 mm/year and a soil moisture deficit of about 472 mm/year (Goudie and Selly 2011; Alvarez et al. 2010).

The average annual temperature is 10.6 °C, ranging from 15 °C to 35 °C in summer and from 0 °C to 15 °C in winter, when there are 12–20 days of frost. The winds are strong, especially in the spring of the southern hemisphere (UNEP/ WCMC 2009).

The prevailing current from the north brings warm water to a part of the peninsula, at the same time that cold waters coming from the south Atlantic runs to other sectors. Nevertheless, the waters of the enclosed gulfs and bays are shallow and warm (UNEP/WCMC 2009). Because of the aridity, there are no streams, and freshwater is present in low levels (Secretaria de Turismo de Argentina 1998).

28.2.2 Geology and Geomorphology

Geologically, Patagonia corresponds to a passive margin, resulting from the division of the south Gondwana 140 million years ago (Matos 1992).

The rocks are mostly sedimentary, but there are evidences of volcanic activity on the isthmus that connects the island to the continent (UNEP/WCMC 2009). The

Fig. 28.2 Punta Piramide, modeled in Miocene sandstones, with elephant seals on the right. (Picture reproduced by permission of Maria Leoni Gaffet, PNOR)

peninsula unfolds over the Atlantic Ocean, having as a main landscape the plains, whose plateau environment is broken by the closed depressions of the salt pans (Goudie and Selly 2011; Yorio et al. 1988).

Its roughly 400 km of shoreline include the gulfs, both covering several thousand square kilometers (UNESCO WHC 2014; see Fig. 28.1).The coastal area is characterized by the presence of rocky seacliffs (Fig. 28.2), beaches, dunes, marshes, and mudflats (UNESCO WHC 2016).

The seacliffs can be 100 m high and are sustained by Miocene sandstones (UNEP/WCMC 2009). They make headlands that alternate with sand beaches and pebble beaches. On the eastern end is the Caleta Valdés, a cove of 35 km length, with some islets on the Northern inner point (Yorio et al. 1988). The Caleta is entirely covered by pebbles, and its inside offers shelter to elephant seals, sea lions, and whales (Yorio et al. 1998).

The Península Valdés has the most important aeolian deposits of the Patagonia. Four main types of active sand dune were identified: compound linear, compound transverse, compound dome-shaped, and complex dome-shaped. Because of the windy climate, dune mobility is high (del Valle et al. 2008).

Fig. 28.3 A southern right whale in a shallow area close to the beach, in the barren coast of Península Valdés. The beaches are made by pebbles and other coarse sediments. (Picture reproduced by permission of Maria Leoni Gaffet, PNOR)

28.2.3 Biodiversity

The waters of the two gulfs that frame the peninsula are calm and shallow, making them suitable for the breeding and colonization of three threatened marine mammals: the southern sea lion, the southern elephant seal, and the southern right whale (Secretaria de Turismo de Argentina 1998).

The site shelters the planet's northernmost population of southern elephant seals and the world's largest concentration of southern right whales (UNEP/WCMC 2009) (Fig. 28.3). It contains the globally most important breeding grounds of the southern right whale (UNESCO WHC 2014). There are also dolphins, among them orcas (Secretaria de Turismo de Argentina 1998).

The area exhibits an exceptional example of adaptation of hunting techniques by the orca to the local coastal conditions (UNESCO WHC 2016). They have developed a particular approach to hunting, pursuing the prey (young sea lions or elephant seals) into the shallows and then grabbing them in their jaws (International Herald Tribune 2009) (Fig. 28.3).

Besides the mammals, numerous marine birds breed in the area. There are 181 bird species, with 66 migratory (Goudie and Selly 2011). For this reason, Península Valdés and the Patagonia coast in general have been identified as one of the most important wetlands in the world (Blanco and Canevari 1995).

One particular type of bird is the Magellanic penguin, which ranges on the coast from Península Valdés to Tierra del Fuego. Five reproductive colonies have been identified at Península Valdés (Secretaria de Turismo de Argentina 1998). Seven other species of marine and coastal birds form nesting colonies on 12 distinct sites scattered throughout the peninsula (UNEP/WCMC 2009).

The flora is mainly represented by bushy plains and grassy plains, of the desert steppe type. From north to south, the bushy plains fade into herbaceous plains, helped by a loose sandy soil, free from excessive salinity (Yorio 1998). There are 18 communities of plants, with 130 species from 41 families, with 38 species endemic to Argentina (Goudie and Selly 2011).

28.3 Conservation Challenge

The southern right whale is a species that had severely suffered from commercial whaling. The conservation efforts in Península Valdés have been playing an important role in their recovery (UNESCO WHC 2014).

The southern sea lion was also heavily hunted for oil and skins on the peninsula, legally until 1953 and illegally into the 1970s. The populations have responded to the conservation measures with impressive recoveries (UNESCO WHC 2014).

Despite ongoing sheep grazing and related competition between livestock and native fauna and flora, the property continues to support diverse communities of local vegetation and wildlife (UNESCO WHC 2016).

Tourism has complex environmental impacts in the property, because of the uncontrolled whale watching, that can result in disturbances of breeding populations. Besides, pollution from sewage treatment facilities, fish processing plants, and industry needs adequate environmental management (UNESCO WHC 2014).

Global warming and climate change may threaten the ecosystem, which is already under stress due the low levels of freshwater. Sea level rise may produce erosion on the seacliffs and headlands, challenging the faunas' habitats.

The pollution, marine traffic, and overfishing in important feeding habitats of southern sea elephants result in increasing concerns to the conservation of the site. There also has been an abrupt and still unexplained increase in right whale mortality rates on its shores, raising major concerns in the scientific community (IUCN 2017). The overall situation is ranked by the last conservation outlook from the World Heritage sites (Osipova et al. 2017) as being of significant concern and presenting a trend of deterioration.

References

Alvarez M, Weiler N, Hernández M (2010) Linking geomorphology and hydrodynamics: a case study from Península Valdés, Patagonia, Argentina. Hydrogeol J 18(2):473–486

Blanco D, Canevari P (1995) Situación actual de los chorlos y playeros migratorios de la zona costera patagónica (Prov. de Rio Negro, Chubut y Santa Cruz). Informes Técnicos del Plan de Manejo Integrado de la Zona Costera Patagónica 3:1–26

del Valle HF, Rostagno CM, Coronato FR, Bouza PJ, Blanco PD (2008) Sand dune activity in northeastern Patagonia. J Arid Environ 72(4):411–422

Goudie A, Selly M (2011) World Heritage Desert landscapes: potential priorities for the recognition of desert landscapes and geomorphological sites on the world heritage list. IUCN, Gland

International Herald Tribune (2009) Tides of time, protected promontory is a haven for marine mammals in placid Patagonian seas. 15 December 1999, New York, USA

IUCN (2017) Peninsula Valdes. IUCN World Heritage Outolook. https://www.worldheritageoutlook.iucn.org/node/1088. Accessed 30 Dec 2017.

Matos RMD (1992) The northeast Brazilian rift system. Tectonics 11(4):766–791

Osipova E, Shadie P, Zwahlen C, Osti M, Shi Y, Kormos C, Bertzky B, Murai M, Van Merm R, Badman T (2017) IUCN world heritage outlook 2: a conservation assessment of all natural world heritage sites. IUCN, Gland

Secretaria de Turismo de Argentina (1998) Nomination of properties for the inclusion of Península Valdés on the world heritage list. Report, Bueno Aires

UNEP/WCMC (United Nations Environmental Programme/World Conservation Monitoring Centre) (2009) Península Valdés Argentina. Report, Cambridge, England

UNESCO WHC (United Nations Educational, Scientific and cultural organization world heritage Centre) (2014) World Heritage Committee Thirty-eighth session. Report, Paris

UNESCO WHC (United Nations Educational, Scientific and Cultural Organization World Heritage Centre) (2016) Península Valdés. http://whc.unesco.org/en/list/937. Accessed 29 June 2016

Yorio P (1998) Región 13: Costa Argentina. In: Canevari P, Blanco DE, Bucher EH, Davison I (eds) Los Humedales de la Argentina. Secretaria de Recursos Naturales y Desarrollo Sustentable, Buenos Aires

Yorio P, Tagliorette A, Harris G (1988) Informe preliminar sobre las areas protegidas de la zona costera patagonica. Informes tecnicos del plano de manejo integrado de la zona costera patagonica. Fundacao Patagonia Natural, Puerto Madryn

Yorio P, Tagliorette A, Harris (1998) Informe preliminar sobre las áreas protegidas de la zona costera patagónica. Informes Técnicos del Plan de Manejo Integrado de la Zona Costera Patagónica. Fundación Patagonia Natural, Puerto Madryn

Chapter 29
Atlantic Forest Southeast Reserves, Brazil

Abstract Atlantic Forest Southeast Reserves are in the states of São Paulo and Paraná, southeastern Brazil. The area comprises the Serra do Mar coastal mountains chain and associated groups of peaks, hills, deep valleys, and coastal areas. The site is formed by 25 protected areas and contains one of the largest and best preserved domains of the Brazilian Atlantic Forest, one of the most threatened biomes of the world. Part of the area is included in the international Mata Atlântica Biosphere Reserve. From mountains covered by dense forests with numerous waterfalls and an abundance of orchids and bromeliads and karst phenomena to coastal islands and estuaries with vast mangroves, the site offers a natural environment of great beauty with tremendous terrestrial and marine biodiversity.

29.1 Introduction

Atlantic Forest Southeast Reserves are in the states of São Paulo and Paraná, in southeastern Brazil (Fig. 29.1). The area comprises the Serra do Mar coastal mountain chain and its associated group of high peaks, small hills, deep valleys, and coastal areas, covered by the rainforest "Mata Atlântica" (Atlantic Forest) (UNESCO WHC 2016).

With an area of 4700 km^2, the site is formed by 25 federal- and state-protected areas. There are also private properties designated as natural heritage private reserves. Ownership in the buffer zone is a mixture of private and public (UNEP/WCMC 2008). It became a World Heritage site in 1999 (UNESCO WHC 2016).

The management of the state-protected areas is the responsibility of the state environmental agencies: the Environmental Institute of Paraná (IAP) and the Foundation for the Conservation and Forest Production of the State of São Paulo (Forest Foundation) (UNEP/WCMC 2008).

The Chico Mendes Institute for Biodiversity Conservation is responsible for the federal areas. These institutions conduct public policies related to protection, sustainable use of natural resources, research, and environmental education (UNESCO WHC 2016).

V. Claudino-Sales, *Coastal World Heritage Sites*, Coastal Research Library 28,
https://doi.org/10.1007/978-94-024-1528-5_29

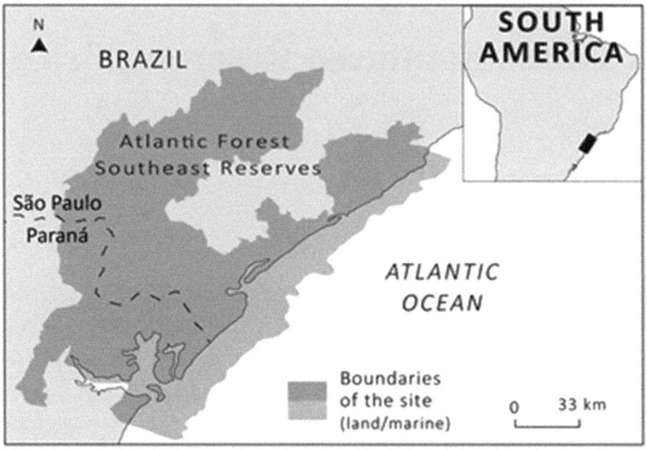

Fig. 29.1 Location of the 25 protected areas that form the Atlantic Forests Southeast Reserves World Heritage site, in Brazil

The region has the oldest traces of colonization of Brazil and is located near two major cities of the country, São Paulo and Curitiba. In the area, there is the presence of indigenous peoples and other traditional groups such as "quilombolas" (communities formed by descendants of former slaves) and "Caiçaras" (coastal communities formed from the miscegenation between Indians, whites and blacks) (UNESCO WHC 2016).

29.2 Natural Settings

29.2.1 Climatic Conditions

The region has a subtropical humid climate in Parana, and the coast of São Paulo is tropical. The site is on the border of tropical and temperate climatic areas. Average temperature in the hottest month is 22 °C (February) and 18 °C in the coldest (July). Average annual precipitation is 1200–1500 mm. Rains are well distributed throughout the year, and there is not a marked dry season, though there is less precipitation in the months of the southern winter (Cavalcanti 2016).

29.2.2 Geology and Geomorphology

The site represents an area of mountains covered with a continuous Atlantic Forest connected with a coastal ecosystem. The altitude goes from 1400 m at the top of the Serra do Mar mountain chain to sea level (UNEP/WCMC 2008).

Fig. 29.2 Comprida Island, a barrier island on the coast of São Paulo State, Brazil, facing the Serra do Mar Mountain. Mangroves suffer from Holocene sea level variations in this narrow island. (Picture reproduced by permission of Tacio Philip Sansonovski/Shutterstock Stock Photos)

The mountains are represented by the Serra do Mar. The Serra do Mar consists of igneous and metamorphic rocks of ages between Archean and early Paleozoic. This chain separates the Brazilian plateau from the coastal lowlands (Serviço Geológico do Paraná 2016).

The origin of the Serra do Mar is related to the separation of Gondwana, which in this area started 140 million years ago. Sixty five million years ago, uplift occurred in the eastern edge of the South American continent, producing uplifting and subsidence of large blocks by faults. The Serra do Mar corresponds to one of these elevated areas (Serviço Geológico do Paraná 2016; Almeida and Carneiro 1998).

The Serra do Mar escarpment was located dozens of kilometers from the current shoreline. Due to erosional processes, the scarps retreated to its present position (Almeida and Carneiro 1998).

In the Serra do Mar, there are more than 300 karstic caves. Among them, there is the "Casa de Pedra" (rocky house), which has the largest portico in the world at 215 m high, and Santana Cave, highly decorated with speleothems (UNESCO WHC 2016). The intensity of the rains and the high humidity in the region explain the exceptionally large size and degree of development of the caves and associated concretions such as stalactites, stalagmites, curtains, and pillars (UNEP/WCMC 2008; Weber 1998)

In the coastal area, in the State of São Paulo, two sites are part of the 25 preserved areas of the heritage site: Comprida Island and Cardoso Island. Comprida Island is a barrier island 72 km long, with a width of 2 to 5 km and altitude of less than 7 m (Fig. 29.2). It is separated from the mainland by lagoon channels colonized by mangrove (Gandolfo et al. 2001).

Fig. 29.3 Ilha do Mel Island, in the State of Paraná, Brazil, with igneous rocks creating an isthmus, with headlands and hills in the shoreline and pocket beaches in the coastal area. (Picture reproduced by permission of Rui Morel Carneiro)

Cardoso Island has an area of approximately 151 km^2, being composed of igneous complex represented by syenitic rocks. On both islands, the Quaternary deposits are represented by Holocene and Pleistocene sediments, resulting from Quaternary variations of sea level (Cheliz 2011).

In São Paulo, the site also comprises part of the estuarine lagoon complex of Iguape-Cananéia, which includes a great variety of wetlands, from the flooded plains of the Ribeira de Iguape River to saline waters of the lagoon complex itself (Governo do Estado de São Paulo 2016).

In the coastal area of Paraná State, there is the island Ilha do Mel Ecological Station. The island has two domains: Precambrian to Mesozoic igneous and metamorphic rocks, which form hills and headlands (Fig. 29.3), and Quaternary sand deposits, which form the coastal plains (Cano et al. 2013). The evolution of Ilha do Mel could be reconstructed since the Late Pleistocene (Angulo and Souza 2005).

Both coastal areas of Paraná and São Paulo are occupied by mobile sand dunes and mangrove forests (Cheliz 2011; Angulo and Souza 2005; Gandolfo et al. 2001). These coastal lands are separated from the ocean by an archipelago that runs parallel to the shoreline (UNEP/WCMC 2008).

29.2.3 *Biodiversity*

The Atlantic Forest once extended more than 1,300,000 km^2. Today, little remains of the rainforest. The protected areas that were created to save its remains represent only 7% of the total of the original area (Ministério do Meio Ambiente 2016).

The site represents the most important ecological corridor of the highly diverse and endangered Atlantic rainforest. This forest, historically partly isolated, has evolved into a complex biome with a multitude of endemic species (UNESCO WHC 2016).

More than 450 species of trees per hectare can be found in some spots, which indicate that the diversity of woody plants in the region is larger than in the Amazon rainforest. Along the coast, there are extensive areas of mangroves and shrubland (UNEP/WCMC 2008).

Mammals include 120 species, probably the largest number in Brazil. The bird fauna is very diverse, with 350 recorded species. The system of caves gives shelter to an endemic microfauna, mainly arthropods and species of blind fish (UNEP/WCMC 2008).

29.3 Conservation Challenges

The site is still barely affected by the process of fragmentation, one of the greatest threats to the biome. Access difficulties, due to its geographical characteristics of rugged mountains and deep valleys with extensive wetlands, contribute to its conservation (UNESCO WHC 2016).

The main threats to the site are the construction of roads, installation of power lines, and urbanization. These infrastructures generate problems of increased fragmentation, disturbance, and penetration corridors for certain invasive species, as well as reduced natural connectivity (UNESCO WHC 2016).

Global warming is also a threat to the Atlantic Forest, considering that it potentially would reduce hydric resources, change the rain regime, and disturb the microclimates of the protected reserves. The rising of sea level could harm the preservation of the unstable coastal features such as barrier islands, beaches, and dunes.

The establishment of protected areas within the remaining Atlantic Forest was an important step to prevent the irreversible loss of a unique and exceptionally diverse forest ecosystem. However, many of the conservation units that make up the property are very small and vulnerable to outside influences. At the same time, recent budgetary cuts in federal budgets in Brazil for protected areas raise serious concerns over the already limited resources to address these issues (IUCN 2017).

For these reasons, the last conservation outlook related to World Heritage sites found that the area's conservation presents significant concerns, associated with all the threats mentioned above (Osipova et al. 2017).

References

Almeida FFM, Carneiro CDR (1998) Origem e Evolução da Serra do Mar. Revista Brasileira de Geociências 28(2):135–150

Angulo RJ, Souza MC (2005) Geologia e Geomorfologia. In: Marques MCN, Britez RN (eds) História Natural e Conservação da Ilha do Mel. Editora UFPR, Curitiba

Cano NF, Turbiani-Filho IT, Gennari RF, Munita CS, Souza MC, Angulo RJ, Watanabe S (2013) TL dating of sediments from Ilha do Mel, Brazil. Quat Int 306:137–145

Cavalcanti IFA (2016) Tempo e clima no Brasil. Oficina de Textos, São Paulo

Cheliz PM (2011) Ilha do Cardoso - Interfácie entre suportes ecológicos e oscilações quaternárias dos mares em uma ilha continental. XIII Congresso da Associação Brasileira de Estudos do Quaternário ABEQUA. Armação dos Búzios, Rio de Janeiro

Gandolfo OCB, Souza LAP, Tessler MG, Rodrigues M (2001) Estratigrafia rasa da Ilha Comprida (SP): um exemplo de aplicação do GPR. Revista Brasileria de Geofísica 19(3):25–36

Governo do Estado de São Paulo (2016) Parque Estadual Lagamar de Cananéia. http://www.ambiente.sp.gov.br/parque-lagamar-de-cananeia/sobre-o-parque/. Accessed 25 July 2016

IUCN (International Union for Conservation of Nature) (2017) Atlantic forest Southeast reserves. IUCN World Heritage Outlook. https://www.worldheritageoutlook.iucn.org/node/1090

Ministério do Meio Ambiente (2016) Mata Atlântica. http://www.mma.gov.br/biomas/mata-atlantica. Accessed 26 July 2016

Osipova E, Shadie P, Zwahlen C, Osti M, Shi Y, Kormos C, Bertzky B, Murai M, Van Merm R, Badman T (2017) IUCN world heritage outlook 2: a conservation assessment of all natural world heritage sites. IUCN, Gland

Serviço Geológico do Paraná Mineropar (2016) Geologia do Paraná. http://www.mineropar.pr.gov.br/arquivos/File/Paineis_geologicos/SerradoMar2_portugues.pdf. Accessed 2 July 2016

UNEP/WCMC (United Nations Environmental Programme/World Conservation Monitoring Centre) (2008) Southeast Atlantic Forest reserves, Brazil. Report, Cambridge, England

UNESCO WHC (United Nations Educational, Scientific and Cultural Organization World Heritage Site) (2016) Atlantic Forest South-East Reserves. http://whc.unesco.org/en/list/893. Accessed 26 July 2016

Weber W (1998) Geologia e Geocronologia da Ilha do Cardoso, Sudeste do Estado de São Paulo. Dissertação de Mestrado, Instituto de Geociências, Universidade de São Paulo, São Paulo

Chapter 30
Discovery Coast Atlantic Forest Reserves, Brazil

Abstract The Discovery Coast Atlantic Forest Reserves, located on the east coast of Brazil, consist of eight separate protected areas containing representative remnants of the rainforest that covered the coast of Brazil (Atlantic Forest), and also by a type of coastal shrubland vegetation called "resting," associated with the Atlantic Forest. The rainforests of Brazil's Atlantic coast are the world's richest in terms of biodiversity. The landscape is characterized by coastal tablelands that form sea-cliffs, as well as by sandy beaches, beach rocks, mangrove swamps, lagoons, and coral reefs. The Discovery Coast was the first landscape sighted by the Portuguese when they arrived in Brazil in 1500 and has an important meaning to the national heritage.

30.1 Introduction

The Discovery Coast is in the central east coast of Brazil, on by the Atlantic Ocean. It has an area of 1119 km^2 (UNESCO WHC 2016). It covers 12 municipalities in Bahia State and 4 in the Espírito Santo State. The site has two separate areas, one north of the Espírito Santo and another south of Bahia, covering the total of eight nature reserves and its adjacent areas (IPHAN 2014) (Fig. 30.1).

The Discovery Coast was the first landscape seen by the Portuguese when they arrived in Brazil in 1500. Because of its historical, geographical, cultural, and scenic values, the area has an important meaning to the national heritage.

The luxurious Atlantic forest found in the Discovery Coast is described in the charter of discovery of the 15th Century as having infinite fertility, extending "to where the eye can see" (Costa 2009). It remained almost intact for more than 400 years, until the opening of a coastal road that caused violent exploitation of the woods, especially the "pau brasil" (red wood), which gave its name to the country (Costa 2009).

The Atlantic Forest originally extended by approximately 1,300,000 km^2. Little remains of that lush forest: today the remnants of the native vegetation are reduced to about 22% of its original cover. With the intention to save what is still standing,

V. Claudino-Sales, *Coastal World Heritage Sites*, Coastal Research Library 28,
https://doi.org/10.1007/978-94-024-1528-5_30

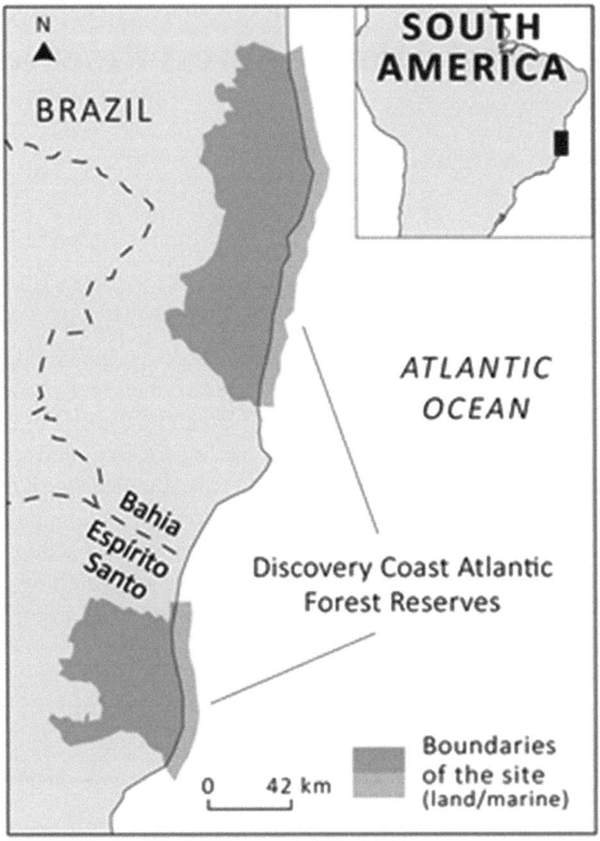

Fig. 30.1 Location of the two segments of the Discovery Coast Atlantic Forest Reserves World Heritage site, in the states of Bahia and Espírito Santo

the various protected areas were created. These protected areas represent just 7% of the total area covered by the forest (Ministério do Meio Ambiente 2016).

In the Discovery Coast, the Mount Pascoal area – the first landform observed by the Portuguese – was the first place to be transformed into a protected area (a national park) by the Brazilian government, in 1961. The other protected areas were established in the 1980s and 1990s (UNEP/WCMC 2008). In 1991, the entire Atlantic Forest was defined as "Mata Atlântica Biosphere Reserve" by UNESCO. The ensemble of the eight sites and adjacent areas of the Discovery Coast became a World Heritage site in 1999 (UNESCO WHC 2016).

The site holds a series of cultural features, mostly representing examples of the first European occupation of the New World. Numerous archaeological sites, both prehistoric and historic, can be found in the area. The ruins of Brazil's first church lay at the top of a cliff to the north of Porto Seguro, Bahia. An old Tupi native civilization town has been unearthed (UNEP/WCMC 2008).

The land tenure is characterized by a mixture of public in state and federal protected areas and private (UNEP/WCMC 2008). The area is well populated and has Indian reservations of the Pataxó people (IPHAN 2014). The management of the site is the responsibility of the Chico Mendes Institute for Biodiversity Conservation (ICMBio), an autonomous federal agency under the Ministry of the Environment (UNESCO WHC 2016).

The site attracts thousands of tourists throughout the year, and tourism is one of the most important economic activities in the area. The most visited of the municipalities is the city of Porto Seguro, in Bahia State, with its paradise-like beaches.

30.2 Natural Settings

30.2.1 Climatic Conditions

The climate is tropical, hot with no dry seasons. The annual rainfall is about 2000 mm, concentrated between April and June. The annual average temperature is over 20 °C, with a maximum above 30 °C in summer (Projeto Costa do Descobrimento 2000).

The weather is influenced by southern trade winds, as well as by periodic intrusions of cold winds coming from Patagonia. Relative humidity is around 80% throughout the year. Dominant winds come from the east, being stronger from September to November (UNEP/WCMC 2008).

30.2.2 Geology and Geomorphology

The area is part of the passive continental margin of Brazil, resulting from the breakup of Gondwana 120 million years ago (Matos 1992).

Coastal tablelands predominate in the coastal zone. The tablelands are sustained by Miocene continental and marine friable sediments, and they are largely dissected by rivers (Dominguez et al. 2002). The tablelands reach the coastline, where they form white and red seacliffs, sometimes covered by vegetation (Fig. 30.2). The seacliffs, with heights of 40 m, characterize the coastal profile (Lima and Dominguez 2015).

There are also coastal plains. The plains are Quaternary in age and are formed by marine deposits and lagoon deposits. They occupy the lower segments of the coastal area, bordering the shoreline and the valleys cut in the tablelands (Amorim and Oliveira 2010). In the coastal plain, there are Pleistocene and Holocene marine terraces, resulting from the Quaternary fluctuations of sea level. Sandy beaches, beach rocks (Fig. 30.2), mangrove swamps, and coral reefs also occur (Dominguez et al. 2002).

Fig. 30.2 Seacliffs with Atlantic coastal vegetation in Miocene sediments and sandy beach in Praia do Espelho beach, Bahia, framed by Holocene beach rocks. (Picture reproduced by permission of Adriana Alvim Oliveira Carvalho, webpage http://adrianaalvim.blogspot.com.br)

River sediments are made up of gravel and sand. At the river mouths and lagoons, there are deposits of silt and clay, rich in organic matter. Yellow/reddish podzol soils exist on the plateau. In the coastal plains, the soil is litholic and sandy. Marine sediments penetrate into the riverbeds, a long distance from the coast (UNEP/WCMC 2008).

In the hinterland, a hilly and mountainous landscape appears. It is sustained by metamorphic rocks of the crystalline basement. It forms Mount Pascoal, an inselberg and the most elevated form of the region at 586 m (Lima and Dominguez 2015). The area is also subjected to neotectonism, with tectonic Quaternary forces tilting blocks and faulting the Miocene formation (Lima and Dominguez 2015).

30.2.3 Biodiversity

The Atlantic rainforests of the Discovery Coast are considered to be the world's richest in terms of the number of species of trees per hectare – a study by the New York Botanical Garden found 458 different species of trees in 1 ha of forest in Bahia (UNEP/WHCM 2008).

Fig. 30.3 Coastal plain with low forest and shrub vegetation ("restingas") in the beach Praia dos Coqueiros, Bahia. (Picture reproduced by permission of Adriana Alvim Oliveira Carvalho, webpage http://adrianaalvim.blogspot.com.br)

The biodiversity is characterized by mixing regional endemic Atlantic and Amazonian forest, suggesting that there may have been galleries connecting them (IPHAN 2014). The principle vegetation type is highly diverse primary tropical moist broadleaf forest with tall (20–30 m in height) canopy trees. The area contains perhaps the largest number of trees of "pau brasil" (red wood) left on Earth (UNEP/WCMC 2008).

In the Mount Pascoal there are rare species of orchids and species of precious woods (IPHAN 2014). Along the river valleys, there is a gallery forest. On sandy coastal soils, there is shrub vegetation (named "restingas" in Portuguese) which, depending on the distance to the shore, ranges from humid prairies to low forests (UNEP/WCMC 2008) (Fig. 30.3).

It is estimated that in the Atlantic Forest, there are about 20000 plant species (about 35% of the existing species in Brazil), including several endemic and endangered species (Ministério do Meio Ambiente 2016). These numbers are larger than in some continents (17000 species in North America and 12500 in Europe), making the Atlantic Forest region a high priority for the conservation of global biodiversity (Ministério do Meio Ambiente 2016).

Regarding the fauna, the Atlantic Forest is home to 849 species of birds, 370 species of amphibians, 200 species of reptiles, 270 mammals, and 350 species of fish. Marine fauna is also rich, especially around reefs (Ministério do Meio Ambiente 2016).

30.3 Conservation Challenges

The beaches suffer from popularity among Brazilian tourists, and environmental problems follow, such as the accumulation of and inadequate waste disposal and the trampling of the reef areas and its removal by tourists to take as souvenirs. Development is also threatening nature and Indian reserves (Rocha et al. 2011).

Global warming and climate change will increase erosion on the beaches. It has already resulted in frequent storms in the Atlantic Forest, in an area that is already subjected to flooding and linear erosion (Amorim and Oliveira 2013). Climate change may also threaten the high diversity of the Atlantic Forest: the number of species should decline under changing climate conditions (Leme et al. 2014).

The main challenges for the protection of the site seem to be the ability of the managers to develop a campaign for environmental education and to obtain more resources for the implementation of recommendations and management plans (UNESCO WHC 2016).

The establishment of the various protected areas was a highly needed step to save the last remnants of one of the most fragmented and degraded forest regions on the planet, but the protected areas require more coherent and encompassing management. At a time when better connectivity is urgently needed and recognition of a need to restore parts of the Atlantic Forest is emerging, the trends in the landscape surrounding the site keep deteriorating. Unless land use dynamics can be influenced to promote more environmentally friendly activities, further deterioration seems inevitable due to biological isolation (IUCN 2017).

For these reasons, the state of conservation outlook of the World Heritage sites published by the International Union for Conservation of Nature (IUCN) at the end of 2017 (Osipova et al. 2017) found that the Discovery Coast Atlantic Reserves site's conservation presents a significant concern.

References

Amorim RR, Oliveira RG (2013) Zoneamento ambiental, subsfidio ao planejamento no uso e ocupação das terras da costa do descobrimento. Mercator 12(29):211–231

Amorim RR, Oliveira RC (2010) Uso e ocupação das mesoformas da região Costa do Descobrimento (Bahia, Brasil). VI Seminário Latino-Americano de Geografia Física, Coimbra, Portugal

Costa JPO (2009) Sitios del patrimonio natural de Brasil. Journal of Cult Herit Stud 22(2):184(14)

Dominguez JML, Martin L, Bittencourt (2002) A costa do descobrimento, a geologia vista das caravelas. In: Schobbenhaus C, Campos DA, Queiroz ET, Winge M, Berbert-Born MLC (eds) Sitios geológicos e Paleontológicos do Brasil. DNPM/CPRM, Brasília

IPHAN (Instituto do Patrimônio Histórico e Artístico Nacional) (2014) Costa do Descobrimento: Reservas da Mata Atlântica (BA/ES). http://portal.iphan.gov.br/pagina/detalhes/37. Accessed 30 June 2016

IUCN (International Union for Conservation of Nature) (2017) Discovery Coast Atlantic Forest Reserves Conservation Outlook. http://www.worldheritageoutlook.iucn.org/explore-sites/wdpaid/198292. Accessed 19 December 2017

Leme P, Melo A, Loyola R (2014) Climate change threatens protected areas of the Atlantic Forest. Biodivers Conserv 23(2):357–368
Lima CCU, Dominguez JML (2015) Discovery coast: the Brazilian landscape first sighted by Europeans. In: Vieira BC, Salgado A, Santos L (eds) Landscapes and landforms of Brazil. Springer, Dordrecht
Matos RMD (1992) The northeast Brazilian rift system. Tectonics 11(4):766–791
Ministério do Meio Ambiente (2016) Mata Atlântica. http://www.mma.gov.br/biomas/mata-atlantica. Accessed 1 July 2016
Osipova E, Shadie P, Zwahlen C., Osti M, Shi Y, Kormos C, Bertzky B, Murai M, Van Merm R, Badman T (2017) IUCN world heritage outlook 2: a conservation assessment of all natural world heritage sites. IUCN Gland
Projeto Costa do Descobrimento (2000) Costa do Descobrimento Programa de Avaliacao de Potencialidades Minerais. CBPM (Companhia Baiana de Pesquisa Mineral), Salvador
Rocha MC, Paixão MM, Castro LLC, Vieira S (2011) O Ecoturismo sob o olhar da comunidade local: Praia do Espelho, APA Caraiva/Trancoso, Porto Seguro (BA). Revista Brasileira de Ecoturismo 4(4):579–590
UNEP/WCMC (United Nations Environmental Programme/World Conservation Monitoring Centre) (2008) Discovery Coast Atlantic Forest Reserves, Brazil. Report, Cambridge, England
UNESCO WHC (United Nations Educational, Scientific and Cultural Organization World Heritage Site) (2016) Discovery Coast Atlantic Forest Reserves. http://whc.unesco.org/en/list/892. Accessed 1 July 2016

Part VIII
World Coastal Heritage in Islands, Gulfs and Reefs: Arctic Ocean

Chapter 31
Natural System of Wrangel Island Reserve, Russia

Abstract The site consists of two islands – Wrangel Island and the small Herald Island – in the Arctic Ocean, 500 km above the Arctic Circle, northwest of the Bering Strait. The islands are an ecosystem and marine area that was not glaciated during the last Quaternary Ice Age, resulting in exceptionally high levels of biodiversity. Wrangel Island has the world's highest density of polar bears and also has reindeer, musk ox, walrus, snow goose, and other polar species. The area is so isolated and so far north that it was the last place on Earth that woolly mammoths survived, with the last dying out around 1700 BC. Wrangel and Herald islands formed part of the land bridge between Asia and North America at the intersection of two major continental systems and of plant and animal migration routes between Asia and America.

31.1 Introduction

Wrangel Island, in the Arctic Ocean, northwest of the Bering Strait, Russia, is 150 km long and 126 km wide. The World Heritage site of Natural System of Wrangell Island Reserve also includes the small 26 km^2 uninhabited Herald Island 64 km to the east (Fig. 31.1).

The property has an area of 19163 km, with 7620 km^2 of land and a marine area of 12 nautical miles around each island, totaling 11543 km^2 (UNESCO WHC 2016). The land around the village of Wrangel Island and an area facing the coast are excluded from the reserve (see Fig. 31.1) (UNEP/WCMC 2008).

In 1975, the islands were designated a federally protected nature reserve by the USSR government. In 1997, the reserve area was enlarged by a 12 nautical mile protective zone. In 1999, the Russian government established a 24 nautical mile protective marine zone around Wrangel Island. The ensemble became a World Heritage site in 2004 (UNEP/WCMC 2008).

Herald Island was first seen in 1849. Wrangel Island was discovered by an American whaler in 1867 and named for the Russian navigator and explorer Ferdinand von Wrangel (UNESCO WHS 2016).

V. Claudino-Sales, *Coastal World Heritage Sites*, Coastal Research Library 28,
https://doi.org/10.1007/978-94-024-1528-5_31

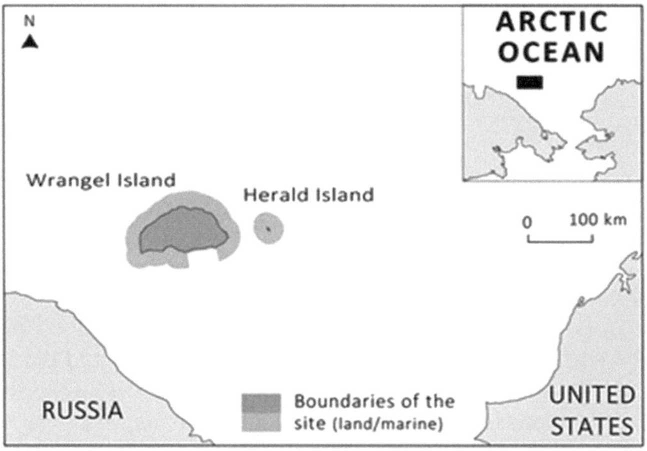

Fig. 31.1 Location of Natural System of Wrangel Island Reserve, in the Arctic Ocean, northwest of the Bering Strait

Evidence of a Neolithic camp inhabited by Paleo-Eskimo hunters of approximately 3400 years ago is found on the south coast of Wrangel Island. Some of their prey species were mammoths, which lived on the island 6000 years later than the official extinction date (Vartanyan et al. 2008).

The land is the property of Russian government, under the federal Ministry of Natural Resources. The Chukotka Autonomous Regional Administration is responsible for the management of both land and marine sectors of the reserve (UNESCO WHC 2016).

31.2 Natural Settings

31.2.1 Climatic Conditions

The climate is Arctic, influenced by Pacific air masses, which contributes to the predominance of high winds constantly blowing from the north and northeast (Directorate of Wrangel Island Reserve 2003). The islands are subjected to cyclones with rapid circular winds. It is an area of blizzards and 2 months of polar night but also constant overcast skies, mists, and fog (UNEP/WCMC 2008).

The mean July temperature is 2.4 °C. The seas around the islands are rarely free from pack ice, and water around the seacliffs is usually iced (Fig. 31.2) (Encyclopaedia Britannica 2016). The frost-free period is about 2–3 weeks. The average annual precipitation is 200 mm, usually accumulating as snow (UNEP/WCMC 2008).

There are noticeable differences in climate between the northern, central, and southern parts of the Wrangel Island. The central and southern portions are warmer,

Fig. 31.2 Seacliff with ice and iced waters in the Arctic Ocean around Wrangel Island. (Picture reproduced by permission of Alexey Bezrukov)

with some of the valleys having semi-continental climates that support a number of sub-Arctic steppe-like meadow species. This is a unique feature in the High Arctic (Stishov et al. 2000).

31.2.2 Geology and Geomorphology

Geologically, Wrangel Island consists of crystalline slates, granites, and gneisses together with alluvial sands. It represents a small but unique exposure of Neoproterozoic basement and its upper Paleozoic and Mesozoic cover within the mostly unexplored East Siberian Shelf (Miller et al. 2010).

Herald Island is formed of gneiss and granite mountains with cliffs up to 250 m rising from the sea on all sides. The ground is underlain with permafrost and a mosaic of tundra and steppe types (Directorate of Wrangel Island Reserve 2003).

Wrangel and Herald Islands are part of a platform formed by Mesozoic folding, which in Precambrian times was part of the Canadian Shield (UNEP/WCMC 2008). There is ample evidence that Wrangel Island has undergone a long evolutionary process since the Mesozoic, uninterrupted by the glaciations that swept most other parts of the Arctic during the Late Quaternary period (Gualtieri et al. 2003).

The absence of glaciations is attributable to contrasting sources of precipitation during the glacial global climates (Stauch et al. 2008). Although the highest part of the island reaches 1096 m at Sovetskaya Mountain, there are no glaciers

Fig. 31.3 Female polar bear with cubs in the snowy hills of Wrangel Island Mountain. (Picture reproduced by permission of Alexey Bezrukov)

(Encyclopaedia Britannica 2016). Glacial morphology and glacial sediments are also absent from the island (Karhu et al. 2001).

Half of the island is covered by a plain of alluvial sediments with karst basins of Holocene age and tundra plains. There are many small lakes, and the northern and southwestern coasts are characterized by numerous Pleistocene sandspits enclosing lagoons (UNEP/WCMC 2008).

The drainage includes approximately 1400 rivers of more than a kilometer in length and 5 longer than 50 km. River waters are characterized by an elevated background value of total mineralization (i.e., total dissolved solids). This is related to an abundance of Late Carboniferous gypsiferous and dolomitic sequences in the mountainous area of the island (Lavrushin and Gruzdev 2012).

31.2.3 Biodiversity

The isolation of the natural systems of the islands at the heart of plant and animal migration streams between Asia and America led them to become refuges with a high degree of speciation (Directorate of Wrangel Island Reserve 2003). The absence of glaciation also contributed to this speciation (UNESCO WHC 2016).

The islands have the world's largest population of polar bear (Fig. 31.3). But despite diversity in the past, there are just seven species of resident terrestrial mammals (UNEP/WCMC 2008).

The world's largest population of the Pacific walrus (80000–100,000) is found on Wrangel Island and on Herald Island. It is the most numerous mammal in the seas off the islands. Ringed seal and bearded seal are also common in the coastal waters. During summer and fall, the waters around the islands are important summer feeding grounds and migration areas for whales, particularly the gray whale (UNEP/WCMC 2008).

Endemism among the avifauna is not pronounced, but the site has high species diversity compared to other arctic areas of about the same size. It is the last and only landfall in Asia for more than a hundred migratory bird species flying north through central Beringia from Pacific North America. A hundred and sixty-nine bird species have been identified on the islands (Stishov et al. 2000).

On Wrangel Island, cold-arid steppe and tundra steppe are widespread, and tundra carbonate and arid saline soils occur, all atypical for the Arctic (UNEP/WCMC 2008). The result is a higher diversity of vascular plants and animals than in most other parts of the Arctic. Remnant species not present elsewhere are found, as well as species far north of their usual range (UNEP/WCMC 2008).

The islands are part of the Arctic tundra subzone, the only one in the entire Bering Sea region of the Arctic. Wrangel Island has an almost typical arctic tundra subzone steppe-like vegetation of low shrubs and herbaceous meadows, underlain by dry permafrost, but it is unique in the richness of species and number of endemic plants (Stishov et al. 2000). Xerophilous herbs on south slopes and dry halophytic communities are also unusual for the Arctic (Directorate of Wrangel Island Reserve 2003).

Generally, the islands' types of tundra range from fields of dry, sparsely vegetated, rounded hummocks, mossy hillocks, sheltered meadows, lush grasslands, and numerous wet and marshy areas interspersed with tundra ponds (Stishov et al. 2000). Four hundred and seventeen species and subspecies of vascular plants have been identified on the islands, more than double that of any other Arctic tundra of comparable size (UNEP/WCMC 2008).

31.3 Conservation Challenges

Climate change and global warming are threatening the site. Average temperatures on Wrangel Island appear to be rising. Extreme weather episodes have been increasing, and summers are becoming wetter (UNEP/WCMC 2008).

Changes in the weather and climate are visible in other levels: ice has been melting earlier in the spring, and the fall freeze-up has been occurring later. The number of ice-free years has also been increasing each decade (UNEP/WCMC 2008).

This warming trend may cause hardship for polar bears and walrus which depend on platforms of ice from which to feed. In addition, it will alter the natural vegetation of the site, with expansion of bush tundra and swamps with higher levels of humidity and temperature. If the climate becomes more continental in character, areas of xerophilous and meso-xerophitic, tundra steppe and meadow steppe may increase (UNEP/WCMC 2008).

The management regime of the site has been maintaining the values of the remote and nearly uninhabited Wrangel Island Reserve, and while these values appear well preserved to date, the long-term conservation outlook of the property mainly depends on the impact of emerging threats, such as the construction of a military base on the island, geophysical prospecting for oil in the immediate site vicinity, and maritime transportation and shipping (IUCN 2017).

For all these reasons, the last conservation outlook for World Heritage sites released by the International Union for Conservation of Nature (Osipova et al. 2017) found that the state of conservation of the site presents a significant concern.

References

Directorate of Wrangel Island Reserve (2003) Natural system of Wrangel Island Reserve for inscription into the world cultural and natural heritage list of UNESCO. Report, Chukotka, Russia

Encyclopaedia Britannica (2016) Wrangel Island, Russia. https://www.britannica.com/place/Wrangel-Island. Accessed 21 September 2016

Gualtieri L, Vartanyan S, Brigham-Grette J, Anderson P (2003) Pleistocene raised marine deposits on Wrangel Island, northeast Siberia and implications for the presence of an East Siberian ice sheet. Quat Res 59(3):399–410

IUCN (International Union for Conservation of Nature) (2017) Natural System of Wrangel Island Reserve Conservation Outlook. http://www.worldheritageoutlook.iucn.org/explore-sites/wdpaid/902356. Accessed 19 December 2017

Karhu JA, Tschudi S, Saarnisto M, Kubik P, Schlüchter C (2001) Constraints for the latest glacial advance on Wrangel Island, Arctic Ocean, from rock surface exposure dating. Glob Planet Chang 31(1):447–451

Lavrushin VY, Gruzdev AR (2012) The salt composition of rivers in Wrangel Island (report). Lithol Miner Resour 47(1):1–17

Miller EL, Gehrels GE, Pease V, Sokolov S (2010) Stratigraphy and U-Pb detrital zircon geochronology of Wrangel Island, Russia; implications for Arctic paleogeography. Am Assoc Pet Geol 94(5):665–692

Osipova E, Shadie P, Zwahlen C, Osti M, Shi Y, Kormos C, Bertzky B, Murai M, Van Merm R, Badman T (2017) IUCN world heritage outlook 2: a conservation assessment of all natural world heritage sites. IUCN Gland

Stishov M, Zheleznov N, Nikiforov V, Botorin A (2000) World heritage nomination for the natural system of Wrangel Island Sanctuary. Report, Chukotka, Russia

Stauch G, Gualtieri L, Thackray GD, Owen LA, Yi C (2008) Late quaternary glaciations in northeastern Russia. Landsc Evol 23(6–7):545–558

UNEP/WCMC (United Nations Environmental Programme/World Conservation Monitoring Centre) (2008) Natural system of Wrangel Island reserve, Chukotka, Russia Federation. Report, Cambridge, England

UNESCO WHC (United Nations Educational, Scientific and Cultural Organization World Heritage Centre) (2016). Natural System of Wrangel Reserve. http://whc.unesco.org/en/list/1023. Accessed 22 September 2016

Vartanyan SL, Arslanov KA, Karhu JA, Possnert G, Sulerzhitsky LD (2008) Collection of radiocarbon dates on the mammoths (Mammuthus primigenius) and other genera of Wrangel Island, northeast Siberia, Russia. Quat Res 70(1):51–59

Part IX
World Coastal Heritage in Islands, Gulfs and Reefs: Atlantic Ocean

Chapter 32
Brazilian Atlantic Islands: Fernando de Noronha and Atol das Rocas, Brazil

Abstract The island of Fernando de Noronha, off the northeast Brazilian coast, was formed by volcanic peaks of a submerged mountain chain. The Atol das Rocas Island, located 150 km west of Fernando de Noronha, is the only atoll in the South Atlantic. They are formed by small islets and islands and accommodate the largest concentration of tropical seabirds to be found in the western Atlantic Ocean. Dolphin Bay, in Fernando de Noronha Island, hosts an exceptional population of dolphin, being the only known place in the world with such a large population of resident dolphins. Fernando de Noronha also contains the only oceanic mangrove in the South Atlantic region. Its surroundings have exceptional underwater landscapes, recognized worldwide as a special place for diving. Atol das Rocas, at low tide, provides a spectacular seascape of lagoons and tidal pools with fish, shellfish, sponges, mollusks, and corals.

32.1 Introduction

Fernando de Noronha and Atol das Rocas are volcanic islands off the northeast Brazilian coast, 4° below the equator, 350 and 150 km from the mainland, respectively (Fig. 32.1).

Part of Fernando de Noronha Island was designated as an environmental protection area in 1986. A marine national park was created in 1988. In 1989, the entire area was declared as an environmental protection area (Reis and Hayward 2013). The Rocas Atoll, the only atoll in the South Atlantic, was established as a national biological reserve in 1979 (ICMBio 2016a). It was the first marine conservation unit established in Brazil and the first marine area protected by law (Kikushi 1994). Both areas became a World Heritage site in 2001 (UNESCO WHC 2016).

The World Heritage site in Fernando de Noronha has the same area as its marine park, covering the majority of the main island and the offshore islands and islets. It has 112.7 km^2, of which 85% is marine (UNESCO WHC 2016).

Tourism is a very important activity in Fernando de Noronha Island. Diving is one of the most popular activities by tourists. A hot oceanic current coming from Africa allows diving to up to 30 m without the need of special clothes (Fernando de

V. Claudino-Sales, *Coastal World Heritage Sites*, Coastal Research Library 28,
https://doi.org/10.1007/978-94-024-1528-5_32

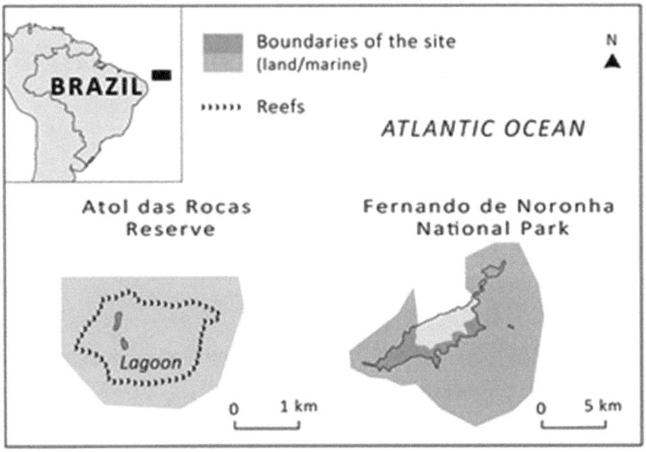

Fig. 32.1 Location of Fernando de Noronha (right) and Atol das Rocas (left) islands off the northeast Brazilian coast

Noronha Official Site 2016), what makes the sport very accessible to tourists in general.

Fernando de Noronha Archipelago was discovered by Portuguese navigators in the sixteenth century. The islands have numerous archaeological sites. It was incorporated into the state of Pernambuco in 1988 (Fernando de Noronha Official Site 2016). The archipelago is a state district, the only one in Brazil, and is administered directly by the state of Pernambuco (Ministério do Meio Ambiente 2008).

In Rocas Atoll, the World Heritage site is formed by the area of its biological marine reserve, which covers 7.5 km^2 (UNESCO WHC 2016). No public access is allowed, and the reserve is solely used for protection of species and scientific research. It was first charted on a sixteenth-century map, with the first detailed chart dating from 1852. The atoll remains largely undisturbed from human activity (UNEP/MCMC 2011).

The Federal Republic of Brazil is the owner of the Atol das Rocas Island. It is managed by the Chico Mendes Institute for the Conservation of Biodiversity, linked to the Ministry of the Environment (ICMBio 2016a).

32.2 Natural Settings

32.2.1 Climatic Conditions

The climatic conditions are those of a drier tropical climate with strong oceanic influences. There are two well-defined seasons, a dry season and a rainy season (Marques et al. 2007).

The rainy season goes from February to July. The annual average precipitation is 1300 mm. The heaviest rains come between March and June, sometimes reaching 190 mm in 24 hours (Fávaro et al. 2006).

The mean annual temperature is 25.4 °C. The dominant winds come from SE, with average speeds of 6.6 m/s. Winds are stronger between July and August (Manso et al. 2011).

32.2.2 Geology and Geomorphology

The archipelago of Fernando de Noronha is part of the remains of an extinct volcanic ocean building, which develops from the surface to marine depths of 4000 m (Almeida 2000). The archipelago is made up of a substrate of pyroclastic rocks deposited in subaerial environments, crossed by a wide range of eruptive rocks which, after a hiatus, were covered by basaltic lava spills (Manso et al. 2011).

In addition to the volcanic rocks, sedimentary rocks are also found, originated by the erosion of the sedimentary volcanic edifice. These rocks represent only 7.5% of the total area of the archipelago (Almeida 1955). The origin of the Fernando de Noronha volcano has been estimated between 1.8 and 12.3 million years (Almeida 2000).

The coastline of Fernando de Noronha Island is complex, with crests and high cliffs (Fig. 32.2). The altitude goes from sea level to 323 m (UNEP/WCMC 2011). The north-facing shores are relatively calm, while the south shores face the predominant currents and winds and are largely rocky shores with significant wave action (Manso et al. 2011).

The Atol das Rocas is one of the smallest atolls in the world (Soares 2009). It developed over the past 7000 years, on the flattop of a seamount, in the same submarine mountain that forms Fernando de Noronha Archipelago (Almeida 1955). It probably developed above a karstic surface of Pleistocene age, where depth ranged from 15 to 30 m (Kikushi 1994).

It is an atoll with elliptical shape. Its greater axis, oriented EW, is about 3.7 km long, and the minor axis, oriented NS, is about 2.5 km long. Despite its small dimensions, the reef front, reef flat, and a lagoon can be clearly distinguished and subdivided into discrete features (Pereira et al. 2013; Kikushi 1999). These features are affected by bioerosion, bioturbation, cementation of the biogenic carbonate sand, production of biodetritic sediments, and bioconstruction of calcareous algae, vermetids, corals, and foraminifers (Soares 2009).

Birds, fish, turtles, corals, zoanthids, algae, mollusks, polychaetes, sipunculids, foraminifers, and microorganisms generate the unique biogeological characteristics of the atoll (Soares 2009). A shallow lagoon, with maximum depth of 6 m, is seen on the northeastern side of the atoll. It connects with the open sea through the northern channel (Kikushi 1994). Cays and sandy beaches also occur (Pereira et al. 2010) (Fig. 32.3). Atol das Rocas at low tide provides a spectacular seascape of lagoon and tidal pools with fish, shellfish, sponges, mollusks, and corals (UNESCO WHC 2016).

Fig. 32.2 Volcanic rocks create peaks and cliffs with sandy pocket beaches in the Peak Beach, Fernando de Noronha Island. (Picture reproduced by permission of Kcris Ramos/Shutterstock Stock Photos)

Fig. 32.3 Vegetated cay with sandy beaches in the Rocas Atoll, equatorial South Atlantic. (Picture reproduced by permission of Dr. Valdir do Amaral Vaz Manso, Federal University of Pernambuco)

32.2.3 *Biodiversity*

Fernando de Noronha Archipelago is rich in endemism but very poor in relation to the diversity of its flora and fauna, as might be expected from its isolation (Batistella 1996).

The original vegetation of the islands has clear affinities with the highly threatened Atlantic rainforest. It has been termed insular Atlantic rainforest, but only 5% of its original cover remains (UNEP/WCMC 2011). A small area of mangroves occurs, representing the only oceanic mangrove community in the South Atlantic. More than 400 plant species have been recorded, including 3 endemic species (UNEP/WCMC 2011).

Fernando de Noronha is one of the most important areas for the reproduction of seabirds in both the North and South Atlantic Ocean (ICMBio 2016a, b). It is also considered a nursery for threatened animal groups such as cetaceans and chelonians (ICMBio 2016b). A resident population of the spinner dolphin, numbering up to 1200 individuals, is present in the surrounding waters. Some 95 fish species have been recorded, including 2 endemic ones (UNEP/WCMC 2011).

The vegetation in Rocas Atoll is herbaceous and salt resistant. A few individuals of coconuts and casuarinas were introduced by fishermen (ICMBio 2016a). It hosts more than 150,000 seabirds. Reef-building corals are widespread (UNEP/WCMC 2011).

32.3 Conservation Challenges

The number of tourists is a concern in Fernando de Noronha Island, considering its pristine environment and its limited amount of freshwater (Montenegro et al. 2008). By law, the number of tourists allowed per day is limited to 450, but it seems that there are in average 1400 in high peak seasons (Ministério do Meio Ambiente 2016).

Tourism has been causing negative impacts (Feitosa and Gomez 2013), such as trampling of the soil, introduction of nonnatives species, and biological disturbances (ICMBio 2016b). Pollution is also a major concern, considering that there is a lack of elementary public services, such as effective waste treatment (Reis and Hayward 2013).

Other threats are overfishing in both Fernando de Noronha and Atol das Rocas islands and degradation in the marine environment due to exotic species (IUCN 2017; UNESCO WHC 2016) and climate changes. Global warming can be deadly to corals, due the elevation of the sea level, and also may alter weather conditions, increasing the dry season and the deficit of water in the volcanic islands.

Despite these threats, the International Union for Conservation of Nature (IUCN), in its last conservation outlook related to World Heritage sites, found that the general level of threat to the property is relatively low (Osipova et al. 2017). Nevertheless, they also found that resources and monitoring tools are insufficient to control the threats to the property's values (IUCN 2017). The state of conservation of the site was found to be good, with some concerns (Osipova et al. 2017).

References

Almeida, FFM (2000). Arquipelago de Fernando de Noronha. Registro de um monte vulcanico do Atlantico Sul. In: Schobbenhaus C, Campos DA, Queiroz ET, Winge M, Berbert-Born M (eds), Sitios Geológicos e Paleontológicos do Brasil. http://sigep.cprm.gov.br/sitio033english.htm. Accessed 23 Oct 2016

Almeida FFM (1955) Geologia e Petrologia do Arquipélago de Fernando de Noronha. Departamento Nacional de Produção Mineral/Diretoria de Geologia e Recursos Minerais, Monografia XIII, Rio de Janeiro, Brazil

Batistella M (1996) Especies vegetais dominantes do Arquipelago de Fernando de Noronha: grupos ecologicos e reparticao especial. Acta Botanica Brasileira 10(2):223–235

Fávaro DIT, Oliveira SMB, Damatto SR, Menor EA, Moraes AS, Mazzilli BP (2006) Caracterização química e mineralógica do material inconsolidado do fundo da Lagoa da Viração, Arquipélago de Fernando de Noronha, Pernambuco. Geologia USP 5(2):1–11

Feitosa MJS, Gomez CRP (2013) Aplicação do Tourism Ecological Footprint Method para avaliação dos impactos ambientais do turismo em ilhas: um estudo em Fernando de Noronha. Revista Brasileira de Pesquisa em Turismo 7(2):220–238

Fernando de Noronha Official Site (2016) Fernando de Noronha. http://www.noronha.pe.gov.br. Accessed 23 Oct 2016

IMCBio (Instituto Chico Mendes para Conservação da Biodiversidade) (2016a) Rebio Atol das Rocas. Rapport, Brasília, Brazil

IMCBio (Instituto Chico Mendes para Conservação da Biodiversidade) (2016b) Plano de Manejo da APA de Fernando de Noronha. http://www.icmbio.gov.br/portal/images/stories/servicos/_APA_Fernando_de_Noronha_-_Encarte_3.pdf. Accessed 24 Oct 2016

IUCN (Internation Union for Conservation of Nature) (2017) Brazilian Atlantic Islands Fernando de Noronha and Atol das Rocas. http://www.worldheritageoutlook.iucn.org/explore-sites/wdpaid/900631. Accessed 19 Dec 2017

Kikushi RKP (1999) Rocas Atoll, Southwestern Equatorial Atlantic, Brazil. I: Schobbenhaus C, Campos DA, Queiroz ET, Winge M, Berbert-Born M (eds), Sitios Geológicos e Paleontológicos do Brasil. http://sigep.cprm.gov.br/sitio033english.htm. Accessed 23 Oct 2016

Kikushi RKP (1994) Geomorfologia, Estratigrafia e Sedimentologia do Atol das Rocas(Rebio/Ibama/RN), Atlântico Sul Ocidental Equatorial. Dissertação de Mestrado, Universidade Federal da Bahia, Salvador

Manso VMV, Menor EA, Valenca LMM, Neumann VHM, Pereira NS, Junior CFAS, Silva ER (2011) Morfodinâmica de Praias Setentrionais da Ilha de Fernando do Noronha. Revista da Gestão Costeira Integrada 11(3):327–339

Marques FA, Ribeiro MR, Bittar SMB, Tavares-Filho AN, Lima JFWF (2007) Caracterização e classificação de Neossolos da Ilha de Fernando de Noronha (Pe). Revista Brasileira de Ciências do Solo 31(6) https://doi.org/10.1590/S0100-06832007000600032

Ministério do Meio Ambiente (2008) Estudo e determinação da capacidade de suporte e seus indicadores de sustentabilidade com vistas à implantação do plano de manejo da área de proteção ambiental do arquipélago de Fernando de Noronha. Ministério do Meio Ambiente, Brasília

Ministério do Meio Ambiente (2016) Governo criara mosaico de UCs marinhas. www.mma.gov.br/index.php/comunicacao/agencia-informa?view=blob&id=2810. Accessed 12 June 2016

Montenegro SMGL, Montenegro AAA, Silva VL, Sobrinho MAM, Correa MM (2008) Water quality evaluation of the springs in Fernando de Noronha Island. Ambiente e Agua 3(3):114–127

Osipova E, Shadie P, Zwahlen C, Osti M, Shi Y, Kormos C, Bertzky B, Murai M, Van Merm R, Badman T (2017) IUCN world heritage outlook 2: a conservation assessment of all natural world heritage sites. IUCN Gland

Pereira NS, Manso VAV, Macedo RJA, Dias JMA, Silva AMC (2013) Detrital carbonate sedimentation of the Rocas Atoll, South Atlantic. Annais da Academia Brasileira de Ciências 85. https://doi.org/10.1590/S0001-37652013000100005

Pereira NS, Manso VAV, Silva AMC (2010) Mapeamento geomorfologico e morfodinamica do Atol das Rocas. Revista de Gestao Costeira Integrada 10:331–345

Reis A, Hayward P (2013) Pronounced particularity: a comparison of governance structures on Lord Howe Island and Fernando de Noronha. I Stud J 8(2):285–298

Soares MO (2009) Geobiologia do Atol das Rocas. Tese de Doutorado. Departamento de Geociências, Universidade Federal do Rio Grande do Sul, Porto Alegre

UNEP/WCMC (United Nations Environmental Programme/World Conservation Monitoring Centre) (2011) Brazilian Atlantic Islands, Brazil. Report, Cambridge, England

UNESCO WHC (United Nations Educational, Scientific and Cultural Organization World Heritage Centre) (2016) Brazilian Atlantic Islands: Fernando de Noronha and Atol das Rocas Reserves. http://whc.unesco.org/en/list/1000. Accessed 26 Oct 2016

Chapter 33
Gros Morne National Park, Canada

Abstract Gros Morne National Park, on the west coast of Newfoundland Island, Eastern Canada, illustrates the process of plate tectonics and continental drift. In the area, deep ocean crust and rocks of the Earth's mantle are exposed, contributing to the understanding of the geological evolution of ancient mountain belts. Shaped by crusher glaciers, the landscape is an outstanding wilderness environment of spectacular freshwater fjords and glacier-scoured headlands in an ocean setting with coastal features such as beaches, dunes, bogs, forests, barren cliffs, alpine plateau, waterfalls, and pristine lakes. The evolution of the North Atlantic basin and glacial activities can be determined from the ancient landscape. There is also an unusually complete paleontological sequence which has been proposed as the world stratotype for the Cambrian-Ordovician boundary.

33.1 Introduction

Gros Morne National Park is located on the western coast of Newfoundland Island, 150 km from the mainland. It faces the Gulf of Saint Lawrence, in Eastern Canada, and has an area of 1805 km^2 (Fig. 33.1).

The park was established in 1973, under a federal/provincial agreement between the governments of Canada and of Newfoundland and Labrador. It was formally established under the National Parks Act in 1986 (UNEP/WCMC 2005). The park became a World Heritage site in 1987 (UNESCO WHC 2016) and is managed by Parks of Canada. The land tenure pertains to the federal government (Parks of Canada 2016).

There are a number of archaeological sites in the park, where human habitation can be traced back to 4500 years ago. Europeans settled there in the late eighteenth century, developing fishing and commercial logging (UNEP/WCMC 2005). Despite developments and modernization, the area retains a distinctive language and cultural tradition (Moorhead 1971). Today, they are no residents in the park (UNESCO WHS 2016). Its boundaries exclude eight coastal settlements, where the population carries out fishing, subsistence agriculture, logging, hutting, and tourism (UNEP/WCMC 2005).

V. Claudino-Sales, *Coastal World Heritage Sites*, Coastal Research Library 28,
https://doi.org/10.1007/978-94-024-1528-5_33

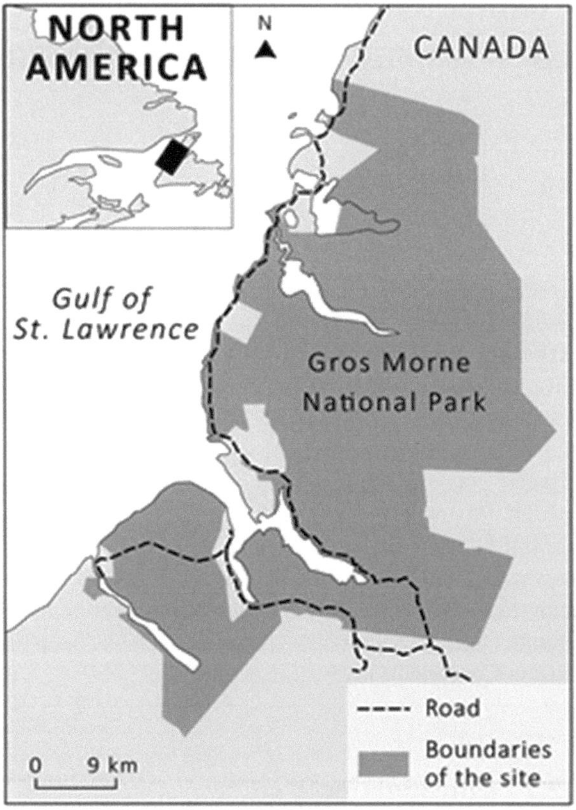

Fig. 33.1 Location of Gros Morne National Park, Eastern Canada

Visitors can hike through wild, uninhabited mountains and camp by the sea. Boat tours bring visitors under the towering cliffs of a freshwater fjord carved out by recent glaciers. Watching wild animals is also part of a visit to the national park (Parks of Canada 2016). A number of campsites are located in the park, and hotels and other services are available in the adjacent communities (UNEP/WCMC 2005).

33.2 Natural Settings

33.2.1 Climatic Conditions

The climate in Gros Morne is cool, being a wet maritime type at sea level and subarctic at higher altitudes. It is influenced by the adjacent ocean and by the strong southwesterly winds from the Gulf of St. Lawrence (Bouchard and Hay 1976). The weather is very unstable, and there are great differences between the conditions in the coastal plain and in the higher elevations of the Long Range Mountains (Parks of Canada 2016).

The mean annual temperature is 3 °C, with a maximum of 15.5 °C in July and minimum of −8.4 °C in February (UNEP/WCMC 2005). Mean annual precipitation is 1397 mm of rain and 3281 mm of snow, with 10–30 days of fog (Moorhead 1971). Sea ice forms in winter and spring and autumn are delayed by the oceanic influence (Parks of Canada 2016).

33.2.2 Geology and Geomorphology

Gros Morne National Park comprises two distinct physiographic components: coastal lowland and the frontal slopes and upland areas of the Long Range Mountains, which has a maximum altitude of 806 m (UNEP/WCMC 2005).

The park illustrates some of the world's best examples of plate tectonics. Within a relatively small area are classic situations related to earth-building forces (UNESCO WHS 2016), with examples that took place in several stages of the geological table time and of the cycle of Wilson (Miall 2011).

Around 1.2 billion years ago, during the Precambrian, the ancient core of what is now eastern North America collided with another continent, forming a vast mountain range. All that remains today of this mountain chain are the deeply eroded granites and gneisses of the Long Range Mountains (Parks of Canada 2010).

By 570 million years ago, in the Late Precambrian, the supercontinent rifted apart, and the resulting basin became the Iapetus Ocean. Some of the rocks of Gros Morne National Park were part of the continental margin on the western side of this new ocean (Parks of Canada 2010). About 500 million years ago, the Iapetus Ocean began to close along a subduction zone. Some of the oceanic crust became trapped within the subduction zone and then thrusted slowly over the continental margin (Parks of Canada 2016). The tablelands present in the Gros Morne National Park are formed by these ophiolites (Szponar et al. 2013).

Continental collision continued for 100 million years, creating the Appalachian Mountains, which border the Atlantic Ocean in eastern North America, including the Newfoundland and Labrador Province (Parks of Canada 2010). Since the Devonian (360 million years), the area has remained relatively stable, only suffering erosion and some uplift. The last two million years was characterized by repeated glaciation and deglaciation and associated sea level changes, which shaped the park scenery that we see today (Parks of Canada 2010) (Fig. 33.2).

The glaciation left alpine plateau and deposits on the flanks of the tablelands, as well as cirque and moraines (Osborn et al. 2007). It also resulted in spectacular scenery, with fjords, freshwater fjords, glacial valleys, waterfalls, pristine lakes, and seacliffs (UNESCO WHC 2016).

The coastal plain is bordered to the east by the Long Range fault scarp and extends inland for 13 km and along the coast for about 55 km. It is composed largely of limestone and slopes gently seaward in a series of steps, defined by faulting in the underlying rock (Moorhead 1971) (Fig. 33.3).

Fig. 33.2 Boulders produced by glaciers at the foot of faulted cliffs on the boundary of a wetland in Gros Morne National Park. (Picture reproduced by permission of Judy McIlrath, University of South Florida)

Fig. 33.3 Sheer seacliffs modeled in limestones in the lowland coast of Gros Morne National Park, Eastern Canada. The area is important to seabirds, as seen in the picture. (Picture reproduced by permission of Judy McIlrath, University of South Florida)

The freshwaters range from torrential streams and waterfalls in the mountains to ponds and deep lakes in the lowlands. There are acidic waters in the crystalline areas and basic and high in Ca^{+} waters in the limestone areas (Steele and Green 1973).

The park is lithologically diverse, with areas of Ordovician sedimentary rocks overlying Precambrian granite and gneiss, Paleozoic gabbros, and volcanic and lower Paleozoic sedimentary rocks. Exposed oceanic crust and mantle, as well as a section of ancient Mohorovicic discontinuity, are visible (UNEP/WCMC 2005).

There is also an unusually complete paleontological sequence which has been proposed as the world stratotype for the Cambrian-Ordovician boundary (Parks of Canada 2010).

33.2.3 Biodiversity

The great variety in topography and geological structure within a relatively small area provides habitats for a great variety of plants and vegetation (Fernald 1933). The vegetation ranges from tundra on the tops of the mountains through persistent snow patches and talus slopes on the flanks of the hills to forest, bog, and fen in the coastal lowlands (Steele and Green 1973).

Thirty-six distinct vegetation types and communities are present, with 711 vascular species and 401 bryophytes. It represents about 60% of Newfoundland's insular flora. There are also over 400 species of lichens (UNEP/WCMC 2005).

The glaciations periodically scoured the area of its plant and animal life. The animals that are now considered native made their way over in the last 1500 years, coming from the mainland by swimming, flying, or walking across the pack ice (Parks of Canada 2010). Since the arrival of Europeans, there have been both deliberate and accidental introductions of animal species. Even today, new species are being brought to the island or find their own way across the water and ice (Parks of Canada 2010).

In the Newfoundland and Labrador Province, faunal diversity is markedly reduced compared to the mainland (Moorhead 1971). There are 30 species of mammals, such as moose, brown bear, polar bear, whales, and seals. The avifauna comprises 239 arctic, boreal, and pelagic species, with strays from the mainland, the northwest Atlantic, and Europe (UNEP/WCMC 2005).

Newfoundland waters harbor no true freshwater fish. Of the fish species found on the island, most spend at least part of their lives in saltwater. Common species are Atlantic salmon, trout, and Arctic char (Parks of Canada 2010).

33.3 Conservation Challenges

The geological features and the glacially derived topography are in near-pristine conditions in Gros Morne National Park. There is effective legislation to protect the property from development and to accommodate visitors without adverse effects (UNESCO WHC 2016).

Nevertheless, a moose population introduced over 100 years ago has expanded and degrades native forests, creating visual and ecological impacts in the park (UNESCO WHC 2016). Climate change and global warming can also be a threat, with erosion increasing in the seacliffs due the rising of sea level and changes in the climate, perhaps altering the glacial marks on the rocks by weathering. Changes in the ecosystems and hydrology are also expected with global warming.

The International Union for Conversation of Nature (IUCN) published a conservation outlook on the World Heritage sites at the end of 2017, in which it found that in general, the state of conservation of the site is good, with some concerns about the native forest, the management of snowmobiling in the park, and the ongoing potential threat from oil and gas development in the region. While the park's outstanding universal geological values and the majority of scenic values are not being impacted at this point in time, biodiversity values remain at risk (IUCN 2017; Osipova et al. 2017).

References

Bouchard A, Hay S (1976) The vascular flora of the Gros Morne National Park coastal plain in Newfoundland. Rhodora 78(814):207–260

Fernald ML (1933) Recent discoveries in the New-foundland flora. Rhodora 35:47–63

IUCN (Internationl Union for Conservation of Nature) (2017) Gros Morne National Park Conservation Outlook. http://www.worldheritageoutlook.iucn.org/explore-sites/wdpaid/17759. Accessed 19 Dec 2017

Miall AD (2011) A Phanerozoic time chart for Canada. Geosci Can 38(3):13

Moorhead S (1971) Gros Morne National Park Study. Report No. 1. Analysis of Existing Factors and Constraints. Strong Moorhead Sigsby Ltd, Toronto, Ontario

Osborn G, Spooner I, Gosse J, Clark D (2007) Alpine glacial geology of the Tablelands, Gros Morne National Park, Newfoundland. Can J Earth Sci 44(6):819–834

Osipova E, Shadie P, Zwahlen C, Osti M, Shi Y, Kormos C, Bertzky B, Murai M, Van Merm R, Badman T (2017) IUCN world heritage outlook 2: a conservation assessment of all natural world heritage sites. IUCN Gland

Parks of Canada (2010) Gros Morne National Park of Canada. http://www.pc.gc.ca/eng/index.aspx. Accessed 14 Nov 2016

Parks of Canada (2016) Gros Morne National Park. https://www.pc.gc.ca/en/pnnp/grosmorne. Accessed 22 June 2016

Szponar N, Brazelton WJ, Schrendk MO, Bower DM, Steele A, Morril PL (2013) Geochemistry of a continental site of serpentinization, the Tablelands Ophiolite, Gros Morne National Park: a Mars analogue. Icarus 224(2):286–296

Steele DH, Green JM (1973) Gros Morne National Park, Newfoundland. Biol Conserv 4(5):392–394

UNEP/WCMC (United Nations Environmental Programme/World Conservation Monitoring Centre) (2005) Gros Morne National Park Newfoundland and Labrador Canada. Report, Cambridge, England

UNESCO WHC (United Nations Educational, Scientific and Cultural Organization World Heritage Centre) (2016) Gros Morne National Park. http://whc.unesco.org/en/list/419. Accessed 14 Nov 2016

Chapter 34
Ilulissat Icefjord, Denmark

Abstract Ilulissat Icefjord is on the west coast of Greenland, 250 km north of the Arctic Circle. It is a tidal fjord covered with massive floating ice, an outstanding example of an actively calving ice sheet. It is one of the fastest and most productive ice streams in the world, moving at 40 m a day and annually calving over 35 km^3 of ice. The Greenland ice cap, from which Ilulissat Icefjord is one of the most impressive examples, is the only remnant in the Northern Hemisphere of the continental ice sheets from the last Quaternary Ice Age. The oldest ice there is estimated to be 250,000 years and is known for providing important information on past climate. The combination of ice, rock, and sea, with the dramatic sounds produced by the moving ice, is a memorable spectacle.

34.1 Introduction

Ilulissat Icefjord is 250 km inside the Arctic Circle, approximately 1000 km up the west coast of Greenland, in the Bay of Disko (Fig. 34.1).

The World Heritage site has an area of 4024 km^2, with 3199 km^2 of glacier ice, 397 km^2 of land, 386 km^2 of fjord, and 42 km^2 of lakes. The site encloses the fjord, the glacier, and its immediate watershed (UNEP/WCMC 2005).

The area is protected and conserved by an established framework of government legislation and protective designations and by local planning policies. The first preservation legislation dates from 1980. The land tenure is public, and the area is managed by the Municipality of Ilulissat, Greenland (UNEP/WCMC 2005). The icefjord and its ensemble became a World Heritage site in 2004 (UNESCO WHC 2016).

Greenland has been inhabited for 4500 years, and the World Heritage property includes archaeologically valuable sites. Today, the local people are Inuit, and their economy is almost entirely dependent on fishing and hunting. Nevertheless, there are no inhabitants living within the boundaries of the site. The government discouraged tourism until the mid of the twentieth century. It has only been actively promoted since the beginning of the years 1990s (UNEP/WCMC 2005).

V. Claudino-Sales, *Coastal World Heritage Sites*, Coastal Research Library 28,
https://doi.org/10.1007/978-94-024-1528-5_34

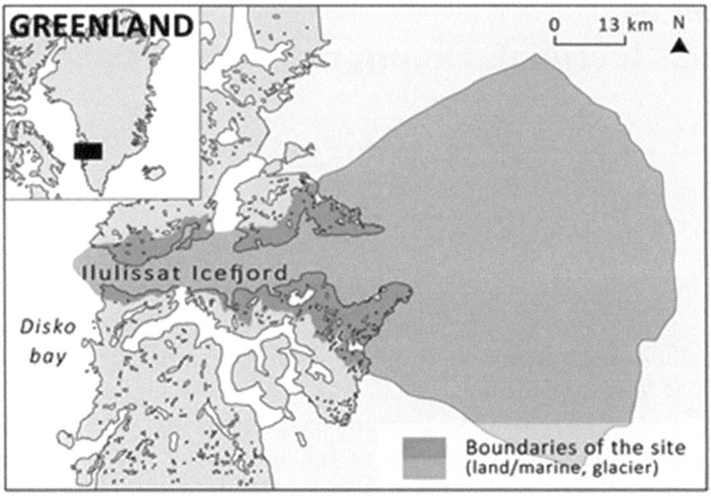

Fig. 34.1 Location of Ilulissat Fjord, in the Bay of Disko, at the Municipality of Ilulissat, Greenland, Denmark

The scenery, the Inuit culture, and the history attract a third of all Greenland's visitors to the Ilulissat and Disko areas. However, it is the icefjord that is the main attraction (UNEP/WCMC 2005).

Long-standing traditions of scientific research in unspoiled natural environments, historic expeditions, and public debates on climate change have directed much attention to the uniqueness of the polar regions, and the tourism that now takes place in Ilulissat is an example of this (Smed 2014). Ilulissat offers easy access for a close view of the calving glacier front. It has been studied for more than 250 years, which has helped develop an understanding of climate change and ice cap glaciology (UNESCO WHC 2016).

34.2 Natural Settings

34.2.1 Climatic Conditions

The climate is cold and continental (Mikkelsen and Ingerslev 2004). The maximum temperature is 8 °C, reached in August. The minimum temperature is −18 °C, in February. The precipitation is around 300 mm/year, with about 25.4 cm/year as snow. The average wind speed is about 4.1 m/s (Weather and Climate 2016).

Calm days are quite frequent, especially during the summer, but the storms can be violent. Stormy weather occurs in connection with passages of low pressure systems, or in connection with føhn winds, which are relatively warm and dry winds with a westerly direction from the inland ice to the outer coast (Mikkelsen and Ingerslev 2004).

Fig. 34.2 Ilulissat Glacier reaching the sea and calving a large amount of ice, creating an important drain for the Greenland continental ice cap. (Picture reproduced by permission of Els Slots, webpage http://www.worldheritagesite.org)

The area has sunless winters and nightless summers for 2–3 months (UNEP/WCMC 2005). The average sea surface temperature in Ilulissat in recent years is 0.6 °C (Sea temperature 2016).

34.2.2 Geology and Geomorphology

The most productive outlet of the inland ice in Greenland is the Ilulissat Glacier (Schumann et al. 2012). Ilulissat Glacier is a continuously fast-moving surge-type glacier that presently drains about 7% of the Greenland ice sheet (Rignot and Kanagaratnam 2006), which represents 35 km^3 of ice being calved annually (UNESCO WHC 2016).

The Ilulissat Icefjord is the sea mouth of Sermeq Kujalleq, one of the few glaciers through which the continental ice of the Greenland ice cap reaches the sea (Mikkelsen and Ingerslev 2004). It is the fastest-moving glacier in the world outside Antarctica and the most prolific ice-calving tidewater glacier in Greenland, producing a constant procession of icebergs (UNEP/WCMC 2005) (Figs. 34.2 and 34.3).

The Ilulissat Glacier has eroded a subglacial trough that is incised 1500 m into the base rock on land (Clarke and Echelmeyer 1996). The icefjord is the seaward continuation of this trough with a maximum depth of around 1000 m (Schumann et al. 2012).

Fig. 34.3 Icebergs resulting from the Ilulissat Glacier calving, in the mouth of the Ilulissat Icefjord. (Picture reproduced by permission of Judy McIlrath, University of South Florida)

The glacial history of the Greenland ice sheet shows an uplift rate of the continental crust for Ilulissat of +1.6 mm/year, in response to glacial melting adaptations of the lithosphere since the end of the last Quaternary Ice Age (Dietrich et al. 2015).

West Greenland is dominated by Precambrian crystalline basement consisting of rocks such as gneiss, granite, and schist, which represent old fold belts. The bedrock of the Ilulissat Icefjord area consists of gray gneiss with minor intercalations of amphibolite (Mikkelsen and Ingerslev 2004).

The altitude of the ice sheet contour is around 1200 m (UNEP/WCMC 2005). Large well-developed moraine ridges can be seen winding over the landscape in the World Heritage site, and there are also kame terraces, "roches moutonées," marginal deltas, and glacial lakes (Mikkelsen and Ingerslev 2004).

34.2.3 Biodiversity

The flora of the area is a low-arctic type, typical of the nutrient-poor silicaceous soil. Colonization of the margins of retreating ice also provides examples of plant succession. The main plant communities are heath, fell field, snow patch, herb slope, willow scrub, fen, riverbank, seashore, and aquatic (UNEP/WCMC 2005).

The marine fauna associated with Ilulissat Icefjord is also linked to the glacial conditions. The icebergs at the mouth of the fjord and the high tides (ca. 2 m) generate an environment with strong turbulence and upwelling masses. In addition, subglacial erosion adds nutrients to the water. These factors attract a number of predators such as big fish, seals, and whales (Mikkelsen and Ingerslev 2004).

The terrestrial fauna in the site is basically formed by birds and mammals. The land birds are rather few and occur in low densities, and the seabirds occur mainly in areas with open water (Mikkelsen and Ingerslev 2004). The most common mammals are the Arctic fox, the Arctic hare, and the reindeer (Meldgaard 1986).

34.3 Conservation Challenges

Generally, the conservation of the protected area is good, and there are no serious problems regarding the World Heritage values of the site. However, climate change and some other threats could seriously affect the site's values in the future (IUCN 2014).

Freshwater runoff from the Greenland ice sheet has increased since 1960 (Mernild et al. 2015). Atmospheric warming explains half of this contemporary mass loss (Enderlin et al. 2014). The other half is due to the calving of icebergs.

The contribution from the calving of icebergs to mass loss has accelerated in recent decades because of the rapid ice retreat, ice speedup, and dynamic thinning of outlet glaciers (van den Broeke et al. 2009). It appears that the ice area reduction and ice speedup were initiated by a switch to higher ocean temperatures along the west Greenland coast in the late 1990s (Hansen et al. 2012; Rignot et al. 2012) and to changes in climate (Hanna et al. 2013).

Besides climate changes, there are other threats, such as the impact from tourism, fishing, boat traffic, and helicopter flights (IUCN 2017). For these reasons, the conservation outlook about World Heritage sites published by the International Union for Conservation of Nature at the end of 2017 (Osipova et al. 2017) considered that the state of conservation of Ilulissat Icefjord is good but presents some concerns.

References

Clarke TS, Echelmeyer K (1996) Seismic-reflection evidence for a deep subglacial trough beneath Jakobshavns Isbrae, West Greenland. J Glaciol 43(141):219–232

Dietrich R, Rulke A, Scheinert M (2015) Present-day vertical crustal deformations in West Greenland from repeated GPS observations. Geophys J Int 163:865–874

Enderlin EM, Howat IM, Jeong S, Hoh MJ, van Angele JH, vanden Broeke MR (2014) An improved mass budget for the Greenland ice sheet. Geophys Res Lett 41:866–872

Hanna E, Jones JM, Cappelen J, Mernild SH, Wood L, Steffen K, Huybrechts P (2013) The influence of North Atlantic atmospheric and oceanic forcing effects on 1900–2010 Greenland summer climate and icemelt/runoff. Int J Climatol 33:862–880

Hansen M, Nielsen T, Stedmon C, Munk P (2012) Oceanographic regime shift during 1997 in Disko Bay, western Greenland. Limnol Oceanogr 57:634–644

IUCN (International Union for Conservation of Nature) (2014) Ilulissat state of conservation. Rapport, Gland, Switzerland
IUCN (International Union for Conservation of Nature) (2017) Ilulissat Icefjord Conservation Outlook. http://www.worldheritageoutlook.iucn.org/explore-sites/wdpaid/902373. Accessed 19 Dec 2017
Meldgaard M (1986) The Greenland caribou – zoogeography, taxonomy, and population dynamics. Bioscience 20:88–98
Mernild SH, Holland DM, Holland D, Rosing-Asvid A, Yde JC, Liston GE, Teffen KS (2015) Freshwater flux and spatiotemporal simulated runoff variability into Ilulissat Icefjord, West Greenland, linked to salinity and temperature observations near Tidewater glacier margins obtained using instrumented ringed seals. J Phys Oceanogr 45:1426–1446
Mikkelsen N, Ingerslev T (eds) (2004) Nomination of Ilulissat Icefjord for inclusion in the World Heritage List. Geological Survey of Denmark and Greenland, Copenhagen
Osipova E, Shadie P, Zwahlen C, Osti M, Shi Y, Kormos C, Bertzky B, Murai M, Van Merm R, Badman T (2017) IUCN world heritage outlook 2: a conservation assessment of all natural world heritage sites. IUCN, Gland
Rignot E, Kanagaratnam P (2006) Changes in the velocity structure of the Greenland ice sheet. Science 311:986–988
Rignot E, Fenty I, Menemenlis D, Xu Y (2012) Spreading of warm ocean waters around Greenland as a possible cause for glacier acceleration. Ann Glaciol 53(60):257–266
Schumann K, Völker D, Weinrebe WR (2012) Acoustic mapping of the Ilulissat Ice Fjord mouth, West Greenland. Quat Sci Rev 40:78–88
Sea temperature (2016). Sea superface temperature in Ilulissat. http://seatemperature.net/current/greenland/ilulissat-qaasuitsup-greenland-sea-temperature. Accessed 16 Nov 2016
Smed KM (2014) The culture of nature: destination visitability in Ilulissat, Greenland. North Rev 38:269–284
UNEP/WCMC (United Nations Environmental Programme/World Conservation Monitoring Centre) (2005) Ilulissat Icefjord, Greenland, Denmark. Rapport, Cambridge, England
UNESCO WHC (United Nations Educational, Scientific and Cultural Organization World Heritage Centre) (2016) Ilulissat Icefjord. http://whc.unesco.org/en/list/1149. Accessed 16 Nov 2016
van den Broeke MJ, Bamber J Ettema J, Rignot E, Schrama E, van de Berg WJ, van Meijgaard E, Velicogna I, Wouters B (2009) Partitioning recent Greenland mass loss. Science 326:984–986
Weather and Climate (2016) Ilulissat. https://weather-and-climate.com/average-monthly-precipitation-Rainfall-inches,ilulissat,Greenland. Accessed 16 Nov 2016

Chapter 35
Surtsey, Iceland

Abstract Surtsey is a volcanic island approximately 32 km from the coast of Iceland. It was formed by volcanic eruptions that took place from 1963 to 1967 and has been protected since its birth. The island has been intensively studied for its geology, geomorphology, and ecology since its eruption. The main island is 90% basalt volcanic rock with 29 active volcanic systems, much hydrothermal activity, and an eruption on average every 3 years. Free from human interference, Surtsey is a pristine natural laboratory for scientists, which has been producing long-term information on the colonization process of new land by plant and animal life. It is part of the world's most active tectonic boundaries, the Mid-Atlantic Ridge between the North American and Eurasian plates.

35.1 Introduction

Surtsey is a small island lying 32 km from the mainland, the most remote of a chain of islands off the southwest coast of Iceland. The Surtsey World Heritage site includes the whole island and a surrounding marine reserve (Fig. 35.1).

Surtsey volcano is part of the Vestmannaeyjar archipelago, which consists of 18 islands and a number of skerries. In total, it has 337 km^2 of area, with 14 km^2 of land, 132 km^2 of submerged volcano, and 191 km^2 of ocean (UNEP/WCMC 2007).

Surtsey volcano emerged from the sea in November 1963, remaining active until mid-1967. In 1965, Surtsey Island and surrounding seas were declared a "Strict Nature Reserve" by the Iceland government. In 2006, the reserve was expanded to include the underwater slopes of the volcano and the submarine islets. It became a World Heritage site in 2008 (UNEP/WCMC 2007).

The land is a property of the Iceland government, and it is attached to the municipality of Vestmannaeyjar, which is located at the Heimaey Island, the largest and only inhabited island of the Vestmannaeyjar archipelago (Baldursson and Ingadóttir 2007). The World Heritage and protected site is supervised by the Surtsey Research Centre under the National Environment and Food Agency (UNEP/WCMC 2007).

The formation of a volcanic island is a rare event, occurring on the average of about twice a century. None has been as intensely investigated as Surtsey. Surtsey

V. Claudino-Sales, *Coastal World Heritage Sites*, Coastal Research Library 28,
https://doi.org/10.1007/978-94-024-1528-5_35

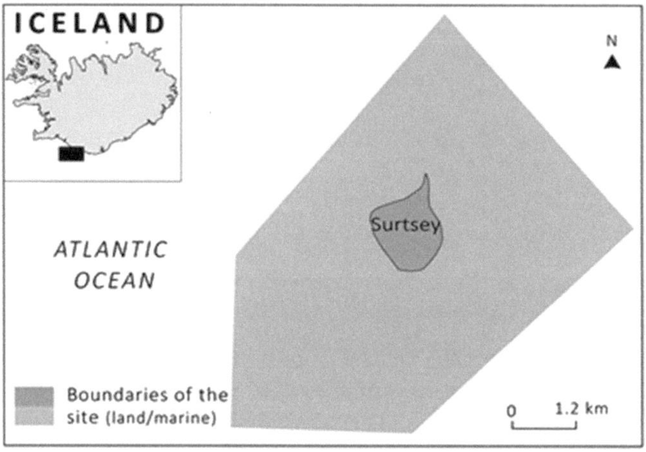

Fig. 35.1 Surtsey World Heritage site, close to the south coast of Iceland, in the Atlantic Ocean

has offered an unparalleled natural laboratory in geology, geomorphology, and ecology and has provided invaluable information on the history of the Earth and its biosphere (Baldursson and Ingadóttir 2007).

35.2 Natural Settings

35.2.1 Climatic Conditions

Surtsey enjoys a much milder climate than the location would imply: a branch of the warm Gulf Stream flows along the coast, greatly moderating the climate conditions. But the Atlantic air in contact with colder Arctic air results in a climate that is marked by frequent changes in storminess and leads to more rainfall than in the northern part of Iceland (Iceland Meteorological Office 2016).

The mean annual temperature is 5.7 °C, and the mean annual precipitation is about 2250 mm (Fridriksson 1975). Winds are strong, prevailing from the east off the mainland and of hurricane strength 15 days a year. Waves are driven from the southwest and can be very high: 16.68 m was recorded at a measuring buoy (Viggósson and Grétarsson 2010).

35.2.2 Geology and Geomorphology

Iceland is an integral part of the Mid-Atlantic Ridge and also its largest aerial component. The discovery of this ridge in the 1950s led to general acceptance of Wegener's 1915 theory of continental drift. Its volcanism results from rising of magma from a hotspot over a magmatic plume (Baldursson and Ingadóttir 2007).

Fig. 35.2 Beach formed by coarse sediments derived from volcanic rocks, colonized by Icelander coastal vascular vegetation and covered in the strand by boulders of basalt. (Picture reproduced by permission of Professor Roger del Moral, University of Washington)

The crest of the ridge, which marks the boundary between the North American and Eurasian tectonic plates, crosses Iceland from southwest to northeast, forming a major rift belt that expands continuously by accreting new land through volcanic activity (Baldursson and Ingadóttir 2007).

The eruption of Surtsey, a marine volcano built on the south insular shelf of Iceland, is a rare example of observed volcanism at this propagating rift (Schipper et al. 2016). The eruption leading to the island's emersion lasted for 3.5 years (Moore et al. 1995).

At the end of the eruption, Surtsey had attained a surface area of 2.65 km^2. The total volume of erupted material was estimated at 1.1 km^3. From this total, about 70% was fragmental material (tephra) and 30% magma (Baldursson and Ingadóttir 2007). Later, the lava evolved into palagonite, due to chemical weathering (Romagnoli and Jakobsson 2015).

The volcano formed on a short submarine ridge, rising 285 m above the former seabed to 155 m above sea level. Its size is approximately 1.33 km by 1.8 km, formed of crescentic tuff cones; two craters and a wide apron of lava to their south; gullied northern talus slopes of tephra, aeolian sand, and debris; and a long spit of lava boulders and coastal sediments to the north (UNEP/WCMC 2007) (Fig. 35.2).

Destructive forces have been active in Surtsey for more than 50 years, from the beginning to present day. These forces produced rapid subaerial and submarine erosion and major morphological changes (Baldursson and Ingadóttir 2007).

Fig. 35.3 Steep seacliffs modeled in basalts on Surtsey Island, with nests of seagulls occupying the reentrances of the cliff. (Picture reproduced by permission of Professor Roger del Moral, University of Washington)

The original area of the island has been greatly reduced to 1.31 km^2 in 2012, and the position of the shoreline has changed considerably due to combined effects of volcanic activity (during 1963–1967) and coastal processes (Romagnoli and Jakobsson 2015).

Coastal processes also formed dunes and soil in the island. In the case of the soil, it also counted with the action of biogenic processes producing organic matter (Stefansdottir et al. 2014).

35.2.3 Biodiversity

When Surtsey rose from the North Atlantic Ocean, it became a unique natural laboratory on how organisms form ecosystems with contrasting structures and functions (UNESCO WHC 2016).

The studies showed that some species benefit from the formation of new habitats (Ingimundardóttir et al. 2014). That is the case of the evolution of the vegetation cover on the island, strongly related to the density of gull nests (Ingimundardóttir et al. 2014). Actually, the amount of nutrients brought onto the island by the seagulls was the critical factor that most influenced ecosystem fluxes and vegetation development on Surtsey (Sigurdsson and Magnusson 2010) (Fig. 35.3).

Nevertheless, except for seaborne pioneers, the development of vegetation in the subarctic climate of Surtsey has been slow, and it remains thin except around the colony of gulls (Baldursson and Ingadóttir 2007). The first gulls landed 2 weeks after land first appeared, and birds have been annually monitored ever since. Eighty-nine species of birds have been recorded: 45 seabirds and 44 terrestrial species (UNEP/WCMC 2007).

Three hundred and five species of invertebrates are recorded, 26% of Iceland's total. A hundred and seventy-four of these species arrived in the first 10 years. Of marine mammals, seals visited early, and by 1983 gray seals were breeding on the northern spit. Whales and dolphins have been seen offshore (UNEP/WCMC 2007).

Marine species have accrued as planktonic algae or as benthic larvae. In the benthos about 80 species of macroalgae grow, and 180 benthic animals have been found in the rocky littoral and sublittoral zones. Along the shores, algae have 60% coverage (UNEP/WCMC 2007).

35.3 Conservation Challenges

Surtsey is a highly controlled and isolated environment, so threats are very limited. The purpose of strictly prohibiting visits is to insure that colonization by plants and animals and the shaping of geological formations will be as natural as possible. The most significant management issue will be to retain the level of control that has characterized the protective history of the island (UNESCO WHC 2016).

Global warming may change the natural dynamics, by increasing the wind speeds and wave heights, as well as changing the levels of rainfall, the rate of erosion of the shore, and the newly created ecosystem dynamics.

However, under the current regime of strict protection, the site's values are well preserved. Nevertheless, other problems exist. Current threats include shipping lanes, large vessels, fishing boats, and dumping of waste at sea (UNESCO WHC 2016).

Vigilance is required to ensure that oil spills, the discharge of sewage or solid waste from visiting recreational boats, fishing boats, or cruise ships, are prevented and no harm occurs to the marine areas and natural environment surrounding Surtsey. Invasive alien species are a major potential threat to the biodiversity of the island (IUCN 2017).

But the last state of conservation outlook by the International Union for Conservation of Nature (IUCN), published at the end of 2017, found that Surtsey's conservation is good without concerns (Osipova et al. 2017).

References

Baldursson S, Ingadóttir A (eds) (2007) Nomination of Surtsey for the UNESCO World Heritage List. Institute of Natural History of Iceland, Reykjavík

Fridriksson S (1975) Surtsey evolution of life in a Volcanic Island. Butterworths, London

Iceland Meteorological Office (2016) Climate of Iceland. http://en.vedur.is/climatology/iceland. Accessed 22 Nov 2016

Ingimundardóttir GV, Weibull H, Cronberg N (2014) Bryophyte colonization history of the virgin volcanic island Surtsey, Iceland. Biogeosciences 11(16):4415–4427

IUCN (International Union for Conservation of Nature) (2017) Surtsey Conservation Outlook. http://www.worldheritageoutlook.iucn.org/explore-sites/wdpaid/903139. Accessed 20 Dec 2017

Moore JG, Sveinn J, Holmjarn JUS (1995) Subsidence of Surtsey volcano, 1967–1991. Bull Volcanol 55(1–2):17–24

Osipova E, Shadie P, Zwahlen C, Osti M, Shi Y, Kormos C, Bertzky B, Murai M, Van Merm R, Badman T (2017) IUCN World Heritage Outlook 2: a conservation assessment of all natural World Heritage sites. IUCN, Gland

Romagnoli C, Jakobsson SP (2015) Post-eruptive morphological evolution of island volcanoes: Surtsey as a modern case study. Geomorphology 250:384–396

Schipper CI, Le Voyer M, Moussallam Y, White JDL, Thordarson T, Kimura JI, Chang Q (2016) Degassing and magma mixing during the eruption of Surtsey Volcano (Iceland 1963–967): the signature of a dynamic and discrete rift propagation event. Bull Volcanol 78(4):33–45

Stefansdottir G, Aradottir AL, Sigurdsson BD (2014) Accumulation of nitrogen and organic matter during primary succession of Leymus arenarius dunes on the volcanic island Surtsey, Iceland. Biogeosciences 11(20):5763–5771

UNEP/WCMC (United Nations Environmental Programme/World Conservation Monitoring Centre) (2007) Surtsey Iceland. Report, Cambridge, UK

UNESCO WHC (United Nations Educational, Scientific and Cultural Organization World Heritage Centre) (2016) Surtsey. http://whc.unesco.org/en/list/1267. Accessed 23 Nov 2016

Viggósson G, Grétarsson S (2010) Landeyjahöfn Ferry Harbour at the South Coast of Iceland. PIANC MMX Congress: 1–20 Liverpool, UK

Chapter 36
Laurisilva of Madeira, Portugal

Abstract Madeira is a volcanic island off the Atlantic northwest coast of Africa, politically part of Portugal. It is characterized by the presence of a primary laurel forest, named Laurisilva, which covered the European continent in the Tertiary and disappeared during the Quaternary climatic changes. Madeira Island is the largest surviving example of this type of vegetation. The forest is largely undisturbed and plays a predominant role in the island's hydrological balance. It is formed by evergreen trees and bushes with dark green leaves. The Laurisilva of Madeira also has important biological diversity. A large proportion of its plants and animals are unique to the laurel forest, and it is larger than and with significant differences from other laurel forest areas in the world.

36.1 Introduction

Madeira Island is in the Atlantic Ocean, 970 km southwest of Portugal and 540 km west of Morocco. The Laurisilva of Madeira World Heritage site is along 35 km of the northern slopes of the island (Regional Government of Madeira 1998). It has 15 km^2, entirely within the 27 km^2 Madeira Nature Park (UNESCO WHC 2016) (Fig. 36.1).

Madeira Island is part of Portugal, but it is an autonomous region with its own government that can discuss and regulate many matters regarding its own territory (Baioni 2011). Hence, they could make their laws about the preservation of the Laurisilva forest.

The Madeira Nature Park was created in 1982. In 1992, the park was declared a "Council of Europe Biogenetic Reserve," a "Special Conservation Zone within the European Union Bird Habitats Directive," and a "Site of Community Interest within the European Union Habitats Directive." The World Heritage site in part of the Madeira Nature Park was declared by UNESCO in 1999 (UNEP/WCMC 2011).

The land is 10% owned by the regional government and 90% by local authority. It is managed by the regional government through the Regional Directorate of Forests under the Regional Secretariat for Agriculture, Forests and Fish (UNEP/WCMC 2011).

V. Claudino-Sales, *Coastal World Heritage Sites*, Coastal Research Library 28,
https://doi.org/10.1007/978-94-024-1528-5_36

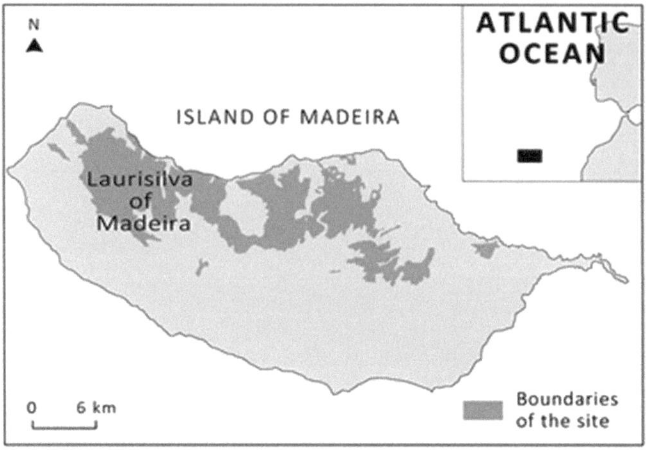

Fig. 36.1 Location of the Laurisilva of Madeira World Heritage site, in the Madeira Island, in the North Atlantic

Madeira Island was claimed by Portuguese sailors in 1419 and settled after 1420 (Regional Government of Madeira 1998). The settlers constructed water channels that run through the forest, following the contours of the landscape. They carry water from the forest to hydropower stations where they provide essential drinking water and irrigation supplies and represent an important element of the occupation of the island and of the forest (UNESCO WHC 2016).

The park lies within an internationally designated Conservation Hotspot, a World Wildlife Fund (WWF) Global 200 Eco-region, a WWF/International Union for Conservation of Nature (IUCN) Center of Plant Diversity, and a BirdLife-designated Endemic Bird Area. Tourism is an important activity on the island and in the World Heritage site (UNEP/WCMC 2011).

36.2 Natural Settings

36.2.1 Climatic Conditions

The climate is temperate and humid, moderated by the surrounding ocean and the prevailing northeasterly winds. It has a relative humidity of 85% and is often cloudy, preserving the conditions in which the Laurisilva once dominated Tertiary Southern Europe (Gouvernement Régional de Madère 1998).

The climatic conditions are determined mainly by the subtropical Azores anticyclone. During the summer, masses of tropical air reach Madeira, while in the winter, when the Azores anticyclone returns to its usual, more southerly position, Madeira is affected by the polar front from the North Atlantic (Caldera et al. 2002).

The topographical barrier formed by the main drainage divide in the center of the island creates a climatic difference between the southern and northern coasts (UNEP/WCMC 2011). The annual precipitation varies from 600–800 mm on the south coast to 1500–2000 mm on the north coast, reaching 3000 mm in the higher areas, where the Laurisilva forest is present (Baioni 2011).

An important aspect of Madeira Island's climate is its persistent nebulous covering, mainly of orographic origin, which normally exists between the 600 and 800 m to 1600 m of altitude. Depending on wind direction and speed, the highest fog water production can be 250 mm/day (Prada and da Silva 2001).

36.2.2 Geology and Geomorphology

The island of Madeira is part of an archipelago of volcanic origin. It is the emerged part of a large volcanic edifice that rises from the surrounding seafloor, from the 2400 m isobath to 1861 m above present sea level (Ferreira 1985). The island is a deeply dissected shield volcano elongated in an E-W direction, as the rcsult of a hotspot that impinged the slow-moving African plate, leaving a possible hotspot track (Ferreira et al. 1988).

The island emerged around 7.0–5.6 million years ago, in the Miocene, essentially through an uplift process. The subaerial shield volcano was built upon the eroded remains of an uplifted seamount. Basal intrusions are a likely uplift mechanism, and their placement is possibly enhanced by the slow motion of the African plate relative to the source of partial melting (Ramalho et al. 2015).

The island is mountainous: 90% of it is over 500 m in altitude, rising up abruptly from the sea (Fig. 36.2). Relief increases in altitude from east to west, and the mountains act as an barrier between the north and south sides, giving them completely different climatic characteristics and different exposures to erosive agents (Baioni 2011).

The most active geomorphic processes are coastal erosion, runoff erosion, chemical weathering of pyroclastic rocks, and landslide activity, this last especially occurring in the south part of the island (Baioni 2011).

36.2.3 Biodiversity

The Paleogene (64-25 million years) was characterized in Southern Europe and North Africa by the influence of warm marine currents connecting the Indian and the Atlantic oceans through the Tethys Sea. This influence ensured the prevalence of a wet-tropical climate in the region (Uriarte 2003). The land at the margins of the Tethys Sea was populated by the Palaeotropical geoflora (Mai 1991), a flora comprising a mixture of conifers, palms, trees of lauroid-shape leaves, and ferns (Barron and Peyrot 2006).

Fig. 36.2 Steep seacliffs on Madeira Island, sustained by basaltic rocks. The Laurisilva forest is degraded in the coastal area due to historical human activities. (Picture reproduced by permission of Vadim Semjonov, website www.2madeira.com)

After the end of the Oligocene (34-23 million years), several complex global-scale tectonic events created the context for a progressive climatic deterioration, culminating with the Quaternary glaciations, which had a heavy impact on the tropical-like flora, extinguishing it from the European continent (Mai 1991).

Today, the major refugia of the Palaeotropical flora are found in Southeast Asia, Korea/Japan, Anatolian Peninsula, the USA (California and a strip located between the Atlantic coast and the Appalachian chain), and the Macaronesian archipelagos (e.g., the archipelagos of the Canaries, Cape Verde, Azores, Selvagen, and Madeira) (Fernandéz-Palacios et al. 2011; Milne 2006).

Most of Madeira Island was forested when it was discovered, but it was soon cleared for sugar cane and for fueling the refineries. Forests are now found between 300 and 1600 m of altitude (Gouvernement Régional de Madère 1998).

The forest in the World Heritage site is a relict: a 90% intact laurel forest, of moist montane evergreen hardwoods with bryophytes, dripping mosses, and lichens characteristic of unpolluted humid ecosystems, supporting a huge and diverse understory of shrubs, ferns, mosses, and liverworts (UNEP/WCMC 2011) (Fig. 36.3). The forest plays a predominant role in the island's hydrological balance (UNESCO WHC 2016).

A hundred fifty plant species are found in the forest, 66 being endemic to Madeira, representing some 15% and 9% of the endemic species of Madeira and Macaronesia, respectively (UNEP/WCMC 2011).

Fig. 36.3 Illustration of the Laurisilva forest of Madeira Island, with shrubs, ferns, and 40-m-tall evergreen trees with flat, dark green leaves. (Picture reproduced by permission of Sandra Mesquita)

The fauna is relatively poor, but most species are of conservation concern, being endemic to the island. Mammals in the forest are represented by only two species. There is 1 reptile, 500 species of invertebrates, and many species of butterflies and mollusks. The Madeira Archipelago has 43 breeding species of birds, and the forest contains 4 of Madeira's 11 important bird areas (Gouvernement Régional de Madére 1998).

36.3 Conservation Challenges

Most of the World Heritage site is believed to have never been felled. It includes trees that may be over 800 years old, which have been growing since before the island was settled (Fernandéz-Palacios et al. 2004). In the past, goats and sheep have caused some damage, but the problem is now finished with the protection and management actions (UNESCO WHC 2016).

The site has strong and effective legal protection under regional, national, and European laws. The difficult access to the core areas of Laurisilva of Madeira has also helped in its protection. The threats arising are related to the presence of invasive species introduced from both agriculture and forestry (UNESCO WHC 2016) and to the recent introduction of cows – cow grazing marks are now common (IUCN 2014).

Fire is also a recent problem and has been increasing over the recent years (IUCN 2014), probably due to global warming. Management authority has developed measures to fight all the threats; however, the implementation of the measures is hindered by lack of resources (IUCN 2014).

Global warming is already producing increases in the temperature (IUCN 2014). If these changes go in the direction of increasing rainfall, the risk of landslides in the mountainous environment will increase. If it goes toward less rain, the temperate forest will suffer from lack of humidity, which can add a high risk of new deterioration.

More recently, the site is facing a number of threats with invasive species being one of the most serious. Recent increase in tourist numbers also poses a management challenge. Significant uncertainties remain over the likely long-term impact of associated increasing water use, whose effects will likely be further affected by climate change (IUCN 2017).

The International Union for Conservation of Nature (IUCN) in its conservation outlook found that the state of conservation of the site was good in 2014 (IUCN 2014). In the last state of conservation outlook, published at the end of 2017, the state of conservation of the site was found to be good with some concerns, which shows a deterioration of the values of the site and an increase in the threats menacing the area (Osipova et al. 2017).

References

Baioni D (2011) Human activity and damaging landslides and floods on Madeira Island. Nat Hazards Earth Syst Sci 11(11):3035–3046

Barron E, Peyrot D (2006) La vegetacion forestal en el Terciario. In: Fernandez S, Fuentes N (eds) Paleoambientes y cambio climatico. Fundacion Seneca/Agencia de Ciencia y Tecnologia de la Region de Murcia, Murcia

Caldera RMA, Groom S, Miller P, Pilgrim D, Nezlin NP (2002) Sea-surface signatures of the island mass effect phenomena around Madeira Island, Northeast Atlantic. Remote Sens Environ 80:336–360
Fernandéz-Palacios JA, Arévalo JR, Gonzaléz-Delgado GJD, Delgado JD, Otto R (2004) Estrategias de regeneración en laurasilva. Makaronesia 6:90–101
Fernandéz-Palacios JA, de Nascimento L, Otto R, Delgado JD, Garcıa-del-Rey E, Arevalo JR, Whittaker RJ (2011) A reconstruction of Palaeo-Macaronesia, with particular reference to the long-term biogeography of the Atlantic island laurel forests. J Biogeogr 38:226–246
Ferreira MP (1985) Evoluçãao geocronológica e paleomagnética das ilhas do arquipélago da Madeira – uma síntese. Memórias e Notícias 99:213–218
Ferreira MP, Macedo C, Ferreira J (1988) K-Ar geochronology in the Selvagens, Porto Santo and Madeira islands (Eastern Central Atlantic): A 30 m.y. spectrum of submarine and subaerial volcanism. Lunar and Planetary Science Conference Abstracts 19:325
Gouvernement Régional de Madère (1998) Candidature de domains à l'inscription au répertoire de Patrimoine Mondial. Report, Funchal, Madeira, Portugal
IUCN (International Union for Conservation of Nature) (2014) Outlook 2014. A conservation assessment of all natural World Heritage sites. Gland, Switzerland
IUCN (International Union for Conservation of Nature) (2017) Laurisilva of Madeira Conservation Outlook. http://www.worldheritageoutlook.iucn.org/explore-sites/wdpaid/198300. Accessed 20 Dec 2017
Mai DH (1991) Palaeofloristics changes in Europe and the confirmation of the Arctotertiary–Palaeotropical geofloral. Revi Palaoeobot Palynol 68:29–36
Milne RJ (2006) Northern hemisphere plant disjunctions: a window on tertiary land bridges and climatic change? Ann Bot 98:465–472
Osipova E, Shadie P, Zwahlen C, Osti M, Shi Y, Kormos C, Bertzky B, Murai M, Van Merm R, Badman T (2017) IUCN world heritage outlook 2: a conservation assessment of all natural world heritage sites. IUCN, Gland
Prada S, da Silva M (2001) Fog precipitation on the Island of Madeira (Portugal). Environ Geol 41(3):384–389
Ramalho R, Brum Da Silveira A, Fonseca P, Madeira J, Cosca M, Cachao M, Fonseca M, Prada S (2015) The emergence of volcanic islands on a slow-moving plate: the example of Madeira Island, NE Atlantic. Geochem Geophys Geosyst 16(2):522–537
UNEP/WCMC (United Nations Environmental Programme/World Conservation Monitoring Centre) (2011) Portugal – Madresilva de Madeira. Report, Cambridge, UK
UNESCO WHC (United Nations Educational, Scientific and Cultural Organization World Heritage Centre) (2016) Laurisilva of Madeira. http://whc.unesco.org/en/list/934. Accessed 12 Dec 2016
Uriarte A (2003) Historia del Clima de la Tierra. Servicio Central de Publicaciones del Gobierno Vasco. Report, Balbao, Spain

Chapter 37
Gough and Inaccessible Islands, the United Kingdom of Great Britain and Northern Ireland

Abstract Gough and Inaccessible Islands are part of a volcanic chain in the South Atlantic Ocean, on the east slope of the Mid-Atlantic Ridge, presenting spectacular seacliffs. Inaccessible Island is part of the Tristan da Cunha-Gough Island group and is located 350 km north-northwest of Gough Island. The two islands are largely pristine and are one of the few temperate oceanic islands free of introduced mammals, as well as one of the world's largest colonies of sea birds. Gough Island has been described as the most important seabird colony in the world. It has never been permanently populated, and the only inhabitants are the scientists working at the meteorological station. Inaccessible Island has been uninhabited except for a 2-year farming settlement 1936–1938.

37.1 Introduction

Gough and Inaccessible Islands are located in the South Atlantic Ocean, 340 km from each other, as part of the Tristan da Cunha archipelago (Fig. 37.1). The archipelago was discovered by the Portuguese explorer Tristão d'Ancunha in 1506 and annexed to Britain in 1816. Today, it is part of the overseas United Kingdom of Great Britain and Northern Ireland territory of Saint Helena, Ascension, and Tristan da Cunha (UNEP/WCMC 2005).

In 1950, the land birds of Gough Island were protected by the British government. In 1976, the island with its islets and territorial waters out to 3 nautical miles was designated a Wildlife Reserve. In 1977, the marine zone was extended to 12 nautical miles. In 1983, the waters within 200 nautical miles of the islands were protected from fishery. In 1995, both islands were proclaimed a World Heritage site by UNESCO. In 1997, Inaccessible Island and its territorial waters out to 12 nautical miles were designated a nature reserve by the British government (UNEP/WCMC 2005).

Both islands are uninhabited. Gough Island has never been permanently populated, and the only inhabitants are the scientists working at the meteorological station. Inaccessible Island has been uninhabited except for a 2-year farming settlement

V. Claudino-Sales, *Coastal World Heritage Sites*, Coastal Research Library 28,
https://doi.org/10.1007/978-94-024-1528-5_37

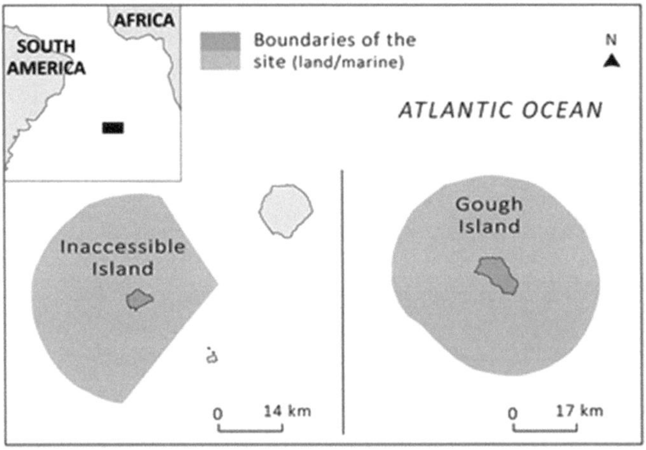

Fig. 37.1 Location of Gough and Inaccessible Islands in South Atlantic Ocean, on the east slope of the Mid-Atlantic Ridge

1936–1938 but has been regularly visited from Tristan for birds, eggs, driftwood, guano, and apples (UNEP/WCMC 2005).

The site has a total area of 79 km^2. Gough Island has 65 km^2 and Inaccessible Island 14 km^2. Both properties are managed by the Tristan da Cunha Conservation Department. The department outlines the environmental management commitments of the United Kingdom government and the Government of Tristan da Cunha and serves as a framework policy to guide the development of the management and plans (UNESCO WHC 2016).

37.2 Natural Settings

37.2.1 Climatic Conditions

Lying in the region of the subtropical front, the islands have a temperate oceanic climate with strong prevailing westerly winds and oceanic currents. Mean annual precipitation is 3154 mm, with mean monthly temperatures varying between 8.9 °C in August and 14.5 °C in February (Jones et al. 2003).

The temperature of the seas surrounding Inaccessible Island increases by some 4° in summer, influenced by the movement of the subtropical convergence from north to south of the island (UNEP/WCMC 2005).

37.2.2 Geology and Geomorphology

Gough Island and Inaccessible Island are part of a chain of South Atlantic volcanic seamounts placed on the east slope of the Mid-Atlantic Ridge (Class and le Roex 2008). The islands represent a succession of basaltic lava flows interbedded with pyroclastic horizons and cut by dykes (Clif et al. 1991).

Gough Island is a small, mountainous island formed as a result of Neogene volcanism (Maund et al. 1988). The recent volcanism on this island is linked to age-progressive volcanism along the Walvis Ridge on the African Plate and its mirror image, the Rio Grande Rise, on the South American Plate, and to the vast Paraná and smaller Etendeka flood basalt provinces, which were adjunct prior to the continental breakup of Gondwana (Class and le Roex 2008).

Inaccessible Island is the surviving sector of a stratovolcano that was centered about 2 km southwest of the present island. The volcano has been severely reduced by marine erosion, but the upper surface of lavas and cinder cones remains largely unmodified. It has an estimated age of 1 million years (Clif et al. 1991).

Spectacular cliffs of each island tower above the ocean. They are composed principally of thin basaltic lava flows with trachytes and pyroclastic deposits (Clif et al. 1991). The altitude goes from seabed to 910 m on Gough Island and to 600 m on Inaccessible Island (UNEP/WCMC 2005).

In Gough Island, the seacliffs are dissected by deep, steep-sided valleys separated by narrow, serrated ridges. The western side of the island consists of rounded slopes. Boulder beaches are found beneath the cliffs, and there are numerous offshore islets and stacks (UNEP/WCMC 2005) (Fig. 37.2).

On Inaccessible Island, an undulating plateau rises from 100 m in the east to over 500 m in the west above a line of steep cliffs and is punctuated by waterfalls (UNESCO WHC 2014) (Fig. 37.3).

37.2.3 Biodiversity

Isolation has kept the islands relatively free from direct human influence, resulting in reasonably pristine habitats (Jones et al. 2003). They represent two extraordinary uninhabited temperate oceanic islands that have remained relatively undisturbed, making them of special conservation significance. The islands' undisturbed nature makes them particularly valuable for biological research (UNESCO WHC 2016).

Gough Island's climate supports an altitudinal succession of vegetation types, from coastal tussock grassland through lowland fern-bush, wet heath, peat bog, and montane moorland communities (Wace 1961). Through this variety, the flora exhibits low species diversity, with a predominance of ferns, lichens, and bryophytes, typical of southern cold temperate islands (Moore 1979). Nevertheless, there are 12 endemic plant species in Gough Island and 8 endemic plant species in Inaccessible Island (UNESCO WHC 2014). Pollen analysis has indicated that the flora has been relatively stable for the last 43,000 years (Bennett et al. 1989).

Fig. 37.2 Steep cliffs in the west coast of Gough Island facing several islets. There is a coastal temperate vegetation cover in the high plateaus of the island. (Picture reproduced by permission of Professor Peter Ryan, University of Cape Town)

Fig. 37.3 West coast of Inaccessible Island, with steep seacliff and boulder beaches, where a penguin and a seal rest in tranquility. (Picture reproduced by permission of Professor Peter Ryan, University of Cape Town)

The spectacular seacliffs on each island host some of the most important seabird colonies in the world, presenting some 22 species of seabirds, several of which only breed there (UNESCO WHC 2016). These include albatrosses, petrels, and penguins. Gough Island is home to two endemic species of land birds. Inaccessible Island also boasts four endemic species of land bird (UNESCO WHC 2014).

About 48% of the world's population of northern rockhopper breeds on Gough (Jouventin 1982), and Atlantic petrel is endemic. Gough is also a major breeding site of the great shearwater. Actually, the island has been described as the most important seabird colony in the world (Bourne 1981).

Subantarctic seal and southern elephant seal are the only two native breeding mammals in Gough Island. Two other marine mammals are found within the reserve, the southern right whale and the dusky dolphin; the rarely seen Shepherd's whale has also been reported. Reptiles, amphibians, freshwater fish, and native terrestrial mammals are absent from the island, although the introduced house mouse is widespread and abundant (UNEP/WCMC 2005).

On Inaccessible Island, the terrestrial fauna, like the flora, has relatively low diversity but a high degree of endemism. There are no land mammals, reptiles, amphibians, or freshwater fish. A number of birds and insects are flightless, and there are no alien species except for invertebrates and house mice. The two marine mammals found on land are the subantarctic seals and southern elephant seals (UNEP/WCMC 2005).

37.3 Conservation Challenges

Invasive alien species are the primary threat to the native biodiversity on the islands (Dawson et al. 2015). House mice were introduced to Gough Island by sealers during the nineteenth century and display unusual phenotypes, including exceptionally large body size and carnivorous feeding behavior (Gray et al. 2014). Actually, house mice in Gough Island are the largest wild house mice on record (Parmenter et al. 2016).

The International Union for Conservation of Nature (IUCN) listed the Tristan albatross of Gough Island as critically endangered, due to mouse predation (Wanless et al. 2009). The house mice are in reality a significant predator of seabird chicks and will, if uncontrolled, gradually reduce the biological value of the site (Parmenter et al. 2016).

Sagina, an aggressive alien plant, was accidentally introduced during the 1990s, and a few other introduced plant species, such as New Zealand flax, could also degrade the integrity of the property if current control measures prove inadequate (UNESCO WHC 2016).

Global warming and climate changes signify threats to the World Heritage site. Based on meteorological data collected on Gough Island, there was already a significant rise in temperature of ca. 0.6 °C between 1963 and the year 2000 (Jones et al. 2003). The increase of temperature can change the amount of precipitation and damage the fauna niches, interfering in the ecology of the islands.

The International Union for Conservation of Nature (IUCN) in its last conservation outlook published in the end of 2017 (IUCN 2017) found that the values of the site are deteriorating. Several species of globally threatened seabird are declining, including Tristan albatross (by ca. 1% per year for the last 30–50 years). Population trends for Atlantic petrel, MacGillivray's prion, and several other bird populations on the island are also likely declining. The population of Gough bunting (critically endangered) is declining and has been estimated to go extinct within the next 10–20 years unless mice are eradicated. The state of conservation of the site was found as presenting significant concerns in this last outlook conservation report (Osipova et al. 2017).

References

Bennett KD, Gribnitz KH, Kent LE (1989) Pollen analysis of a quartenary peat sequence on Gough Island, South Atlantic. New Pathologist 113(3):417–422
Bourne W (1981) Fur seals return to gough islands. Oryx 16:46–47
Class C, le Roex AP (2008) Ce anomalies in Gough Island lavas – trace element characteristics of a recycled sediment component. Earth Planet Sci Lett 265:475–486
Clif RA, Baker PE, Mateer NJ (1991) Geochemistry of Inaccessible Island volcanics. Chem Geol 92:251–260
Dawson J, Oppel S, Cuthbert RJ, Holmes N, Bird JP, Butchart SHM, Spatz DR, Tershy B (2015) Prioritizing islands for the eradication of invasive vertebrates in the United Kingdom overseas territories. Conserv Biol 29(1):143–153
Gray MM, Wegmann D, Hassi RJ, White MA, Gabriel SI, Searle JB, Cuthbert RJ, Ryan PG, Payseur BA (2014) Demographic history of a recent invasion of house mice on the isolated Island of Gough. Mol Ecol 23(B):1923–1939
IUCN (International Union for Conservation of Nature) (2017) Gough and Inaccessible Islands Conservation Outlook. http://www.worldheritageoutlook.iucn.org/explore-sites/wdpaid/93767. Accessed 20 Dec 2017
Jones AG, Chown SL, Ryan PG, Gremmen NJM, NJM GKJ (2003) A review of conservation threats on Gough Island: a case study for terrestrial conservation in the Southern Oceans. Biol Conserv 113:75–87
Maund JG, Rex DC, Le Roux AP, Reid DL (1988) Volcanism on Gough Island: a revised stratigraphy. Geol Mag 152:175–181
Moore DM (1979) Origins of temperate islands floras. In: Bramdwell D (ed) Planst and islands. Academic Press, London
Osipova E, Shadie P, Zwahlen C, Osti M, Shi Y, Kormos C, Bertzky B, Murai M, Van Merm R, Badman T (2017b) IUCN World Heritage Outlook 2: a conservation assessment of all natural World Heritage sites. Gland, Switzerland
Parmenter MD, Gray MM, Hogan CA, Ford IN, Broman KW, Vinyard CJ, Payseur BA (2016) Genetics of skeletal evolution in unusually large mice from Gough Island. Genetics 204(4):1559–1572
UNEP/WCMC (United Nations Environmental Programme/World Conservation Monitoring Centre) (2005) Gough and Inaccessible Islands United Kingdom. Report, Cambridge, UK
UNESCO WHC (United Nations Educational, Scientific and Cultural Organization World Heritage Centre) (2014) World Heritage Committee thirty-eighth session. Report, Doha, Qatar
UNESCO WHC (United Nations Educational, Scientific and Cultural Organization) (2016) Gough and Inaccessible islands. http://whc.unesco.org/en/list/740. Accessed 22 Dec 2016
Wace NM (1961) The vegetation of Gough Island. Ecol Monogr 31:337–367
Wanless RM, Ryan PG, Altwegg R, Angel A, Cooper J, Cuthbert R, Hilton GM (2009) From both sides: dire demographic consequences of carnivorous mice and longlining for the critically endangered Tristan albatrosses on Gough Island. Biol Conserv 142(8):1710–1718

Chapter 38
Saint Kilda, the United Kingdom of Great Britain and Northern Ireland

Abstract Saint Kilda is a tiny archipelago that lies off the west coast of mainland Scotland, in the North Atlantic Ocean. Formed from an ancient volcano associated with the opening of the North Atlantic in the Early Tertiary, it has jagged landscape with towering seacliffs, the highest on Europe. The combination of oceanic influences and local geology has created a marine environment of unparalleled richness. With nearly one million seabirds present at the breeding season, Saint Kilda supports the largest seabird colony in the northeast Atlantic. Overlaying the spectacular natural landscape, there is a rich cultural landscape that bears testimony to millennia of human occupation in extreme conditions. Age-old traditions and land uses shaped the landscape, in such a way that the site is also a cultural World Heritage site.

38.1 Introduction

Saint Kilda is formed by a group of four small and isolated islands on the North Atlantic Ocean, 165 km west of the mainland of northern Scotland. The site has 242 km^2 of area, with terrestrial and marine segments (Fig. 38.1).

In 1957, the island was declared a national nature reserve by the Scotland government. In 1979, it was made an ancient monument and architectural area. In 1981, the island became a national scenic area. In 1984, it became a site of special scientific interest, due to its geology (UNEP/WCMC 2005).

Saint Kilda was inscribed by UNESCO as a World Heritage site in 1986, in recognition of its natural heritage. In July 2005, further recognition of the islands cultural heritage was awarded, making it one of the few places in the world with dual World Heritage listing for both natural and cultural significance (The National Trust for Scotland 2016).

The archipelago has been inhabited since the prehistoric times, confirmed by the discovery of Neolithic ceramics and Bronze Age pots. There are also structures of the Iron Age, of the time of Viking invasions and of early Christian carving. In fact, the archipelago is believed to have been continuously occupied for some 2000 years (Government of Scotland 2004) (Fig. 38.2).

V. Claudino-Sales, *Coastal World Heritage Sites*, Coastal Research Library 28,
https://doi.org/10.1007/978-94-024-1528-5_38

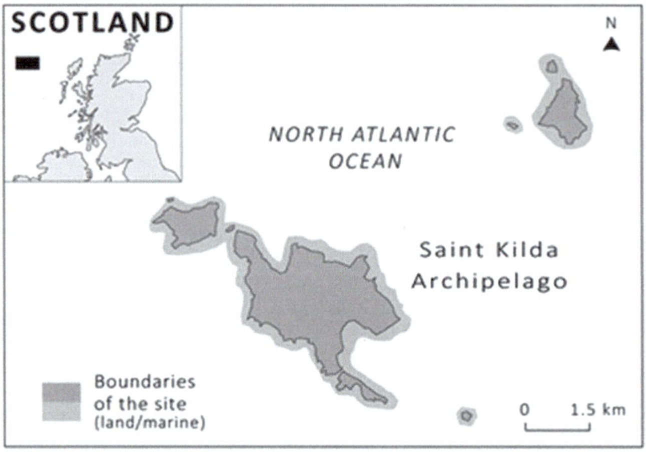

Fig. 38.1 Location of Saint Kilda archipelago, World Heritage site, in the Northern Atlantic Ocean

Fig. 38.2 Ruins of a medieval village, with hundreds of turf-covered stone storage huts ("cleits") and a stone wall. The coast is characterized by the presence of mountainous landscape, with steep cliffs. (Picture reproduced by permission of the Historic Scotland, webpage www.historic-scotland.gov.uk)

In historic times, a small population – about 180 at the end of the seventeenth century – rented the land from a distant landlord. They lived in a small settlement with seasonal dwellings. Until the late nineteenth century, rents were paid in kind, especially barley, oats, fish, produce from cattle and sheep, and specially seabirds (The National Trust of Scotland 2016).

Saint Kilda is a major seabird breeding ground, and the islanders took gannets, fulmars, and puffins for food, feathers, and oil. The birds were taken during the breeding season. Islanders climbed the cliffs, and the birds were caught by hand or with a fowling rod or a snare. It is relatively unusual that inhabitants exploited seabird colonies for food, enabling a population to be sustained on a limited and naturally poor, soil resource (Meharg et al. 2006).

Gradually, the Saint Kildans lost their self-sufficiency. In 1930, the remaining 36 islanders requested evacuation to the mainland. In 1931, Saint Kilda was sold to the Marquis of Bute, a keen ornithologist. He bequeathed the archipelago to the National Trust for Scotland in 1957 (Fleming 2000).

Today, the National Trust for Scotland (a non-governmental organization, working as an independent charity supported by its membership), the Scottish Natural Heritage (a governmental department), and the Ministry of Defense of the UK work together on conservation and research on the archipelago (Fleming 2000).

38.2 Natural Settings

38.2.1 Climatic Conditions

The climate in Saint Kilda is oceanic, with extremely high humidity and strong winds in the winter months. These average 110 km/h and can gust over 185 km/h, carrying salt spray as high as 60 m. The tidal range averages 2.9 m, and the ocean swell generates 2 m waves most of the year (UNEP/WCMC 2005).

The average rainfall is 1400 mm, evenly spread. The mean temperature in January is 5.6 °C and in July 11.8 °C. Studies of the pollen profile reveal 6000 years of climatic change in an environment relatively undisturbed by man (UNEP/WCMC 2005).

38.2.2 Geology and Geomorphology

The Saint Kilda Archipelago is a remnant of a Tertiary igneous complex. Resistant ultrabasic rocks give rise to a very rugged, craggy topography. No "in situ" sedimentary or metamorphic rocks have been found, although the surrounding continental shelf consists of gneiss (Sutherland et al. 1984).

Fig. 38.3 Reverse seacliff profiles, with probable vestiges of folded structures controlled by faults and eroded by glaciations. Sea caves develop in the foot of the cliffs. (Picture reproduced by permission of Historic Scotland, webpage www.historic-scotland.gov.uk)

The archipelago was formed from an ancient volcano associated with the opening of the North Atlantic in the Early Tertiary. The North Atlantic Ocean evolved as the continental crust between the Western Britain Isles and Greenland. It was stretched and thinned, and then a plume impingement resulted in large volcanism. This process happened during the Paleocene and Early Eocene, around 63–52 million years ago (Government of Scotland 2005).

The present landforms and landscapes in part result from the modifications that occurred during Quaternary glaciations. The archipelago comprises a coastline of 35 km, almost completely cliffed, with a maximum altitude of 420 m, the highest in Europe (Government of Scotland 2005).

There are vertical, composite, and reversed cliff profiles, as well as geos, sea caves, arches, tunnels, blowholes, and narrow inlets at all stages of formation. Cobble and boulder beaches occur, forming steep storm beach ridges above water mark (Government of Scotland 2005) (Fig. 38.3).

Scenically, every element appears vertical, except the smooth amphitheater of Village Bay with its relict historic landscape (see Fig. 38.2). Exposure to some of the greatest wave heights and strongest wind speeds in Europe plays a major role in shaping the coastal ecology (UNESCO WHC 2016).

38.2.3 Biodiversity

The heath and bogland vegetation is typical of the archipelago. The effect of salt spray hundreds of meters inland results in *Plantago*-dominated swards on cliff tops. The fertilizing effects of vast numbers of nesting seabirds create a rich turf. More than 184 species of vascular plants have been recorded from the islands, 170 species of fungi, 194 lichens, and 160 bryophytes, but no trees (UNEP/WCMC 2005).

Pollen analysis shows that the islands were partially covered with birch-hazel scrub between 5200 and 6400 years ago. Subsequent climatic change led to the extinction of tree cover and the expansion of halophytic communities (UNEP/WCMC 2005).

With nearly one million seabirds present at the height of the breeding season, Saint Kilda supports the largest seabird colony in the northeast Atlantic and without parallel in Europe (UNESCO WHC 2016). Effectively, Saint Kilda is Europe's most important seabird colony with the world's largest colony of northern gannets nesting on Boreray and the sea stacks (Government of Scotland 2005).

Saint Kilda is an outstanding example of the ecological colonization of a remote island and of the genetic divergence caused by the isolation of small populations. The endemic Saint Kilda wood mouse survived the human depopulation. The terrestrial avifauna of 10 species is impoverished, but the terrestrial invertebrate fauna includes 200 species of fly, 150 beetles, and over 280 lepidoptera (UNEP/WCMC 2005).

Soay sheep are now free ranging. Gray seals and harbor porpoises now breed on the islands, and bottlenose dolphins breed offshore. Ten species of cetaceans are seen. The marine invertebrate fauna of all three zones, intertidal, infralittoral, and sublittoral, is extremely diverse (UNEP/WCMC 2005).

38.3 Conservation Challenges

The archipelago's remote location, the difficulty of accessing it, and human activities almost entirely centered in one of the four islands have significantly contributed to retaining of its integrity (UNESCO WHC 2005).

However, both natural and cultural attributes are threatened. Climatic conditions and coastal erosion are attacking the abandoned houses, cleits, and other archeological remains. Large-scale offshore developments could pose a potential threat to the pristine setting of the islands. Introduction of invasive species poses a significant threat to the natural heritage. The most severe potential threat comes from variations in the marine ecosystem, especially the plankton, caused by climate change (UNESCO WHC 2016; ICUN 2014).

Climatic changes in Saint Kilda Archipelago have already produced warm, wet, and windy winters (Stenseth et al. 2002). It will probably also change the amount of precipitation and the weathering conditions, as well as seacliff erosion, in a way that may interfere in the ecology of the archipelago. Nevertheless, the state of conservation of the site was considered good, in the last conservation outlook published by the International Union for Conservation of Nature at the end of 2017 (IUCN 2017; Osipova et al. 2017).

References

Fleming A (2000) Saint Kilda: family, community, and the wider world. J Anthropol 19(3):348–368

Government of Scotland (2005) Revised nomination of Saint Kilda for inclusion in the World Heritage site list. Report, Edinburgh

IUCN (International Union for Conservation of Nature) (2017) Saint Kilda Conservation Outlook. http://www.worldheritageoutlook.iucn.org/explore-sites/wdpaid/902368, Accessed 20 Dec 2017

IUCN (International Union for Conservation of Nature) IUCN World Heritage Outlook (2014) A conservation assessment of all natural World Heritage sites. Report, Gland, Switzerland

Meharg A, Deacon C, Edwards K, Donaldson M, Davidson D, Spring C, Scrimgeour CM, Feldmann J, Rabb A (2006) Ancient manuring practices pollute arable soils at the Saint Kilda World Heritage site, Scottish North Atlantic. Chemosphere 64(11):1818–1828

Osipova E, Shadie P, Zwahlen C, Osti M, Shi Y, Kormos C, Bertzky B, Murai M, Van Merm R, Badman T (2017) IUCN World Heritage Outlook 2: a conservation assessment of all natural World Heritage sites. IUCN, Gland

Stenseth NC, Mysterud A, Ottersen G, Hurrell JW, Chan KS, Lima M (2002) Ecological effects of climate fluctuations. Science 297(5585):1292–1296

Sutherland DG, Ballantyne CK, Walker MJC (1984) Late Quaternary glaciation and environmental change on Saint Kilda, and their palaeoclimatic significance. Boreas 3:18–29

The National Trust of Scotland (2016) Saint Kilda. http://www.kilda.org.uk. Accessed 26 Dec 2016

UNEP/WCMC (United Nations Environmental Programme/World Conservation Monitoring Centre) (2005) Saint Kilda United Kingdom of Great Britain and Northern Ireland. Report, Cambridge, UK

UNESCO WHC (United Nations Educational, Scientific and Cultural Organization World Heritage Centre) (2016) St Kilda. http://whc.unesco.org/en/list/387. Accessed 27 Dec 2016

Part X
World Coastal Heritage in Islands, Gulfs and Reefs: Indian Ocean

Chapter 39
Aldabra Atoll, Seychelles

Abstract Aldabra Atoll is located in the Indian Ocean. The atoll comprises four large coral islands surrounded by a coral reef enclosing a lagoon, which is linked to the ocean by channels and reef passages. Geomorphological processes have produced a varied and rugged topography which has undergone several emergences and submergences during its natural evolution, creating one of the Earth's largest raised coral atolls. Aldabra is a classic coral atoll which has been built up from the seabed. Due to its isolation and to the difficulty of access, the atoll has been protected from human influence and has the world's largest population of giant tortoise. It forms an ideal natural laboratory for the study of tropical marine corals and lagoon habitats such as mangroves and seagrasses.

39.1 Introduction

Aldabra Atoll is approximately 420 km northwest of Madagascar, 700 km east of the East African mainland, and 1200 km southwest of Seychelles, in the Indian Ocean. It is part of the Republic of Seychelles, which is a member of the African Union. The atoll comprises four main coral islands separated by narrow passes. It has an area of 350 km^2, with 188 km^2 terrestrial, 20 km^2 mangrove, and 142 km^2 marine environment (UNEP/WCMC 2008) (Fig. 39.1).

In 1976, when Seychelles was still a British territory, the atoll was designated a strict nature reserve. In the same year, the Royal Society of London acquired a 14-year lease from the British government. By 1980, Seychelles had become an independent country and the lease on the atoll was assumed by the Seychelles Islands Foundation, a public charitable trust. In 1981, the area was designated as a special reserve, and in 1982, it was declared a World Heritage site (UNEP/WCMC 2008).

Due to its remoteness and inaccessibility, the atoll has remained largely untouched by human influence for most of its existence. The Government of Seychelles owns the reserve, which is managed by the Seychelles Island Foundation. Tourism in the area is limited and carefully controlled (UNESCO WHC 2016).

V. Claudino-Sales, *Coastal World Heritage Sites*, Coastal Research Library 28,
https://doi.org/10.1007/978-94-024-1528-5_39

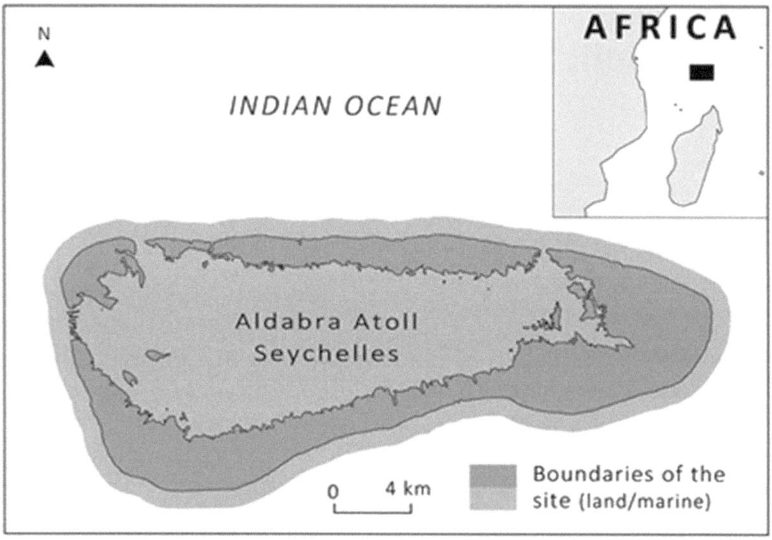

Fig. 39.1 Aldabra Atoll, located in the Indian Ocean. The atoll is part of the territory of the Republic of Seychelles

39.2 Nature Settings

39.2.1 Climatic Conditions

The climate is tropical, with a mean annual temperature of 27 °C. The monthly mean maximum is in December (31 °C) and mean minimum in August (22 °C). The average rainfall is about 1100 mm, although the rainfall varies greatly between years (Farrow 1971). Rain most commonly falls as heavy, violent showers of short duration and limited areal extent (Hnatiuk et al. 1976).

The climate is heavily influenced by northwest monsoon winds from November to March, which bring the heaviest rainfall. Southeastern trade winds blow the remainder of the year, which is drier. The area is occasionally affected by cyclones (UNEP/WCMC 2008).

39.2.2 Geology and Geomorphology

The raised reefs of Aldabra Atoll are the summits of ocean mountains approximately 4000 m high, rising from a fairly flat seafloor between 4000 and 4200 m deep (Stoddart et al. 1971). It consists of an annular rim surrounding by a shallow lagoon (see Fig. 39.1), most of it being about 8 m above sea level. Jagged rocky

Fig. 39.2 Mangroves in the shallow lagoon of Aldabra Atoll, growing up on rocky substrate and raised reefs covered by a thin layer of silt and sand. (Picture reproduced by permission of Seychelles Island Foundation)

cliffs about 3 m high form much of the coastline. Few beaches occur, and behind them, large dunes have developed (Hnatiuk et al. 1976).

The surface geology shows a series of Pleistocene depositional and erosional events, consisting of alternating shallow-water marine limestones (Taylor 1978) dating from 125,000 years (Thomson and Walton 1972).

The marine deposits form most of the present land area. The lower deposits consist largely of a fine calcarenite characterized by abundant calcareous red algae but with few corals and mollusks. The higher deposits occupy most of the outer rim of the atoll (Thomson and Walton 1972). The erosional events are characterized by the presence of marine planation. In the absence of tectonic activity, these events must be interpreted as resulting from glacioeustatic rise and fall of Quaternary sea level (Taylor 1978).

There is no evidence of any extensive marine deposit postdating the limestones. In such a context, present-day conditions of Aldabra atoll may be the result of the most recent erosional event, what probably took place during Holocene sea level rising to its present position around 5000–3000 years BP (Taylor 1978).

The shallow lagoon is almost completely fringed by mangroves which are growing upon rock or silt substrates (Fig. 39.2). The rest of the lagoon floor, which rarely exceeds 5 m deep, is generally rock bottomed with a thin cover of sand or silt; the lagoon contains many smaller islands (Macnae 1971).

Along the southerly and easterly coasts, the seaward platform of the atoll consists of algal-covered rocky pavement, but in the northerly and westerly ends the wider platform is covered by sand dunes. The seaward slope is mainly rocky, with extensive coral growth only in the western end (Barnes et al. 1971).

Fig. 39.3 Giant tortoise in Aldabra Atoll. The atoll is home to the largest population of these turtles in the world. (Picture reproduced by permission of Seychelles Island Foundation)

Aldabra is one of the largest atolls in the world – it is 35 km long and 15 km wide (UNEP/WCMC 2008). It is entirely surrounded by a fringing coral reef, especially abundant in the narrow channels developed through it into the lagoon (Taylor 1978).

39.2.3 Biodiversity

Aldabra Atoll contains one of the most important natural habitats for studying evolutionary and ecological processes. Geomorphologic actions have produced a rugged topography which supports a variety of habitats with a relatively rich biota for an oceanic island and a high degree of endemism (UNESCO WHC 2016).

The vegetation is unusually rich and varied for an isolated atoll. The terrestrial flora includes 178 species of indigenous flowering plants, of which about 8% (15 species) are thought to be endemic to Aldabra. Many of these plants are considered to be threatened. The most common type of vegetation is the mangrove (see Fig. 39.2) (UNEP/WCMC 2008).

The atoll is home to the largest giant tortoise population in the world, with more than 100,000. They represent the last survivors of a life form once found on other Indian Ocean islands, but Aldabra is the only remaining habitat (UNESCO WHC 2016). The island group is also one of the few world areas where reptiles dominate the terrestrial fauna (UNEP/WCMC 2008) (Fig. 39.3).

Aldabra is unusual for a tropical island of its size in having no introduced avifauna. Of the 13 land birds, 2 species and 11 sub-species are endemic. The endemic bird species include the last of the western Indian Ocean flightless birds - the endemic Aldabran white-throated rail. Endemic insects form some 38% of the estimated 1000 species. Two hundred and ten species of fish have been reported. The only indigenous mammals are the fruit bat (Ahamada 2004).

The atoll's remoteness has preserved it from most threats, but during the 1997–98 El Niño, it suffered extensive bleaching, especially in the shallower areas. Mortality was particularly high in the branching corals. However, by 2004 live reefs covered up to 58% of areas shallower than 10 m and up to 52% in deeper ones (UNEP/WCMC 2008).

39.3 Conservation Challenges

Due to its remoteness, lack of fresh water, dense scrub, and difficult terrain, Aldabra's ecosystem has survived relatively intact, although turtles, tortoises, shells, mangroves, and some birds were once extensively exploited (UNEP/WCMC 2008).

Nevertheless, since the transformation of the atoll to a preserved area in the 1970s, the population of different species has recovered: turtles show an increase of 500–800% (Mortimer et al. 2011), and there is an increase of bird population in the atoll (De Crommenacker et al. 2016).

While the remoteness of the site has limited human interference, it also poses tremendous logistical challenges. The managers also have to deal with the constant risk of the arrival of invasive species and oil exploration in the surroundings and, especially, with the results of climate change (UNESCO WHC 2016).

Climate changes are already registered in Aldabra Atoll. Rainfall declined in the last four decades, corresponding to general regional declines. There is a sensitivity of Aldabra's vegetation to rainfall changes, highlighting the potential impact of increasing water stress in East Africa on the region's endemic ecosystems (Shekeine et al. 2015). The increase of the acidity of sea water, the increase of El Niño events, the rise of the sea level, and the rising of seawater temperature are threatening the survival of the corals of the atoll.

For these reasons, the last outlook about the state of conservation of the World Heritage sites, published at the end of 2017 by the International Union for Conservation of Nature (IUCN 2017), found the site to be classified as good with some concerns (Osipova et al. 2017) .

References

Ahamada S (2004) Status of the coral reefs of the Southwest Indian Ocean island states. In: Wilkinson C (ed) Status of coral reefs of the world. Australian Institute of Marine Science, Cape Ferguson

Barnes J, Bellamy DJ, Jones DJ, Whitton BA, Drew EA, Kenyon L, Lythgoe JN, Rosen BR (1971) Morphology and ecology of the reef front of Aldabra. Symp Zool Soc Lond 28:87–114

De Crommenacker J, Richards H, Onezia C, Mahoune T, Haupt P, Accouche W, Fleisher-Dogley F, Bunbury N (2016) Long-term monitoring of landbirds on Aldabra Atoll indicates increasing population trends. Bird Conserv Int 26(3):337–349

Farrow GE (1971) The climate of Aldabra Atoll. Philos Trans R Soci B 260:67. https://doi.org/10.1098/rstb.1971.0007

Hnatiuk RJ, Woodel SRJ, Bourn DM (1976) Giant tortoise and vegetation interactions on Aldabra Atoll – part 2: coastal. Biol Conserv 9:305–316

IUCN (Internationl Union for Conservation of Nature) (2017) Aldabra Atoll Conservation Outlook. http://www.worldheritageoutlook.iucn.org/node/993. Accessed 22 Dec 2017
Macnae N (1971) Mangroves of Aldabra. Philos Trans R Soc Lond B 268:237–247
Mortimer J, Von Brandis R, Liljevik A, Chapman R, Collie J (2011) Fall and rise of nesting turtles (Chelonia mydas) at Aldabra Atoll, Seychelles: positive response to four decades of protection (1968-2008). Chelonian Conserv Biol 10(2):165–176
Osipova E, Shadie P, Zwahlen C, Osti M, Shi Y, Kormos C, Bertzky B, Murai M, Van Merm R, Badman T (2017) IUCN world heritage outlook 2: a conservation assessment of all natural world heritage sites. IUCN, Gland
Shekeine J, Turnbull LA, Cherubini P, Jong R, Baxter R, Hansen D, Bunbury N, Fleischer-Dogley F, Schaepman-Strub G (2015) Primary productivity and its correlation with rainfall on Aldabra Atoll. Biogeosci Discuss 12(2):981–1013
Stoddart DR, Taylor JD, Fosberg FR, Farrow GE (1971) Geomorphology of Aldabra Atoll. Philos Trans R Soc B 260(386):31–65
Taylor JD (1978) Faunal response to the instability of reef habitats: Pleistocene molluscan assemblages of Aldabra Atoll. Paléo 26:2–30
Thomson J, Walton A (1972) Redetermination of chronology of Aldabra Atoll by $^{230}Th/^{234}U$ dating. Nature 240:145–146. https://doi.org/10.1038/240145a0
UNEP/WCMC (United Nations Environmental Programme/World Conservation Monitoring Centre) (2008) Aldabra Atoll Seychelles. Report, Cambridge, UK
UNESCO WHC (United Nations Educational, Scientific and Cultural Organization World Heritage Centre (2016) Aldabra Atoll. http://whc.unesco.org/en/list/185. Accessed 30 Dec 2016

Chapter 40
Socotra Archipelago, Yemen

Abstract Socotra Archipelago is in the northwest Indian Ocean, near the Gulf of Aden. It comprises four islands and two rocky islets which appear as a prolongation of the Horn of Africa. These isolated and almost inviolate islands lie between three biogeographic regions, African, Oriental, and Palearctic, forming a living laboratory which has preserved intact unique endemic ecosystems, a bizarre xerophytic flora, and a traditional culture living in balance with the environment. The coastal landscapes are varied: seacliffs, wave-cut platforms, fossil reef, cobble beaches, sandy beaches, raised beaches, and lagoons. Limestones are present, forming an extensive desert platform. There are also granitic mountains with maximum altitudes of about 1500 m.

40.1 Introduction

Socotra Archipelago lies in the northwest Indian Ocean near the mouth of the Gulf of Aden. Socotra, the largest island, is 250 km long and is located 450 km south of the coast of Yemen. The site comprises four islands and two rocky islets which appear as a prolongation of the Horn of Africa (UNEP/WCMC 2008). Politically, it is part of the territory of Yemen (Fig. 40.1).

The site is of universal importance because of its biodiversity, with rich and distinct flora and fauna (UNESCO WHC 2017). But the archipelago also has a geologically interesting evolution, considering that it is an island of continental origin, related to the Arabian basement as part of Gondwanaland (Whitehouse et al. 2008).

These isolated and until recently almost unviolated islands lie between three biogeographic regions, African, Oriental, and Palearctic, and are a biological museum with a traditional culture living in balance with nature (UNEP/MCM 2008). The islands are frequently referred to as "the Galapagos of the Indian Ocean."

Due to its remarkable fauna and flora, the Socotra Archipelago has been the focus of a certain number of ecological studies dating back to the beginning of the nineteenth century (Brown and Mies 2012). Its environmental protection started in the end of the 1990s.

V. Claudino-Sales, *Coastal World Heritage Sites*, Coastal Research Library 28,
https://doi.org/10.1007/978-94-024-1528-5_40

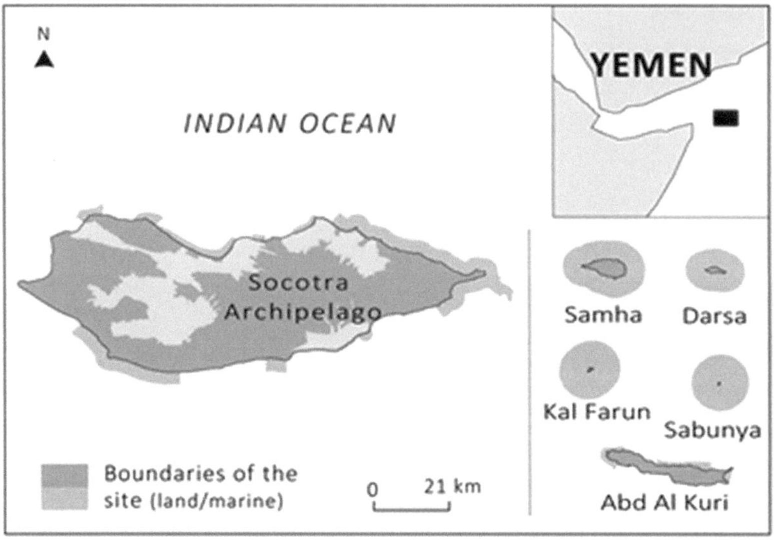

Fig. 40.1 Location of Socotra Archipelago, facing the mouth of the Gulf of Aden, in the Indian Ocean

In 1996, the Socotra Island Protected Area and the Socotra Biodiversity Project were created by the government of Yemen. In 2000, a national park and a nature sanctuary were endorsed. In 2003, it was declared a UNESCO Biosphere Reserve. The archipelago became a World Heritage site in 2008 (UNEP/WCMC 2008). It is also a WWF Global 200 Eco-region, a WWF/IUCN Centre of Plant Diversity, and a BirdLife-designated Endemic Bird Area (UNEP/WCMC 2008).

The terrestrial nature sanctuaries and national parks included in the World Heritage property encompass about 75% of the total land area. They protect all the major vegetation types and areas of faunal values (UNESCO WHC 2017).

Nearly all inhabitants of Socotra, estimated at nearly 50,000, live on the main island. They are of Arab, Somali, Greek, and South Asian origins. The language is Soqotri, an unwritten Semitic language related to other modern South Arabian languages (New World Encyclopedia 2015).

The chief product of the island is fish, but the production of the traditionally managed small-scale fishery is severely declining (Zajonz et al. 2016). They also produce dates and tobacco, and some residents raise cattle and goats. Tourism has boosted the economy in recent decades (New World Encyclopedia 2015).

40.2 Natural Settings

40.2.1 Climatic Conditions

Socotra Archipelago has a tropical arid climate, characterized by pronounced seasonal, altitudinal, spatial, and inter-annual variability (Mies 1999). Around 80% of the islands receive less than 200 mm of annual precipitation on average (Mies 2000), but in the mountains it may rise to 1000 mm (UNEP/WCMC 2008).

The precipitation is controlled by the seasonal changes of the Intertropical Convergence Zone (ITCZ), which gives rise to SW monsoons in summer and to NE monsoons in winter. But cloud cover and mist are a regular feature in many parts of the area, which probably means a supplementary source of humidity (Brown and Mies 2012).

The winds associated with the SW monsoons are generally much stronger than those of the NE monsoon, with a relatively calm period occurring in fall and spring (Mies 1999). When the monsoons are at their most intense, wind speeds are more than 50 km/h on average, sometimes rising to 160 km/h (Mies 2000). Desiccating winds strongly influence the vegetation, and rough seas resulting from the strong winds make fishing and communication with the outside world very difficult for up to 4 or 5 months (UNEP/WCMC 2008).

Mean monthly temperature is usually below 28 °C in most locations, but in summer it can be over 40 °C. At the same time, at the highest elevations, temperatures can drop to 10 °C in some days, although frost is unknown (Brown and Mies 2012).

40.2.2 Geology and Geomorphology

Socotra is a typical example of an ancient continental island or a continental fragment. It has an igneous and metamorphic Precambrian basement, and represents a block of the Gondwanaland, linked to the Arabian Plate.

The block separated from the Arabian Plate by a rifting event which created the Gulf of Aden and the Red Sea. This took place between the Oligocene and the Miocene, around 34–23 million years ago (Whitehouse et al. 2008; Krupp et al. 2002). When it was still a part of Gondwana, the archipelago was linked to the Horn of Africa, from which it is a continuation (Lukashov 2013).

Most parts of Socotra were subjected to repeated marine transgressions until the Miocene. As a consequence, Precambrian basement is overlain unconformably by Cretaceous and Tertiary limestones (Brown and Mies 2012). In the Tertiary, an uplifting of part of the granite basement created the mountains present in the area, with altitudes of the order of 1505 m (Whitehouse et al. 2008).

Fig. 40.2 Limestone plateau in the form of a karren in Socotra Island, colonized by the particular type of vegetation that characterizes the area. (Picture reproduced by permission of Harry Jans, website www.jansalpines.com)

From a geomorphological point of view, the main island is characterized by the presence of three units: the Haghier Mountains, modeled in the granites of the basement uplifted in the Tertiary; a limestone plateau with karstic features of arid climates, which occupy most of the area (Fig. 40.2); and coastal plains with sea-cliffs, wave-cut platforms, fossil reef, cobble beaches, sandy beaches, raised beaches, sandy dunes, and lagoons (Tardelli and Baldini 2000) (Fig. 40.3).

40.2.3 Biodiversity

The Socotra Archipelago is considered to be one of the most remote and most biodiverse rich and distinct islands in the world. Despite its small size, it is one of hotspot biodiversity on Earth with respect to its endemic fauna and flora, at both species and generic levels (Brown and Mies 2012). Nevertheless, despite this richness, the taxonomy of most of the groups is still not clear, and their origin and evolution remain unknown (Sindaco et al. 2012).

Fig. 40.3 Sandy beaches and limestones cliffs on Socotra Island's coastal area. The particular vegetation of the site is present on the limestone, creating a spectacular landscape. (Picture reproduced by permission of Harry Jans, website www.jansalpines.com)

But it is believed that the long geological isolation of the islands and its climate, characterized by heat and drought, are responsible for the existence of this unique and spectacular endemic flora (see Figs. 40.2 and 40.3). Surveys have revealed that of a total of nearly 900 species, 37% are endemic (Miller and Morris 2004; Elie 2002).

The marine biodiversity around Socotra is also rich, characterized by a mixture of species. The islands support unique coastal and coral-associated fish assemblages in spite of the limited biogenic reef frameworks (Kemp 1998). Fish diversity is the highest among comparable Arabian ecoregions, and fish biomass productivity is also high, by Indian Ocean standards (Zajonz et al. 2016).

As with many isolated island systems, bats are the only mammals native to Socotra. Reptiles constitute the most relevant vertebrate fauna of the archipelago, with 90% of the 30 species and 45% of the 12 genera being found nowhere else in the world (Vasconcelos et al. 2016; Sindaco et al. 2012).

40.3 Conservation Challenges

In 2008, the Yemeni government approved decrees providing for increased protection for the islands, limiting damage from road building, promoting sustainable ecotourism, reducing the effects of grazing by uncontrolled livestock, and securing funding for a management structure of the archipelago (UNEP/WCMC 2008).

Although inhabited for 6 millennia, Socotra only began to open up to the outside world in 1990. With conservation interventions starting prior to major developments, and building on centuries-old low-intensity resource management, the island has been in a unique position to practice preemptive conservation (Scholte et al. 2011).

Nevertheless, botanists rank the Socotra flora among the ten most endangered island flora in the world, vulnerable to introduced species (such as goats), modernization, and climate change (New World Encyclopedia 2015). Major causes for biodiversity loss are habitat fragmentation and degradation, pollution, impact of tourism, unsustainable resource use, the loss of traditional land management, and illegal trade in biota, a worrying phenomenon that further increases the pressures on Socotra's ecosystems (Van Damme and Banfield 2011).

The International Union for Conservation of Nature (IUCN), in a conservation outlook published in the end of 2017, found that the island is undergoing rapid development that brings about unprecedented pressures and threats. At the same time, they found that critical conditions associated with political turmoil could negatively affect the archipelago. This outlook pointed out that current and potential threats to Socotra's values are increasing rapidly. Infrastructure development, tourism, and unsustainable natural resource management following the deterioration of traditional management are already affecting the island (IUCN 2017).

The status of some values, particularly ecosystems and endemic flora, has begun to deteriorate, with further deterioration predicted, following rapid ongoing socioeconomic changes, justifying the assessment of the current state of World Heritage values to be of high concern (Osipova et al. 2017).

References

Brown G, Mies BA (2012) Vegetation ecology of Socotra. Springer, London

Elie SD (2002) A historical genealogy of socotra as an object of mythical speculation, scientific research & development experiment. American Institute for Yemeni Studies 44

IUCN (International Union for Conservation of Nature) (2017) Socotra Archipelago Conservation Outlook. http://www.worldheritageoutlook.iucn.org/node/1154. Accessed 23 Dec 2017

Kemp JM (1998) The zoogeography of the coral reef fishes of the Socotra archipelago. J Biogeogr 25:919–933

Krupp F, Apel M, Hairiri K (2002) Conservation and sustanable use of biodiversity of Socotra Archipelago. Marine Habitat, Biodiversity and Fisheries Surveys and Management. Final Overall Report, Frankfurt, Germany

Lukashov AA (2013) Morphostructural evolution of the Gulf of Aden's southern margin. Geomorfologiya 1:35–43

Mies BA (1999) On the comparison of the flora and vegetation of the islands groups of Socotra and Macaronesia. Boletim do Museu Municipal de Funchal 4:455–471

Mies BA (2000) Flora und Vegetationsökologie der Insel Soqotra. Essener Ökologische Schriften 15:1–270

Miller AG, Morris M (2004) Ethnoflora of the Soqotra Archipelago. Royal Botanic Garden, Edinburgh

New World Encyclopedia (2015). Socotra Archipelago. http://www.newworldencyclopedia.org/entry/Socotra/. Accessed 7 May 2017

Osipova E, Shadie P, Zwahlen C, Osti M, Shi Y, Kormos C, Bertzky B, Murai M, Van Merm R, Badman T (2017) IUCN world heritage outlook 2: a conservation assessment of all natural world heritage sites. IUCN, Gland

Scholte P, Al-Okaishi A, Suleyman AS (2011) When conservation precedes development: a case study of the opening up of Socotra archipelago, Yemen. Oryx 45:401–410

Sindaco R, Metallinou M, Pupin F, Fasola M, Carranza S (2012) Forgotten in the ocean: systematics, biogeography and evolution of the Trachylepis skinks of the Socotra Archipelago. Zool Scr 41(4):346–362

Tardelli M, Baldini RM (2000) Botanical report in the island of Socotra. Port Act Biol 19:443–453

UNEP/WCMC (United Nations Environmental Programme/World Conservation Monitoring Centre) (2008) Socotra Archipelago Yemen. Report, Cambridge, UK

UNESCO WHC (United Nations Educational, Scientific and Cultural Organization World Heritage Centre) (2017) Socotra Archipelago. http://whc.unesco.org/en/list/1263. Accessed 7 May 2017

Van Damme K, Banfield L (2011) Past and present human impacts on the biodiversity of Socotra Island (Yemen): implications for future conservation. Zool Middle East 54:31–88

Vasconcelos R, Montero-Mendieta S, Simó-Riudalbas M, Sindaco R, Santos X, Fasola M, Llorente G, Razzetti E, Carranza S (2016) Unexpectedly high levels of cryptic diversity uncovered by a complete DNA barcoding of reptiles of the Socotra Archipelago. PLoS One 11(3):e0149985

Whitehouse M, Al Khirbash S, Pease V (2008) Socotra; a missing part of the Arabian basement puzzle. 33rd International Geological Congress, IUGS, Oslo

Zajonz U, Lavergne E, Klaus R, Krupp F, Aideed MS, Saeed FN (2016) The coastal fishes and fisheries of the Socotra archipelago, Yemen. Mar Pollut Bull 105:660–675

Part XI
World Coastal Heritage in Islands, Gulfs and Reefs: Pacific Ocean

Chapter 41
Fraser Island, Australia

Abstract Fraser Island lies along the eastern coast of Australia. At 122 km, it is the largest sand island in the world. The island has majestic remnants of tall rainforest growing on sand and half the world's perched freshwater dune lakes. It offers an outstanding example of ongoing biological, hydrological, and geomorphological processes. The immense sand dunes are part of the longest and most complete age sequence of coastal dunes systems in the world and are still evolving. It boasts the world's largest unconfined aquifer on a sand island. The combination of subtropical rainforests and lakes, shifting sand dunes, spectacular blowouts, and 220 km of white sand beaches with long, uninterrupted sweeps of ocean beach and strikingly colored sand cliffs make it an exceptional site.

41.1 Introduction

Fraser Island lies just off the east coast of Australia, stretching along the southern coast of Queensland, separated from it for some 60 km by the narrow Great Sandy Strait which opens into the wide Hervey Bay. Fraser Island is 122 km long, 7–25 km wide, and 1840 km^2 in area. It is the largest sand island in the world (Fig. 41.1).

Its history of protection is unique, since it became an Aboriginal reserve in 1860. In 1908, the center of the island was transformed into a forestry reserve, but the forests were strongly logged anyway, altering its structure, floristic composition, and relative species abundance (Sinclair and Morrison 1990). Together with mining, these activities challenged the environment. Nevertheless, it took place mainly in the southern part of the island, so it did only minor damage to the ensemble of the mobile dunefield (Donders et al. 2006).

In 1923, the island was declared a state reserve. Fraser Island Fish Habitat Area was created in 1983. In 2000, the Great Sandy Regional Area was established, including Fraser Island (UNEP/WCMC 2008). The area became a World Heritage site in 1992. It lies within a WWF Global 200 Eco-region and is a BirdLife-designated Endemic Bird Area. It adjoins a Ramsar wetland under the Ramsar Convention of Wetlands, which is a refuge for humpback whales and dugongs (UNESCO WHC 2017).

V. Claudino-Sales, *Coastal World Heritage Sites*, Coastal Research Library 28,
https://doi.org/10.1007/978-94-024-1528-5_41

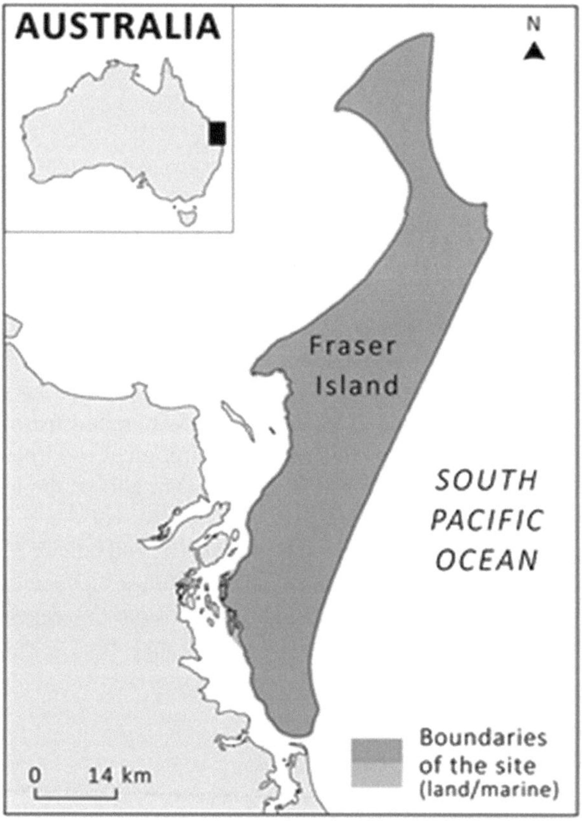

Fig. 41.1 Location of Fraser Island World Heritage site, off the eastern coast of Australia

The World Heritage site includes all the Fraser Island and several small islands off the island's west coast. The Government of Australia, the state of Queensland, and a small area in private ownership have the land tenure. The site is managed by the Queensland Parks and Wildlife Service (QPWS) of the state Environmental Protection Agency (UNESCO WHC 2017).

The property reveals Aboriginal occupation of at least 50,000 years (Sinclair 1997). Nevertheless, the Aboriginal life was disrupted soon after European settlement. The island contains many sites of spiritual, archeological, and social significance, such as artifact scatters, fish traps, campsites, and more than 200 shell middens (UNEP/WCMC 2008). Today, there are only few centuries of permanent residents (UNESCO WHC 2017).

41.2 Natural Settings

41.2.1 Climatic Conditions

Fraser Island is exposed to the open ocean currents of the Coral Sea on the east coast and the waters of Hervey Bay on its western margin and is positioned to receive moisture from the Indo-Australian monsoon and from southeast trade winds (Dontz et al. 2015).

Climatic conditions are maritime subtropical with mean annual temperatures ranging from 14.1 °C minimum to 28.8 °C maximum. Rainfall is high, reaching 1800 mm on the highest dunes in the center of the island (Sinclair and Morrison 1990).

Winds are not strong, and the average is 8 m/s during most months. In October and November, they can be stronger, with speeds of 17 m/s. The dominant direction of the winds is NW (Meteoblue 2017). The area is subject to the action of tropical cyclones (Levin 2011).

41.2.2 Geology and Geomorphology

Fraser Island is the result of Quaternary deposition of sand swept north from the continental shelf and of continual erosion along a fluctuating coastline. It is a continuation of the similar Cooloola sand mass on the mainland to the south (DASETT 1991) and was once connected to mainland Australia (Howard et al. 2016). The sand, often tinted by iron-rich minerals, derives from granites, sandstones, and metamorphic rocks in river catchments to the south and from the seafloor (UNEP/WCMC 2008).

Fraser Island's dunes have formed over a long time period of dune emplacements, beginning at least 730,000 years ago (Lees 2006). It represents the world's oldest recorded dune sequence, with episodes of major dune-building forming at least eight overlapping dune systems (QPWS 2006). The dunes reach an altitude greater than 200 m. Mobile dunefields have a width of 10–20 km and are mostly vegetated by eucalyptus forests (Levin 2011) (Fig. 41.2).

The longest sandy island in the world also has some solid rock. Igneous rocks outcrop at Indian Head and Waddy Point in the north (DASETT 1991).

The hydrology of the sand masses is notable for unique strata and the almost unaltered catchments of the region (QPWS 2003). The aquifers consist of an extensive regional freshwater lens within the sand mass and aquifers perched over more or less impervious organically bound sands. Over 100 lakes are scattered among the dunes (UNEP/WCMC 2008). The 40 perched, window, and barrage dune lakes are unusual

Fig. 41.2 High mobile, semifixed, and fixed dunefields on Fraser Island. The typical vegetation on the rainforest is eucalyptus. (Picture reproduced by permission of Noam Levin)

in their number, size, elevation, depth, and acidic oligotrophic waters and comprise more than half the world total of such lakes (UNESCO WHC 2017) (Fig. 41.3).

Some of the lakes, formed in wind-scoured depressions that become impermeable from the collection of wind-hardened organic debris, are estimated to be up to 300,000 years and contain in their organic sediments a continuous record of changes to the island's hydrology and vegetation through Quaternary glacial and interglacial cycles (UNEP/WCMC 2008).

The island also has several barrage lakes formed when a wall of landward migrating (transgressive) sand dunes has blocked a watercourse or groundwater springs (DASETT 1991; Sinclair and Morrison 1990). Actually, the island has the world's largest unconfined aquifer on a sand island (UNESCO WHC 2017).

The island also has a particular combination of subtropical rainforests and lakes, shifting sand dunes, spectacular blowouts, and white sand beaches up to 250 km, with uninterrupted sweeps of ocean beach and strikingly colored sand cliffs, which makes Fraser Island an exceptional site (UNEP/WCMC 2008).

The process of soil formation on the island is also unique and results from the successive overlaying of dune systems. The depth of the podzol there (25 m thick) far exceeds known depths of podzols anywhere in the world (UNESCO WHC 2017).

Fig. 41.3 A coastal lake on Fraser Island, formed because a wall of landward migrating (transgressive) sand dunes has blocked groundwater spring. (Picture reproduced by permission of Noam Levin)

41.2.3 Biodiversity

The island has a surprising variety of vegetation, ranging from coastal heath to subtropical rainforest, and is the only place in the world where tall rainforest (trees up to 50 m) grow on sand dunes over 200 m in elevation (UNEP/WCMC 2008). It is a region geologically old enough to have continuously supported rainforest in refugia (Howard et al. 2016).

There is an east-west zonation and succession of plant communities according to the salinity, water table, age, and nutrient status of dune sands and the exposure, frequency, and intensity of fires (Sinclair and Morrison, 1990). In the leached old dunes, the forests are replaced by woodlands, shrubs, and low heaths by a process of retrogressive succession which is of considerable scientific interest (UNEP/WCMC 2008).

Compared with the mainland, the island is noted for its low number of introduced species. The native plant communities contain a large number of specialized habitats that support a very diverse fauna, although diversity within habitats is low and few species are endemic to the sandy coastal heaths (DASETT 1991; Sinclair and Morrison 1990).

Sixty-six mammal species, 17 amphibian, 78 reptile, and 352 bird species are present (QPWS 2005). The mammals include 12 marsupials, and also the dingo, a canine which is the purest genetic strain remaining in eastern Australia (Twyford 1996).

41.3 Conservation Challenges

There are some threats requiring ongoing attention, such as inappropriate fire, invasive plants and animals, and lake water quality deterioration (UNESCO WHC 2017). It seems that the high threats are linked to the degradation due to the high number of annual visitors and climate change.

Dingo attacks on tourists have been reported, which is provoking debate about the need to kill the animals (Carter and Palmer 2017). The situation indicates a lack of balance between nature and society in the World Heritage site. Another environmental problem linked to the presence of tourists is related to the use of the dunes as camping sites. Trucks used to transport the campers damage the dunes. It was estimated that 20% of the foredunes were eroding and the shoreline was retreating. Crabs and other marine animals were also declining because of these practices (Thompson and Schlacher 2008).

In relation to climate change, studies show a significant decrease in tropical cyclone frequency and intensity in eastern Australia since the early 1980s, leading to rapid dune stabilization (Levin 2011). This situation changes the nature of the site's dunefields and ecosystems.

In the last years, Fraser Island has been viewed as a model in participatory conservation management. However, pressures from tourism and recreational use, as well as climate change, will require continuing monitoring and increased management efforts to ensure preservation of the site's values in the long-term (IUCN 2017). The state of conservation of the site in the conservation outlook published by the International Union for Conservation of Nature (IUCN) in the end of 2017 found that the level of conservation of the island is good, with some concerns (Osipova et al. 2017).

References

Carter J, Palmer J (2017) Dilegams of transgression: ethical responses in a more-than-human world. Cult Geogr 24:213–230

DASETT (1991) Nomination of Fraser Island and the Great Sandy Region by the Government of Australia for inclusion in the World Heritage List. Department of the Arts, Sport, Environment, Tourism and Territories, Sidney

Donders TH, Wagner-Cremer F, Visscher H (2006) Late Pleistocene and Holocene subtropical vegetation dynamics recorded in perched lake deposits on Fraser Island, Queensland, Australia. Palaeogeogr Palaeoclimatol Palaeoecol 241:417–439

Dontz AM, Moss PT, Sloss CR, Petherick LM, Mccallum A, Shapland F (2015) Understanding past climate variation and environmental change for the future of an iconic landscape – K'gari Fraser Island, Queensland, Australia. Aust J Environ Manag 22:105–123

Howard MG, McDonald JF, Forster PI, Kress WJ, Ericsoon D, Faith DP, Shapcott A (2016) Patterns of Phylogenetic diversity of subtropical rainforest of the Gread Sandy region, Australia indicate long term climatic refugia. PLoS One 11(4):e0153565 https://doi.org/10.1371/journal.pone.0153565

IUCN (International Union for Conservation of Nature) (2017) Fraser Island Conservation Outlook. http://www.worldheritageoutlook.iucn.org/node/1056. Accessed 23 Dec 2017
Lees B (2006) Timing and formation of coastal dunes in northern and eastern Australia. J Coast Res 22:78–89
Levin N (2011) Climate-driven changes in tropical cyclone intensity shape dune activity on Earth's largest sand island. Geomorpholgoy 125:239–252
Meteoblue (2017) Climate Fraser Island. https://www.meteoblue.com/en/weather/forecast/model-climate/fraser-island/. Accessed 10 May 2017
Osipova E, Shadie P, Zwahlen C, Osti M, Shi Y, Kormos C, Bertzky B, Murai M, Van Merm R, Badman T (2017) IUCN world heritage outlook 2: a conservation assessment of all natural world heritage sites. IUCN, Gland
QPWS (Queensland Parks and Wildflie Service) (2005). Coastal Sandmasses of southern Queensland (Fraser Island) (2003). Environmental Protection Agency, Government of Queensland, Australia
QPWS (Queensland Parks and Wildlife Service) (2003) Fraser Island World Heritage Area. Environmental Protection Agency, Government of Queensland, Australia
Sinclair J (1997) Discovering Fraser Island & Cooloola. Australian Environmental Publications, Gladesville
Sinclair J, Morrison R (1990) Fraser Island and Cooloola. Weldon Publishing, Sydney
Thompson L, Schlacher T (2008) Physical damage to coastal dunes and ecological impacts caused by vehicle tracks associated with beach camping on sandy shores: a case study from Fraser Island, Australia. J Coast Conserv 12:67–82
Twyford K (1996) Vertebrate Fauna of Fraser Island world heritage area. Central Coast Region, Department of Environment and Heritage, Sydney
UNEP/WCMC (United Nations Environmental Programme/World Conservation Monitoring Centre) (2008) Fraser Island, Queensland, Australia. Report, Cambridge
UNESCO WHC (United Nations Educational, Scientific and Cultural Organization World Heritage Centre) (2017) Fraser Island. http://whc.unesco.org/en/list/630. Accessed 10 May 2017

Chapter 42
Great Barrier Reef, Australia

Abstract The Great Barrier Reef is the world's largest coral reef ecosystem and one of the most complex and biodiverse natural systems on Earth. It is on the northeast coast of Australia and is of remarkable variety and beauty. The barrier reef has 400 types of coral, 1500 species of fish, and 4000 types of mollusk and is the habitat for species such as the dugong ("sea cow") and the large green turtle. It stretches from the low water mark along the mainland coast up to 250 km offshore and is characterized by vast shallow inshore areas and outer reefs, going beyond the continental shelf to oceanic water for over 2000 m deep. Within the Great Barrier Reef, there are many islands, sometimes with more than 1000 m of altitude. The landscapes and seascapes there provide spectacular maritime scenery.

42.1 Introduction

The Great Barrier Reef extends for 2300 km along the northeastern coast of Australia. Covering an area of 348,000 km^2, it goes across a contiguous latitudinal range of 14°, from 10°S to 24°S, and extends beyond the low water mark 60–250 km from the mainland to oceanic water for over 2000 m deep (UNESCO WHC 2017) (Fig. 42.1). The complete reef system extends to the Papua New Guinea coast and is one of a few living structures visible from space (UNEP/WCMC 2008).

In 1934 and again in 1943, national parks were created in the area of the Great Barrier Reef, and in 1974, the first marine national park was installed. The Great Barrier Reef Marine Park was created in 1979, which initially preserved only one part of the reefs. Between 1981 and 1989, other segments were included in the park (UNEP/WCMC 2008). The present marine park abuts around 99% of the total of the area occupied by the barrier reef (UNESCO WHC 2017; GBRMPA 2014).

The reef system was defined as a World Heritage in 1981. The World Heritage property has an area of 348.700 km^2, being divided into four different management areas. The site includes islands that are excluded from the marine park (Hopley et al. 2012). The park also lies within a WWF Global 200 Ecoregion, a WWF/IUCN Centre of Plant Diversity, and contains two wetlands established by the Ramsar Wetland Convention (UNEP/WCMC 2008).

V. Claudino-Sales, *Coastal World Heritage Sites*, Coastal Research Library 28,
https://doi.org/10.1007/978-94-024-1528-5_42

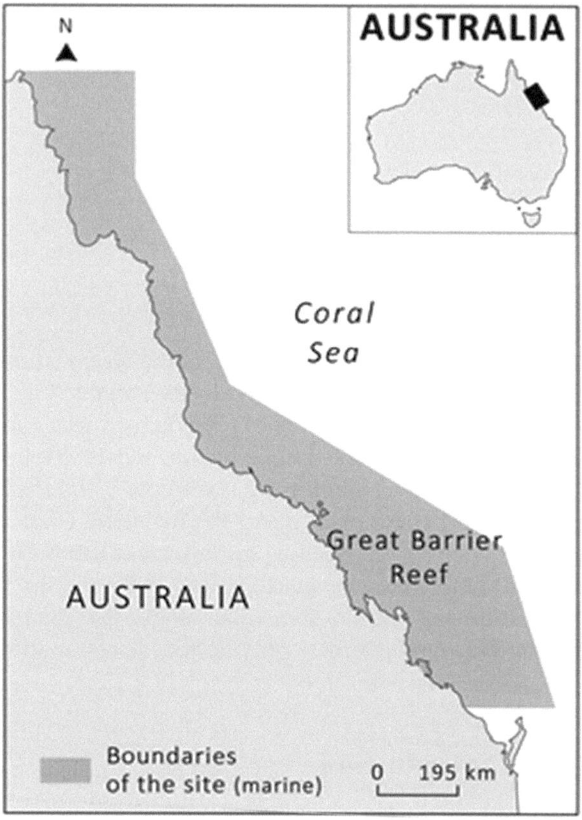

Fig. 42.1 Great Barrier Reef World Heritage site, in the northeast coast of Australia

The marine park is managed by the Great Barrier Reef Marine Park Authority, an independent Australian government agency (UNESCO WHC 2017). The land tenure is shared by the state of Queensland and the federal government. Some land is held privately (UNEP/WCMC 2008).

The cultural heritage of the site is important, considering that aboriginal occupation of the coast probably dates back to the earliest human occupation of Australia (ca. 50,000 years ago), and the Great Barrier Reef is essential in both the historic and contemporary culture of the coastal tribes (UNEP/WCMC 2008).

Currently, people living in aboriginal communities in the Great Barrier Reef area have access to the marine and inshore resources (GBRMPA 2001). Over 70 coastal clan groups maintain strong cultural relationships with the area, and about 11 native title claims are registered over parts of the property (GBRMPA 2012).

The Great Barrier Reef site is and has always been managed as a multiple use area. Uses include a range of commercial and recreational activities, with tourism as the most important of those economic activities (UNESCO WHC 2017). More than two billion people visit the Great Barrier Reef every year; it is worth 15–20 billion/year dollars to the Australian economy and provides approximately 64,000 full-time jobs (Brodie and Pearson 2016).

The Great Barrier Reef was first described by James Cook in 1770. It became important to the Australian society only in the 1920s, when the Great Barrier Reef Committee was established (Hopley et al. 2012). The site offers unparalleled opportunities for scientific research, and researchers have been active in the area since the formation of the committee. Today, it is one of the most studied reefs in the world (UNEP/WCMC 2008).

42.2 Natural Settings

42.2.1 Climatic Conditions

The area lies between continental Australia and the open ocean of the South Pacific, and its tropical climate is strongly influenced by both. The region has two distinct seasons, a winter period of mild temperatures and low rainfall and a summer period of warm temperatures and higher rainfall, under the influence of the monsoon and irregular tropical cyclones and depressions (Great Barrier Reef Foundation 2017).

The wind pattern for most of the year is dominated by the southeast trades, but from January to March, northwesterlies prevail over the north of the reef under the influence of the intertropical monsoonal front (UNEP/WCMC 2008).

42.2.2 Geology and Geomorphology

The site covers the most extensive coral reef ecosystem in the world, extending from warm temperate waters just south of the Tropic of Capricorn to tropical waters in the Torres Strait halfway to Papua New Guinea (UNESCO WHC 2017).

It comprises some 3400 individual reefs, including 760 fringing reefs, providing the most spectacular marine scenery on earth – beneath the ocean surface, there are an abundance and diversity of shapes, sizes, and colors, with a spectacular coral assemblage of hard and soft corals (Fig. 42.2).

There are also 300 coral cays, 150 mangrove islands, and 618 continental islands, with altitudes going to over 1000 m (GBRMPA 2001) (Fig. 42.3). The reef is about 15,000 years old and evolved in association with marine regressions and transgressions (Hopley et al. 2012). The area has been exposed and flooded by at least four glacial and interglacial cycles, with reefs growing on the continental shelf (UNESCO WHC 2012).

But only 7% of the site is made up of coral reefs. The rest of the area is formed by an extraordinary variety of marine habitats, characterized by the presence of shallow inshore areas such as seagrass, mangroves, sand, and algal and sponge gardens, besides offshore interreef communities (Great Barrier Reef Foundation 2017).

Fig. 42.2 Corals and tropical fish in the Great Barrier Reef: a typical local seascape. (Picture reproduced by permission of Brian Kinney/Shutterstock Stock Pictures)

Fig. 42.3 Mountains, river mouth, sand bars, and barriers at the Whitehaven Beach, Whitsundays Island, Great Barrier Reef. (Picture reproduced by permission of Tanya Puntti/Shutterstock Stock Pictures)

The water circulation is very complex, governed by the properties of the sea, evaporation, southeast trade winds, upwellings forced by strong tidal currents through narrow reef passages, mangrove swamps, and land runoff (UNEP/WCMC 2008). Twenty-six river catchments drain into the park, and freshwater runoff can have significant physical and biological effects (Kelleher et al. 1989).

Tides are generally semidiurnal with diurnal inequality toward the north, becoming almost diurnal in Torres Strait. The maximum tidal range is about 3 m along most of the coast, increasing to 6–9 m in some areas (GBRMPA 2012).

42.2.3 Biodiversity

The latitudinal and cross-shelf diversity, combined with diversity through the depths of the water column, makes for a unique array of ecological communities, habitats, and species and is one of the richest and most complex natural ecosystems on Earth (UNESCO WHC 2017).

The form and structure of the individual reefs show great variety. Two main classes can be defined: platform reefs, resulting from radial growth, and wall reefs, resulting from elongated growth (Kelleher et al. 1989).

There are 1500 species of fish, 6 turtles, and 15 sea snakes. The immense diversity of marine life encompasses 5000–8000 species of mollusks, 630 echinoderms, 1930 crustaceans, 6000 amphipod crustaceans, 1500 sponges, 350 ascidians, 500 bryozoans, and more than 500 polychaete (GBRMPA 2001).

The site includes feeding grounds for one of the world's largest populations of dugong ("sea cow"), 7 whales, 19 species of dolphin, and 28 species of sharks and rays. There are nesting beaches of world significance for green turtles, and the estuarine crocodile can be found along the whole coast (UNEP/WCMC 2008). Seasonally, between 1.4 and 1.7 million birds gather in the area, and 118 species of butterfly are recorded (GBRMPA 2001).

42.3 Conservation Challenges

Assailed by multiple threats, the Great Barrier Reef seems to be deteriorating, considering that half of its coral cover has been lost over the last three decades (Great Barrier Reef Foundation 2017).

Some of the main threats to the World Heritage values are coastal development; the increasing fishing; litter from tourism and recreational use; biodiversity loss from oil pollutants; predation by red foxes, feral cats, and pigs; and cyclic outbreaks of crown of thorns plagues (Wilson and Verlis 2017; De'ath et al. 2012; Fabricius et al. 2010; UNEP/WCMC 2008). Nevertheless, the biggest threats to the survival of the Great Barrier Reef ecosystem are climate change and catchment water quality along the coastal area (GBRMPA 2012).

Water quality has been declining due to a variety of causes including increasing urbanization and agriculture along the coast and modifications to all the major rivers which flow into the Great Barrier Reef lagoon since European settlement (Deng et al. 2014; McCulloch et al. 2003). In addition, there has been increased tourism, and while sewage treatment works at all the resorts have been upgraded to increase levels of treatment, water quality has continued to decline (Chazottes et al. 2017).

On the other hand, climate change and global warming are responsible for the rising of seawater temperature, producing bleaching, a phenomenon affecting coral by which they lose their natural color and die (e.g., Ainsworth et al. 2016; Hughes et al. 2015). At the same time, global warming is producing ocean acidification, which reduces the capacity of the corals to build their calcium carbonate-based skeletons. Climate change is also accelerating rates of sea level rise, changing the amount of sunlight in the reefs, and changing weather patterns, which is a threat to the physical integrity of the reefs (GBRMPA 2012).

A number of values for which the property was added to the World Heritage List are deteriorating, with a further dramatic decline occurring during the 2016 and 2017 coral bleaching events (IUCN 2017). The last conservation outlook published by the International Union for Conservation of Nature (ICUN) found the level of conservation of the site to present significant concerns (Osipova et al. 2017).

References

Ainsworth TD, Heron SF, Ortiz JC, Mumby PJ, Grech A, Ogawa D, Eakin CM, Leggat W (2016) Climate change disables coral bleaching protection on the Great Barrier Reef. Science 352(6283):338–342

Brodie J, Pearson RG (2016) Ecosystem health of the Great Barrier Reef: time for effective management action based on evidence. Estuar Coast Shelf Sci 183:438–452

Chazottes V, Hutchings P, Osorno A (2017) Impact of an experimental eutrophication on the processes of bioerosion on the reef: One Tree Island, Great Barrier Reef, Australia. Mar Pollut Bull 118:125–130

De'ath AG, Fabricius KE, Sweatman HPA, Puotinenm ML (2012) The 27- year decline in coral cover on the Great Barrier Reef and its causes. Proc Natl Acad Sci U S A 109:17995–17999

Deng W, Wei G, McCulloch M, Xie L, Liu Y, Ti Zeng T (2014) Evaluation of annual resolution coral geochemical records as climate proxies in the Great Barrier Reef of Australia. Coral Reefs 33:965–977

Fabricius KE, Okaji K, De'ath G (2010) Three lines of evidence to link out-breaks of the crown of thorns seastar Acanthaster planci to the release of larval food limitations. Coral Reefs 29:593–605

GBRMPA (Great Barrier Reef Marine Park Authority) (2001) Great Barrier Reef Outlook. Report, Queensland

GBRMPA (Great Barrier Reef Marine Park Authority) (2014) Great Barrier Reef Outlook. Report, Queensland

GBRPMD (Great Barrier Reef Marine Park Authority) (2012) Great Barrier Reef Outlook. Report, Queensland

Great Barrier Reef Foundation (2017). Great Barrier Reef. https://www.barrierreef.org/the-reef/the-facts. Accessed 15 May 2017

Hopley D, Smithers SG, Parnell K (2012) The geomorphology of the Great Barrier Reef: development, diversity and change. Cambridge University Press, London

Hughes TP, Day J, Brodie J (2015) Securing the future of the Great Barrier Reef. Nat Clim Chang 5:508–511

IUCN (International Union for Conservatino of Nature) (2017) Great Barrier Reef Conservation Outlook. http://www.worldheritageoutlook.iucn.org/node/974. Accessed 23 Dec 2017

Kelleher G, Childs R, Quilty P (1989) Great Barrier Reef area: a protected seascape. Rapport, Great Barrier Reef Marine Park Authority, Queensland

McCulloch M, Fallon S, Wyndham T, Hendy E, Lough J, Barnes D (2003) Coral record of increased sediment flux to the inner Great Barrier Reef since European settlement. Nature 421:727–730

Osipova E, Shadie P, Zwahlen C, Osti M, Shi Y, Kormos C, Bertzky B, Murai M, Van Merm R, Badman T (2017) IUCN world heritage outlook 2: a conservation assessment of all natural world heritage sites. IUCN, Gland

UNEP/WCMC (United Nations Environmental Programme/World Conservation Monitoring Centre) (2008) Great Barrier Reef, Queensland, Australia. Rapport, Cambrigde, UK

UNESCO WHC (United Nations Educational, Scientific and Cultural Organization World Heritage Center) (2012) World Heritage Comittee. Thirty-six session. Report. Paris, France

UNESCO WHC (United Nations Educational, Scientific and Cultural Organization World Heritage Centre) (2017) Great Barrier Reef. http://whc.unesco.org/en/list/154. Accessed 15 May 2017

Wilson SP, Verlis KM (2017) The ugly face of tourism: marine debris pollution linked to visitation in the southern Great Barrier Reef, Australia. Mar Pollut Bull 117:239–246

Chapter 43
Lord Howe Island Group, Australia

Abstract Lord Howe Island Group is in the South Pacific, 800 km from Sidney. The islands are a remarkable example of an isolated oceanic island born of volcanic activity and demonstrating the nearly complete stage in the destruction of a shield volcano. They have a grand topographic relief, with spectacular and scenic landscapes within a small area, characterized by the presence of sheer mountain slopes rising abruptly from the ocean and a broad arc of hills enclosing a lagoon, where the world's most southerly true coral reef exists. The islands support extensive colonies of nesting seabirds, and iconic species include endemics such as the world's largest stick insect. Many species are at their ecological limits, and unique assemblages of temperate and tropical forms cohabit.

43.1 Introduction

Located in the Tasman Sea, South Pacific Ocean, Lord Howe Island Group forms an isolated group of oceanic islands of volcanic origin. It is characterized by the presence of a main island and many surrounding islets and rocks: the wider island group includes several offshore islets, the Ball's Pyramid group of rocks 25 km southeast, and associated coral reefs and seas. The site has an area of 12 km^2 (UNEP/WCMC 2008) (Fig. 43.1).

In 1982, the Australian government created a preserve area on the Lord Howe Island. In 1998, a state marine park was declared by the state of New South Wales around Lord Howe Island and Ball's Pyramid. In 2000, a marine national park was created out from each main island. The World Heritage site was announced in 1982, including the islands' land surface, the marine area, and the associated coral reef covers (UNEP/WCMC 2008).

The islands were discovered in the end of the seventeenth century by a British ship and were named after the First Lord of the Admiralty. A small permanent settlement was established in 1834, with the settlers making a living by hunting and fish-

V. Claudino-Sales, *Coastal World Heritage Sites*, Coastal Research Library 28,
https://doi.org/10.1007/978-94-024-1528-5_43

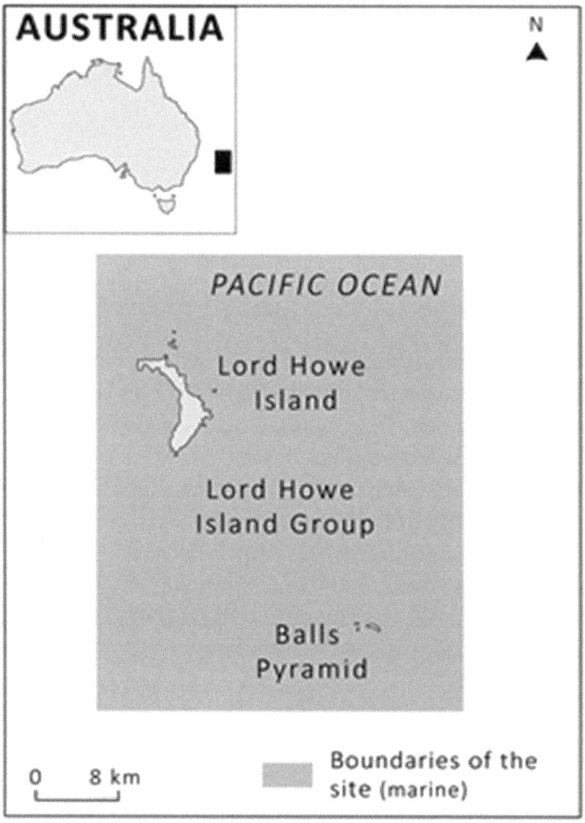

Fig. 43.1 Location of Lord Howe Island Group, in the South Pacific. The islands are politically part of Australia

ing and cultivating the land. With numerous fluctuations over the years, the settlement slowly expanded and consolidated, developing a distinctive social structure and culture (Davey 1986).

The community lives in a narrow isthmus of lowland country in the north-central part of Lord Howe Island. The island is an interesting example of restricted island settlement, although the World Heritage nomination was not made on cultural grounds. Tourist is today one important economic source of living for the settlers (Environment of Australia 2017).

The Lord Howe Island Act of 1953 from the state of New South Wales, Australia, established the "Lord Howe Island Board" a statutory authority responsible for managing the island. The board is composed of three nonlocal members appointed by the governor of NSW and 4 islanders elected among the 380 residents (Reis 2013).

43.2 Natural Settings

43.2.1 Climatic Conditions

Climate is humid subtropical with a mean temperature of 16 °C in August and 23 °C in February. Both diurnal and seasonal temperature ranges are about 7 °C. Mean annual rainfall in the lowlands is almost 1700 mm, with a pronounced maximum in winter and a mean rainfall of 100 mm in February. The southerly part of Lord Howe Island is generally wetter due to orographic effects. Relative humidity is high at 75–78% (UNEP/WCMC 2017).

Wind levels average 7 m/s in August and 5 m/s in January and March. The alternating influences of warm and cool ocean currents create a transition zone between temperate and tropical regions that contributes to an unusual mix of tropical, subtropical, and temperate marine fauna and flora and a high level of endemism (Environment of Australia 2017).

43.2.2 Geology and Geomorphology

Lord Howe Island is a remnant of a much larger Late Miocene basaltic shield volcano and is surrounded by a wide, near-horizontal shelf of water about 50 m deep (Dickson et al. 2004).

The volcanic evolution of the island began with deposition of tuff, which provided a subaerial foundation from which lava constructed a large shield volcano, about 6–9 million years ago. A caldera formed when the summit of the shield volcano collapsed, and a breccia was deposited from a vent within or close to the caldera. Eruptions ceased about 4–6 million years ago (McDougall et al. 1981). The islands now form the peak of a large volcanic seamount some 65 km long by 24 km wide, called the Lord Howe Rise. They mark the successive movement of the Australian tectonic plate over a hotspot in the upper mantle below (McDougall et al. 1981).

The coastal geomorphology around Lord Howe Island is spectacular and diverse. Precipitous basalt cliffs up to 800 m high plunge into the sea or have talus slopes at their base. Lower cliffs formed in a range of lithologies (Dickson et al. 2004). Superimposed on these cliffs are a suite of localized erosional landforms including sea caves, gulches, arches, and blowholes (Dickson 2002). There are also eolianites, which were deposited in the Middle and Late Pleistocene and which consist of calcarenite with isolated beach facies (Brooke et al. 2003).

Fig. 43.2 Lord Howe Island, enclosing a lagoon and limited by two volcanic pics to the south. (Picture reproduced by permission of Jack Shick Photography, webpage http://lordhoweisland-tours.net)

Fig. 43.3 North segment of Lord Howe Island, showing two volcanic pics that emerge abruptly from the seafloor, a characteristic of the island group. (Picture reproduced by permission of Jack Shick Photography, webpage http://lordhoweislandtours.net)

Lord Howe Island, the main island, measures 10 km north to south by 2 km wide, in a shallow crescent shape. It encloses a coral reef lagoon on its west and is anchored in the south by two mountainous volcanic plugs rising sheer from the sea (Figs. 43.2 and 43.3). Ball's Pyramid is a steep triangular stack of bare rock which rises a sheer 650 m (Dickson et al. 2004).

43.2.3 Biodiversity

The Lord Howe Island Group supports an extraordinary array of terrestrial and marine ecosystems and landscapes, a diverse flora and fauna, and a large number of endemic species and communities (Pickard 1983).

Forty-seven percent of the vascular plant species are endemic, and 50% of the pre-European species of land birds were endemic species or subspecies (Brown and Baker 2009).

The area is also exceptional for seabirds and is reputed to have more seabird species breeding and in higher numbers than anywhere else in Australia (Mcallan et al. 2004). More than 1600 terrestrial invertebrate species have been recorded in the islands, with up to 60% of the species being endemic (Cassis et al. 2003). No indigenous native mammals are known except for a small population of the large forest bat (Davey 1986).

The waters around Lord Howe Island provide an unusual mixture of temperate and tropical organisms, including many species at their distributional limits (UNEP/WCMC 2008).

At least 500 species of fish have been recorded, in 107 families of which 4% are unrecorded elsewhere except for the waters around Norfolk Island and Middleton Reef to the north; 400 are inshore species, and 15 are endemic (Davey 1986).

In relation to the corals, three distinctive community types were identified within the site – coral reefs, macroalgal beds, and an offshore/open coast community. The distribution of these community types is strongly related to wave exposure (Graham et al. 2010).

Lord Howe Island Group is largely covered in rainforest and palm forest, and most of the site (87%) has retained its original vegetation (Hunter 2002). There is a wide variety of vegetation types: lowland, montane, valleys, ridges, and areas exposed to maritime influence. Variable exposure to wind and salt spray appear to be the main determinants of vegetation occurrence, structure, and floristics (Pickard 1983).

There are 241 native species of vascular plants, in 25 plant associations, 20 alliances, and 14 sub-formations. There are 105 endemic species which dominate 14 of the associations, 16 of which are considered rare, endangered, or vulnerable (Pickard 1983). There are 160 naturalized introduced plant species, mostly, but not exclusively, in the lowland settlement area (UNEP/WCMC 2008).

43.3 Conservation Challenges

Threats on the Lord Howe Island Group vary in severity and geographical extent. The main threats include the introduction of new pests, diseases and fauna species, weed invasion, habitat clearing and modification, trampling, browsing and grazing by domestic stock, and climate change (Brown and Baker 2009). Fishing and increasing tourism are also pressuring the environment (UNESCO WHC 2017).

Nevertheless, the site has good management, and its values should remain preserved (IUCN 2017). The International Union for Conservation of Nature, in its last conservation outlook published at the end of 2017, found that the level of conservation of the site is good, without any concerns (Osipova et al. 2017).

References

Brooke BP, Woodroffe CD, Murray-Wallace CV, Heijnis HJ, Jones BG (2003) Quaternary calcarenite stratigraphy on Lord Howe Island, southwestern Pacific Ocean and the record of coastal carbonate deposition. Quat Sci Rev 22:859–880

Brown D, Baker L (2009) The Lord Howe Island biodiversity management plan: an integrated approach to recovery planning. Ecol Manage Restor 10:570–578

Cassis G, Meades L, Harris R (2003) Lord Howe Island terrestrial invertebrate biodiversity and conservation. Australian Museum, Sydney

Davey A (1986) Plan of management: Lord Howe Island Permanent Park preserve. New South Wales National Parks and Wildlife Service, Sydney

Dickson ME (2002) The development of rock coast morphology on Lord Howe Island, Australia. PhD thesis: School of Geosciences. University of Wollongong, Wollongong

Dickson ME, Kennedy DM, Woodroffe CD (2004) The influence of rock resistance on coastal geomorphology around Lord Howe Island, Southwest Pacific. Earth Surf Process Landf 29:629–643

Environment of Australia (2017) Australian Heritage Database. http://www.environment.gov.au/. Accessed 16 May 2017

Graham EJ, Davey D, Kelly G, Mawbey RB, Parsons K (2010) Biogeographical and ecological context for managing threats to coral and rocky reef communities in the Lord Howe Island Marine Park, Southwestern Pacific. Aquat Conserv Mar Freshwat Ecosyst 20:378–396

Hunter J (2002) Vegetation and habitat of significance within the settlement area of Lord Howe Island. South Wales National Parks and Wildlife Service, Coffs Harbour

IUCN (International Union for Conservation of Nature) (2017) Lord Howe Island Group Conservation Outlook. http://www.worldheritageoutlook.iucn.org/node/990. Accessed 23 Dec 2017

Mcallan IAW, Curtis B, Hutton I, Cooper HM (2004) The birds of Lord Howe Island Group: a review of records. Aust Field Ornithol 21(supplement):1–84

McDougall I, Embleton BJJ, Stone DB (1981) Origin and evolution of Lord Howe Island, Southwest Pacific Ocean. J Geol Soc Aust 28:155–176

Osipova E, Shadie P, Zwahlen C, Osti M, Shi Y, Kormos C, Bertzky B, Murai M, Van Merm R, Badman T (2017) IUCN world heritage outlook 2: a conservation assessment of all natural world heritage sites. IUCN, Gland

Pickard J (1983) Vegetation of Lord Howe Island. Cunninghamia 1:133–266

Reis A (2013) Pronounced particularity: a comparison of governance structures on Lord Howe Island and Fernando de Noronha. Island Stud J 8:285–298

UNEP/WCMC (United Nations Environment Programme/World Conservation Monitoring Centre) (2008) Lord Howe Island group, new South Wales, Australia. Report, Cambridge, England

UNESCO WHC (United Nations Educational, Scientific and Cultural Organization World Heritage Centre) (2017). Lord Howe Island Group. http://whc.unesco.org/en/list/186. Accessed 17 May 2017

Chapter 44
Shark Bay, Australia

Abstract Shark Bay is on the East Indian Ocean in Western Australia. It is composed of a marine segment and a coastal segment. In the marine segment, there are numerous islands, a bank of seagrass that forms the biggest beds of that type in the world, and a large population of sea cow. In the littoral, microbial activity creates living deposits of stromalitos, dome-shaped deposits that are among the oldest forms of life on Earth, contributing to a singular seascape in the intertidal area. The hydrologic structure of Shark Bay has produced a basin where marine waters are hypersaline and contributed to the presence of extensive beaches consisting entirely of shells. Shark Bay is also one of the few marine areas in the world dominated by carbonates not associated with reef-building corals.

44.1 Introduction

Shark Bay is at the westerly point of Australia. It covers an area of 2200 km^2, of which about 70% are marine waters (UNESCO WHC 2016). It averages 100 km from the mainland (UNEP/WCMC 2008). The coastline of Shark Bay World Heritage site is more than 1500 km long (Department of Environment of Australia 2006) (Fig. 44.1).

The property is composed of marine parks, marine nature reserves, terrestrial nature reserves, public land, and pastoral land. The largest of these preserves is the Shark Bay Marine Park. The oldest is from 1957. The entire area became a World Heritage site in 1991 (UNEP/WCMC 2008).

The Shark Bay has been home to aboriginal people for more than 40,000 years (Playford et al. 2013). After 1850, it was occupied by guano miners, pearlers, fishermen, sandalwood cutters, and pastoralists. Pearling was the main industry from 1850s to the 1940s. The fishing declined with the introduction of regulations introduced to prevent overexploitation of fish stocks (UNEP/WCMC 2008).

In the 1960s, a coastal highway was sealed and the area for the first time opened to tourists. In 1985, the road was paved and tourist numbers increased dramatically (UNEP/WCMC 2008). Today, tourism in Shark Bay is a very important element of the economy. The Western Australian Government is responsible for the management of the Shark Bay World Heritage Area (Department of Environment of Australia 2006).

V. Claudino-Sales, *Coastal World Heritage Sites*, Coastal Research Library 28,
https://doi.org/10.1007/978-94-024-1528-5_44

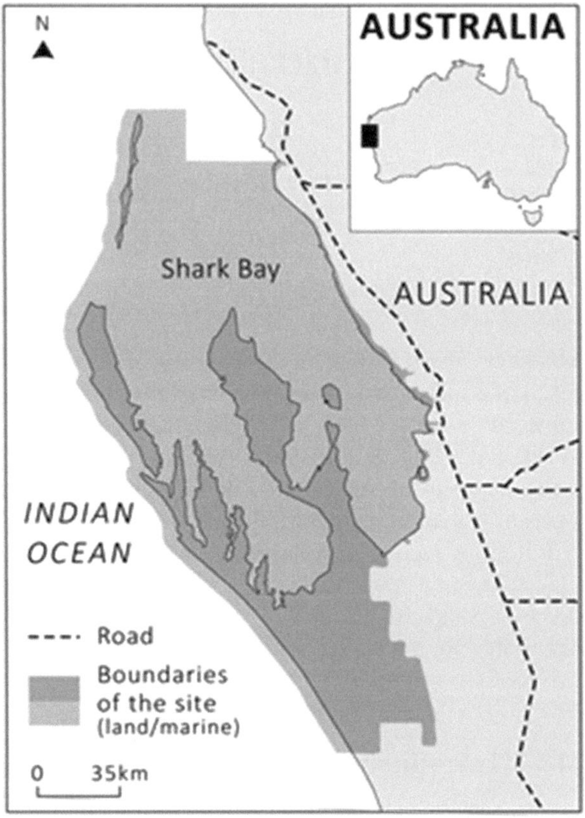

Fig. 44.1 Location of Shark Bay, in Western Australia

44.2 Natural Settings

44.2.1 Climatic Conditions

Shark Bay climate is semiarid to arid, characterized by hot dry summers and mild winters. Summer temperatures range between 20 and 35 °C. Winter temperatures are between 10 °C and 20 °C. The annual precipitation is low, ranging from 200 mm to 400 mm. Annual evaporation is high, between 2000 mm and 3000 mm in the east (UNEP/WCMC 2008).

The area suffers the interference of the Ningaloo Niño, which represents an episodic occurrence of anomalously warm ocean (Feng et al. 2015), caused by changes in the poleward flowing Leeuwin Current that flows along the western coast of Australia due to both local and remote wind forcing (Benthuysen et al. 2014).

The Ningaloo Niño develops during the spring in the South Hemisphere, having peaks in summer and decaying in fall (Feng et al. 2015). It influences the precipitation (Tozuka et al. 2014).

Fig. 44.2 Seacliffs in faulted Cretaceous limestones on the coast of Shark Bay. (Picture reproduced by permission Dragi Markovic and Department of Environment of Australia)

44.2.2 Geology and Geomorphology

Shark Bay is a large semi-enclosed bay on the coast of the Indian Ocean lying behind a chain of barrier islands. The area basement rock is Late Cretaceous limestone and chalk (UNEP/WCMC 2008). There are also Holocene carbonates.

The outstanding marine feature of the bay is the steep salinity gradient. Water exchange with the ocean is restricted, except by the presence of ocean water brought by the Leeuwin Current that flows along the coast of Western Australia. The Holocene hypersaline carbonate system is represented by microbial deposits (stromatolites) and a coquina beach ridge system (Jahnert et al. 2012).

The salinity ranges from oceanic (35–40 ppt) to hypersaline (56–70 ppt) (UNEP/WCMC 2008). The loss of freshwater through evaporation causes salinity levels in Shark Bay to be higher than the Indian Ocean adjacent, leading to its classification as an inverse estuary (Hetzel et al. 2013).

In the coastal area, the limestones appear especially as seacliffs, forming a striking feature of the geomorphology of the area. They are present for some 200 km long and up to 250 m high (Government of Western Australia 2016). The cliffs are considered a fault-line scarp, resulting from normal faulting in Late Pleistocene or Early Holocene (Playford et al. 2013) (Fig. 44.2).

The living stromatolites of Shark Bay constitute one of the largest and most diverse occurrences in the world. The microbial carbonate system developed in response to environmental changes, from near-open marine into restricted embay-

Fig. 44.3 Living stromatolites in the intertidal area of Shark Bay coast, Western Australia. (Picture reproduced by permission of Monica Johansen/Shutterstock Stock Photos)

ment with hypersalinity and high evaporation. The deposits started to be built around 2000 years BP (Jahnert and Collins 2011). They form dome-shaped deposits and are among the oldest forms of life on Earth (UNESCO WHC 2016), responding for a singular seascape in the intertidal area (Fig. 44.3).

The coquina beach system overlies the Pleistocene coastal formation and is prograding seaward, over the supratidal-intertidal microbial deposits, as a consequence of Late Holocene sea level fall, and also because of the high volume of bivalve shells available. The process has gone on for 4500 years (Jahnert et al. 2012).

The hydrologic structure of Shark Bay has produced a basin where marine waters are hypersaline and contributed to the presence of extensive beaches consisting entirely of shells. Shark Bay is also one of the few marine areas in the world dominated by carbonates not associated with reef-building corals (UNESCO WHC 2016).

44.2.3 Biodiversity

Shark Bay is of great botanical and zoological importance due to the long isolation of its island and peninsular ecosystems. In addition, it is in the limits of the transition between temperate and tropical zones, and because of this, many species are at the limits of their geographical range (UNEP/WCMC 2008).

On the terrestrial segment, the flora is adapted to a semiarid and arid environment. More than 620 species are recorded for the region, at least 51 being endemic (UNEP/WMCW 2008).

The marine flora is characterized by the presence of vast seagrass beds, which with 4800 km^2 are the largest and richest in the world (UNESCO WHC 2016). The seagrass system works as a channel-bank complex. Its construction seems to be related to Holocene sea level fluctuations (Bufarale and Collins 2015). Over the last 5000 years, seagrasses have modified the physical, chemical, biological, and geological environment of the bay, forming marine barriers which forbid the penetration of ocean waters, creating, together with the low rainfall and high evaporation, the hypersaline conditions that characterize the area (UNESCO WHC 2016; UNEP/WCMC 2008).

The region has a rich avifauna, and over 230 species or 35% of Australia's bird species have been recorded. It is also noted for the diversity of its amphibians and reptiles, supporting nearly 100 species. Of the 26 species of endangered Australian mammals, five are found in Shark Bay's islands (Department of Environment of Australia 2006).

The bay is known for its marine fauna: 323 species of fish have been recorded, of which 83% are tropical, 11% are warm temperate, and 6% are cool temperate (UNEP/WCMC 2008).

The population of 14,000 sea cows is claimed to be 12.5% of the world population. Wild Indo-Pacific dolphins regularly come to be fed at the islands. The only other similar long-term human-wild dolphin interactions are at Banc d'Arguin in Mauritania (see Sect. 1.1) and in northern Brazil (UNEP/WCMC 2008).

44.3 Conservation Challenges

The ecosystems in Shark Bay appear relatively unaltered by human impact. Some potential threats could be from improved technology in producing drinking water, increased tourism and residential density, upgrading of road access, agricultural developments, expansion of gypsum mining, and the introduction of intensive aquaculture and fishing technologies (UNESCO WHC 2016).

Climate change could also impact the complex ecosystem. The area can suffer desertification, considering its semiarid conditions, having diminished the levels of annual rainfall. The hypersaline environment can increase the amount of salt and change the ecological characteristics of the environment. Sea level rising due to global warming can bring more erosion to the coastline.

Nevertheless, the site has a good conservation outlook, thanks to the strength of its management in the face of anthropogenic impacts, its relative inaccessibility, and the appropriateness of the boundaries of the area (IUCN 2017). The last conservation outlook published by the International Union for Conservation of Nature (IUCN) in the end of 2017 found that the level of conservation of the site is good without any concerns (Osipova et al. 2017).

References

Benthuysen J, Feng M, Zhong L (2014) Spatial patterns of warming off Western Australia during the 2011 Ningaloo Niño: quantifying impacts of remote and local forcing. Cont Shelf Res 91:232–246

Bufarale G, Collins LB (2015) Stratigraphic architecture and evolution of a barrier seagrass bank in the mid-late Holocene, Shark Bay, Australia. Mar Geol 359:1–21

Department of Environment of Australia (2006) Shark Bay, Western Australia. https://www.environment.gov.au/system/files/pages/758ea283-6b9d-41bc-a29a-1276e1c7fced/files/shark-bay-factsheet.pdf. Accessed 23 June 2016

Feng M, Hendon HH, Xie SP, Marshall AG, Schiller A, Kosaka Y, Caputi N, Pearce A (2015) Decadal increase in Ningaloo Niño since the late 1990s. Geophys Res Lett 42(1):104–112

Government of Western Australia (2016) The geology of Shark Bay. http://www.dmp.wa.gov.au/Geological-Survey/The-geology-of-Shark-Bay-1539.aspx. Accessed 22 June 2016

Hetzel Y, Pattiaratchi C, Lowe R (2013) Intermittent dense water outflows under variable tidal forcing in Shark Bay, Western Australia. Cont Shelf Res 66:36–48

IUCN (International Union for Conservation of Nature) (2017) Shark Bay conservation outlook. http://www.worldheritageoutlook.iucn.org/node/1050. Accessed 23 Dec 2017

Jahnert RJ, Collins LB (2011) Significance of subtidal microbial deposits in Shark Bay, Australia. Mar Geol 286(1):106–111

Jahnert RJ, de Paula O, Collins LB, Strobach E, Pevzner R (2012) Evolution of a coquina barrier in Shark Bay, Australia by GPR imaging: architecture of a Holocene reservoir analog. Sediment Geol 281:59–74

Osipova E, Shadie P, Zwahlen C, Osti M, Shi Y, Kormos C, Bertzky B, Murai M, Van Merm R, Badman T (2017) IUCN world heritage outlook 2: a conservation assessment of all natural world heritage sites. IUCN, Gland

Playford PE, Cockbain AE, Berry PF, Roberts AP, Haines HW, Brooke BP (2013) The geology of Shark Bay. Geol Surv W Aust Bull 146:1–146

Tozuka T, Kataoka T, Yamagata T (2014) Locally and remotely forced atmospheric circulation anomalies of Ningaloo Niño/Niña. Clim Dyn 43:2197–2244

UNEP/WCMC (United Nations Environmental Programme/World Conservation Monitoring Centre) (2008) Shark Bay Western Australia. Report, Cambridge, England

UNESCO WHC (United Nations Educational, Scientific and Cultural Organization) (2016) Shark Bay, Western Australia. http://whc.unesco.org/en/list/578. Accessed 23 June 2016

Chapter 45
Tasmanian Wilderness, Australia

Abstract These scenic parks and reserves are mixed cultural and natural World Heritage site. It comprises a contiguous network of reserve lands extending over much of west-central and southwest Tasmania. The site is geologically and geomorphologically complex, representing the product of at least three major glaciations, and covers an area of more than 1 million hectares. It constitutes one of the last expanses of temperate rainforest in the world. The altitude goes from sea level to 1617 m, and the area is characterized by the presence of steep gorges. Remains found in caves attest to human occupation of the area for more than 35,000 years. The Aborigines who lived there represent the longest isolated human group in world history, surviving hundreds of generations without outside influence.

45.1 Introduction

Tasmanian Wilderness is located on the island of Tasmania, a state of Australia. It comprises a contiguous network of reserve lands extending over much of west-central and southwest Tasmania. The total area of the site is 13.836 km^2, about 20% of the area of the island (UNEP/WCMC 2005) (Fig. 45.1).

The site is formed by four national parks, three forest reserves, three state reserves, two protected archaeological sites, two historic sites, four conservation areas, one game reserve, three commonwealth freehold, and private lands. The preservation of these sites by the government started in 1922 and evolved until 1991 (UNEP/WCMC 2005). In 1977, one of the national parks was designated a Wetland of International Importance under the Ramsar Wetland Convention. It became a natural World Heritage site in 1982 and a cultural World Heritage site in 1993 (UNESCO WHC 2017).

V. Claudino-Sales, *Coastal World Heritage Sites*, Coastal Research Library 28,
https://doi.org/10.1007/978-94-024-1528-5_45

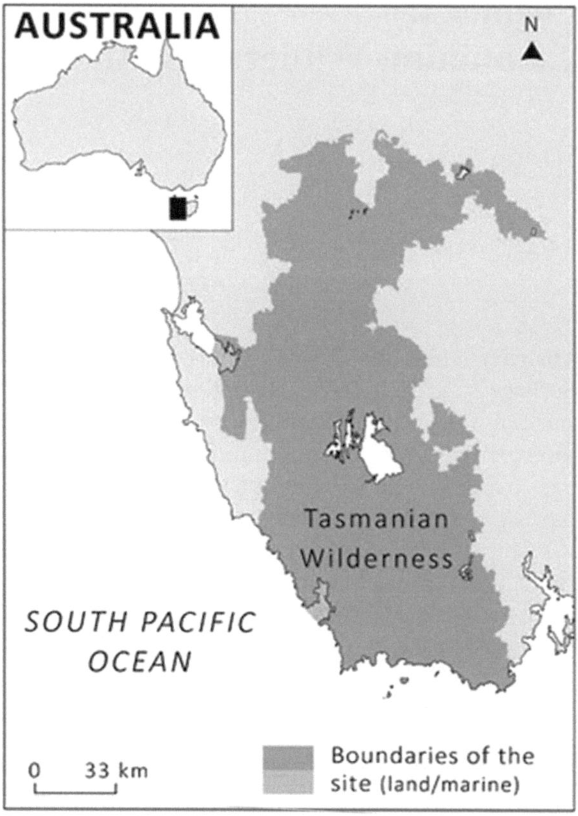

Fig. 45.1 Tasmanian Wilderness in Tasmania, Australia

The Tasmanian Aborigines were, until the advent of the Europeans two centuries ago, the longest isolated human group in world history, surviving some 500 generations without outside influence. Archaeological data indicates the existence of a Pleistocene Ice Age hunter-gatherer society on the island, which existed from at least 35,000 years and possibly as long as 70,000 years (Australian Government 2017a). Aboriginal people used fire as a sophisticated tool to continually manage and modify the landscape in Tasmania. They left many archaeological registers, from extensive and impressive rock markings to isolated artifacts. They were removed from the island in the 1830s (Aboriginal Heritage Tasmania 2017).

The site is an important tourist destination in Australia, with many recreational offerings, such as walking, camping, boating, fishing, kayaking, rafting, and climbing (The Official Site for Australian Travel and Tourism 2017). All the land is Crown property administered by the Ministry for Tourism, Parks and Heritage, through the Parks and Wildlife Service of Tasmania, which is responsible for more than 99% of the total area (UNEP/WCMC 2005).

45.2 Natural Settings

45.2.1 Climatic Conditions

Tasmania has a temperate maritime climate, characterized by wet, windy, cloudy, and cool conditions. Rainfall ranges from about 1800 mm to more than 3400 mm a year (Bosworth 1979). Average annual temperatures range between the low teens in winter and the low twenties in summer (Australian Government 2017b).

In summer, wind speeds are around 8 m/s, and the speed can double in winter. During the storms, wind speeds can be around 27 m/s (100 km/h). The dominant direction is NW (University of Tasmania 2017).

45.2.2 Geology and Geomorphology

In contrast with mainland Australia, Tasmania is rugged and forested and is geologically and geomorphologically complex. The western half of the island is underlain by fold structures, the eastern half by fault structures. The area contains Australia's greatest array of landscapes and geological types, including rocks from all but one geological period (PWST 2017).

Precambrian formations are widespread and consist of quartzite, schist, phyllite, conglomerate, dolomite, siltstone, and sandstone. The more resistant sequences form most of the ranges (Fig. 45.2), with altitudes of 1617 m. The less resistant rocks underlie many of the valleys and plains. The area was subjected to two orogenies (UNEP/WCMC 2005).

The landscape reveals features associated with the separation of the Australian and Antarctic plates during the latter stages of the breakup of Gondwana. The final stages of the breakup of Gondwana during the Tertiary resulted in widespread faulting, producing a landscape very similar to that occurring today (PWST 2017).

The dramatic landscapes of the Tasmanian Wilderness reveal rich insights into the forces that shape the Earth's surface. The rugged mountain landscapes are largely the legacy of at least three major glaciations during the Pleistocene. Thousands of highland lakes and tarns, cirques, and U-shaped valleys and gorges have been carved by the massive erosional force of glaciers (Australian Government 2017a) (Fig. 45.3).

Depositional features such as outwash gravels and moraines which hold back the waters of lakes are typical features of the landscape. Extensive areas of limestone, in places up to 2 kilometers thick and considered to be the best developed of its kind in the world, are a major feature of the site. Fluctuating sea levels during the glaciations also provided conditions for the development of a wide range of coastal features (PWST 2017).

Fig. 45.2 Mountainous coastal landscape sustained by Precambrian quartzite in the Tasmanian Wilderness, with sandy beaches and seacliffs in its feet. (Picture reproduced by permission of Judy McIlrath, University of South Florida)

45.2.3 Biodiversity

The site is one of the three largest remaining temperate wilderness regions in the world. Because of the isolation, the fauna is of world importance. It includes an unusually high proportion of endemic species and ancient relict groups. The varied topography, geology, soils, and vegetation together with the harsh and variable climatic conditions have combined to create a wide array of habitats and a correspondingly diversely adapted fauna (UNEP/WCMC 2005).

Within the forests are the descendants of some of the most ancient of Australia's plants. Some species date back over 60 million years. The ancestors of many rainforest species evolved on the ancient continent of Gondwana (PWST 2017). There are four main vegetation types in the protected areas: closed temperate rainforest, open eucalypt forest, and woodland and alpine and subalpine communities. There are also coastland, wetlands, and grasslands. These occur in a unique mosaic of Antarctic and Australian elements which has much in common with the cool temperate regions of South America and New Zealand (UNEP/WCMC 2005).

The region provides pristine habitats for a range of plants and animals that are found nowhere else in the world, including many rare and endangered species. For a number of animals which have become extinct on mainland Australia in recent times, the area offers a last refuge (PWST 2017).

Fig. 45.3 Glacial-carved gorge in the sandstones of the Tasmanian Wilderness. (Picture reproduced by permission of Judy McIlrath, University of South Florida)

Twenty-seven mammals are present in the site. Four are endemic to Tasmania, including the world's largest extant carnivorous marsupial. Thirteen of the 150 bird species recorded are also endemic, and there are 11 reptile species. The rainforest invertebrate fauna includes many groups of Gondwanan descent. Within aquatic habitats, the freshwater crustaceans are of global significance, as are many amphipods, isopods, crayfish, and shrimps which are relictual Gondwanan fauna (PWST 2017).

45.3 Conservation Challenges

Natural and cultural heritage specialists have studied the main threats and pressures that can cause impacts on the values of the World Heritage site (e.g. Xiaojiang 2016). They are (PWST 2017) the illegal cutting and removal of timber, the illegal introduction of trout into trout-free lakes, unlawful lighting of fires, removal of mineral specimens, unauthorized track cutting into remote areas, and the spreading of weeds.

The effects of climate change, global warming, and related sea level rise impose degradation and loss of alpine environments and associated communities. It is also increasing coastal erosion with associated loss of significant coastal aboriginal midden sites, producing alteration to natural rates and magnitudes of change in the region's drainage system (Australian Government 2017a).

The last conservation outlook published by the International Union for Conservation of Nature (IUCN) at the end of 2017 found that the level of conservation of the Tasmanian Wilderness World Heritage site is good but presents concerns, such as those cited above (Osipova et al. 2017).

References

Aboriginal Heritage Tasmania (2017) Aboriginal Culture Heritage. http://www.aboriginalheritage.tas.gov.au/aboriginal-cultural-heritage. Accessed 21 May 2017

Australian Government (2017a) World heritage places – Tasmanian wilderness. Department of Environment and Energy. http://www.environment.gov.au/heritage/places/world/tasmanian-wilderness. Accessed 21 May 2017

Australian Government (2017b) Bureau of Meteorology. Climate in Tasmania. http://search.bom.gov.au/s/search.html?query=climate+in+tasmania&collection=bom&Submit=Search. Accessed 22 May 2017

Bosworth P (1979) A collation of meteorological information for South-West Tasmania. South West Tasmania Resources Survey Occasional Paper 2

Osipova E, Shadie P, Zwahlen C, Osti M, Shi Y, Kormos C, Bertzky B, Murai M, Van Merm R, Badman T (2017) IUCN world heritage outlook 2: a conservation assessment of all natural world heritage sites. IUCN, Gland

PWST (Parks & Wildlife Service Tasmania) (2017) World heritage values. http://www.parks.tas.gov.au/index.aspx?base=7450. Accessed 22 May 2017

The Official Site for Australian Travel and Tourism (2017) Tasmanian wilderness. http://www.australia.com/en-us/search.html?q=tasmanian+wilderness. Accessed 22 May 2017

UNEP/WCMC (United Nations Environmental Programme/World Conservation Monitorin Centre) (2005) Tasmanian wilderness Australia. Report, Cambridge, UK

UNESCO WHC (United Nations Educational, Scientific and Cultural Organization World Heritage Centre) (2017) Tasmanian Wilderness. http://whc.unesco.org/en/list/181. Accessed 22 May 2017

University of Tasmania (2017) Wind Forecast. https://wind.willyweather.com.au/tas/hobart/university-of-tasmania.html. Accessed 28 Dec 2017

Xiaojiang Y (2016) Protect the Tasmanian wilderness. Science 351(6275):824–825

Chapter 46
Malpelo Fauna and Flora Sanctuary, Colombia

Abstract Malpelo Fauna and Flora Sanctuary is an isolated oceanic habitat and a large marine protected area 506 km off Colombia's Pacific Coast. The area represents the largest no-fishing zone in the Eastern Tropical Pacific and lies within an oceanic system known as the Eastern Tropical Pacific Marine Wilderness, unified by the movements of major oceanic currents and countercurrents related to the winds. The site is a major source of nutrients resulting in large aggregations of marine biodiversity. The rugged underwater topography reaches a depth of 3400 m. The varied underwater mountain seascape and the excellent state of conservation make the site an "ocean oasis." It has some of the best diving in the world and is terrestrially barren and unlikely to be developed.

46.1 Introduction

Malpelo Island stands isolated in the Pacific Ocean, 506 km west of the Pacific coast of Colombia. It has an area of 8575 km^2, with a land area of 3.5 km^2 and a marine area of 8571 km^2. Fewer than 3% of World Heritage sites are predominantly marine, and Malpelo is one of them (UNESCO WHC 2017) (Fig. 46.1).

In 1995, Malpelo Island was designated a Flora and Fauna Sanctuary by the Colombian Government. The protected area was extended in 1996, 2002, and 2005. The International Maritime Organization declared the sanctuary a Particularly Sensitive Sea Area in 2005, and the Bird Life International designed the area of importance to the preservation of birds in the same year. It was made a World Heritage site in 2006 (UNEP/WCMC 2006).

Malpelo was one of the first Pacific oceanic islands to be discovered by Europeans, from a Spanish galleon in 1530 (Lonsdale and Fornari 1980). It has no harbors or running water so was probably never used by passing sailors. The island

V. Claudino-Sales, *Coastal World Heritage Sites*, Coastal Research Library 28,
https://doi.org/10.1007/978-94-024-1528-5_46

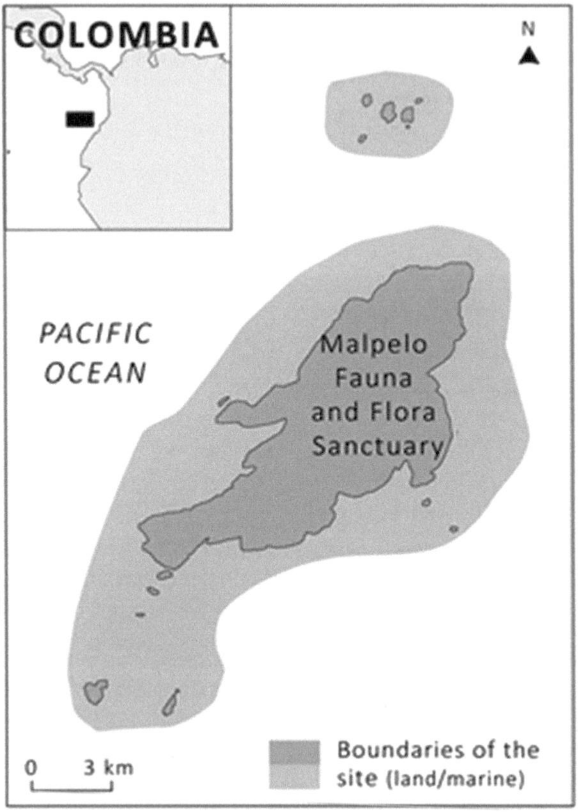

Fig. 46.1 Malpelo Island, located off the Pacific coast of Colombia

is uninhabited except for six soldiers, rotated biannually, plus ten more from a patrol boat. Fishermen visit illegally from the mainland (UNEP/WCMC 2006).

Diving is an important activity in the marine park and is practiced only with authorization of the manager. The varied underwater mountain seascape and the excellent state of conservation make the site an "ocean oasis." It is terrestrially barren and unlikely to be developed (UNESCO WHC 2017).

Malpelo Flora and Fauna Sanctuary is under the administration of the system of Natural National Parks of Colombia, linked to the Ministry of Environment, with administrative headquarters in the city of Santiago de Cali. The land tenure is of the Colombian Government (PNNC 2017).

A large part of the property's marine secrets remain to be discovered, presenting a great potential for research in an almost pristine natural environment (UNESCO WHC 2017).

46.2 Natural Settings

46.2.1 Climatic Conditions

The island's climate is tropical and dictated by the cycles of the surrounding marine currents and the intertropical convergence zone. Trade winds converging from the northeast and southeast push the warm air up, producing heavy rainfall along the fluctuating equatorial front (Villegas 2007).

Malpelo is subject to the variable winds and various sea currents of the region. From December to May, conditions are calm, cold, and clear, and from June to November, they are warm and rainy. The temperatures generally are not over 27 °C (UNEP/WCMC 2006).

46.2.2 Geology and Geomorphology

Malpelo Island lies within an oceanic system known as the Eastern Tropical Pacific Marine Wilderness, an area that extends from the California Peninsula to Peru, unified by the movements of major oceanic currents and countercurrents related to the winds (UNEP/WCMC 2006). This marine conservation network also includes World Heritage properties in Costa Rica, Ecuador, and Panama (UNESCO WHC 2017).

The island appears as the elevated segment of the Malpelo Ridge, which is the oldest part of a volcanic trail left by passage of the Cocos Plate over a "hotspot" centered near the Cocos-Nazca microplates spreading axis (Blakley 1974). Malpelo Island is formed by a series of massive lava flows with intercalated volcanic breccia and tuff that dips monotonically to the northwest and is cut by steep fault zones (Lonsdale and Klitgord 1978). It is an aseismic block that rises from abyssal depths of 3400 m to altitudes of ca 370 m above sea level. It was formed between 17 and 20 million years ago (Lonsdale and Fornari 1980).

The island makes an isolated seamount. It is wet but barren, rugged, and inhospitable (Fig. 46.2), in relatively cool waters, with steep submarine walls and sheer cliffs, tunnels, and caves encrusted with algae and invertebrates (PNNC 2017).

46.2.3 Biodiversity

The wide range of oceanic influences such as the large number of marine currents adds to the diversity of habitats and species in the site. The island and its satellite rocks boast a limited but highly specialized terrestrial biodiversity characterized by a high degree of endemism (UNESCO WHC 2017).

Fig. 46.2 Marine birds in a steep seacliff with barren rocks in Malpelo Fauna and Flora Sanctuary. (Picture reproduced by permission of Dr. Fernando A. Zapata, Universidad del Valle)

For the terrestrial flora, scientific data show that 25 species of lichens, 2 species of vascular plants, and 1 moss were identified in the island. Lichens were the most diverse group found, including records of four new genera (González-Román et al. 2014).

The terrestrial fauna is adapted to the barren conditions and the deposits of guano which are the basis of the island's ecology, both on land and under sea. The terrestrial fauna includes 5 reptiles and 61 species of birds (with 24,034 individuals of masked booby, the second largest colony of this bird in the world: Pitman et al. 1995); dragonflies have been discovered in the last decade (Bermudez and Lopez-Victoria 2009).

Malpelo Island and the surroundings waters are a conservation gem and contribute to the maintenance, dispersal, and replenishment of benthic larvae of corals (Fig. 46.3), fish and mollusks, and other marine life (UNESCO WHC 2017; Hausdorf et al. 2012).

The island provides food for an abundance of marine mammals (17 species of marine mammals have been reported) and turtles, and schools of large pelagic fish and sharks, which become more obvious when the winter thermocline rises to 15 m (UNEP/WCMC 2006). High levels of fish biomass and functional richness suggest that Malpelo is one of the most pristine and vulnerable sites on the Eastern Tropical Pacific (Quimbayo et al. 2017).

Fig. 46.3 Corals growing in the volcanic substrate of Malpelo Ridge, in the Eastern Tropical Pacific Ocean, on the Malpelo Fauna and Flora Sanctuary World Heritage site. (Picture reproduced by permission of Dr. Fernando A. Zapata, Universidade del Valle)

46.3 Conservation Challenges

The vast majority of the site is a "marine wilderness," constituting the largest no-fishing zone in the Eastern Tropical Pacific. It is free of alien invasive species and represents a baseline for untouched assemblages. For this reason, it is a priority area for conservation at the national and international level (Quimbayo et al. 2017).

The remoteness of the island means a high degree of natural protection. Nevertheless, the area is under increasing pressure imposed by several threats, such as illegal fishing, marine traffic, tourism, and climate change (UNESCO WHC 2017). Global warming is putting pressure in the coral community and may change the overall ecosystem of the sanctuary in the future.

The International Union for Conservation of Nature (IUCN) published a report about the state of conservation of the World Heritage sites at the end of December 2017. The document found that the conservation outlook for Malpelo Flora and Fauna Sanctuary is positive, compared to the situation of many marine protected areas in the world. This is because of the remoteness of the property, the existence of a large no-take area, and the encouraging conservation efforts by governmental and non-governmental actors, including foundations, research institutions, and tourism operators (IUCN 2017).

Illegal fishing can be considered the largest current threat to the site, with both illegal artisanal and illegal industrial fishing occurring within and around the marine protected zone, including by foreign vessels (IUCN 2017). For these reasons, the state of conservation of the site was found to be good, with some concerns (Osipova et al. 2017).

References

Bermudez RC, Lopez-Victoria M (2009) First records of dragonflies (Odonata: Anisoptera) of Malpelo Island, Colombia. Rev Colomb Entomol 35:286–287

Blakely R (1974) Geomagnetic reversals during the Miocene. J Geophys Res 79:2987–2985

González-Román RD, López-Victoria M, Silverstone-Sopkin PA (2014) Terrestrial flora of Malpelo Island, Colombia, eastern tropical Pacific. Rev Biol Trop 62(1):327–336

Hausdorf B, Kroll O, López-Victoria M (2012) The land snails of Malpelo Island, Colombia. J Molluscan Stud 78:157–165

IUCN (International Union for Conservation of Nature) (2017) Malpelo flora and fauna sanctuary conservation outlook. http://www.worldheritageoutlook.iucn.org/explore-sites/wdpaid/902899. Accessed 24 Dec 2017

Lonsdale P, Fornari D (1980) Submarine geology of Malpelo ridge, Panama Basin. Mar Geol 36:65–83

Lonsdale P, Klitgord KD (1978) Structure and tectonic history of the eastern Panama Basin. Geol Soc Am Bull 89:981–999

Osipova E, Shadie P, Zwahlen C, Osti M, Shi Y, Kormos C, Bertzky B, Murai M, Van Merm R, Badman T (2017) IUCN world heritage outlook 2: a conservation assessment of all natural world heritage sites. IUCN, Gland, Switzerland

Pitman RL, Spear LB, Force MP (1995) The marine birds of Malpelo Island. Colon Waterbirds 18:113–119

PNNC (Parques Naturales Nacionales de Colombia) (2017) Santuario de Fauna y Flora Malpelo. http://www.parquesnacionales.gov.co/portal/es/ecoturismo/region-pacifico/santuario-de-flora-y-fauna-malpelo/. Accessed 24 May 2017

Quimbayo J, Mendes T, Kulbicki M, Floeter S, Zapata F (2017) Unusual reef fish biomass and functional richness at Malpelo, a remote island in the tropical eastern Pacific. Environ Biol Fish 100:149–162

UNEP/WCMC (United Nations Environmental Programme/World Conservation Monitoring Centre) (2006) Malpelo fauna and flora sanctuary. Report, Cambridge

UNESCO WHC (United Nations Educational, Scientific and Cultural Organization World Heritage Centre) (2017) Malpelo fauna and flora sanctuary. http://whc.unesco.org/en/list/1216. Accessed 24 May 2017

Villegas B (2007) Colombia National Parks. Villegas Editors, Bogota

Chapter 47
Cocos Island National Park, Costa Rica

Abstract Cocos Island ("Isla del Coco"), 530 km off the Pacific coast of Costa Rica, is the only island in the Eastern Tropical Pacific to be colonized by tropical rainforest. It is the only landmark of a line of submerged volcanoes formed over the margin of the Cocos and Pacific tectonic plates. The relief and coastline are rugged, and many rocks and cliffs rise almost vertically from a narrow shore. It harbors a rare and complex mosaic of land and sea environments, including forested mountains, rivers, waterfalls, estuaries, rocky beaches, bays, coral reefs – the most extensive in the East Pacific – and large pelagic habitats. It is also a magnificent diving site and one of the best places in the world to see large species such as sharks, rays, tuna, and dolphins. Is it known as the "sharkiest place in the sea."

47.1 Introduction

Cocos Island (also known as "Isla del Coco"), in the Pacific Tropical Eastern, is 532 km far away from the Pacific coast of Costa Rica. The area of the island is 24 km^2, and the surrounding marine ecosystems have 1974 km^2. The total area is 1998 km^2. It represents the only oceanic island of Central America (Bolaños et al. 2012) (Fig. 47.1).

In 1978, Cocos Island was designated a national park by the Costa Rican Government. The park was extended in 1984, 2002, and 2011, to include all marine ecosystems within 22 km of the island (Earle 2012). It was declared a wetland of importance under the Ramsar Wetland Convention in 1998 and became a World Heritage site in 1997 (UNEP/WCMC 2008).

In 2003, the United Nations Foundation and the UNESCO World Heritage Center launched the Marine Conservation Corridor of the Pacific Tropical Eastern, a 211 million hectare area that encompasses the marine protected areas of Cocos (Costa Rica), Gorgona and Malpelo (Colombia), Coiba (Panamá), and the Galápagos (Ecuador) islands and the oceanic waters that link them (USGS 2017). In this area, fishing is not allowed.

Cocos Island is owned by the Government of Costa Rica, which took official possession of it in 1869. It is managed by the National Park Service of the Ministry

V. Claudino-Sales, *Coastal World Heritage Sites*, Coastal Research Library 28,
https://doi.org/10.1007/978-94-024-1528-5_47

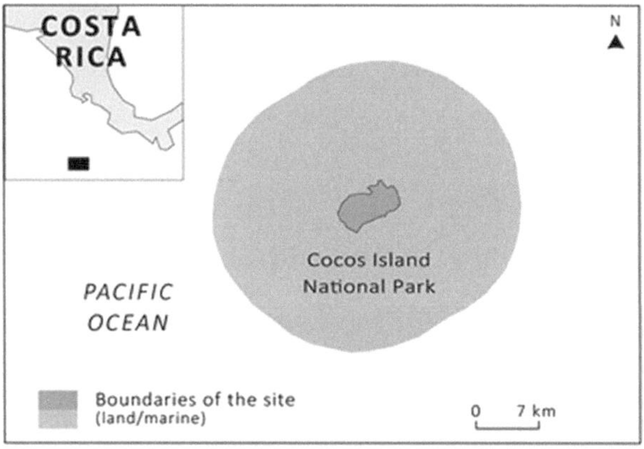

Fig. 47.1 Location of Cocos Island, Costa Rica, in the Pacific Tropical Eastern

of Environment and Energy (UNEP/WCMC 2008). The history of exploration of Cocos Island and the surroundings dates from the sixteenth century. The uninhabited island has always attracted seafarers, pirates, whalers, and fishermen and, more recently, scientists and divers (Earle 2012; Cortéz 2008).

47.2 Natural Settings

47.2.1 Climatic Conditions

The tropical climate of the island is controlled by the movements of the intertropical convergence zone (ITCZ). The ITCZ produces constant nebulosity, abundant precipitation, and persistent winds, which come from southeast. The annual average rain is very high, almost 7000 mm; the mean temperature is 25.5 °C all over the year (USGS 2017; Alfaro 2008).

The climate is strongly influenced by a complex system of ocean currents, but especially by the warm eastward-flowing North Equatorial Countercurrent. The island is also periodically affected by the southern oscillation of El Niño (UNEP/WCMC 2008).

47.2.2 Geology and Geomorphology

Cocos Island is the only point above sea level of the Cocos Ridge, a submarine feature that runs from Costa Rica to the Galápagos. It is a landmark of this line of submerged volcanoes formed over the margin of the Cocos and Pacific tectonic plates (Rojas and Alvoredo 2012).

Cocos Island is part of a series of young volcanoes on an old hotspot trace (Castillo et al. 1988). The submarine shield volcano is of complex evolution and erupted several times since the Tertiary (15 million years) until the Lower Pleistocene (2 million years: Rojas and Alvoredo 2012). It was created at the Galapagos Islands and transported to the northeast toward Central America by Cocos Plate motion, where it subducts beneath the Osa Peninsula in southern Costa Rica (Rojas and Alvoredo 2012). Being the only landmass of the Cocos Plate that emerges above sea level, Cocos Island is the only place where Cocos Plate motion can be measured. Surveys show that it is converging with the Caribbean Plate at a rate of 78 mm/year (Protti et al. 2012).

Seamounts characterize the surrounding waters. They have depths of 300 m from the sea surface and heights up to 2700 m from the marine bottom. The island and the seamounts are the result of thermal anomalies that emerged from the mantle through the fissures of the oceanic crust and that have an age of between 2.2 and 1.5 million years (Lizano 2012).

Cocos Island is characteristic of an accumulation of lava flows, dykes, and pyroclastic rocks. The basic feature of the coastal morphology consists of elevated cliffs up to 183 m high, almost vertical, formed by the volcanic rocks. Columnar disjunction (rock prisms) of the lava flows has favored the formation of these living cliffs, which are attacked by intense regressive erosion and undermining caused by waves (Rojas and Alvoredo 2012) (Fig. 47.2).

The almost total absence of sandy beaches in Cocos Island is due to the intense activity of waves, which do not allow the accumulation of material, and to the existence of relatively deep waters in the proximity of the seacliffs (Rojas and Alvoredo 2012).

Beyond cliffs, the island harbors a rare and complex mosaic of land environments, including forested mountains, rivers, waterfalls, estuaries, and rocky beaches (Fig. 47.3). The maximum elevation of the peaks is about 630 m above sea level. Another characteristic feature of the coastal morphology of the island is the presence of caverns, marine tunnels, and submerged arches (Lockwood and Benumof 2000).

47.2.3 Biodiversity

Cocos Island Marine Park, because of the distance from the mainland, is an important repository of species unique to the island. It is the only island in the Pacific Tropical Eastern with an undisturbed humid tropical forest. The vegetation is dense and exuberant due to the heavy rainfall and the condensation which is intensified by the rugged relief (Friends of Isla del Coco National Park Foundation (2017).

Like other oceanic islands, it has a relatively impoverished flora of 235 vascular plants, 70 (nearly 30%) being endemic. In addition to vascular plants, there are 74 species of ferns (5 endemic), 56 species of moss, 106 liverworts, at least 85 fungi, and 99 lichens (3 endemic) (UNEP/WCMC 2005). There are also five species of orchids (Bogarin et al. 2011).

Fig. 47.2 Rocket of basaltic rocks on Cocos Island, with columnar disjunction characteristic of differential cooling of lava flows. The basis of the volcanic pillar is deeply attacked by wave erosion. (Picture reproduced by permission of Dr. Jorge Cortés Núñez, Universidade de Costa Rica)

There are three main plant associations on the island: a coastal, mostly herbaceous plain with some marshlands, an undisturbed very moist montane forest, and above 500 m, cloud forest. The species are predominantly neotropical (UNEP/WCMC 2008). For the marine flora, macroalgae represent a large percentage of the benthic reef environments and play important ecological roles (Fernández 2008).

There is low diversity of terrestrial fauna. The five terrestrial mammals are introduced: pigs, goats, deer, cats and rats. Seven species of freshwater fish, 100 species of birds, 5 reptiles and 362 insects have been recorded (UNEP/WCMC 2008).

The marine biodiversity is characterized by the presence of at least 1688 species, with the gastropods (383 species), bony fishes (354 species), and crustaceans (at least 263 species) being the most species-rich groups. Forty-five species are endemic to Isla del Coco National Park (2.7% of the total). The number of species per kilometer of coastline and by square kilometer of seabed shallower than 200 m deep are the highest recorded in the Eastern Tropical Pacific (Cortés 2012).

The island is a magnificent diving site and one of the best places in the world to see large species such as sharks, rays, tuna, and dolphins. Is it known as the "sharkiest place in the sea." It also presents the most diverse coral reefs of the entire Eastern Tropical Pacific (Earle 2012).

Fig. 47.3 Vegetated rocky coast with seacliffs and small estuary in Bahia Wafar, Cocos Island. (Picture reproduced by permission of Dr. Jorge Cortés Núñez, Universidad de Costa Rica)

47.3 Conservation Challenges

On land, the main threats are invasive alien species of both flora and fauna. The introduced mammals have complex effects on the ecosystems. Temporary settlers brought plants like guinea grass and coffee. The later has since been invading the understory of the forest (UNESCO WHC 2017).

In the waters, illegal fishing is common, being concentrated on a seamount within the park (Romero and Cinner 2016). The illegal fishing is producing degradation and diminution of biodiversity. A decrease in relative abundance for six species has been documented, including the iconic scalloped hammerhead shark (−45%), whitetip reef shark (−77%), mobula ray (−78%), and manta ray (−89%) (White et al. 2015).

Another important threat to the island is climate change, which could increase the temperature and level of the sea water and then modify all of the environment, reducing the biodiversity. Phenomena such as El Niño, which already reaches the island, will probably increase in the next decades, impacting the fauna and flora, marine and terrestrial.

The International Union for Conservation of Nature in its last conservation outlook related to the World Heritage sites published at the end of 2017 found that Cocos Island National Park state of conservation presents significant concerns, with illegal fishing and climate changes the most important threats to the integrity of the area (Osipova et al. 2017).

References

Alfaro EJ (2008) Ciclo diario y anual de variables troposfericas y oceanicas en la Isla del Coco, Costa Rica (Pacifico Oriental). Rev Biol Trop 56(supplement):99–11

Bogarín D, Warner J, Powell M, Savolainen V (2011) The orchid flora of Cocos Island National Park, Puntarenas, Costa Rica. Bot J Linn Soc 166:20–39

Bolaños RED, Alfaro EJ, Quesada AU (2012) Diccionario histórico-toponímico de la Isla del Coco, Costa Rica. Rev Biol Trop 60:1–14

Castillo P, Batiza B, Vanko D, Malavassi E, Barquero J, Fernández E (1988) Anomalously young volcanoes on old-spot traces: geology and petrology of Cocos Island. Geol Soc Am Bull 100:1400–1414

Cortéz J (2008) Historia de la investigación marina de la Isla del Coco, Costa Rica. Rev Biol Trop 56:1–18

Cortéz J (2012) Marine biodiversity of an eastern tropical Pacific oceanic island, Isla del Coco, Costa Rica. Rev Biol Trop 60:131–185

Earle AS (2012) Foreword. Rev Biol Trop 60:XI–XII

Fernández C (2008) Flora marina del Parque Nacional Isla del Coco, Costa Rica, Pacífico tropical oriental. Revi Biol Trop 56:57–69

Friends of Isla del Coco National Park Foundation (2017) Isla del Coco. http://www.cocosisland.org/. Accessed 28 May 2017

Lizano OG (2012) Rasgos morfológicos alrededor de la Isla del Coco y de sus montes submarinos vecinos, Pacífico de Costa Rica. Rev Biol Trop 60:43–51

Lockwood JP, Benumof BT (2000) Investigaciones de las cuevas marinas de la isla del Coco. Report, San Jose, Costa Rica

Osipova E, Shadie P, Zwahlen C, Osti M, Shi Y, Kormos C, Bertzky B, Murai M, Van Merm R, Badman T (2017) IUCN world heritage outlook 2: a conservation assessment of all natural world heritage sites. IUCN, Gland, Switzerland

Protti M, González V, Freymueller J, Doelger S (2012) Isla del coco, on cocos plate, converges with Isla de San Andrés, on the Caribbean plate, at 78mm/yr. Rev Biol Trop 60:33–41

Rojas W, Alvarado GE (2012) Marco geológico y tectónico de la Isla del Coco y la región marítima circunvecina, Costa Rica. Revi Biol Trop 60:15–32

Romero JC, Cinner JE (2016) Optimizing enforcement and compliance in offshore marine protected areas: a case study from Cocos Island, Costa Rica. Oryx 50:18–26

UNEP/WCMC (United Nations Environmental Programme/World Conservation Monitoring Centre) (2005) Cocos Island, Costa Rica. Raport, Cambridge, United Kingdom

UNEP/WCMC (United Nations Environmental Programme/World Conservation Monitoring Centre) (2008) Cocos Island National Park, Costa Rica. Report, Cambridge, England

UNESCO WHC (United Nations Educational, Scientific and Cultural Organization World Heritage Centre) (2017) Cocos Island National Park. http://whc.unesco.org/en/list/820. Accessed 28 May 2017

USGS (United States Geological Survey) Isla del Coco fishes https://coastal.er.usgs.gov/cocos/. Accessed 28 May 2017

White ER, Myers MC, Flemming JM, Baum JK (2015) Shifting elasmobranch community assemblage at Cocos Island – an isolated marine protected area. Conserv Biol 29:1186–1197

Chapter 48
Galápagos Islands, Ecuador

Abstract Galápagos Islands are isolated in the Pacific Ocean, 1100 km west of Ecuador, at the confluence of three ocean currents. The islands are subject to volcanic and seismic activity and are a natural museum of geological, ecological, and evolutionary processes. Their varied climates, oceanographic regimes, and extreme isolation have produced one of the world's highest concentrations of endemic species, including unusual animals such as the land and marine iguanas, giant tortoises, and the finch that inspired Darwin's theory of evolution following his visit in 1835. No other site in the world can offer the experience of diving with such a diversity of marine life. It was the first site to be declared World Heritage by UNESCO.

48.1 Introduction

Situated in the Pacific Ocean some 1000 km from Ecuador mainland, the site is made by 19 islands and a surrounding marine reserve. It has 1406 km^2 of area, with 766 km^2 terrestrial and 1330 km^2 marine. The marine reserve includes all the water within a zone 64 km wide (UNESCO WHC 2017) (Fig. 48.1).

The preservation of the islands started in the first decades of the last century. In 1936, the Galápagos National Park was established by the Ecuadorian Government, and it now covers 97% of the total land mass. In 1968, the effective management of the islands began. In 1971, the marine reserve was created, which is one of the largest in the world (UNESCO WHC 2017). In 1984, the islands were designated a UNESCO Biosphere Reserve, and in 2002 an area on one of the islands was declared a wetland of importance by the Ramsar Wetland Convention (UNEP/WCMC 2008).

The islands were declared a World Heritage site in 1978, the first one to be announced by UNESCO. In 2007, the site was listed as a World Heritage site in danger due to the increasing commercial pressure of tourism, mainland immigration, and introduction of exotic flora and fauna (UNEP/WCMC 2008). It was taken off the list of sites in danger in 2010 (The Guardian 2010).

Galápagos Islands were discovered by the Incas in the middle of the fifteenth century. They were annexed by Ecuador in 1832. In 1835, Charles Darwin visited the area, and his observations support his theory of evolution of the species. They were

V. Claudino-Sales, *Coastal World Heritage Sites*, Coastal Research Library 28,
https://doi.org/10.1007/978-94-024-1528-5_48

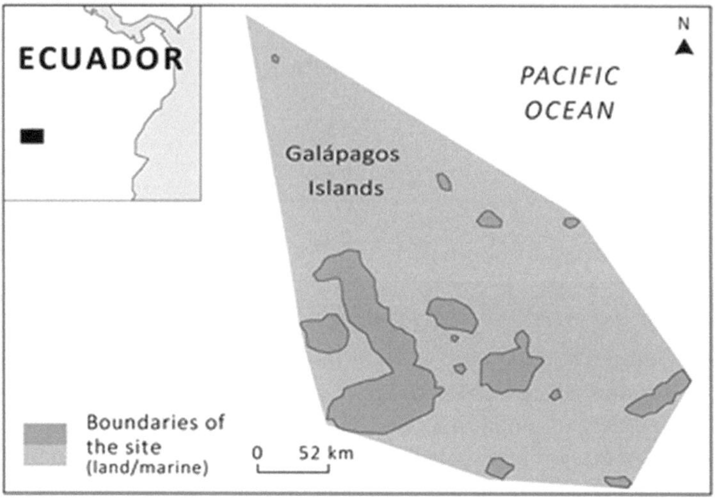

Fig. 48.1 Location of Galápagos Islands and the surroundings waters, in the East Pacific

used as a stop-off by sailors, buccaneers, and whalers who introduced foreign species and hunted down tortoises and seals. Despite this attention, the islands remained largely unsettled until the second half of the last century (UNESCO WHC 2017).

Today, 4 of the 19 islands are inhabited, and the total population, in the Ecuadorian census of 2010, is around 25,000 (Galapagos Conservancy 2017). Tourism, including diving, is the most important economic activity in the islands. The site is property of the Ecuadorian Government and is managed by a tripartite commission formed by the Ministry of the Environment, the National Fisheries Department, and the Navy (UNESCO WHC 2017).

48.2 Natural Settings

48.2.1 Climatic Changes

The climate in Galápagos is equatorial. It is strongly influenced by three main currents, the Humboldt Current, which brings cold and nutrient-rich waters, the warm Panama Current, and the Equatorial Undercurrent, which brings nutrient-rich deep waters to the surface. The first two cool the islands and keep them from being too hot even it if is in the equator. The temperature averages 22 °C many months of the year (University of Minnesota Duluth 2000).

There are two primary seasons in Galápagos, the wet and the dry. The wet season, also the hot season, lasts from June to November. The dry season is characterized by cool temperatures and a fairly persistent fog that envelopes the highlands. The mean annual precipitation is 356 mm, while at 200 m above sea level, it is

1100 mm. But the rainfall is variable: approximately every 4 years, El Niño creates a major warm water flow during this season, bringing heavy rainfall (UNEP/WCMC 2008).

48.2.2 Geology and Geomorphology

The Galápagos Islands are a cluster of volcanic islands created by eruptive activity at the Galápagos hotspot, which occurs as the result of a mantle plume beneath the Nazca Plate (Villagómez 2007). The last eruption of one of those volcanos was in 2015 (NASA 2015).

The seismic Carnegie Ridge, from which Galápagos Islands represent landmarks, is the hotspot track extending eastward on the Nazca Plate from the Galápagos Islands to the subduction zone at the Ecuadorian coast (Harpp 2005). Generally, the ages of the islands range from younger in the west to older in the east, where they are up to ~5.5 million years (White 1993).

The composition of the lavas erupting in Galápagos is predominantly basaltic, leading to shield volcanoes dominating the morphology and the landscape. The flanks of these volcanoes are dotted with cones and associated flows, and many other basaltic eruptions and flow features such as lava tunnels and erratic blocks are found around the islands (Toulkeredis 2011).

The Galápagos Islands archipelago consists of 13 major islands, 6 smaller islands, and 107 small islets (Villacis and Carrillo 2013). Each of the larger islands consists of one or more large shield volcanoes, with the largest island, Isabela, containing six. The altitudes of the central highlands of these volcanoes often ascend to over 1000 m, with Wolf Volcano on Isabela Island reaching the highest in Galápagos at 1707 m (Kelly and Salazar 2016).

The summits are studded with parasitic vents a few tens of meters high. Long stretches of shoreline are only slightly eroded, but in many places faulting and marine erosion have produced steep cliffs and lava, coral, or shell-sand beaches. The terrain is generally composed of uplifted marine lava flows which form an uneven surface, strewn with a deep layer of rounded or angular boulders (Fig. 48.2). Soils are very poor. Freshwater is limited (UNEP/WCMC 2008).

48.2.3 Biodiversity

The varied climates, ongoing volcanism, and extreme isolation of Galápagos Islands have produced one of the highest concentrations of endemic species in the world (UNESCO WHC 2017).

Endemic flora of the Galápagos Islands has adapted to thrive in harsh environmental conditions (Pailles et al. 2017). The islands probably harbor 7000–9000 species of plants. Seven genera are endemic and 19 show adaptive radiation, mainly

Fig. 48.2 Lava flows and erratic blocks on the coastal area of one of the islands of Galapagos Archipelago. The boulders are suffering biogenic erosion, resulting from the presence of seabirds in the area. (Picture reproduced by permission of Gary Lehman)

from nearby mainland species (UNEP/WCMC 2008). The native vegetation occurs in one to three zones on the lowland islands, coastal, arid (80–120 m), and transitional (100–200 m), and on mountainous islands, there are closed forest (200–500 m), mossy open forest and mossy scrub (500–1000 m), and summit pampa above 1000 m (Coffey et al. 2011).

The Galápagos is a distinct biogeographic province the endemic biota of which includes 24 vertebrate and 2000 invertebrate species. There are few indigenous mammals but 35 endemic reptiles. Best known are the giant Galápagos land tortoise, with 20 subspecies on different islands. Others include Galápagos land iguanas and marine iguana which is the world's only sea-going lizard and feeds on seaweed (Fig. 48.3). The native avifauna includes 57 residents, of which 28 (49%) are endemic and 31 are regular migrants (UNEP/WCMC 2008).

The marine environment has some 2900 species formed in the convergence of ocean currents. There are some 460 species of fish, at least 51 species (11%) being endemic. There are 12 species of sharks and 6 species of rays. The interaction between the terrestrial and marine environment is particularly important for the marine iguana and for 27 of the islands' 57 bird species, especially the Galápagos penguin and many species of finch (UNEP/WCMC 2008).

Fig. 48.3 Land iguana, one of the most representative elements of Galapagos fauna. (Picture reproduced by permission of Gary Lehman)

Dolphins and the endemic sea lions and fur seals are abundant. Several species of whales are regularly seen. There are 650 species of sea shells, 200 sea stars and urchins, and 120 crabs. Due to the cool waters of the Humboldt Current during 4 to 6 months per year, the Galápagos is a marginal environment for coral reefs; 120 species, however, are found in its warmer waters (UNEP/WCMC 2008). The archipelago and its immense marine reserve is known as the unique "living museum and showcase of evolution" (UNESCO WHC 2017).

48.3 Conservation Challenges

Galápagos Islands are often cited as an example of the conflicts that are emerging between resource conservation and economic development in island ecosystems, as the pressures associated with tourism threaten nature (Walsh and Mena 2016). Increasing tourism and growth of the population are real pressures on the environment (UNESCO WHC 2017).

Human impacts on the biodiversity are evident. The present flora is without modern analogue before 1930, and exotic species are invading the environment. Approximately 750 nonnative plants were introduced over the last 500 years of human presence in the archipelago, and there are an additional 62 vascular plants where native status remains unclear (Coffey et al. 2011). In such a situation, the habitats are continuously moving toward a novel ecosystem (Restrepo et al. 2012).

The introduction of new species also threatens the harmony of the endemic fauna. Rats, goats, and wild dogs are among these species brought to the islands by humans, and they destroy the habitats and many species' populations (UNESCO WHC 2017). Climate changes are also a threat to the endemic biodiversity of the islands. Warmer wet seasons and more intense convection are producing a transport of lowland pollen to the uplands, with the possibility of more changes in the flora (Restrepo et al. 2012). Extreme weather events have also taken a toll in the wildlife and environment (UNEP/WCMC 2008). Adverse effect of these events is perilous to marine life.

The International Union for Conservation of Nature in its report about the conservation outlook of the World Heritage sites published at the end of 2017 found that the main globally significant values of the site are in overall good condition as a result of good management and governance regime. However, continued pressure to the site's natural phenomena, natural beauty, ecological processes, and biodiversity, endemic and threatened species is increasing (IUCN 2017). The report found that the state of conservation of Galapagos presents significant concerns (Osipova et al. 2017).

References

Coffey EED, Froyd CA, Willis KJ, Katherine J (2011) When is an invasive not an invasive? Macrofossil evidence of doubtful native plant species in the Galápagos Islands. Ecology 92:805–812

Galapagos Conservancy (2017) About galapagos. https://www.galapagos.org/about_galapagos/about-galapagos/. Accessed 1 June 2017

Harpp K (2005) The Cocos and Carnegie aseismic ridges: a trace element record of long-term plume-spreading center interaction. J Petrol 46:109–133

IUCN (International Union for Conservation of Nature) (2017) Galapagos island conservation outlook. http://www.worldheritageoutlook.iucn.org/node/957. Accessed 24 Dec 2017

Kelly DF, Salazar R (2016) Geosites in the Galapagos Islands used for geology education programs. Geoheritage 9(3):351–358. https://doi.org/10.1007/s12371-016-0190-3

NASA (National Aeronautics and Space Administration) (2015) Eruption of wolf volcano, Galapagos Islands. https://www.nasa.gov/image-feature/eruption-of-wolf-volcano-galapagos-islands. Accessed 1 June 2017

Osipova E, Shadie P, Zwahlen C, Osti M, Shi Y, Kormos C, Bertzky B, Murai M, Van Merm R, Badman T (2017) IUCN world heritage outlook 2: a conservation assessment of all natural world heritage sites. IUCN, Gland, Switzerland

Pailles Y, Ho S, Pires IS, Tester M, Negrao S, Schmockel SM (2017) Genetic diversity and population structure of two tomato species from the Galapagos Islands. Front Plant Sci 8:138. https://doi.org/10.3389/fpls.2017.00138

Restrepo A, Colinvaux P, Bush M, Correa-Metrio A, Conroy J, Gardener MR, Jamarilo P, Mirian S-K, Overpeck J (2012) Impacts of climate variability and human colonization on the vegetation of the Galápagos Islands. Ecology 93:1853–1866

The Guardian (2010) UN withdraws Galápagos from world heritage danger list. https://www.theguardian.com/world/2010/jul/29/galapagos-withdrawn-heritage-danger-list. Accessed 31 May 2017

Toulkeridis T (2011) Volcanic Galápagos. Centro de geología, volcanología y geodinámica/Escuela Politécnica del Ejército, Quito

UNEP/WCMC (United Nations Educational Programme/World Conservation Monitoring Centre) (2008) Galapagos Islands National Park and marine reserve, Ecuador. Report, Cambridge, England

UNESCO WHC (United Nations Educational, Scientific and Cultural Organization World Heritage Centre) (2017) Galápagos Islands. http://whc.unesco.org/en/list/1. Accessed 1 June 2017

University of Minnesota Duluth (2000) Climate and meteorology of the Galápagos Islands. https://www.geol.umd.edu/~jmerck/galsite/research/projects/ader/. Accessed 1 June 2015

Villacis B, Carrillo D (2013) The socioeconomic paradox of Galápagos. In: Walsh SJ, Mena CF (eds) Science and conservation in the Galápagos Islands: frameworks and perspectives. Springer, London

Villagómez D (2007) Upper mantle structure beneath the Galápagos archipelago from surface wave tomography. J Geophys Res 112:B07303. https://doi.org/10.1029/2006JB004672

Walsh S, Mena C (2016) Interactions of social, terrestrial, and marine sub-systems in the Galapagos Islands, Ecuador. Proc Natl Acad Sci U S A 113:14536

White W (1993) Petrology and geochemistry of the Galápagos Islands: portrait of a pathological mantle plume. J Geophys Res 98:19,533–19,563

Chapter 49
Lagoons of New Caledonia, France

Abstract The French Pacific Ocean archipelago of New Caledonia is one of the most extensive reef systems in the world and the world's most diverse concentration of reef structures. It ranges from extensive double barrier systems, offshore reefs, and coral islands to nearshore reticulate reef formations. The lagoons feature an exceptional diversity of coral and fish, presenting a continuum of habitats from mangroves to seagrasses and tropical mountain forests. The reef complex provides a variety of different kinds of oceanography exposure, including both warm and cold currents. It is one of the most ancient and complex ecosystem types.

49.1 Introduction

New Caledonia is the southwestern most of the archipelagos of the Pacific, 1700 km west from Australia (Queensland) and 1400 km north-northwest of New Zealand. It is French territory. La Grande Terre, one of the islands of archipelago, is the ocean's third largest oceanic island. The site has 15,743 km^2 of area (UNESCO WHC 2017) (Fig. 49.1).

The preservation of the archipelago started in the 1970s. In 1970, an area inserted in one of the lagoons was declared a marine reserve. In the 1990s, several other marine reserves were established. Six marine clusters, including lagoons and atolls, were designated a World Heritage site in 2008 (UNEP/WCMC 2008).

The island's name was given by Cook in 1774. It was taken by France in 1853. Indigenous people (the Melanesian Kanak) have lived there for 3500 years, and Polynesians came 800 years ago. They live in tribal clans with exclusive rights to the resources of their own communities preserved by customary systems of protective management (CIA 2017).

The site lies within a Conservation International-designated Hotspot, a WWF Global 200 Eco-region, a WWF/IUCN Centre of Plant Diversity, and a BirdLife-designated Endemic Bird Area. Authority in the archipelago derives from the French government and the government of the territory and includes the participation of the Kanak. Tourism is an increasing activity (UNESCO WHC 2017; UNEP/WCMC 2008).

V. Claudino-Sales, *Coastal World Heritage Sites*, Coastal Research Library 28,
https://doi.org/10.1007/978-94-024-1528-5_49

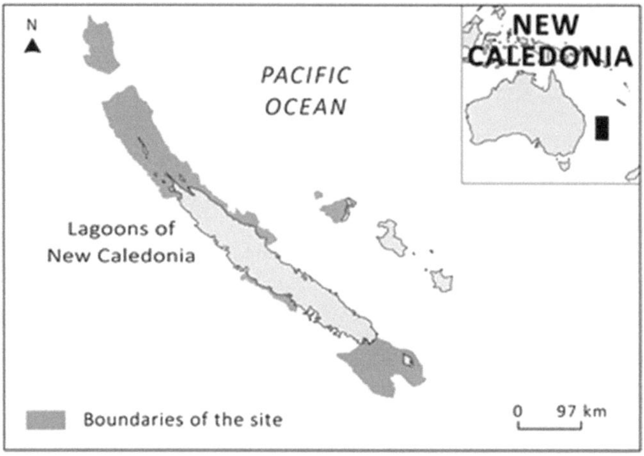

Fig. 49.1 Lagoons of New Caledonia World Heritage site, in the Southern Pacific

49.2 Natural Settings

49.2.1 Climatic Conditions

The sites extend over five degrees of latitude, from warm temperate to subtropical and tropical climates. The seasons alternate between wet, with tropical storms, cyclones, and heavy rains from December to April, and a fresh dry season from July to September (UNEP/WCMC 2008).

The average maximum temperature is around 28 °C, and the minimum average, 18 °C. The average annual rainfall is ca 1200 mm. The average wind speed is 5 m/s (Weather and Climate 2016).

49.2.2 Geology and Geomorphology

The geological evolution of New Caledonia is very complex and can be divided into many successive phases. The Gondwanan phase (Permian-Early Cretaceous) is marked by subduction along the SE Gondwanaland margin. During the Late Cretaceous-Eocene, marginal rifting isolated New Caledonia. A new subduction appeared to the east of New Caledonia at the Paleocene-Eocene boundary and eventually ended with Late Eocene obduction. Finally, during the post-Eocene phase, New Caledonia emerged (Cluzel et al. 2012).

New Caledonia formed from part of a continental fragment that began separating from Australia ca 83 million years ago as the Tasman Sea began to form (Neall and Trewick 2008; Brothers and Lillie 1988). In the Paleocene, this part experienced a

Fig. 49.2 Raised limestones, eroded by wave action and wind in the island of Ouvéa, New Caledonia archipelago. (Picture reproduced by permission of Sbedaux/Shutterstock Stock Photos)

lengthy submersion in deep water, as evidenced by marine sedimentation (Pelletier 2006; Aitchison et al. 1998; Paris 1981).

During the Eocene, its continental crust was tectonically active, and obduction placed a layer of oceanic crust over the submerged continental crust. It emerged during an Oligocene lithosphere extension phase, uplifting with a cover of lithospheric ultramafic mantle rocks (Schellart et al. 2006; Crawford et al. 2003). Owing to these geological constraints, New Caledonia is a remarkable paleogeography model as it presents a combination of continental and oceanic features.

It has a backbone of complex and dissected mountains, with tablelands and peaks reaching elevations of more than 1600 m. The most impressive features are, nevertheless, the coral reefs, which are the third most extensive in the world and form the world's most diverse assemblage of reef structures in one location: 60% of the archipelago's surface is formed of fringing reefs, single barrier reefs, double barrier reefs (which are rare), lagoon-enclosing atolls, raised atolls and limestones (Fig. 49.2), sand cays, oceanic banks, and coral islets (UNESCO WHC 2017; UNEP/WCMC 2008).

Some of the lagoons are deep; others are shallow and dry, in many places lined by fringing reef and mangroves. Some are underlain by a perched lens of fresh water which has created karstic erosion of caves and potholes. They have a variety of substrates derived from terrestrial sediments or from degraded coral and shells. The system hosts an exceptional diversity of landforms, corals (Fig. 49.3), and their associated ecosystems, both continental and oceanic in character (Government of Caledonia 2006).

Fig. 49.3 Corals on the bottom of the lagoon of New Caledonia district, New Caledonia. (Picture reproduced by permission of the Institut de Recherche pour le Développement –IRD, France/C Geoffray)

49.2.3 Biodiversity

It was common to say that the biodiversity derived from Gondwanaland (e.g., UNEP/WCMC 2008; Government of New Caledonia 2006). Nevertheless, it has been already considered that the successive geological events that happened in this part of the Earth after the separation of Gondwana in the Cretaceous have had important consequences for the evolution of the biota. Effectively, more than 20 million years of total submersion, from the Paleocene to the Eocene, makes it difficult to retain the notion that a Gondwanan biota has survived locally (Pelletier 2006; Murienne et al. 2005).

There are five main terrestrial habitats. The most extensive is evergreen wet tropical lowland forest, now covering one-fifth of its original area. Other vegetation types are the very fragmented sclerophyllous forest, with 1% of its original extent, low- to mid-level maquis, high-altitude wet maquis forest of edaphic origin, anthropogenic savanna, and marsh (Government of New Caledonia 2006). The remoteness of the islands further favored the appearance of endemic species (Ladiges 2008). Along 50% of the coast, mangroves of 16 species occur. The marine vegetation is seagrasses and algal beds, which in the extreme south resemble temperate species (UNEP/WCMC 2008).

The great variety of habitats, the latitudinal and altitudinal ranges, the temperature, salinity and sedimentation gradients between east and west coasts, and upwelling currents, all contribute to an estimated total of some 15,000 animal species of which about 10,500 are marine (Grandcolas et al. 2008). The fish are found in the reefs, estuaries and mangrove swamps, and soft-bottomed bays and lagoons, and the crustaceans and mollusks are abundant (Government of New Caledonia 2006).

Whales, dolphins, and dugongs (sea cow) are common. The sea cow population is one of the largest in Oceania and the third in the world. There are many species of reptiles, especially marine turtles, and the site is also an avian hotspot and Endemic Bird Area. Before western importations, the land had no mammals except for bats (UNEP/WCMC 2008).

49.3 Conservation Challenges

The site is currently protected by fisheries legislation. Enhanced surveillance and monitoring are required to address potential impacts from illegal fishing, agriculture, aquaculture, and mining. Tourism is likely to increase in the near future and needs to be well managed (UNESCO WHC 2017).

The most dangerous threat to the site is global warming. The increase of sea water temperature, the rising of sea level, and the increasing of oceans' acidity are situations that demand strong actions from the managers to maintain the resilience of the reefs and lagoons.

In a report published at the end of 2017 by the International Union for Conservation of Nature (IUCN) related to the state of conservation of the World Heritage sites, the organization found that the conservation of the site as a whole is very satisfactory. With a coordinating body of the 13 management committees, with good representation of local stakeholders in the management committees and with a strong participation of local communities in the committees, the overall management of the property is effective (IUCN 2017). Nevertheless, the report found the state of conservation of the site to be good, but with some concerns (Osipova et al. 2017).

References

Aitchison JC, Ireland TR, Clarke GL, Cluzel D, Davis AM, Meffre S (1998) Regional implications of U/Pb SHRIMP age constraints on the tectonic evolution of New Caledonia. Tectonophysics 299:333–343

Brothers RN, Lillie AR (1988) Regional geology of New Caledonia. In: Nairn AEM, Stehli FG, Uyeda S (eds) The ocean basins and margins, vol 7B. Plenum Press, New York, pp 325–374

CIA (Central Intelligence Agency) (2017) Australia-Oceania: New Caledonia the world factbook. https://www.cia.gov/library/publications/the-world-factbook/geos/nc.html. Accessed 11 June 2017

Cluzel D, Maurizot P, Collot J, Sevin BC, Keith M, Jell PA (2012) An outline of the geology of New Caledonia; from Permian-Mesozoic southeast Gondwanaland active margin to Cenozoic obduction and supergene evolution. 34th International Geological Congress 35:72–86, Brisbane

Crawford AJ, Meffre S, Symonds PA (2003) Tectonic evolution of the Southwest Pacific and analogous geological evolution of the 600 to 220ma Tasman fold belt system. Geol Soc Aust Spec Publ 22:377–397

Government of New Caledonia (2006) Les Lagons de Nouvelle-Calédonie. Diversité Récifale et Écosystèmes Associés. Nomination dossier, Noumea

Grandcolas J, Murienne J, Robillard T, Desutter-Grandcolas L, Jourdan H, Guilbert E, Deharveng L (2008) New Caledonia: a very old Darwinian Island? Philos Trans R Soc B 363:3309. https://doi.org/10.1098/rstb.2008.0122

IUCN (International Union for Conservation of Nature) (2017) Lagoons of New Caledonia conservation outlook. http://www.worldheritageoutlook.iucn.org/explore-sites/wdpaid/903134. Accessed 24 Dec 2017

Ladiges PY (2008) Synthesizing biotic patterns and geology for New Caledonia. J Biogeogr 35:2151–2152

Murienne J, Grandcolas P, Piulachs MD, Belles X, D'Haese C, Legendre F, Pelles R, Guilbert E (2005) Evolution of a shaky piece of Gondwana: is local endemism recent in New Caledonia? Cladistics 21:2–7

Neall VE, Trewick SA (2008) The age and origin of the Pacific islands: a geological overview. Philos Trans R Soc B 363:3293–3308

Osipova E, Shadie P, Zwahlen C, Osti M, Shi Y, Kormos C, Bertzky B, Murai M, Van Merm R, Badman T (2017) IUCN world heritage outlook 2: a conservation assessment of all natural world heritage sites. IUCN, Gland, Switzerland

Paris JP (1981) Géologie. In: Atlas de le Nouvelle Calédonie et dépendances. ORSTOM (l'Office de la Recherche Scientifique et Technique Outre-Mer.), Paris

Pelletier B (2006) Geology of the New Caledonia region and its implications for the study of the New Caledonian biodiversity. Documents Scientifiques et Techniques IRD, II 7:17–30, Noumea

Schellart WP, Lister GS, Toy VG (2006) A late cretaceous and Cenozoic reconstruction of the Southwest Pacific region: tectonics controlled by subduction and slab rollback processes. Earth-Sci Rev 76:191–233

UNEP/WCMC (United Nations Environmental Programme/World Conservation Monitoring Centre) (2008) Lagoons of New Caledonia: reef diversity and associated ecosystems, New Caledonia, France. Report, Cambridge, England

UNESCO WHC (United Nations Educational, Scientific and Cultural Organization World Heritage Centre) (2017) Lagoons of New Caledonia: reef diversity and associated ecosystems. http://whc.unesco.org/en/list/1115. Accessed 11 June 2017

Weather and Climate (2016). New Caledonia, climate noumea. https://weather-and-climate.com/average-monthly-Rainfall-Temperature-Sunshine,Noumea,New-Caledonia. Accessed 11 June 2017

Chapter 50
Ogasawara Islands, Japan

Abstract The Ogasawara Islands are in the northwestern Pacific Ocean, 1000 km south of the main Japanese Archipelago. The landscape is dominated by subtropical forest types and scrublands surrounded by steep cliffs. The islands provide precious information on the evolution of the Earth and organisms. They represent the only place in the world where a formation process of an oceanic island arc can be observed from its birth, juvenile phase, to the ongoing young phase. There is also species diversification to adapt to various types of environments. It has unique ecosystems, serving as habitats for a large number of globally important endangered species.

50.1 Introduction

The Ogasawara Islands, also known as Bonin Islands, are a group of subtropical islands in the northwestern Pacific Ocean, located approximately 1000 km south of main island of Japan (Fig. 50.1). Administratively, they form one of the villages of Tokyo (Government of Japan 2010a).

The site covers a surface area of 79 km^2, being mostly terrestrial. It extends for about 400 km from north to south and includes more than 30 small islands. Residential and inhabited areas of the islands are not included in the World Heritage site (UNESCO WHC 2017).

The Ogasawara Islands were discovered in 1593 by a Japanese man, but the occupation started only in the nineteenth century. In 1944, all island residents were forcibly evacuated to the mainland, and the occupation started again 25 years later. In 1979, the village government was established. Today, two islands are populated, with around 2100 inhabitants in 2011 (Ogasawara Village 2011).

The first actions focusing in conservation date from 1969, based on the Act on Special Measures for the Ogasawara Islands Reconstruction, from the Japanese Government. Now, the archipelago is protected as a wilderness area, a national park, a natural monument, a forest ecosystem reserve, and a national wildlife protection area (Government of Japan 2010a). It became a World Heritage site in 2011 (UNESCO WHC 2017).

V. Claudino-Sales, *Coastal World Heritage Sites*, Coastal Research Library 28,
https://doi.org/10.1007/978-94-024-1528-5_50

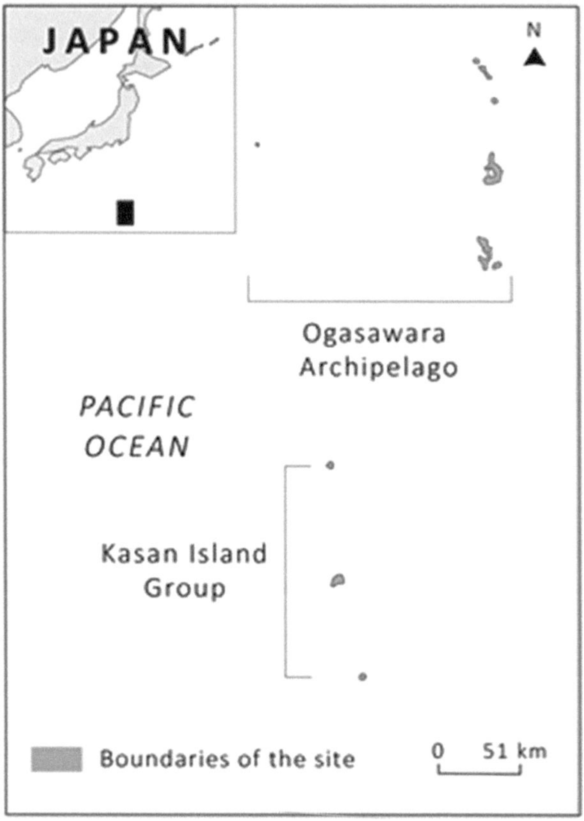

Fig. 50.1 Ogasawara Archipelago World Heritage site, located 1000 km from Tokyo, in Japan

Land ownership is retained by the Forestry Agency, which accounts for about 80% of the islands. The rest is state owned by the Ministry of Finance, the Ministry of the Environment, and the Tokyo Metropolitan Government, as well as privately. It is managed by Tokyo Municipality in association with the village government. The main industries on the islands are tourism, agriculture, and fisheries (Government of Japan 2010b).

The only way to get to Ogasawara Islands is by boat. There are luxury cruises and ferries leaving from Tokyo every week. The number of ferries and tourists more than doubled 1 year after the inscription of the islands in the list of World Heritage sites (Japan Times 2012).

50.2 Natural Settings

50.2.1 Climatic Conditions

The archipelago is in a warm subtropical climatic zone. Because both the annual and diurnal temperature ranges are small and the level of humidity is high, the climate there is considered to be maritime. The types range from a humid subtropical climate to tropical savanna climate (Ogasawara Village 2012).

The mean annual temperature is 23 °C. The mean temperature of the coldest month (February) is 17.7 °C, and the mean of the hottest month (August) is 27.6 °C. The mean annual precipitation is 1277 mm (Government of Japan 2010a).

The amount of evaporation, of the order of 1380 mm annually, exceeds that of precipitation. In consequence, the soil is very dry. There is no record of typhoons in the islands (Government of Japan 2010b).

50.2.2 Geology and Geomorphology

In the context of subduction, when the magma from the mantle rises to the surface and gives birth to a line of volcanoes in the overriding plate, it creates a volcanic arc, which is typically situated a few hundred kilometers behind the oceanic trench. The Ogasawara Archipelago represents an emergent section of the Izu-Ogasawara-Mariana Trench arc and forearc basement that extends for more than 1500 km (Kanayama et al. 2012).

Islands arc magmatism leads to the formation of andesitic middle crust under the Earth's surface, the same chemical composition of continental crust. The phenomenon, also taking place in Ogasawara Islands, demonstrates the very beginning of continent formation (Tatsumi et al. 2008).

Ogasawara Islands are formed by submarine volcanoes that rose above the sea surface due the subduction of the Pacific Plate. The subduction, involving the Pacific Plate and the Philippine Sea Plate, took place in the Eocene (ca 48 million years) (Kanayama et al. 2014). Quaternary volcanoes, including active ones, comprise the present front of the arc (Government of Japan 2010a). The oceanic islands consist of submarine effusive rocks of Eocene age unconformably overlain by Oligocene to Miocene foraminifer limestone (Umino 1985) (Fig. 50.2).

The arc is one of the most studied in terms of geophysics, geology, and petrology, because of its scientific importance as a typical example of an oceanic island arc that can be analyzed in all phases of growth, from its birth to present day (Government of Japan 2010b).

Fig. 50.2 Outcropping of effusive rocks (boninite) formed after magma solidified during submarine volcanic activity. (Picture reproduced by permission of Knilia/Shutterstock Stock Photos)

Most of the Izu-Ogasawara-Mariana arc lies under the deep ocean, making it impossible to directly obtain geological records. However, the Ogasawara Islands are an exceptional subaerial exposure of these geological strata, making the islands the only site in Earth where these processes can be directly analyzed in detail (Ogasawara Village 2012). The islands are surrounded by seacliffs and basaltic outcrops, which provide excellent geological observation (Government of Japan 2010a) (Fig. 50.3).

The area is also a place to study the ongoing evolution from an oceanic arc to a continent. This process is still under way beneath the Izu-Ogasawara-Mariana arc, as demonstrated by the recently active volcano chains that characterize one of the three marine clusters that form the site (Kodaira et al. 2008). Most of the islands have steep shorelines, often with seacliffs ranging from 50 to 100 m in height. The islands are also fringed with coral reefs. The highest point is 916 m (Government of Japan 2010a).

50.2.3 *Biodiversity*

Ogasawara Islands have never been connected to a continent. For this reason, many of their animals and plants have undergone unique evolutionary processes. This has led to the islands' nickname as "The Galápagos of the Orient" (Government of Japan 2010a).

Fig. 50.3 Exposure of ophiolite in Ogasawara Islands, reworked by wave action, creating bridges-type geomorphic features. (Picture reproduced by permission of Knilia/Shutterstock Stock Photos)

The islands form a distinct subtropical moist broadleaf forest ecoregion, with a high degree of biodiversity and endemism. The islands are home to about 500 plant species, of which 43% are endemic (Government of Japan 2010a).

The forests are of three main types:

1. Mesic, subtropical rainforest, found in the moist lowland areas with deep soils. The forests have a closed canopy with a height of about 15 m.
2. Dry forest, found in drier lowland and upland sites with shallower soils. It is also a closed-canopy forest, with a 4-to-8-m canopy.
3. Scrub forest, found on windy and dry mountain ridges and exposed sea cliffs. These forests have the highest species diversity on the islands (Government of Japan 2010b).

The fauna is characterized by extremely disharmonic biological communities, as is typical of oceanic islands (Ono 1994). Specific groups of animals have evolved independently, forming a unique fauna (Government of Japan 2010a). As for the terrestrial fauna, there is only one mammal species, two reptile species, and many species of birds and endemic insects, but no amphibians. For the marine fauna, 23 species of mammal in 6 families have been identified, including the endangered blue whale and sei whale. For other aquatic animals like fish, the variety is not too large, considering the distance from land masses (Government of Japan 2010a).

50.3 Conservation Challenges

The islands have a short history of human occupation and this helped to preserve nature (Okutomi et al. 1983), but there is always a risk of the arrival of invasive species (UNESCO WHC 2017). Global warming can interfere in the life of the corals and may increase the months of droughts, changing the flora and fauna habitats.

In a report about the state of conservation of the World Heritage sites published in 2014 by the International Union for Conservation of Nature (2014), the state of conservation on this site was found to be good but presenting some concerns. The concerns were linked especially to the increasing of construction of tourism infrastructures and tourism impact, as well as logging (IUCN 2014).

The new conservation outlook published in 2017 (IUCN 2017) found that the situation remains the same, though with a new high threat situation, related to the presence of invasive species, which represent by far the most serious threat to the ongoing ecological processes on the islands (IUCN 2017; Osipova et al. 2017).

References

Government of Japan (2010a) Nomination of Ogasawara Islands for inscription on the World Heritage list. Report, Tokyo

Government of Japan (2010b) Ogasawara management plan. Report, Tokyo, Japan

IUCN (International Union for Conservation of Nature) (2014) State of Conservation of World Heritage sites. Report, Gland, Switzerland

IUCN (International Union for Conservation of Nature) (2017) Ogasawara Islands conservation outlook. http://www.worldheritageoutlook.iucn.org/node/1164. Accessed 24 Dec 2017

Japan Times (2012) Japan's Ogasawara Islands: one year after UNESCO. https://japantimes.co.jp/life/2012/11/18/travel/japans-ogasawara-islands-one-year-after-unesco/#.WOIYSL. Accessed 16 June 2017

Kanayama K, Umino S, Ishizuka O (2012) Eocene volcanism during the incipient stage of Izu–Ogasawara Arc: Geology and petrology of the Mukojima Island Group, the Ogasawara Islands. Island Arc 288:316

Kanayama K, Umino S, Ishizuka O (2014) Shallow submarine volcano group in the early stage of island arc development: geology and petrology of small islands south off Hahajima main island, the Ogasawara Islands. J Asian Earth Sci 85:1–25

Kodaira ST, Takahashi N, Yamashita M, Kaneda Y (2008) Seismic imaging of a possible paleoarc in the Izu-Bonin Intraoceanic arc and its implication for arc evolution processes. Geochem Geophys Geosyst 9. https://doi.org/10.1029/2008GC002073

Ministry of Environment of Japan (2017) Ogasawara National Park. https://www.env.go.jp/en/nature/nps/park/ogasawara/. Accessed 14 June 2017

Ogasawara Village (2011) World Natural Heritage Ogasawara Islands. Pamphlet, Chichijima

Ogasawara Village (2012) Ogaswara village outline. https://en.vill.ogasawara.tokyo.jp/outline/. Accessed 17 June 2017

Okutomi K, Iseki T, Hioki Y, Kitayama K, Sumihiro H (1983) Vegetation of Ogasawara. In: Ogasawara Island Natural Environment Status Report (translated). Bureau of Environment, Tokyo

Ono M (1994) Creature of isolated islands: the Galapagos of Ogasawara. Iwanami Shotten, Tokyo

Osipova E, Shadie P, Zwahlen C, Osti M, Shi Y, Kormos C, Bertzky B, Murai M, Van Merm R, Badman T (2017) IUCN World Heritage outlook 2: a conservation assessment of all natural World Heritage sites. IUCN, Gland

Tatsumi Y, Shukuno KN, Takahashi N, Kodaira S, Kogiso T (2008) Structure and growth of the Izu-Bonin-Mariana arc crust: role of crust-mantle transformation and the transparent Moho in arc crust evolution. J Geophys Res 113. https://doi.org/10.1029/2007JB0055121

Umino S (1985) Volcanic geology of Chichijima, the Bonin Islands (Ogasawara Islands). J Geol Soc Jpn 91:505–523

UNESCO WHC (United Nations Environmental Scientific and Cultural Organization World Heritage Centre) (2017) Ogasawara Islands. http://whc.unesco.org/en/list/1362. Accessed 16 June 2017

Chapter 51
Shiretoko, Japan

Abstract Shiretoko is a peninsula on the northernmost island of Japan. It is an admirable example of the interaction of marine and terrestrial ecosystems and habitats, largely influenced by the formation of sea ice at the lowest latitude in the Northern Hemisphere. The sea is enriched by melting sea ice in spring and by a warm current in summer which produces and maintains an explosive growth of phytoplankton, the basis of extremely abundant life. Many rivers connect the sea with a wealth of pristine vegetation from the coast to volcanic mountain tops; one volcano has periodically erupted pure molten sulfur, and there are fumaroles and hot springs in the range. Both on land and in the sea, northern and southern species intermingle, resulting in great diversity of fishes and birds, as well as dense populations of bears and sea lions.

51.1 Introduction

The Shiretoko Peninsula, in the northeast of Hokkaido, the northernmost island of Japan, is about 24 km wide and protrudes some 70 km in the Sea of Okhotsk. The site is 21 km wide, narrowing to an average of 10 km. It extends for 50 km, being one-third sea and two-thirds land, spreading in an area of 711 km^2 (UNEP/WCMC 2005) (Fig. 51.1).

In 1964, the Shiretoko National Park was established by the Japanese Government. It was expanded in 2004, with marine boundaries extending 3 km from the shoreline and corresponding to the depth of 200 m. In 1980, a wilderness area was established, in 1982 the Shiretoko National Wildlife Protection Area was created, and in 1990 the Shiretoko Forest Ecosystem was delimited. The islands became a World Heritage site in 2005 (UNEP/WCMC 2005).

The land is 95% property of the National Forestry Agency with a few areas owned by the state directly, by Hokkaido Prefecture, by Shari and Rausu towns, and also by private owners. The site is managed by a regional liaison committee (UNESCO WHC 2017).

The peninsula, inhabited by the Ainu, was always inaccessible, but the ruins of one earlier village have been found. It was settled between 1912 and 1967,

V. Claudino-Sales, *Coastal World Heritage Sites*, Coastal Research Library 28,
https://doi.org/10.1007/978-94-024-1528-5_51

Fig. 51.1 Location of the Shiretoko Peninsula World Heritage site, northern Japan

nevertheless farming was given up because of the harsh conditions. There are two small towns, Shari and Rausu, and no one lives in the protected area. There is one major road crossing the peninsula, between the towns. Tourism and fishery are the chief activities (UNESCO WHC 2017; Nature Conservation Bureau 2004).

There are various nature-experience programs available at the site, such as night tours for animal watching under starlit skies, daytime treks for alpine plant spotting while surrounded by singing bird, or nature watching enjoying waterfalls and wild animals while walking along the trail (Japan National Tourism Organization 2017).

51.2 Natural Settings

51.2.1 Climatic Conditions

The climate is subpolar, with a marked difference in the weather of the east and west coasts, which are separated by volcano peaks. The east coast has 1600 mm/year of rainfall with an average maximum temperature in August of 18.9 °C; the average minimum is in February, with a temperature of −10.9 °C. The winter is also characterized by heavy snowfall (Nature Conservation Bureau 2004).

On the west coast, the annual rainfall is 813 mm/year; the average maximum temperature in August is 21.8 °C, and the average minimum February temperature is −16.1 °C. Storms from autumn to winter are common, generating high waves and storm surge (Noda 2005). There are icy winds coming from Siberia, which freezes surfaces (Nature Conservation Bureau 2004).

51.2.2 Geology and Geomorphology

The peninsula is formed of a range of volcanic peaks over 1550 m high. Mount Rausu, with 1661 m of altitude, is the highest. One of the volcanoes, Mount Iou, has periodically erupted pure molten sulfur; the last eruption was in 1936, and there are still fumaroles and hot springs in the range (UNEP/WCMC 2005).

The geology of the area consists of Tertiary volcanoclastic rocks and Tertiary to Quaternary lavas (Goto et al. 2002). The area is volcanically and tectonically active, associated with a geological arc (the Inner Kuril Arc), which is linked to the subduction of the Pacific Plate beneath the North American Plate (Noda 2005).

The seismicity that characterizes the area generates landslides, which come from the volcano peaks. The peaks create rugged geomorphic features, in such a way that no river is longer than 10 km (Noda 2005) (Fig. 51.2). Besides the volcanic features, sedimentary and marine features are also present. Effectively, marine terraces have developed on the peninsula. They are situated 80 m above sea level, and correspond to the interglacial stage 5e, ca 125,000 years (Okumura 1996).

The western coast has steep volcanic cliffs up to 120 m high. Forty-four small rivers, nine of which have been dammed, run down to the sea where ice forms in winter. It is the most southerly sea ice in the Northern Hemisphere (latitudes between 43° 56′38″ and 44° 21′10″ N) (Fig. 51.3). The formation of ice in such a latitude results from the enclosed nature of the Sea of Okhotsk, the huge flow of freshwater from the rivers and the icy winter winds from Siberia. In summer, the warm Soya Current flows along the surface of the sea (Nature Conservation Bureau 2004).

51.2.3 Biodiversity

Shiretoko is one of the richest integrated ecosystems in the world, encompassing both terrestrial and marine areas. The high productivity of the ecosystems is largely influenced by the formation of seasonal sea ice. The phytoplankton resulting from the ice melting sustains the food sources for marine and terrestrial species (UNESCO WHC 2017). Besides, warm currents, numerous rivers, vertical mixing, and seasonal upwelling provide nutrient support to the marine ecosystems (Sakurai 2007).

Fig. 51.2 Aerial photo of the tip of Shiretoko Peninsula (Shiretoko Cape), showing a very rugged and dissected landscape, with evidence of snowfall and ice formation. On the bottom of the picture, well-developed marine terraces are present. (Picture reproduced by permission of Shiretoko Nature Foundation, web page www.shiretoko.org.jp)

Fig. 51.3 Ice on the sea and on the volcano peaks in winter, on the west coast of Shiretoko Peninsula. (Picture reproduced by permission of Shiretoko Nature Foundation, webpage www.shiretoko.org.jp)

The complex undulating topography and contrasting weather on either coast create a variety of habitats where many southern as well as northern species grow. More than 817 species of plants are recorded, of which 233 are alpine species from the northern Pacific or Siberia (Nature Conservation Bureau 2004).

About 90% of the land is covered with nearly pristine natural vegetation. Much of it is taken by temperate broadleaf and mixed forests, which are characteristic of deciduous southern species. Their leaves turn to brilliant reds, oranges, and golds in autumn (UNEP/WCMC 2005; Sheppard 2005).

The undisturbed richness of the peninsula's resources enables animals to feed on prey. The quality of the habitat is so high that the home territories of the brown bear are the smallest – and one of the most dense in the world. There are 33 other species of terrestrial mammals, including fox and deer. Seven species of reptile and three amphibians are known, perhaps originally attracted by the volcanic sources of geothermal heat (Nature Conservation Bureau 2004).

The terrestrial habitats also support 2500 species of insects, 1850 moths, and 500 beetles. There are 42 species of freshwater fish. The peninsula is also internationally important as a stopover point for migratory birds(IUCN 2005); rare and endangered bird species found within the site, such as sea eagle and the white-tailed eagle, are legally protected by national laws (UNESCO WHC 2017).

The rich marine water sustains 28 species of marine mammals, among them whales, seals, and sea lions. There are 223 species of marine fish, 150 of which live in on the continental shelf in waters less than 200 m deep. Seventy percent of these fish are northern, but 14% are also southern species, living mostly in the warmer upper layer of the sea (UNEP/WCMC 2005).

51.3 Conservation Challenges

Tourism in an increasing activity in the site, and it needs attention to not cause pollution and stress on the biodiversity. Other issues impacting the area are the fishery industry on the marine ecosystem, the impact of river constructions, the impact on vegetation of grazing of the densely populated deer, and the conflicts between population and tourists and brown bears (UNESCO WHS 2017).

Global warming and climate change are also concerns. Many factors in recent years have indicated alteration in the Shiretoko ecosystem associated with these situations, such as decreases in the amount of seasonal sea ice, changes in the location of fishing grounds, and the appearance of non-local species (Makino and Sakurai 2012).

In 2014, the International Union for Conservation of Nature (IUCN), in its IUCN World Heritage Conservation Outlook, classified the state of conservation of the site as being "good with concerns" (IUCN 2014). The concerns related to the above situations. At the end of 2017, the IUCN published a new conservation outlook of the World Heritage sites, and the situation did not change for the Shiretoko (IUCN 2017; Osipova et al. 2017).

References

Goto Y, Kono T, Tsuchiya N (2002) Geochronological and geochemical constraints on the heat source of thermal activity in the Rausu geothermal field, Shiretoko Peninsula, Hokkaido, Japan. J Mineral Petrol Sci 97:79–84

IUCN (International Union for Conservation of Nature) (2005) World heritage nomination. IUCN Technical evaluation report, Shiretoko

IUCN (International Union for Conservation of Nature) (2014) World heritage conservation outlook. Report, Gland, Switzerland

IUCN (International Union for Conservation of Nature) (2017) Shiretoko conservation outlook. http://www.worldheritageoutlook.iucn.org/node/1138. Accessed 25 Dec 2017

Japan National Tourism Organization (2017) World heritage in Japan. Shiretoko. http://www.jnto.go.jp/eng/indepth/scenic/worldheritage/c_1_shiretoko.html. Accessed 18 June 2017

Makino M, Sakurai Y (2012) Adaptation to climate-change effects on fisheries in the Shiretoko World Natural Heritage area, Japan. ICES J Mar Sci 69:1134–1140

Nature Conservation Bureau (2004) Nomination of the Shiretoko for inscription on the World Heritage List. Ministry of the Environment, Tokyo

Noda A (2005) Texture and petrology of modern river, beach and shelf sands in a volcanic back-arc setting, northeastern Japan. Island Arc 14:687–707

Okumura K (1996) Tephrochronology, correlation, and deformation of marine terraces in eastern Hokkaido, Japan. Geogr Rep Tokyo Metrop Univ 31:19–26

Osipova E, Shadie P, Zwahlen C, Osti M, Shi Y, Kormos C, Bertzky B, Murai M, Van Merm R, Badman T (2017) IUCN world heritage outlook 2: a conservation assessment of all natural world heritage sites. IUCN, Gland

Sakurai Y (2007) An overview of the Oyashio ecosystem. Deep Sea Res II 54:2526–2542

Sheppard D (2005) World heritage nomination technical evaluation, Shiretoko (Japan). International Union for Conservation of Nature (IUCN), Gland

UNEP/WCMC (United Nations Environmental Programme/World Conservation Monitoring Centre) (2005) Shiretoko Hokkaido, Japan. Report, Cambridge

UNESCO WHC (United Nations Educational, Scientific and Cultural Organization World Heritage Centre) (2017) Shiretoko. http://whc.unesco.org/en/list/1193. Accessed 18 June 2017

Chapter 52
Phoenix Islands Protected Area, Kiribati

Abstract The Phoenix Islands Protected Area is an expanse of marine and terrestrial habitats in the South Pacific. The protected area represents one of the three island groups of the Republic of Kiribati. It is one of the largest marine protected areas in the world and the largest and deepest World Heritage site. The greater part of the marine protected area is comprised of ocean floor with a water column averaging more than 4000 m deep with a maximum of 6147 m. It is also is the largest marine conservation effort of its kind by a least developed country (LDC). The protected area has one of the world's largest oceanic coral ecosystems and 14 underwater seamounts which probably are extinct volcanos. It also offers important habitats for seabird nesting.

52.1 Introduction

Phoenix Islands, one of the world's most remote island chains, are located in the South Pacific, near the equator, midway between Australia and Hawaii, within the globally biologically important area called Polynesian/Micronesian hotspot. It is one of the three island groups of the Republic of Kiribati and has a total area of 408,250 km^2, almost all marine (UNESCO WHC 2017) (Fig. 52.1).

Kiribati first declared the creation of Phoenix Islands Protected Area in 2006. In 2008, the country adopted formal regulations more than doubling the original size of the area, making it at the time the largest marine protected segment in the world. In 2010, the protected area became a World Heritage site and is still the largest and deepest of all of them and the largest in the Pacific Ocean (Phoenix Island Organization 2017).

The management of the site is made by Phoenix Islands Protected Area Conservation Trust, a non-governmental organization established under the laws of the Republic of Kiribati, with support from Conservation International and New England Aquarium, two other non-governmental organizations (Conservation International 2017).

To enable effective management of the protected area and the conservation measures needed, the trust has pioneered a market-based contractual approach, in effect

V. Claudino-Sales, *Coastal World Heritage Sites*, Coastal Research Library 28,
https://doi.org/10.1007/978-94-024-1528-5_52

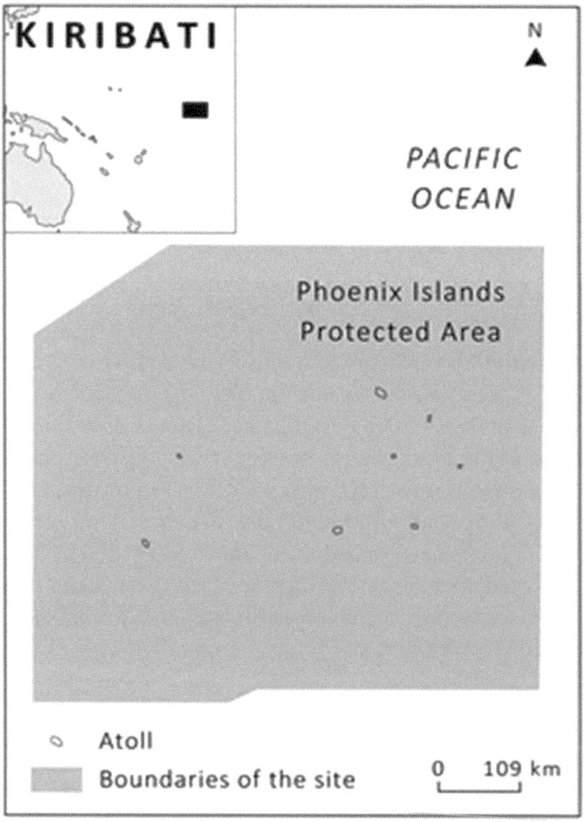

Fig. 52.1 Location of Phoenix Islands, Kiribati, in South Pacific Ocean

since January 2015, on the heels of the closure to all commercial fishing, the first for developing nations (UNESCO WHC 2017).

The history of the area is diverse and spans centuries. There are remnants of unique cultural history reflecting ancient Polynesian and Micronesian exploration and settlement of the Pacific, nineteenth century whaling, guano discovery and extraction, and the area's role in World War II and in the Cold War (Government of Kiribati 2014).

The site is made up of 8 atolls, 2 submerged reefs, 14 seamounts, and extensive open ocean and ocean floor habitat (UNESCO WHC 2017; see Fig. 52.1). The isolation of the islands inhibits any mass visitation, and tourism is not important as an economic activity, as usually is the case in relation to World Heritage sites. Just one of the eight atolls is inhabited, with only a few dozens of people (Phoenix Islands Organization 2017).

The most common way to get access to the islands is by flights leaving from Fiji or Hawaii to South Tarawa, Kiribati's capital, and from there by boat to the site. Due to the isolation, there has been minimal human impact in the area, and it is in a near pristine state, creating a truly oceanic wilderness (Government of Kiribati 2014).

52.2 Natural Settings

52.2.1 Climatic Conditions

The climate of Kiribati is tropical warm with different seasons. The dry seasons are between December and February and from June to September. The rainy seasons last from February to the end of May and from September to the end of November. Temperatures range between 26 and 32 °C year-round, with the water temperature sitting on a comfortable year-round 28–29 °C (Weather online 2017).

Rainfall in Kiribati is affected by the movement of the South Pacific convergence zone and the intertropical convergence zone. Its heavy rainfall is caused by air rising over warm water where winds converge, resulting in thunderstorm activity (Government of Kiribati 2013).

52.2.2 Geology and Geomorphology

As a vast expanse of largely pristine mid-ocean environment, replete with a suite of largely intact uninhabited atolls, Phoenix Islands are a superlative natural phenomenon of global importance (Phoenix Islands Organization 2017).

Phoenix Islands Protected Area contains an outstanding collection of large extinct submerged volcanos, rising directly from the deep seafloor with an average depth of more than 4500 m and a maximum depth of more than 6000 m. There are 14 recognized seamounts that do not penetrate to the surface. The collection of atolls and reef islands represent coral reef capping on eight other volcanic mountains that approach the surface (UNESCO WHC 2017).

The eight islands of the Phoenix Group vary in size from 17.5 to 1.03 km across. Three major reef types are present in the islands: outer reef slopes, lagoon reefs, and channel patch reefs. Channel reefs are negligible and limited to small patches. Lagoon reefs are well developed, growing on relict reef structures and controlled by circulation of water. Outer reefs have the most vibrant growth of corals (Obura 2011). The majority of the atolls is barely more than 6 meters above sea level and surrounded by barrier reefs and white beaches, mostly sandy (Figs. 52.2 and 52.3), with a broad expanse of abyssal plain (Government of Kiribati 2014).

52.2.3 Biodiversity

Due to its extreme isolation, Phoenix Islands Protected Area occupies a unique position in the biogeography of the Pacific Ocean, creating ecosystems appropriated for migratory and pelagic/planktonic species (UNESCO WHC 2017).

Fig. 52.2 Sandy beach surrounded by coconuts in an atoll of the Phoenix Islands Protected Area, Kiribati. (Picture reproduced by permission of Brian J. Skerry/National Geographic Creative)

Fig. 52.3 A lagoon bordered by limestones and reefs in Phoenix Islands, Kiribati. (Picture reproduced by permission of Nava Fedaeff/Shutterstock Stock Photo)

It embraces the full range of marine environments and displays high levels of marine abundance, with habitats for shark, reef fishes, sea turtles, sea birds, corals, giant clams, and coconut crabs, many of which have been depleted elsewhere. It is an area of unique submerged water currents and hence is believed to play a significant role in the dispersal of marine animals and larva throughout the world (Government of Kiribati 2014).

With its rich biota, as a known breeding site for numerous nomadic, migratory, and pelagic marine species, Phoenix Islands make an outstanding contribution to ongoing ecological and biological processes in the evolution and development of global marine ecosystems and communities of animals (UNESCO WHC 2017).

Coral species diversity of the islands is moderately high. While lacking in some corals dominant in other major reef areas, the reefs have interesting species assemblages, with prolific growth of some species normally subdominant in other reef areas (Obura et al. 2011).

There are 200 species of coral (Government of Kiribati 2014). The total known fish fauna of the Phoenix Islands stands at 516 species (Allen and Bailey 2011), and there are also 18 marine mammals and 44 bird species (UNESCO WHC 2017).

52.3 Conservation Challenges

Since the creation of the marine protected area around and within Phoenix Islands, the reefs have shown benefits, and the area is becoming a real refuge for open ocean pelagic and deep-sea marine life (Rotjan et al. 2014).

Nevertheless, there are some emerging threats, so that isolation can no longer be relied upon to conserve Phoenix Islands. The most important is the threat associated with global warming and climate change, which implies sea level rise, ocean acidification, and climate alterations, putting the existence of the Phoenix Islands in jeopardy (Conservation International 2017).

The number of occurrences of coral bleaching is increasing, killing corals and decreasing the abundance of coral fish. In addition, coastal erosion is also increasing, destroying habitats (Mangubhai et al. 2014; Vargas-Angel et al. 2011), and increasing rainfall and floods are killing trees and destroying soils, in such a way that birds are leaving the islands (Government of Kiribati 2013). The management of the site has to deal with these problems, which are predicted to increase.

The International Union for Conservation of Nature (IUCN) published a World Heritage Conservation Outlook report at the end of 2017 and found that the Phoenix Islands site is in a good state of conservation but presents some concerns. The concerns are related to the above cited problems (Osipova et al. 2017).

References

Allen G, Bailey S (2011) Reef fishes of the Phoenix Islands. Atoll Res Bull 589:103–115

Conservation International (2017) Phoenix Islands Protected Area. http://www.conservation.org/projects/Pages/phoenix-islands-protected-area.aspx. Accessed 24 June 2014

Government of Kiribati (2013) Kiribati climate change. http://wwwclimategovki/met/ Accessed 24 June 2017

Government of Kiribati (2014) Kiribati for travelers. http://www.kiribatitourism.gov.ki/practicalinformation/. Accessed 24 June 2017

Mangubhai S, Strauch A, Obura D, Stone G, Rotjan R (2014) Short-term changes of fish assemblages observed in the near-pristine reefs of Phoenix Islands. Rev Fish Biol Fish 24:505–518

Obura D (2011) Coral reef structure and zonation of the Phoenix Islands. Atoll Res Bull 589:84–98

Obura D, Stone G, Mangubhai S, Bailey S, Yoshinaga A, Holloway C, Barrel R (2011) Baseline marine biological surveys on the Phoenix Island, July 2000. Atoll Res Bull 589:1–61

Osipova E, Shadie P, Zwahlen C, Osti M, Shi Y, Kormos C, Bertzky B, Murai M, Van Merm R, Badman T (2017) IUCN world heritage outlook 2: a conservation assessment of all natural world heritage sites. IUCN, Gland

Phoenix Islands Organization (2017) Phoenix Island Protected Area. http://www.phoenixislands.org/. Accessed 23 June 2017

Rotjan R, Jamieson R, Carr B, Kaufman L, Mangubhai S, Obura D, Pierce R, Rimon B, Ris B, Sandin S, Shelley P, Sumaila U, Taei S, Tausig H, Teroroko T, Thorrold S, Wikgren B, Toatu T, Stone G (2014) Establishment, management, and maintenance of the Phoenix Islands protected area. Adv Mar Biol 2014:289–324

UNESCO WHC (United Nations Educational, Scientific and Cultural Organization World Heritage Centre) (2017) phoenix islands protected area. http://whc.unesco.org/en/list/1325. Accessed 24 June 2017

Vargas-Angel B, Looney EE, Vetter OJ, Coccagna E (2011) Severe, widespread El Niño and associated coral bleaching in the Phoenix Islands (2011). Bull Mar Sci 87:623–638

Weather Online (2017) Weather in Kiribati. http://www.weatheronline.co.uk/reports/climate/Kiribati.htm. Accessed 24 June 2017

Chapter 53
Archipiélago de Revillagigedo, Mexico

Abstract Located in the eastern Pacific Ocean, 390 km southeast of the southern tip of the Baja California Peninsula, the Archipiélago de Revillagigedo is made up of four remote islands. It is part of a submerged volcano mountain range, representing the peaks that are emerging above sea level. Both the landscape and seascape exhibit impressive active volcanos, arches, cliffs, and isolated rock outcrops. The area lies along the junction of the California and Equatorial currents, generating a complex and highly productive transition zone rich in marine life recognized as very important for wide-ranging species. The islands support abundant populations of shark, rays, whales, turtles, rays, and large pelagic fish and are popular for diving.

53.1 Introduction

The Archipiélago de Revillagigedo is in the Eastern Tropical Pacific Ocean, 390 km southwest of the southern tip of the Baja California Peninsula and 720–970 km west of the Mexican mainland. The site covers 6366 km^2 and includes a marine protected area extending 12 nautical miles around each of the islands (Fig. 53.1).

The Archipiélago de Revillagigedo was declared a natural protected area and Mexican Biosphere Reserve in 1994. The reserve was also designated as a Wetland of International Importance under the Ramsar International Wetland Convention in 2004 (IUCN 2016). It became a World Heritage site in 2016 (UNESCO WHC 2017).

In 1957, an outpost was established in one of the long-uninhabited islands, and trees were planted for timber and fruit. The expedition found about 8000 sheep on the island, the residue of a temporary Australian settlement of the mid-1800s. The Revillagigedo Islands are rich in sulfur, fish, and guano. Except for two military bases, the inlands are uninhabited (Encyclopedia Britannica 2017).

The state of Colima, Mexico, was granted the sovereignty of the Revillagigedo Islands, and all parts of the archipelago are hence state-owned and state-controlled. The site is managed by the Natural Commission of Natural Protected Areas (CONANP), an agency of Mexico's Secretariat of the Environment and Natural

V. Claudino-Sales, *Coastal World Heritage Sites*, Coastal Research Library 28,
https://doi.org/10.1007/978-94-024-1528-5_53

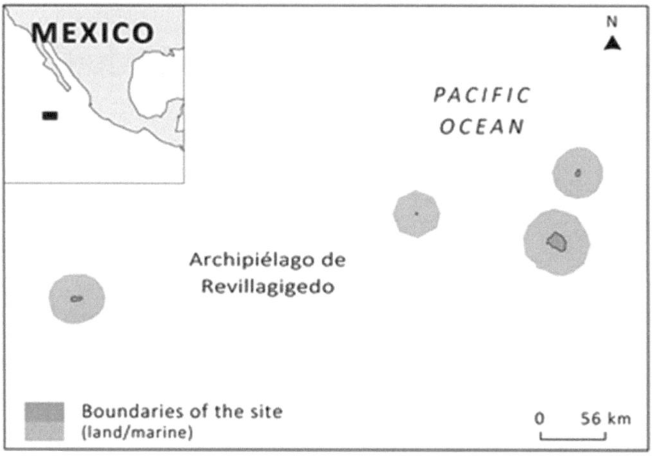

Fig. 53.1 Location of the four islands that form the Archipiélago de Revillagigedo, Mexico

Resources (SEMARNAT). The property is managed in collaboration with the Mexican Navy (IUCN 2016).

The islands are very remote, and only large, powerful vessels can reach them, in a journey of about 30 hours from Baja California. For its great biodiversity, Revillagigedo is known as the "Mexican Galapagos" and represents a spot area for divers (The Yucatan Times 2017).

53.2 Natural Settings

53.2.1 Climatic Conditions

The climate is hot and subhumid. At the top of the highest peak, because of the altitude, the climate is temperate and subhumid. Tropical cyclones form 15–24 times a year, and the area is also reached by hurricanes (University of Delaware 2014).

The average annual temperature is warm, higher than 22 °C. The temperature of the coldest month is above 18 °C. There are rains in summer and winter. The wettest month is September. The islands have an average value of 76.5% humidity (El Clima 2017).

53.2.2 Geology and Geomorphology

The Archipiélago de Revillagigedo is part of a Quaternary submarine volcanic mountain range 4000 m high, with peaks of some volcanos emerging above sea level. Due to their volcanic origin, depths around the islands increase abruptly at distances of around 10–12 km from the shorelines (Taran et al. 2010).

The World Heritage site consists of four islands. The largest, Socorro, which rises to an elevation of 1130 m, is 39 km long and 14 km wide. San Benedicto, 64 km north of Socorro, and Clarión, 400 km west of Socorro, are the two other large islands. Roca Partida is a small island 100 km west of Socorro (Encyclopedia Britannica 2017).

Socorro Island is the exposed part of a volcanic edifice rising from the ocean floor. It is the top portion of Evermann volcano which had its last eruption in 1993. The volcano is located at the intersection of the Clarion Fracture Zone (E-W) and the Mathematician Ridge System (N-S) and has erupted voluminous pantellerite to trachyte ash flows and lava flows which make up 90% of the rocks exposed on the island (Carballido-Sanchez 1994).

Socorro is reported as the only silicic peralkaline volcanic island in the Pacific Ocean. The summit region of the Evermann volcano is composed of a group of rhyolite domes. Moderate fumarolic activity and diffuse degassing are present, making it an active volcano (IUCN 2016).

On San Benedicto Island, the Barcena volcano erupted in 1948 and 1952, denuding the island of all flora and fauna. Roca Partida is the crest of a submarine stratovolcano. Clarión Island, the western most of the islands in the group, was formed by volcanic eruptions during the Miocene and Late Eocene. The landscape on the archipelago exhibits arches, cliffs, and isolated rock outcrops emerging from the middle of the ocean (Taran et al. 2010) (Fig. 53.2).

53.2.3 Biodiversity

The Archipiélago de Revillagigedo is located in a transitional maritime zone influenced by the cold California Current mixed with the warm waters of the North Equatorial Current, resulting in a unique set of biological and ecological processes (IUCN 2016). The isolation and degree of conservation of the archipelago support evolutionary systems which provide a high degree of endemicity in both terrestrial and marine realms (UNESCO WHC 2017).

The vegetation in the islands has its peculiarities. In addition to coastal halophytes, it is characterized, especially in Socorro Islands, by the presence of various primary vegetation types such as shrub vegetation, grassland, prairie, tropical dry forest, tropical rainforest, and lower montane cloud forest (Flores-Palacios et al. 2009).

The marine ecosystems of the archipelago are mostly undisturbed and harbor abundant populations of the largest pelagic fauna in the world. There are large numbers of giant manta rays, tunas, turtles, sharks, and whales (Fig. 53.3) (IUCN 2016).

Among the marine fauna, 251 fish species with 10 endemic species have been recorded, together with a high diversity of invertebrates, especially crustaceans and mollusks. In the four islands, corals are present but form no reefs. They represent 25 species related to five families. More than half of this total of coral species is found only in the Eastern Pacific and more than a third exclusively in the islands that form the archipelago (Mejia 1998).

Fig. 53.2 Stacks and isolated cliffs on volcanic rocks worked by waves and wind near Socorro Island in Revillagigedo Archipelago. (Picture reproduced by permission of Doublex/Shutterstock Stock Photos)

The terrestrial fauna is characterized by the massive presence of birds. Practically all of the terrestrial breeding birds on the islands are endemic, with only two or three recent as introductions. Four endemic species of reptiles, including two lizards and two snakes, and nine invertebrates have also been recorded (UNESCO WHC 2017).

Fig. 53.3 Breaching young humpback whale near Socorro Island, Archipiélago de Revillagigedo. (Picture reproduced by permission of Jacek Jasinsk/Shutterstock Stock Photos)

53.3 Conservation Challenges

The Archipiélago de Revillagigedo, being isolated and largely uninhabited, is submitted to relatively low threats. The greatest problem in the islands is related to the presence of introduced species (UNESCO WHC 2017).

Major conservation successes by the Mexican government have seen the eradication of pigs and sheep from Clarión in 2002 and sheep from Socorro by 2010. Nevertheless, rabbits and iguanas remain on Clarión, and cats and mice on Socorro. However, no rats have ever reached the Archipiélago de Revillagigedo, which is extremely rare and fortuitous in a tropical island system (IUCN 2016).

To date, tourism has been restricted by the Mexican government to a set number of diving boats, and no people are allowed onshore without a permit. Fishing is restricted through the marine area zoning system (though there are concerns regarding policing and instances of sport fishing) (UNESCO WHC 2017).

Global warming in the isolated islands can be a serious problem, considering the limited area and resources to mitigate the situation. Coastal erosion, death of corals, and alterations in the climate characteristics with different levels of precipitation may be the issues associated with climate change.

The International Union for Conservation of Nature (IUCN) published a report about the state of conservation of the World Heritage sites at the end of 2017 in which the conservation of the Archipiélago de Revillagigedo appears as good but presenting some concerns (Osipova et al. 2017). The concerns are related to the situations pointed out above (IUCN 2017).

References

Carballido-Sanchez, EAN (1994) The geology and petrology of Socorro Island, Revillagigedo archipelago, Mexico. PhD thesis, Geology Department. Tulane University, Los Angeles

El Clima (2017) Archipiélago de Revillagigedo: clima, flora y fauna. http://www.elclima.com.mx/archipielago_de_revillagigedo_clima_flora_y_fauna.htm. Accessed 29 June 2017

Encyclopedia Britannica (2017) Revillagigedo Islands. https://www.britannica.com/place/Revillagigedo-Islands. Accessed 29 June 2017

Flores-Palacios A, Martínez-Gómez JE, Curry RL (2009) La vegetation de Isla Socorro, Archipiélago Revillagigedo, México. Boletín de la Sociedad Botánica del México 84:13–23

IUCN (International Union for Conservation of Nature) (2016) Archipiélago of Revillagigedo, Mexico, world heritage nomination – IUCN technical evaluation. Report, Gland

IUCN (International Union for Conservation of Nature (2017) The Archipiélago de Revillagigedo Conservation Outlook. http://www.worldheritageoutlook.iucn.org/node/1187. Accessed 25 Dec 2017

Mejia JTK (1998) Comunidades coralinas del Archipiélago de Revillagigedo México. PhD Thesis, Museo de Historia Natural. Universidade Autonoma de la Baja California Sur, La Paz

Osipova E, Shadie P, Zwahlen C, Osti M, Shi Y, Kormos C, Bertzky B, Murai M, Van Merm R, Badman T (2017) IUCN world heritage outlook 2: a conservation assessment of all natural world heritage sites. IUCN, Gland

Taran Y, Varley N, Inguaggiato S, Cienfuegos E (2010) Geochemistry of H2 and CH4 enriched hydrothermal fluids of Socorros Island, Revillagigedo archipelago, Mexico, evidence for serpentinization and abiogenic methane. Geofluids 10:542. https://doi.org/10.1111/j.1468-8123.2010.00314.x

The Yucatan Times (2017) Revillagigedo Islands: "The Mexican Galapagos". http://www.theyucatantimes.com/2017/01/revillagigedo-islands-the-mexican-galapagos/. Accessed 29 June 2017

UNESCO WHC (United Nations Educational, Scientific and Cultural Organization World Heritage Site) (2017) Archipiélago de Revillagigedo. http://whc.unesco.org/en/list/1510. Accessed 29 June 2017

University of Delaware (2014) Revillagigedo islands. http://www1.udel.edu/fllt/faculty/suztgula/lectura23.htm. Accessed 29 June 2017

Chapter 54
Islands and Protected Areas of Gulf of California, Mexico

Abstract Located in northwestern Mexico, the site comprises 244 islands, islets, and coastal areas. Almost all major oceanographic processes occurring in the major oceans are also present in the Gulf of California, giving it a great importance for study. The area also represents a natural laboratory for speciation – it contains 39% of the world's total number of species of marine mammals and a third of the world's marine cetacean species. Most of the islands and coastal areas are part of the Sonoran Desert, and there is a dramatic contrast between the rugged and inhospitable islands and coasts and the brilliant reflection of the turquoise waters. High rocky cliffs and sandy beaches in countless forms and colors rim the islands and coasts. The diversity of marine life associated with spectacular submarine terrain and unusual water transparency turn the site into a globally renowned diver's area.

54.1 Introduction

Extending from the Colorado River Delta in the north to 270 km southeast of the tip of the Baja California Peninsula, the serial site comprises 244 islands, islets, and coastal segments. The total area is 18,371 km^2, of which about 25% is terrestrial and the rest marine (UNESCO WHC 2017) (Fig. 54.1).

From north to south, the areas included in the serial site are the Colorado River Delta, the Great Islands, the Archipelago de San Lorenzo and San Pedro Mártir, the Vizcaíno coast, Bahía de Loreto, Cabo Pulmo and Cabo San Lucas, the Islas Marías, and Isla Isabel (UNEP/WCMC 2009).

The protection of the islands and coasts that integrates the serial site started in 1993, with part of the delta of the Colorado River being designated as a UNESCO Biosphere Reserve. The government of Mexico designated the delta as a protected area in 1996. In the decade of 2000, many national parks were created, and 21 sites have been designated Ramsar Wetlands of International Importance under the Ramsar Wetland Convention. The sites were added to the list of World Heritage sites in 2005 (UNEP/WCMC 2009).

All of the marine area and most of the islands are federally owned, with very few in private hands. Private owners usually do not live there and the majority of the

V. Claudino-Sales, *Coastal World Heritage Sites*, Coastal Research Library 28,
https://doi.org/10.1007/978-94-024-1528-5_54

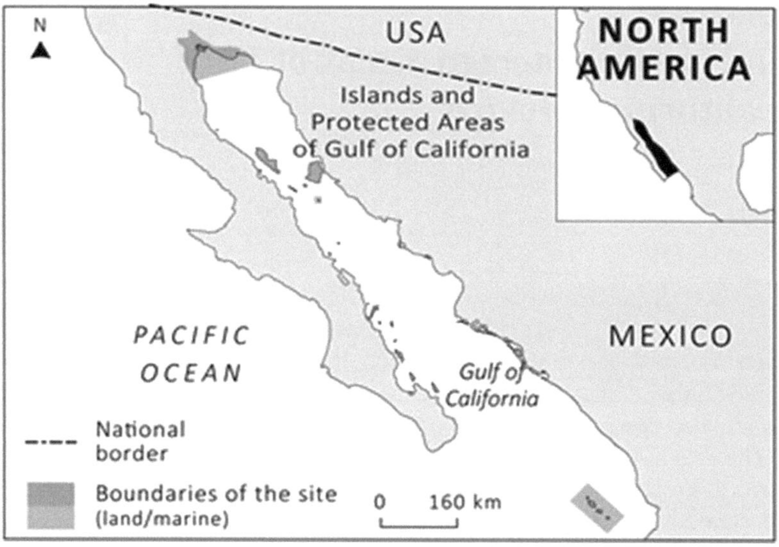

Fig. 54.1 Location of the serial site islands and protected areas of Gulf of California, Mexico

islands are uninhabited. Some contains small camps of fisherman, and one (Shark Island) is owned by the Seri indigenous peoples, who consider it to be sacred (UNESCO WHC 2017).

Most of the islands and marine areas is administered by the "Comisión Nacional de Areas Naturales Protegidas - CONANP" (National Commission of Natural Protected Areas) of the Ministry of the Environment and Natural Resources. Isla Isabel is administered by the Ministry of the Interior and Shark Island by the Seri (UNEP/WCMC 2017).

Archaeologically, little is known of the prehistory of the arid peninsula of Baja California, but hundreds of rock canyons serve as the canvas for traditional wall murals painted by early hunter-gatherer groups. Carbon dating has identified signs of habitation by mollusk-eating humans 40,000 years ago (Rosabel 2005). In 1533, the Spanish conqueror Hernan Cortes accidentally discovered the Baja Peninsula while exploring the Pacific coast of Mexico. In his honor, the Gulf of California is often referred to as the Sea of Cortes. Baja California remained unconquerable until 1697. Franciscan priests and later Dominicans dominated the peninsula until Mexico obtained its freedom from Spain in 1821 (Nix 2010).

The Gulf of California and its islands have been called a natural laboratory for the investigation and was called the "world's aquarium" by Jacques Cousteau. The diversity of marine life associated with spectacular submarine terrain and unusual water transparency turn the site into a globally renowned diver and tourist area (UNESCO WHC 2017).

54.2 Natural Settings

54.2.1 Climatic Conditions

The gulf offers an arid climate with low humidity, high rates of evaporation, and intense solar radiation. The peninsula is surrounded by mountains, so that the climate is more continental than oceanic. The temperature averages 16–24 °C. Summer temperatures can reach 30 °C; in winter nights it can drop to 9 °C. Inland summer temperatures can reach 43 °C (Nix 2010).

The northern section has a Mediterranean pattern of winter rain (50 mm) and summer dryness. In the south the climate is less dry with 190 mm of rain, mainly falling in August and September. At the tip of the gulf, the mean rainfall is between 200 and 317 mm, but tropical storms in late summer and autumn rainfall can bring as much as 700 mm (UNEP/WCMC 2009).

Coastal islands experience mainly land and sea breezes. The islands in the open sea experience northwest winds from November to May and southeast winds from June to October. The area is subject to hurricanes (Nix 2010).

54.2.2 Geology and Geomorphology

The Gulf of California is more than 1050 km wide, going from the mouth of the Colorado River to a line between Cabo San Lucas and Eldorado on the coast of the state of Sinaloa, Mexico. It averages about 175 km wide, widening toward the south. Almost all major oceanographic processes occurring in the major oceans are also present in the Gulf of California, giving it a great importance for scientific study (Rosabel 2005).

Wegener, as part of his scheme of continental drift, proposed that the Gulf of California was formed by separation of the peninsula of Baja California from mainland Mexico. This proposal was ignored for many decades because the concept of continental drift was found unacceptable by geologists, but it is now revived in light of the theory of plate tectonics (Elders et al. 1972).

Effectively, the new research shows that the Gulf of California is a young example of crustal stretching and shearing leading to the birth of a new oceanic basin at a formerly convergent margin. It is a border zone placed between the North American and Pacific plates and also an interface of continental and oceanic tectonics (Duque-Trujillo et al. 2015).

The rifting started 14 million years ago, opening obliquely several basins in the past 6 million years (Wang et al. 2009). These processes make the Gulf of California an oblique rift system with short spreading segments connected by long transform faults (Lizarralde et al. 2007). The increase in rift obliquity and resultant onset of faulting facilitated marine incursion by 6 million years and complete continental rupture at 2–1 million years ago (Darin et al. 2016).

Fig. 54.2 Pocket sandy beach surrounded by barren cliffs in the Island of Espiritu Santo (Holy Spirit Island). It depicts an intense contrast between the reddish volcanic rocks and the turquoise waters of Gulf of California. (Panorama picture reproduced by permission of Giordano J. Cervera)

The present configuration of the gulf shows a modest continental shelf along the east side and precipitous slopes along the western edge which is an almost unbroken mountain wall. The uplift of the western edge was coeval with lithospheric rupture, and the onset of oceanic spreading was driven by the flexural response to lithospheric unloading (Mark et al. 2014).

The Gulf of California has irregular coastlines and forms numerous small bays. The Colorado River Delta extends offshore for as much as 32 km with depths as shallow as 18 m. Its eastern shores are broken by a number of rivers. No rivers feed the gulf on the western edge. Many islands are contained within the gulf (Nix 2010).

The islands within the Gulf of California present a variety of geological formations and structural relationships. Some are geologically similar to the adjacent continent and were probably part of it at one time. They are composed mainly of coarse, crystalline granitic rocks. But some others are composed of lavas, mudflows, tuffs, and other eruptive rocks, sometimes resting on a granitic base. Some others are made of small granitic islets that emerge from a submerged ridge. Isla Coronado includes a Pliocene volcanic cone (Nix 2010).

Most of the islands are barren and without water. Wave action has eroded great seacliffs with few good harbors. High rocky cliffs and sandy beaches in countless forms and colors rim the islands and coasts (Figs. 54.2 and 54.3). The maximum altitude in the islands is 990 m, in Isla Ángel de la Guarda (UNEP/WCMC 2009).

There is a high percentage of biogenic components in the sedimentary deposits of the islands and coasts of the gulf. Due to high-energy waves, coastal deposits on points or islands contain massive biocalcarenite deposits derived from shells, while areas of lower water energy commonly accumulate the crushed debris of coralline red algae. Sheltered areas most likely attract colonization by coral reefs (Ledesma-Vázquez et al. 2012).

54.2.3 Biodiversity

The multitude and diversity of islands in terms of origin, size, and elevation and the harsh conditions, the isolation, and the variations from north to south, as well as the distance to the mainland, have resulted in speciation and endemism (UNESCO WHC 2017).

Fig. 54.3 Cliffs and islets sustained by volcanic rocks in Bahia of Loreto, Baja California, on the coast of the Gulf of California. An important colony of birds harbors in the area. (Picture reproduced by permission of Gerardo Rodriguez Quiroz)

It has to be pointed out that during the formation of the gulf, temperate eastern Pacific waters were trapped in the upper part while the lower part remained tropical. This has created a unique marine ecosystem (the Sea of Cortés ecosystem) which contains a variety of pelagic, benthic, and coastal environments that extend from mangrove to wetlands and coral reefs (UNEP/WCMC 2009). The gulf's immense biodiversity productivity is also due to the abundant phytoplankton and zooplankton, which proliferates on the nutrients brought by upwelling oceanic currents (Rosabel 2005).

The desertic conditions prohibit exuberant flora. The dominant type is that of the floristically rich Sonoran Desert with its many varieties of cactus and other succulents, including some of the tallest cacti in the world, exceeding 25 m. Six hundred and ninety-five vascular plants have been recorded, 28 species or subspecies being endemic (UNEP/WCMC 2009).

The marine environment harbors 626 species of macroalgae in submarine forests. There are 31 species of marine mammals (75% of Mexico's and 39% of the world's total species), 34 species of cetaceans (a third of the world's total), and 891 species of fish in 441 genera, including 90 endemic species and more than 150 rocky and sandy coastal species. About 73% of the fish are tropical. There are also 4848 recorded macroinvertebrates. Impressive numbers of resident and migratory birds are also present (Rosabel 2005).

54.3 Conservation Challenges

The state of conservation of the serial site is generally good. Threats from overfishing, pollution, exotic species, and uncontrolled tourism are significant, but management is rated as relatively effective, and there are no large problems at present, with respect to the health of the marine ecosystem (IUCN 2014).

Global warming and climate change, however, represent a high threat, considering that it can produce an increase in the temperatures of the air and water, resulting in acidification of the sea and mortality of corals and calcifying species. The erosion on the coastlines is probably going to increase, changing the equilibrium of beaches and mangroves and modifying habitats.

The International Union for Conservation of Nature (IUCN) released a report about the state of conservation of the World Heritage sites in 2014, in which the Islands and Protected Areas of Gulf of California level of conservation was found to be good without concerns (IUCN 2014). At the end of 2017, a new report was published (Osipova et al. 2017), in which the situation on the site was shown as being in deterioration, with the problems that threaten the property showing an increasing intensity. The management of the site seems to not have conditions to eliminate all the threats, and the site conservation is presenting significant concerns (IUCN 2017).

References

Darin MH, Bennett SEK, Dorsey RJ, Oskin ME, Iriondo A (2016) Late Miocene extension in coastal Sonora, Mexico: implications for the evolution of dextral shear in the proto-Gulf of California oblique rift. Tectonophysics 693:378–409

Duque-Trujillo J, Ferrari L, Orozco-Esquivel T, López-Martínez M, Lonsdale P, Bryan SE, Kluesner J, Piñero-Lajas D, Solari L (2015) Timing of rifting in the southern Gulf of California and its conjugate margins: insights from the plutonic record. Geol Soc Am 127:702–736

Elders WA, Rex RW, Meidav T, Robinson PT, Biehler S (1972) Crustal spreading in northern California. Science 176:4056

IUCN (International Union for Conservation of Nature) (2014) Islands and protected areas of Gulf of California conservation outlook. Report, Gland, Switerzland

IUCN (International Union for Conservation of Nature) (2017) Islands and protected areas of Gulf of California Conservation Outlook. http://www.worldheritageoutlook.iucn.org/node/1137. Accessed 26 Dec 2017

Ledesma-Vázquez J, Carreño A, Guardado-France R (2012) Biogenic coastal deposits: Isla del Carmo, Gulf of California, Mexico. Facies 58:169–178

Lizarralde D, Gary JA, Brown HE, Fletcher JM, Gonzalez-Fernandez A, Harding AJ, Holbrook WS, Kent GM, Paramo P, Sutherland F, Umhoefer PJW (2007) Variations in the style of rifting in the Gulf of California. Nature 448(7152):466–469

Mark C, Gupta S, Carter A, Mark DF, Gautheron C, Martin A (2014) Rift flank uplift at the Gulf of California: no requirement for astenospheric upwelling. Geology 42:259–262

Nix RK (2010) The Gulf of California. University of Texas. Report, Dallas

Osipova E, Shadie P, Zwahlen C, Osti M, Shi Y, Kormos C, Bertzky B, Murai M, Van Merm R, Badman T (2017) IUCN world heritage outlook 2: a conservation assessment of all natural world heritage sites. IUCN, Gland

Rosabel P (2005) World heritage nomination IUCN summary: islands and protected areas of the Gulf of California (Mexico). International Union for Conservation of Natura (IUCN). Report, Gland, Switzerland

UNEP/WCMC (United Nations Environmental Programme/World Conservation Monitoring Centre) (2009) Islands and protected areas of Gulf of California, Mexico. Report, Cambridge, England

UNESCO WHC (2017) Islands and protected areas of Gulf of California. http://whc.unesco.org/en/list/1182. Accessed 6 July 2017

Wang Y, Forsyth D, Savage B (2009) Convective upwelling in the mantle beneath the Gulf of California. Nature 462(7272):499

Chapter 55
Te Wahipounamu Southwest, New Zealand

Abstract Located in the southwest corner of New Zealand's South Island, the site is spread over a 450 km strip extending 40–90 km inland from the Tasman Sea. The area exhibits many classic examples of the tectonic and climatic processes that have shaped the earth. It has many evidences of the last glacial era and also counts with the world's best intact modern representation of the ancient biota of Gondwana. The landscape shows a sequence of mountains, fjords, marine terraces, rocky coasts, towering cliffs, hanging valleys, cirques, lakes, and waterfalls, all in a pristine state. Temperate rainforest and alpine ecosystems are well represented across wide climatic and latitudinal gradients. Only traces of human influence are evident and then mainly in peripheral areas.

55.1 Introduction

Located in the southwest corner of New Zealand's South Island, Te Wahipounamu is spread over a 450 km strip extending inland 40–90 km from the Tasman Sea, having a seaward boundary at the mean high water mark. This vast park covers 9.6% of the country, representing 26,000 km^2 (UNESCO WHC 2017) (Fig. 55.1).

The preservation of the area started in 1953, when a national park was created at Mount Cook. From 1953 to 1997, many other areas were protected. In 1986, Westland-Mount Cook and Fiordland National Parks were inscribed on the World Heritage List. In 1990, these separated sites were incorporated. The World Heritage site now is one unit, with a few small outliers, composed of 40 different protected areas (UNEP/WCMC 2007). It includes 4 national parks, 2 nature reserves, 3 scientific reserves, 13 scenic reserves, 4 wildlife management reserves, 5 ecological areas, 12 conservation areas, and 1 private reserve (UNESCO WHC 2017).

Te Wahipounamu means the waters of the green stone, valued by the Maori for its beauty and hardness. The area is within the ancestral territories of the Ngai Tahu tribe. The first European to arrive was the Dutch navigator Abel Tasman, in 1642. Sealing by foreigners began in 1792, and soon the seal population had been severely reduced, until protection in 1875. Whalers also established short-lived coastal stations during

V. Claudino-Sales, *Coastal World Heritage Sites*, Coastal Research Library 28,
https://doi.org/10.1007/978-94-024-1528-5_55

Fig. 55.1 Location of Te Wahipounamu Southwest Park, in New Zealand

the 1800s. Gold was discovered in the 1860s but in few years the boom towns were abandoned. The area is the least populated part of New Zealand (DOC 1989).

The site, with one small exception, is Crown (Government and people of New Zealand) owned and is administered by the Department of Conservation. The area has an international profile and international and national tourism are an important activity (Government of New Zealand 2017).

55.2 Natural Settings

55.2.1 Climatic Conditions

The climate is cool temperate oceanic. In summer the temperatures are around 20 °C, in winter between 5 and 10 °C (Meteovista 2017). Rain falls throughout the year. The mountains form a barrier to the prevailing wet westerlies, and the rain is often heavy and prolonged. It is from 3000 to 5000 mm on the coastal lowlands and of 12,000 mm on the western flank of the mountains, where much of it falls as snow (DOC 1989).

The winds have an average speed of 5 m/s. The air humidity is around 80% all over the year. Fog is common in winter, especially around the mountains (World Weather and Climate 2017).

55.2.2 Geology and Geomorphology

The serial site exhibits many classic examples of the natural processes that have shaped the Earth (UNESCO WHC 2017). The tectonic activity there is remarkable: the great Alpine Fault divides the region and marks the contact zone between the Indo-Australian and Pacific continental plates, making it one of only three segments where major plate boundaries are on land (DOC 1989).

This continental collision has uplifted the Southern Alps, a young Cenozoic (~5 million yeas) actively deforming orogen which represents one of the most active mountain belts in the world, presenting nearly 4000 m of altitude within only 30 km from the sea (Beyssac et al. 2016). The area suffers a vertical uplifting of ca 0.3–0.5 mm a year (Beavan et al. 2010) and is submitted to frequent earthquakes (Boese et al. 2013).

The basic shape of the land was set during the Pleistocene glaciations, making it a wilderness of rugged snow-capped mountains, snowfields and glaciers, large alpine lakes, and 14 deeply incised sea fjords (Fig. 55.2). The uplifted mountains have been very deeply sculpted by glaciers, resulting in very high relief with hanging valleys and cirques and containing 28 of the 29 New Zealand peaks above 3000 m, among them Mount Cook, which at 3764 m altitude is the highest peak in New Zealand (UNEP/WCMC 2007).

The immense glacier on Mount Cook flows at 0.65 m a day, the fast-moving glaciers in Westland flow at the rate of 4 m a day, descending into temperate rainforests. Erosion in the mountains is very rapid, with the land showing intense gullying, serrated ridges, rock falls, landslides and a great variety of morainal forms (Reznichenko et al. 2016; Hales and Roering 2005). In Fiordland, a lower massif, post-glacial modification of the topography is much less than in the Southern Alps, the glacial landforms remain almost entirely intact, and mountains are extremely steep (DOC 1989).

Full exposure to Southern Ocean swells has produced a dramatic coastline with irregular high seacliffs, some as high as 2000 m, and many offshore rocks and stacks (Fig. 55.3). On the south coast of Fiordland, 13 marine terraces record a detailed history of uplift over almost a million years, with terraces formed by coastal erosion being found up to 1000 m above sea level (UNEP/WCMC 2007).

The rocks are generally crystalline, dominated by granite, diorite, and gneisses. In the extreme southwest, there are sedimentary rocks. The Dun Mountain ophiolite comprises a slice of oceanic crust and the underlying mantle. The valued New Zealand jade (pounamu) is found as boulders along the coast (DOC 1989).

Fig. 55.2 Fjords in Milford Sound at Fiordland National Park, Te Wahipounamu World Heritage site. (Picture reproduced by permission of Department of Conservation Te Papa Atawhai, Government of New Zealand, web site http://www.doc.govt.nz/)

Fig. 55.3 Rocky coast with cliffs and boulders at Gillespies Beach, Westland National Park, Te Wahipounamu World Heritage site. (Picture reproduced by permission of Bildagentur Zoonar GmbH/Shutterstock Stock Photos)

55.2.3 Biodiversity

The flora and fauna has become the world's best intact modern representation of the ancient biota of Gondwana. The region contains outstanding examples of plant succession after glaciation, with sequences along longitudinal (sea level to permanent snowline), latitudinal (wet west to the dry east), and chronological gradients (fresh post-glacial surfaces to old Pleistocene moraine deposits) (UNESCO WHC 2017).

Two-thirds of the site is covered with southern beech and podocarps, some of which are over 800 years old. The west is covered by temperate rain forest plus wetlands; the drier more continental east has more open forest, generally of mountain beech, shrubs, and short tussock grasslands. The mountains hold 450 km^2 of alpine communities (DOC 1989).

The alpine flora, as a result of isolation in a geologically young habitat, is of very high species diversity and endemicity, considerable adaptive radiation, polymorphism and variation, and great variation in growth forms and has predominantly white flowers, perhaps to attract specialized pollinators which are scarce (UNEP/WCMC 2007).

The southwest is the least modified region of mainland New Zealand and is the core habitat for many indigenous animals, including a number of primitive taxa. The southwest coast has the largest breeding aggregations of New Zealand fur seal, except for the Bounty Islands (DOC 1997).

A number of mammals have been introduced, such as rat, stoat, possum, wild boar, deer, goat, and chamois. There are 16 species of native freshwater fish. The area also contains the largest and most significant populations of forest birds and wading birds in New Zealand, most being endemic. There are two species of flightless birds which were thought to be extinct (kiwi and takahe) and the only alpine parrot in the world. It is also home to the endemic crested penguin (UNEP/WCMC 2007).

55.3 Conservation Challenges

Invasive species are one of the biggest environmental problems in the site. The introduction of alien browsing mammals and predators has resulted in localized extinctions, range reductions, and significant decline in abundance of some indigenous biota. Tourism also places pressure on some of the main attractions (UNESCO WHC 2017).

The other big threat is climate change. Global warming is already changing the environment, especially on the coast, where erosion tends to increase. The problem is also serious in the glaciated areas: between 2000 and 2008, glaciers recessed in extensions of ca 3.7 km, and the melting tends to increase (Dykes et al. 2011). Besides, fragmentation of alpine areas due to global warming could signify the loss of ~80% of the existing flora, what will severely increase extinction risks (Halloy and Mark 2003).

The state of conservation of the site is considered to be good with some concerns (Osipova et al. 2017). It was the same condition found in the conservation outlook published by the International Union for Conservation of Nature (IUCN) in 2014 (IUCN 2014), what shows that the management of the area and the quality of its values remain stable.

References

Beavan J, Denys P, Denham M, Hager B, Herring T, Molnar P (2010) Distribution of present-day vertical deformation across the Southern Alps, New Zealand, from 10 years to GPS data. Geophys Res Lett 37:n/a-n/a

Beyssac O, Cox SC, Vry J, Herman F (2016) Peak metamorphic temperature and thermal history of the southern Alps (New Zealand). Tectonophysics 676:229–249

Boese CM, Stern TA, Townend J, Bourguignon S, Sheehan A, Smith EFC (2013) Sub-crustal earthquakes within the Australia–Pacific plate boundary zone beneath the Southern Alps, New Zealand. Earth Planet Sci Lett 15:212–219

DOC (Department of Conservation) (1989) Nomination of South-West New Zealand (Te Wahipounamu) by the Government of New Zealand for inclusion in the world heritage list. Report, Wellington

DOC (Department of Conservation) (1997) Te Wahipounamu-South West New Zealand world heritage area. https://www.doc.govt.nz/te-wahipounamu. Accessed 6 July 2017

Dykes RC, Brook MS, Robertson CM, Fuller IC (2011) Twenty-first century calving retreat on Tasman Glacier, Southern Alps, New Zealand. Artic Antarctic Alp Res 43:1–10

Government of New Zealand (2017) Te Wahipounamu South West New Zealand World Heritage site. http://www.doc.govt.nz/te-wahipounamu. Accessed 6 July 2017

Hales TC, Roering J (2005) Climate-controlled variations in scree production, Southern Alps, New Zealand. Geology 33:701–704

Halloy SRP, Mark AF (2003) Climate-change effects on alpine plant biodiversity: a New Zealand perspective on quantifying the threat. Artic, Antartic Alp Res 35:248–254

IUCN (International Union for Conservation of Nature) (2014) Te Wahipounamu Conservation Outlook. Rapport, Gland, Switerzland

Meteovista (2017) The climate in New Zealand. http://www.meteovista.com/Oceania/New-Zealand/Te-Wahipounamu-natuurgebied/4955638. Accessed 6 July 2017

Osipova E, Shadie P, Zwahlen C, Osti M, Shi Y, Kormos C, Bertzky B, Murai M, Van Merm R, Badman T (2017) IUCN world heritage outlook 2: a conservation assessment of all natural world heritage sites. IUCN, Gland

Reznichenko N, Davies TRH, Winkler S (2016) Revised paleoclimatic significance of Mueler glacier moraines, Southern Alps, New Zealand. Earth Surf Process Landf 41:196–207

UNEP/WCMC United Nations Environmental Programme/World Conservation Monitoring Centre) (2007) Te Wahipounamu South West New Zealand. Report, Cambridge, UK

UNESCO WHC (United Nations Educational, Scientific and Cultural Organization) (2017) Te Wahipounamu – South West New Zealand. http://whc.unesco.org/en/list/551. Accessed 6 July 2017

World Climate and Weather (2017) New Zealand. https://weather-and-climate.com/average-monthly-Rainfall-Temperature-Sunshine,te-anau,New-Zealand. Accessed 6 July 2017

Chapter 56
Rock Islands Southern Lagoon, Palau

Abstract The Rock Islands of Palau's Southern Lagoon in the Western Pacific are part of a tropical archipelago of 445 small, luxuriant, forested, raised limestone islands uplifted due to volcanism and shaped over time by wind and vegetation. It consists of scattered islands within a lagoon protected by a barrier reef, many of them displaying unique mushroom-like shapes. The site harbors the highest concentration of marine lakes in the world. They sustain a large diversity of plants, birds, and marine life. The terrestrial environment is lush and at the same time harsh, supporting numerous endemic and endangered species. The islands contain a significant set of cultural remains relating to an occupation over some 5000 years that ended in abandonment in the 17th and 18th centuries. For this reason, the area was designated as a mixed cultural and natural site.

56.1 Introduction

Rock Islands Southern Lagoon, in the Western Pacific, is part of the Republic of Palau. The site covers around 1000 km^2, 95% of which is marine (UNESCO WHC 2017) (Fig. 56.1).

The protection of the site started in 1956, when a wildlife preserve was created by the Palau Government. In 1999, the Rock Islands Southern Lagoon were designated a Conservation Zone by Koror State Law. The Southern Lagoon and surroundings became a World Heritage site in 2012, as a mixed cultural and natural property (UNEP/WCMC 2012).

The islands contain a significant set of cultural remains relating to an occupation over some 5000 years that ended in abandonment in the 17th and 18th centuries, when the population moved to larger islands. The villages include the remains of former human occupation in caves, including rock art and burials, defensive walls, terraces, and house platforms (UNESCO WHC 2017).

Many small villages were built of the local stone, the ruins of walls, terraces, and middens of which remain on higher ground along with early cave homes and rock art, though beach sites have often been washed away or reused. The settlements reflect distinctive responses to the local environment, and their abandonment

V. Claudino-Sales, *Coastal World Heritage Sites*, Coastal Research Library 28,
https://doi.org/10.1007/978-94-024-1528-5_56

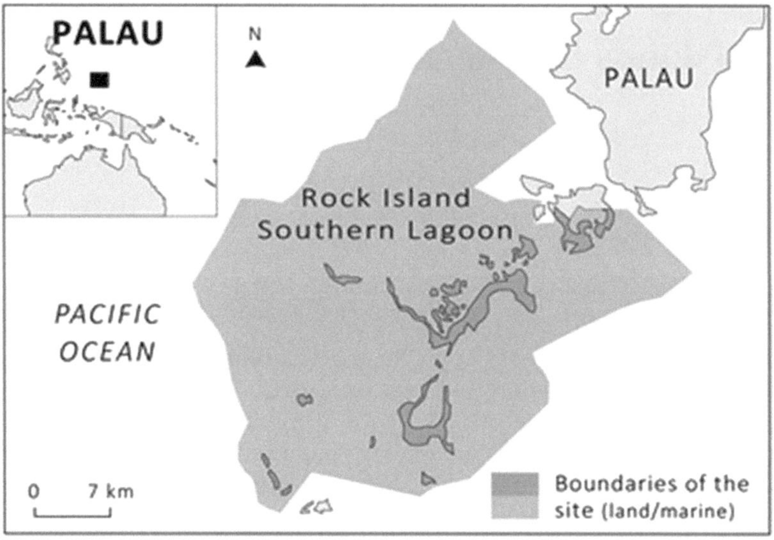

Fig. 56.1 Rock Islands Southern Lagoon, in the Archipelago of Palau, Western Pacific

demonstrates the consequences of population growth impacting and climate conditions on subsistence in a marginal environment (Osborne 1966).

The descendants identify with their ancestral islands through oral traditions that record in legends, myths, dances, and proverbs, and traditional place names the landscape and seascape of their former homes. Palauans continue to use the area and its resources for cultural and recreational purposes. This is regulated through a traditional governance system that remains an important part of national identity. International tourism is increasing every year (Parmentier 1987).

Most of the land in the property is traditionally owned by the Chiefs of Koror State, one of Palau's seven local jurisdictions. The islands are wholly owned and administered by the state of Koror, through the Koror State Department of Conservation and Law Enforcement (UNESCO WHC 2012).

56.2 Natural Settings

56.2.1 Climatic Conditions

The climate of Palau can be classified as tropical maritime, being hot and humid, with all months above 18 °C. Temperatures range between 26 and 32 °C, with the water temperature sitting on a comfortable year-round 28–29 °C. There are two different seasons. The dry season is between December and April; the rainy season lasts from April to December (Bureau of Agriculture 2003).

The annual rainfall is 3760 mm, falling heavier during the storm season between July and October. The average relative humidity is 82%. The islands are affected by the oscillations of El Niño, and there are occasional damaging storms (UNEP/WCMC 2012).

56.2.2 Geology and Geomorphology

The Palau Archipelago comprises 12 inhabited islands and over 700 small islands and islets. The Rock Islands Southern Lagoon represents Neogene volcanoes topped by steep coralline limestones (Neall and Trewick 2008). It comprises a maze of some 445 pristine, mostly small, well-forested raised coral islands, with fringing reefs and patch reefs within an enclosing barrier reef (UNESCO WHC 2012).

The volcanism in Southern Palau is related to the Kyushu-Palau ridge, an active volcanic arc associated with the subduction of the Pacific Plate into the Philippine Sea Plate; it took place mostly in Oligocene (between 37.7 and 20.1million years ago) (Sdrolias et al. 2004; Kobayashi 2000).

The volcanic islands were later overlain by Miocene limestones, what is indicative of subsidence. Rotation of the Pacific Plate gradually led to uplift of the islands, raising Miocene limestones to 220 m above sea level, forming a limestone cap with surrounding raised coral platforms and reefs (Kobayashi 2000; Canfield 1981). Current average rate of uplift is 0.5 mma year (Kobayashi 2000).

The islands are dome-shaped, with wave-undercut shores. Their karstic intricacy includes many coves, caves, fissures and tunnels, sinkholes and channels, inland marine lakes and inner basins, wetlands and beaches, and mangrove swamps. The karstic substrate of the islands is porous, there is little freshwater, and the soil is thin (UNESCO WHC 2017) (Figs. 56.2 and 56.3).

The site has 52 enclosed marine lakes, what represents one-fourth of the world's total. They were formed between 6000 and 15,000 years ago by rising sea levels and are nearly but not quite cut off from the sea. They can be 2 m or 60 m deep, clear, or murky. The walls of some are vertical and tree-shaded, and their waters have many and varying colors which, with the exuberantly forested islands, the varied coastlines, and wide skies, present a large aesthetic interest (UNEP/WCMC 2012). Freshwater is limited, with only one sustained stream flow on the large volcanic island of Babeldaob (UNESCO WHC 2017).

56.2.3 Biodiversity

The predominant vegetation of the rugged karstic substrate of Palau is subtropical moist broadleaf forest ringed by strand forest and a marine flora typical of the tropical northwestern Pacific. The Rock Islands contain 55 (42%) of Palau's 130 endemic plants, 31 (23%) of which are restricted to the islands (Kittalong et al. (2008).

Fig. 56.2 Raised and forested limestone forming a dome-shaped island in Rock Islands Southern Lagoon World Heritage site. (Picture reproduced by permission of Christian Reepmeyer)

Fig. 56.3 Sandy beach surrounded by corals and mangroves in one of the 445 small raised limestone islands of the Southern Lagoon, Palau. (Picture reproduced by permission of Christian Reepmeyer)

The islands harbor the endangered cycad in remnant native forest and endangered endemic palms. The Ngerukuid Islands Wildlife Preserve includes 113 native plants of which 30 are endemic. On ten of its islands, three endemic species form 11% of measured trees (Ohba et al. 2007).

The islands are part of one of the most ecologically diverse ecosystems in the Indo-Pacific, with 32% of the region's coral reef fauna. This contains 746 of the over 1350 species of Palau's fish, at least 13 of 17 reported species of shark, 7 species of giant clam, an estimated 385 species of stony corals in 66 genera, most of Palau's 150 species of soft coral and an endemic nautilus (UNEP/WCMC 2012).

The islands possess all of Palau's mammals and herpetofauna. Among the mammals, there is an isolated but thriving population of dugongs (sea cow). There are also two bats and the endemic flying fox. The islands contain 56 of Palau's 151 bird species and its 9 endemic. The green turtle and seawater crocodile also live in the islands' waters (UNESCO WHC 2012).

56.3 Conservation Challenges

The property has been largely isolated from human interference since pre-European occupation ceased. It is of exceptional conservation value for its low fishing pressure, low human impacts, and limited pollution. Exotic introduced animals have been eradicated from the site, and aliens' species are not a threat now. The site is nevertheless highly vulnerable to uncontrolled tourism activities (UNESCO WHC 2012).

In a general point of view, the state of conservation of the site is good with some concerns (IUCN 2014, 2017). Long-term protection and management requirements for the site include the need to prevent negative impacts from tourism, ensuring visitor numbers are within the capacity of the property, and mitigating adverse effects from development of infrastructure in Kotor.

The management of the site also has to deal with problems from climate change (Osipova et al. 2017; UNESCO WHC 2017). Climate change can be a high threat to the site. The rising of sea level associated with global warming can change the status of the almost cut-off-ocean marine lakes, as well as increase erosion in the uplifted shores. Changes in the humidity can also impose limitations, considering the historic scarcity of freshwater in the area.

References

Bureau of Agriculture (2003) Animal genetic resources report for the Republic of Palau. Report, Ngerulmud, Palau

Canfield JE (1981) Palau: diversity and status of the native vegetation of a unique Pacific Island ecosystem. Newsl Hawaii Bot Soc 20:14–20

IUCN (International Union for Conservation of Nature) (2014) IUCN world heritage outlook 2014. Report, Gland

IUCN (International Union for Conservation of Nature) (2017) Rock Island Southern Lagoon Conservation Outlook. http://www.worldheritageoutlook.iucn.org/node/1173. Accessed 27 Dec 2017

Kittalong H, Demao R, Holm T (2008) Native trees and shrubs of Palau. The Environment Inc, Koror

Kobayashi K (2000) Horizontally-moving subducted slab may generate enigmatic features of the Palau and Yap Trench-Arcs. Proc Jpn Acad 76:133–138

Neall VE, Trewick S (2008) The age and origin of the Pacific Islands: a geological overview. Philos Trans R Soc Lond Ser B Biol Sci 363(1508):3293–3308

Ohba H, Victor S, Golbuu Y, Yukihira H (2007) Tropical marine plants of Palau. Palau International Coral Reef Center, Koror

Osborne D (1966) The archaeology of the Palau Islands: An intensive survey. Bishop Museum Bulletin 230.Honolulu, USA

Osipova E, Shadie P, Zwahlen C, Osti M, Shi Y, Kormos C, Bertzky B, Murai M, Van Merm R, Badman T (2017) IUCN world heritage outlook 2: a conservation assessment of all natural world heritage sites. IUCN, Gland

Parmentier R (1987) The sacred remains: myth, history and polity in Belau. The University of Chicago Press, Chicago

Sdrolias M, Roest WR, Müller RD (2004) An expression of Philippine Sea plate rotation: the Parece Vela and Shikoku Basins. Tectonophysics 394:69–86

UNEP/WCMC (United Nations Environmental Programme/World Conservation Monitoring Centre) (2012) Rock Islands southern lagoon, Palau. Report, Cambridge, England

UNESCO WHC (United Nations Educational, Scientific and Cultural Organization World Heritage Centre) (2012) World Heritage Committee Thirty-six session. Report, Paris, France

UNESCO WHC (United Nations Educational, Scientific and Cultural Organization World Heritage Centre) (2017) Rock Islands Southern Lagoon. http://whc.unesco.org/en/list/1386. Accessed 10 July 2017

Chapter 57
Coiba National Park and Its Special Zone of Marine Protection, Panama

Abstract Coiba Island, in the Eastern Pacific, off the southeast coast of Panama, is on the edge of the Gulf of Panama and 23 km southwest of the mainland. It is the largest island of an archipelago of 9 smaller islands plus 28 islets. It is part of a system of oceanic islands of volcanic origin formed from subduction of the Pacific Plate under the Caribbean Plate, subsequently rising during tectonic movement at the end of the Tertiary. The site contains marine environments that have characteristics of both continental and oceanic influence. It maintains exceptionally high levels of endemism and provides a key ecological link to the Eastern Tropical Pacific for the transit and survival of pelagic fish and marine mammals. It is also the last refuge for a number of threatened animals.

57.1 Introduction

Coiba Island in the Gulf of Chiriquí is on the edge of the Gulf of Panama and 23 km southwest of the mainland. It is the largest island of an archipelago that includes many islands close to the coast. The site has an area of 2700 km^2, the terrestrial sector having 536 km^2, and the marine sector 2165 km^2 (UNESCO WHC 2017) (Fig. 57.1).

The preservation of the site started in 1991, when the national park was created by the Government of Panama. In 2004, the park was enlarged and designated a National Park and Special Zone of Marine Protection. The state is the proprietor of the land, and the site is managed by the National Authority of the Environment (UNEP/WCMC 2012).

Archaeological findings prove that the island was settled until the Spanish conquest in the fifteenth century. From 1919 to 2005, there was a penal colony. About 20% of the original vegetation has been altered by the occupation and its accompanying livestock, such as cattle, horses, pigs, dogs, and buffalo (UNEP/WCMC 2012).

The islands have been isolated long enough for endemic species of flora and fauna to develop, making them a natural laboratory, and there is a well-established biological station. The Smithsonian Tropical Research Institute has surveyed plants and marine invertebrates and with Duke University also surveyed fishermen to develop a

V. Claudino-Sales, *Coastal World Heritage Sites*, Coastal Research Library 28,
https://doi.org/10.1007/978-94-024-1528-5_57

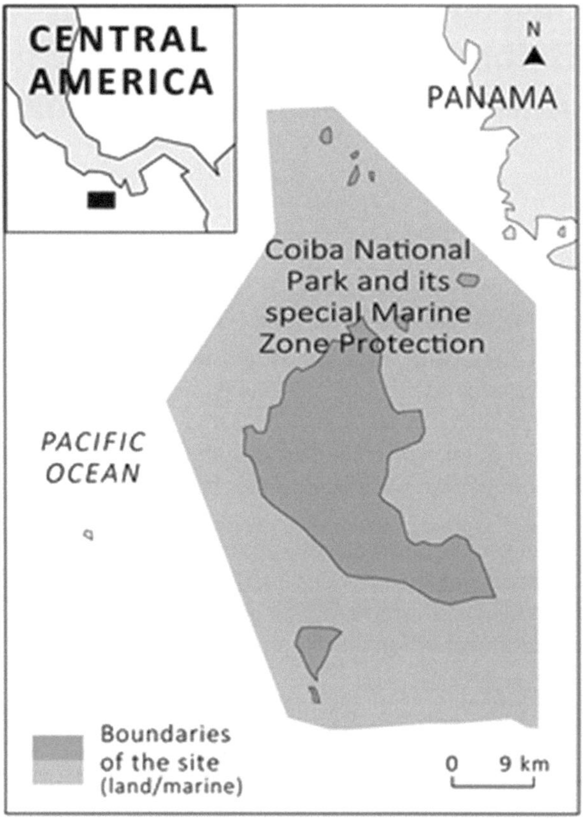

Fig. 57.1 Coiba National Park and its Special Zone of Marine Protection, in the Eastern Pacific, off the southeast coast of Panama

strategy for fish conservation and for limiting illegal fishing. The Spanish Agency for International Cooperation has worked with local communities researching into artisanal fishing, sustainable farming, and their commercial marketing (IUCN 2005).

The Panamanian Tourism Institute has chosen the area as one of its national priorities for tourism and the number of visitors is expected to grow rapidly year by year. Activities include the use of the beaches and coastal areas as well as scuba diving. Talks to school groups and the public are also part of the park's mission (UNEP/WCMC 2012).

57.2 Natural Settings

57.2.1 Climatic Conditions

The climate in the Gulf of Chiquirí is tropical. It lies within the Intertropical Convergence Zone, ITCZ, where the trade winds from northeast and southeast converge to produce heavy rainfall along the fluctuating North Equatorial Front. The average annual rainfall is 3500 mm, and it is strongly seasonal, falling mainly between May and December (Cardiel et al. 1997).

The average daily temperature is a fairly constant 26 °C. The high rainfall results in less salty seawater and seasonally washes sediment from the land that enriches the water. Every 3–7 years, the El Niño occurs, when the sea temperature warms to over 30° over long periods, causing dieback of coral (Brenner et al. 2017).

57.2.2 Geology and Geomorphology

Coiba is the largest island in Central America. The national park includes 9 volcanic islands and 28 islets. The geology of the area is controlled by the subduction of Cocos Plate into the Caribbean Plate, responsible for the Miocene volcanism that took place in an oceanic segment close to the continental shelf (Werner et al. 2003). Limestones were formed later on the top of the seamounts before it was uplifted in the Neogene (Bauch 2006; Meschede and Barckhausen 2001). From base to top, the volcano's structure comprises submarine basaltic lava flows locally interlayered with hemipelagic sediments, basaltic breccias, shallow-water limestones, and subaerial basaltic lava (Buchs et al. 2011) (Fig. 57.2).

Fig. 57.2 One of the numerous islet on Coiba National Park, formed by boulders and blocks of basalts and having carbonates as basement on the shore and in the shallow ocean floor of the surroundings. The islet is colonized by dense tropical coastal vegetation. (Picture reproduced by permission of Ian Woolcock/Shutterstock Stock Photos)

The National Park with the Special Zone of Marine Protection covers 60% of the continental shelf of Panama and 90% of the islands within the Gulf of Chiriquí. The site contains marine environments that have characteristics of both continental and oceanic influence. Coiba itself has 15 rivers and 240 km of coastline. Remote islands with sandy beaches are common (Fig. 57.2). It has the second biggest reef coral of all Pacific Ocean (National Environmental Authority 2005; UNESCO WHC 2017).

The islands are on the northeastern edge of the Pacific Marine Biological Corridor, which extends from Baja California to Peru. It includes World Heritage sites archipelagos of Revillagigedo and Galapagos and the islands of Cocos, Coiba, and Malpelo. This marine corridor, which is part of the Eastern Tropical Pacific Marine Wilderness, is characterized by the convergence of major warm oceanic currents and cool coastal countercurrents (UNEP/WCMC 2012).

57.2.3 Biodiversity

Coiba National Park and its Special Zone of Marine Protection include insular marine coastal and terrestrial island ecosystems. The wide range of environments and habitats is a result of its location, close to the mainland, but despite its proximity to the coast, Coiba has been isolated long enough to allow speciation to occur. The area provides a key ecological link to the Eastern Tropical Pacific (UNESCO WHC 2017; Castroviejo and Ibáñez 2001).

The national park preserves approximately 80% of its original tropical moist forest cover. Two principal life zones on the island are present: very humid tropical forest and pluvial forest. In addition, there are four coastal vegetation types: shoreline with coconut (see Figs. 57.2 and 57.3), mangrove forest, mangrove swamp, and brackish water forest. There are around 2000 species of vascular plants, and 130 lichen species and 148 fungi have been recorded (Ibáñez 2001). The island is still botanically unexplored, and further study would certainly reveal more species (UNEP/WCMC 2012).

The diverse marine habitats include coral reefs, sandy bottoms, and mangrove intertidal flats. Panama's Pacific shore has 91% of the Eastern Pacific reef-building corals, and the comparatively stable marine environment of the sheltered Gulf of Chiriquí contains many of these (Cortés 1997).

Twenty-two coral species were observed, mostly endemic. Coral reefs and coral communities cover about 2% of the marine habitats on Coiba National Park (Guzman et al. 2004). Seven-hundred species and 375 genera (85% of eastern Pacific genera) of fish are known from the gulf. Thirty-three species of shark exist locally, and 20 species of cetaceans have been recorded. There are also four turtle species (Castroviejo and Ibáñez 2001).

Few studies have been made about the terrestrial fauna, but Coiba is known to have 2 endemic mammals, 20 endemic birds, and 1 endemic subspecies of snake. The mammalian fauna is dominated by bats. A hundred and forty-seven species of birds have been recorded, 96 resident on Coiba itself. The island is also a refuge for birds which have almost disappeared from mainland Panama (UNEP/WCMC 2012).

Fig. 57.3 Wide sandy beach in one of the heavily vegetated islands of the Coiba National Park. (Picture reproduced by permission of Chris Roher, webpage http://www.gardeningandthespanish-way-roherbot.blogspot.com)

57.3 Conservation Challenges

Current facilities, funding, and staff seem to be insufficient to control the expected pressures from fishing and tourism. A decrease in the number of sharks, rays, groupers, and snappers have already been recorded (UNEP/WCMC 2012). Impacts on the site with infrastructure development, agriculture, forest cutting, human settlements, and exploitation of mineral resources, while strictly prohibited, remain potential threats (UNESCO WHC 2017).

In 2014, the International Union for Conservation of Nature – IUCN – found that the state of conservation of the site inspired significant concerns (IUCN 2014). IUCN published a new state of conservation outlook at the end of 2017, in which the situation remained the same (Osipova et al. 2017; IUCN 2017).

IUCN listed the series of threats to which the site was submitted: fishing and collecting of aquatic resources, insufficient human resources, impacts of tourism, limiting legal framework, livestock farming and grazing of domesticated animals, insufficient management systems and management plan, and problems with marine transport infrastructure, with the projection of construction of a naval base in the area (IUCN 2017; UNESCO WHC 2017).

As with other world marine areas, the site faces the impacts of global warming and climate change. The situations associated with these problems are coral death, stronger El Niño, frequency of hurricanes, coastal erosion and flooding caused by sea level rise, and modifications on terrestrial and marine habitats, with risk of extinction of parcels of the biodiversity.

References

Bauch T (2006) Carbonate-siliciclastic facies patterns related to the closure of the Central American seaway: a comparison between the Pliocene of Costa Rica and the present-day gulf of Panama and gulf of Chiriquí. PhD thesis, mathematics and Geosciencies faculty, Christians-Albrechts university, Kiel, Germany

Brenner LD, Linsley BK, Dunbar RB (2017) Examining the utility of coral Ba/Ca as a proxy for river discharge and hydroclimate variability in Coiba Island, Gulf of Chiriquí, Panama. Mar Pollut Bull 118:48–56

Buchs DM, Arculus RJ, Baumgartner PO, Ulianov A (2011) Oceanic intraplate volcanoes exposed: example from seamounts accreted in Panama. Geology 39:335–338

Cardiel J, Castroviejo S, Velayos M (1997) Parque Nacional de Coiba: El medio físico. In: Castroviejo S (ed) Flora y fauna del Parque Nacional de Coiba (Panamá). Spanish Agency for International Cooperation Internacional, Madrid

Castroviejo S, Ibáñez A (2001) Origen y análisis de la diversidad biológica de la isla de Coiba. Quercus 188:29–32

Cortés J (1997) Biology and geology of eastern Pacific coral reefs. 8th Symposium, Smithsonian Tropical Research Institute, Balboa, Republic of Panama

Guzman HM, Guevara CA, Breedy O (2004) Distribution, diversity, and conservation of coral communities in the largest marine protected area of the Pacific Panama (Coiba Island). Environ Conserv 31:111–121

Ibáñez A (2001) Estudio de la Composición Forística y Ecología del Bosque Tropical de la Isla de Coiba (Panamá). PhD dissertation, Department of Botany, University of Salamanca, Spain

IUCN (International Union for Conservation of Nature (2017) Coiba National Park and the zone of Marine Protection Conservation Outlook. http://www.worldheritageoutlook.iucn.org/es/explore-sites/wdpaid/902479. Accessed 27 Dec 2017

IUCN (International Union for Conservation of Nature) (2005) World heritage nomination technical evaluation Coiba National Park (Panama), Gland

IUCN (International Union for Conservation of Nature) (2014) IUCN world heritage outlook 2014. Report, Gland, Switzerland

Meschede M, Barckhausen U (2001) The relationship of the Cocos and Carnegieridgs: age constraints from paleogeographic reconstruction. Int J Earth Sci 90:386–392

National Environmental Authority (2005) Proposal for the inscription of the Coiba National Park in the list of world heritage sites of UNESCO. Government of Panama, Panama City

Osipova E, Shadie P, Zwahlen C, Osti M, Shi Y, Kosmos C, Bertzky B, Murrai M, Van Merm R, Badman T (2017) IUCN World Heritage Outlook 2: a conservation assessment of all natural World Heritage sites. IUCN, Gland

UNEP/WCMC (United Nations Environmental Programme/World Conservation Monitoring Centre) (2012) Coiba National Park and its special zone of marine protection Panama. Report, Cambridge, England

UNESCO WHC (United Nations Educational, Scientific and Cultural Organization World Heritage Centre) (2017) Coiba National Park and its special zone of marine protection. http://whc.unesco.org/en/list/1138. Accessed 13 July 2017

Werner R, Hoernle K, Barckhausen U, Hauff F (2003) Geodynamic evolution of the Galapagos hot spot system (Central East Pacific) over the past m.y.: constraints from morphology, geochemistry, and magnetic anomalies. Geochem Geophys Geosyst 4. https://doi.org/10.1029/2003GC000576

Chapter 58
Volcanoes of Kamchatka, Russia

Abstract The Kamchatka Peninsula is in the far east of Russia between the Sea of Okhotsk and the Bering Sea. The heritage complex comprises six separated protected areas in the center and southeast of the peninsula. The pristine wilderness represents one of the most outstanding examples of the volcanic regions in the world. It contains a high density of active volcanoes of many types, as well as a wide range of volcanic features. The peninsula location between a large continental landmass and the Pacific Ocean also exhibits unique characteristics with major concentrations of wildlife, including the world's largest known variety of salmonid fish and exceptional concentrations of sea otter, brown bear, and sea eagle. The contrast of high active volcanoes and glaciers, together with wild rivers and spectacular coastline, creates a dynamic landscape of great beauty.

58.1 Introduction

The Kamchatka Peninsula is in the far east of Russia between the Sea of Okhotsk and the Bering Sea. Covering an area of 38,302 km^2, the heritage site comprises six separated protected areas in the center and southeast of the peninsula (UNESCO WHC 2017) (Fig. 58.1).

The story of preservation of the area dates to 1880 and 1882, when two reserves were established by imperial decree, to protect sable and sea otter. Between 1934 and 1995, nature reserves and nature parks were created protecting wildlife and landscape. In 1996, five preserved areas became a World Heritage site, which was extended in 2001, to include one more. In total, the site is made by six areas: Southern Kamchatka Regional Nature, Southern Kamchatka Wildlife Reserve, Bystrinsky Regional Nature Park, Kronotsky Strict Nature Reserve, Nalychevo Regional Nature Park, and Klyuchevskoy Regional Nature Park (UNESCO WHC 2017; UNEP/WCMC 2011).

The oldest settlement in the peninsula dates from the Paleolithic, some 21,000 years ago, which continued throughout most of the Holocene. The peninsula is part of Beringia and at various times during the Pleistocene when sea level was lower, the Bering land bridge connected Asia with North America, and this land

V. Claudino-Sales, *Coastal World Heritage Sites*, Coastal Research Library 28,
https://doi.org/10.1007/978-94-024-1528-5_58

Fig. 58.1 Volcanoes of Kamchatka in eastern Russia

bridge is postulated to be an important route for human populations coming from Asia to populate the Americas (Goebel 2013).

The Kamchatka Peninsula became part of Russia in 1699, and the first description of Kamchatka was given in 1742 by the explorer S.P. Krasheninnikov. At the beginning of the 18th century, the Itel'meni people settled in the region, depending on fishing. Evens and Koryak people settled later, depending on reindeer herding (UNEP/WCMC 2011).

In modern times, the peninsula was closed to all outsiders due its strategic importance as a military area, until 1990. The major city on the peninsula is its capital, Petropavlovsk-Kamchatsky, which with nearby Elisovo has most of the region's population. The isolation, combined with few roads and settlements, has preserved much of the wilderness. More recently, some areas have been zoned for permanent settlements, but there is strong outside pressure to make greater economic use of these lands (Jones and Solomina 2015).

Tourism is increasing in Kamchatka. The tourist product offered by travel agencies varies from stationary excursions to hunting and salmon fishing, hiking tours, sea cruises on Avacha Bay, and extreme ecotours. Existing hotel capacities allow accepting 15,000 tourists in a tourist season (Kamchatka Organization 2017).

58.2 Natural Settings

58.2.1 Climatic Conditions

The climate is subarctic. It is influenced by the position of the Kamchatka Peninsula at the eastern margin of the Asian landmass, by its mountain ranges, by the cold ocean currents present to the east, and by the Sea of Okhotsk to the west. These result in cold continental conditions arising from the northwest during winter and warm-moist air masses moving across the peninsula, from southeast to northwest during summer (Jones and Solomina 2015).

The maximum average temperature is 15 °C in summer and −8 °C in winter. The maximum amount of precipitation is up to 2600 mm a year close to the coast, but less than 400 mm in the center of the peninsula (Kamchatka Organization 2017). The climate is windy, with winter speeds of 40 m/s. There is often fog, and the area is subject to very heavy snowstorms (UNEP/WCMC 2011).

58.2.2 Geology and Geomorphology

Kamchatka Peninsula is one of the most outstanding volcanic regions in the world, with a high density of active volcanoes, a variety of types, and a wide range of related features, resulting from plate tectonics. The Pacific Plate is actively subducting beneath it at a speed of about of 80–100 mm per year, resulting in one of the more active volcanic areas in the world and the greatest concentration of active volcanism in Eurasia. There are about 300 volcanoes, more than 20 of which are active (Nikulin et al. 2012).

The most effective periods of eruptions occurred in the late Pleistocene and mid-Holocene, but some volcanoes erupt today, some every 5 or 6 years. The eruptions are accompanied by explosions, the throw-outs of gas and ashes and ash falls (Gledhill 2007). Since 1690, some 200 eruptions have been recorded (Ponomareva et al. 2015). The volcanoes are the stratovolcano type, somma volcanos, shield volcanos, and mixed types (Braitseva et al. 1997). Neotectonics is also an important factor shaping the local landscape (Ivanov 2002) (Figs. 58.2 and 58.3).

The Klyuchevskaya range has 3 of the highest and most active volcanoes and 12 volcanic peaks. Klyuchevskoy itself is Eurasia's highest and most active volcano after Etna and is still growing: its 1986 height was 4813 m; in 2002 it was 4835 m. It is a stratovolcano which has erupted 25 times in the last 50 years and still erupts

Fig. 58.2 Eruption of Tolbachik Volcano, which occurs on average every 5 years. This volcano is part of the Klyuchevskoy group of volcanos, which are located in one of the six areas of the World Heritage serial site in Kamchatka Peninsula. (Picture reproduced by permission of Alexander Piragis/Shuttersotck Stock Photos)

regularly, most recently in 2007, creating a wilderness of ash and lava (Kamchatka Organization 2017; UNEP/WCMC 2011).

Due to the extent of recent explosive volcanic activity and the presence of preserved tephra in datable archives, Kamchatka has one of the best resolved Holocene tephra sequences in the world (Ponomareva et al. 2015). Additionally there are many thermal and mineral springs, geysers, calderas, scoriae cones, lava streams, cinder fields, solfataras, mud pots, and many other constantly active volcanic features (Jones and Solomina 2015).

The Pleistocene glaciations shaped a great part of the present geomorphological characteristics of the area, considering that there are many active glaciers creating cirques, hanging valleys, U-shaped valleys, moraines, glacial till, and almost all the types of ice formation common in volcanic areas (Solomina et al. 2007). Most of the glaciers are along the coast of the peninsula and 238 cirques are occupied by them (Barr and Spagnolo 2013). The contrast of high active volcanoes and glaciers creates a dynamic landscape of great beauty (UNESCO WHC 2017).

There is also a very diverse and scenic landscape created by lakes, wild rivers, and spectacular rocky coastline with islands, deltas, and wide swampy estuaries (UNEP/WCMC 2011). Along the coast, it is also possible to find marine deposits created by tsunamis, some of which have been documented as having taken place thousands of years ago (Pinegina and Bourgeois 2001).

Fig. 58.3 The stratospheric Koryak Volcano, in the Kronotsky Strict Nature Reserve, with glaciers descending from the top. The area is also characterized by the exceptional concentration of brown bear, as seen here. (Picture reproduced by permission of DPS/Shutterstock Stock Photos)

58.2.3 Biodiversity

Kamchatka lies in the boreal zone of Eurasia, and about one-third of the area is forested. The rest is occupied with shrub, alpine tundra, and wetlands, besides the volcanic landscapes (Ivanov 2002). The severity of the climate and the active volcanism do not allow a large endemic flora, but there is a diverse Palaearctic flora (Nazarova et al. 2013).

There are clear altitudinal vegetation zones which are strongly climate-dependent. Six vegetation zones are present: taiga, birch forest, riparian forest, wetland, subalpine, and mountain tundra. The total number of plant species is 1168, with 10% endemic (UNEP/WCMC 2011).

Wildlife is abundant on the peninsula. The fauna has a relatively low diversity, exhibiting some of the biogeographic qualities of an island. However, a number of species are abundant, including the world's largest population of brown bears (van Zoelen 2002). There is also a large presence of snow sheep, deer, sable, and wolverine (Jones and Solomina 2015).

Tundra birds and seabirds are numerous. Some 179 bird species have been recorded. There are numerous seabird colonies along the coasts of the reserves, several of which contain a notable portion of the world population of certain species. More than half the world populations of Steller's sea eagle and of Aleutian tern nest on the peninsula (UNEP/WCMC 2011).

The rivers of Kamchatka contain the greatest concentration and diversity of salmonid fish species on Earth and are the only place on the Pacific rim where all the species of Pacific salmon coexist. The rivers are exceptionally unpolluted and rich in biodiversity, which sustains the large populations of terrestrial animals (UNEP/WCMC 2011).

58.3 Conservation Challenges

A number of situations are imposing important threats to the serial site. Mining and the installation of ground transport infrastructure and of renewable energy facilities (construction of a geothermal power station) are the principal. The decline in populations of wild reindeer and snow sheep, due to illegal hunting, is also considerable (UNESCO WHC 2017).

There are no studies related to global warming and climate changes in the area, as well as no information about an eventual melting of glaciers. Nevertheless, there is an increasing occurrence of wildfire in the forests (UNESCO WHC 2017), which very probably is linked to a recent decrease of humidity in the area.

The state of conservation of the serial site was found by the International Union for Conservation of Nature in its outlook about the degree of conservation and situation of the UNESCO World Heritage sites in 2014 (IUCN 2014) to present significant concerns. The same evaluation appeared in the state of conservation report published at the end of 2017 (Osipova et al. 2017; IUCN 2017). It indicated that the threats to the integrity of the site are still challenging the site.

References

Barr ID, Spagnolo M (2013) Palaeoglacial and palaeoclimatic conditions in the NW Pacific, as revealed by a morphometric analysis of cirques upon the Kamchatka Peninsula. Geomorphology 192:15–29

Braitseva OA, Ponomareva VV, Sulerzhitsky LD, Melekestsev IV, Bailey J (1997) Holocene key-marker tephra layers in Kamchatka, Russia. Quat Res 47:125–139

Gledhill D (2007) Kamchatka: a journey and guide to Russia's land of ice and fire. Odyssey Books and Guides, Hong Kong

Goebel T (2013) Archaeological records: global expansion 300,000-8,000 years ago, America. In: Elias S (ed) Encyclopedia of quaternary science. Elsevier, Amsterdam

IUCN (International Union for Conservation of Nature) (2014) IUCN world heritage conservation outlook. Rapport, Gland, Switzerland

IUCN (International Union for Conservation of Nature) (2017) Volcanoes of Kamchatka Conservation Outlook. http://www.worldheritageoutlook.iucn.org/node/1074. Accessed 27 Dec 2017

Ivanov A (2002) The far east. In: Shahgedanova M (ed) The physical geography of northern Eurasia. Oxford University Press, Oxford

Jones V, Solomina O (2015) The geography of Kamchatka. Glob Planet Chang 134:3–9

Kamchata Organization (2017) Volcanoes of Kamchatka. http://kamchatka.org.ru/eng/aboutK/volcanoes.html. Accessed 17 July 2017

Nazarova L, de Hoog V, Hoff U, Dirksen O, Diekmann B (2013) Late Holocene climate and environmental changes in Kamchatka inferred from the subfossil chironomid record. Quart Sci Rev 67:81–92

Nikulin A, Levin V, Carr M, Herzberg C, West M (2012) Evidence for two upper mantle sources driving volcanism in Central Kamchatka. Earth Planet Sci Lett 321-322:14–19

Osipova E, Shadie P, Zwahlen C, Osti M, Shi Y, Kormos C, Bertzky B, Murai M, Van Merm R, Badman T (2017) IUCN world heritage outlook 2: a conservation assessment of all natural world heritage sites. IUCN, Gland

Pinegina TK, Bourgeois J (2001) Historical and paleo-tsunami deposits on Kamchatka Russia: long-term chronologies and long-distance correlations. Nat Hazards 1:177–185

Ponomareva VV, Pornyagin M, Pevzner M, Blaauw M, Klye P, Derkachev A (2015) Tephra from andesitic Shiveluch volcano, Kamchatka, NW Pacific: chronology of explosive eruptions and geochemical fingerprinting of volcanic glass. Int J Earth Sci 104:1459–1482

Solomina O, Wiles G, Shiraiwa T, D'Arrigo R (2007) Multiproxy records of climate variability for Kamchatka for the past 400 years. Clim Past 3:119–111

UNEP/WCMC (United Nations Environmental Programme/World Conservation Monitoring Centre (2011) Volcanoes of Kamchatka Russia federation. Report, Cambridge, England

UNESCO WHC (United Nations Educational, Scientific and Cultural Organization World Herigate Centre) (2017) Volcanoes of Kamchatka. http://whc.unesco.org/en/list/765. Accessed 19 July 2017

van Zoelen A (2002) A world of mountains, yet to conquer: the Kamchatka Peninsula. Mt Res Dev 22:191–193

Chapter 59
East Rennell, Solomon Islands

Abstract Rennell is the southernmost island in the Solomon Island group in the Western Pacific. It is the largest raised coral atoll in the world. An important feature of the island is Lake Tegano, the largest closed lake on a Pacific Island, which was the former lagoon on the atoll. The lake is brackish and contains many rugged limestone islands. The surrounding kart terrain has a dense cover of indigenous forest. Remaining in its natural state, the forest has a rich biodiversity with many endemic species. The island is a stage in the migration and evolution of species in the Western Pacific and, combined with the strong climatic effects of frequent hurricanes, makes a natural laboratory for the study of island biogeography.

59.1 Introduction

East Rennell World Heritage site takes the southern 40% of Rennell Island, the southernmost of the Solomon Islands which lie on the northern edge of the Coral Sea, some 1000 km east of New Guinea. The site has 370 km^2, with terrestrial and marine segments (UNESCO WHC 2017) (Fig. 59.1).

The area was listed as a World Heritage site in 1998. It is protected under a national act passed in 2010 and managed by the Ministry of the Environment. The land is in traditional customary ownership, what ensures the sustainable use of its resources at some level. Nevertheless, the level of protection is not enough, in such a way that the site was added to the UNESCO list of "sites in danger" in 2013 and is still in this condition (UNESCO WHC 2017).

Polynesians lived in Rennell Island between 2000 and 1600 years ago, when people of the Lapita culture appeared. The next settlement was around 130 BC, with another major occupation in about 1000 AD. The present inhabitants say their ancestors landed about 1400 AD and came from Wallis Island north of Fiji (UNEP/WCMC 2012).

A Spanish navigator discovered the archipelago in 1568 and an English captain rediscovered it in 1793. In the 19th century the islands were subjected to missionaries and then to the British government as a protectorate in 1893, which ended in 1978. It has never been very developed because there is limited freshwater and

V. Claudino-Sales, *Coastal World Heritage Sites*, Coastal Research Library 28,
https://doi.org/10.1007/978-94-024-1528-5_59

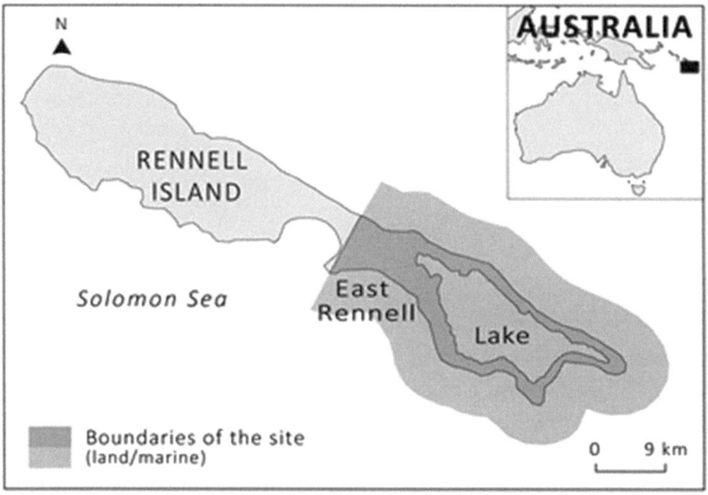

Fig. 59.1 Location of East Rennell World Heritage site, in the Western Pacific

only a few small scattered patches of soil suitable for cultivation. The population is declining through emigration (Wingham 1997).

Tourism is increasing with the offer of facilities, and ecotourism is the most important, due to the primitive conditions of the islands. Fuel is shipped over once a month on a cargo boat that also takes passengers. Freshwater for all uses, including drinking and cooking, relies on basic but effective rainwater catchment systems. For access to the site, there are also small airplanes arriving twice a week. Much of the movement around the island is made by boat around Lake Tegano (UNEP/WCMC 2012).

59.2 Natural Settings

59.2.1 Climatic Conditions

The Solomon Islands have an equatorial, hot, and humid climate. The islands have an average temperature of about 27 °C with only very small changes all year. The area has two distinct seasons – a wet season from November to April and a dry season from May to October. The average rainfall is approximately 2500–3000 mm; however, it varies from west to east across the archipelago and also depends on the island's topography and elevation (Solomon Islands Official Tourism Site 2017).

Southeast trade winds prevail from April to the end of November. The island is in the path of hurricanes and is subject to them between November and April. The hurricanes frequently cause damage to the forests and villages (Wingham 1997).

59.2.2 Geology and Geomorphology

The Solomon Islands is a group of 992 islands mainly of volcanic origin with basement rocks exposed along a spreading mid-ocean ridge (Wells 1989). They correspond to an ophiolite massif coming from an Eocene intra-oceanic subduction in the southeast Pacific; the ophiolitic belt has been disrupted by later transcurrent faults, more recent spreading phenomena and two other subductions, related to the Indo and Australian plates, in the Oligocene and Miocene-Recent (Parrot and Dugas 1980).

Rennell is around 200 km south of the main group, well away from the plate collision zone, presenting very low seismicity. It is geologically younger than most of the other islands and has emerged near the end of the Pleistocene due to tectonic movement along the convergent plate boundary (Cowley et al. 2004). The tectonic movement raised the seabed in a ridge high enough to allow coral to build the island's reefs (Mann and Taira 2004).

The structure and geomorphology of East Rennell Island show that the ridge in which it sits first suffered subsidence and is now in a phase of active uplift. The island was initially deposited as coralline algal limestone and then dolotomized. The dolomitic reef complex was in the sequence overlain by younger undolotomized coral limestone (Wingham 1997).

Rennell is 86 km long by 14 km wide and nearly surrounded by limestone cliffs between 120 and 150 m high. It is the world's largest uplifted coral atoll and one of about 25 such atolls in the Pacific. The island is almost ringed by a narrow fringing reef (UNEP/WCMC 2012) (Fig. 59.2).

In the center of the island, there is Tegano Lake, which is the largest enclosed lake in a Pacific island. It fills the original lagoon at the east end of the atoll. At 27 km long and averaging 9 km wide, the lake occupies nearly a quarter of the island. It is brackish with a salt concentration maintained by a subterranean connection with the sea. At the west end are some 200 rugged limestone islets (Wingham 1997) (Fig. 59.3).

There are no streams or rivers in the island, but there are freshwater springs around the lake edge and in some places emerging from the coastal cliffs. Groundwater flow is directed from the higher surrounding land into the lake (UNEP/WCMC 2012).

There has been no soil enrichment by volcanic dust because of the great distance from the larger volcanic islands and the direction of the prevailing wind and sea currents. For this reason, the soil in the island is thin, derived from weathered coral limestone, and occurring only in small scattered pockets (Wingham 1997).

Fig. 59.2 Coral ringing East Rennell Island in the shallow water, with coastal area characterized by the presence of dense and undisturbed vegetation colonizing the raised limestone cliffs. (Picture reproduced by permission of Paddy Ryan, website http://www.ryanphotographic.com)

59.2.3 Biodiversity

East Rennell evidences significant ongoing ecological and biological processes and is an important site for biogeography. The area is an important stepping stone in the migration and evolution in the Western Pacific (UNESCO WHC 2017).

Around 90% of the island is covered by pristine tropical rainforest with a canopy averaging 20 m high. The rainforest presents a number of adaptations to the effects of the frequent hurricanes which the island gets annually (IUCN 2008).

There are three major vegetation types: low mature forest on the karst ridge of the island's perimeter, tall forest of the island interior, and beach flora on the Lake Tegano margins, together with small patches of mangrove. The differences in vegetation between the karst ridge and the interior forests result from the relatively sheltered island interior where deeper soils occur (Wolff 1958).

The Solomon group contains almost 650 species of vascular plants, with 162 (25% of the total) not occurring on any other island to the east. No endemic tree species have yet been found, but ten endemic plants are recorded, and the lake's waters contains 312 species of algae and diatoms, some endemic (UNEP/WCMC 2012).

The Solomon Islands have a greater diversity of animal species and higher level of endemism than any other Pacific island nation except for Papua New Guinea. They have more restricted range and unique bird species by area than any other

Fig. 59.3 Islet with coastal vegetation surrounded by sandy beach and corals in East Rennell. (Picture reproduced by permission Ethan Danields/Shutterstock Stock Photos)

place on Earth. Rennell has the highest endemism per hectare of all the islands owing to its isolation and relatively light disturbance by human and alien species (UNESCO WHC 2017).

At least 43 species of breeding land and water birds have been identified in East Rennell. The site has 11 species of bats, and there are 5 species of geckos and 27 species of land snails. No amphibians have yet been recorded. A total of 731 insects have been registered (UNEP/WCMC 2012). There is an endemic sea snake (PHCG 2010).

Around the Solomon Islands, 754 species of fish and a total of 1019 reef-dwelling fish have been recorded. Eight whale species and nine dolphin species are found. The Solomon Islands as a group has the second highest diversity of coral species in the world, with more than 494 species (IUCN 2008).

59.3 Conservation Challenges

The site presents a series of vulnerabilities and threats, such as mining, logging, overexploitation of marine resources, and the recent presence of invasive species. The management plan of the site is not clear in relation to specific policies to address these threats and has no timeline or budget. The ability of the traditional owners to protect and manage the natural resources is limited by lack of funding, capacity, and resources (UNESCO WHC 2017).

This context justifies the inclusion of the site in the list of sites in danger. It is worth noting that the isolation of the island and the extreme difficulty of overland travel help the protection of the site but also impact attempts to develop ecotourism, which could be a good and sustainable source of funding to the local society. Climate changes and global warming are also a dangerous threat to the island, which is already suffering with more frequent and intense floods (Solomon Island Official Tourism 2017). Death of corals, greater coastal erosion, and increase of storms and hurricanes are also expected due to these factors.

The last conservation outlook published by the International Union for Conservation of Nature (IUCN) at the end of 2017 (Osipova et al. 2017) found the site to be in critical conditions of conservation, and that additional support should urgently be provided to communities to ensue enhancement of protection and management of the site (IUCN 2017).

References

Cowley S, Mann P, Coffin MF, Shipley TH (2004) Oligocene to recent tectonic history of the Central Solomon intra-arc basin as determined from marine seismic reflection data and compilation of onland geology. Tectonophysics 389:267–307

IUCN (International Union for Conservation of Nature) (2008). State of conservation of report East Rennell (Solomon Islands). Report, Gland, Switzerland

IUCN (International Union for Conservation of Nature) (2017) East Rennell Conservation Outlook. http://www.worldheritageoutlook.iucn.org/node/1087. Accessed 27 Dec 2017

Mann P, Taira A (2004) Global tectonic significance of Solomon Islands and Ontong Java Plateau convergent zone. Tectonophysics 389:137–190

Osipova E, Shadie P, Zwahlen C, Osti M, Shi Y, Kormos C, Bertzky B, Murai M, Van Merm R, Badman T (2017) IUCN world heritage outlook 2: a conservation assessment of all natural world heritage sites. IUCN, Gland

Parrot JF, Dugas F (1980) The disrupted ophiolitic belt of the Southwest Pacific: evidence of an Eocene subduction zone. Tectonophysics 66:349–372

PHCG (Pacific Horizon Consultancy Group) (2010) Solomon Islands state of environment report. Ministry of Environment, Conservation & Meteorology, Honiara, p 2010

Solomon Islands Official Tourism Site (2017) Solomon Islands. http://www.visitsolomons.com.sb/. Accessed 19 July 2017

UNEP/WCMC (United Nations Environmental Programme/World Conservation Monitoring Centre) (2012) East Rennell, Solomon Islands. Report, Cambridge, England

UNESCO WHC (United Nations Environmental, Scientific and Cultural Organization World Heritage Centre) (2017) East Rennell. http://whc.unesco.org/en/list/854. Accessed 21 July 2017

Wells RE (1989) Origin of the oceanic basalt basement on Solomon Islands arc and its relationship to the Ontong Java Plateau – insights from Cenozoic plate motion models. Tectonophysics 165:219–235

Wingham E (1997) Nomination of East Rennell, Solomon Islands for inclusion in the world heritage list – natural sites. Ministry of Foreign Affairs and Trade of New Zealand, Wellington

Wolff T (1958) The natural history of Rennell Island. Danish Science Press, Copenhagen

Chapter 60
Henderson Island, the United Kingdom of Great Britain and Northern Ireland

Abstract Henderson Island, located in the southeastern Pacific Ocean, is a fossil coral atoll and is the world's only raised atoll where the ecology is essentially intact, having been little degraded because of its extreme isolation and rough terrain. The surface is largely a plateau of reef rubble interspersed with areas of sharp coral pinnacles and with extremely rugged dissected karstic limestone round the edges. The central depression is the site of a fossil lagoon which formed before the island rose. The land is surrounded by high limestone cliffs which protect the island from storms. Its fossil corals record changes in sea levels and the climate over half a million years, and its isolation makes it an ideal site for studying the dynamics of island evolution and natural selection.

60.1 Introduction

Henderson, a remote island in the southeastern Pacific 4800 km equidistant from both Chile and New Zealand, is part of the Pitcairn Archipelago, a dependent territory of the UK. It lies 200 km from the largest island, Pitcairn. The site has an area of 37 km^2 (UNESCO WHC 2017) (Fig. 60.1).

Henderson Island was defined as a World Heritage site in 1988. It is not a protected area, and only its isolation ensures its protection. Its land tenure is the Crown. It is managed by the Pitcairn Island Council under the British High Commissioner to New Zealand, who is governor of Pitcairn (UNEP/WCMC 2011). While the governor holds most formal power, much day-to-day administration of the site comes from a commissioner in the Pitcairn Islands Administration office in Auckland, with the elected island council (UNESCO WHC 2017).

The island is uninhabited because of its lack of resources. Nevertheless, there is evidence that it was colonized by a small Polynesian group between the late 8th and 17th centuries (Weisler 1995). Their impact may have been slight except on now-extinct bird species (Steadman and Olson 1985), although there is some disagreement among scientists over this (UNEP/WCMC 2011).

Access to Henderson requires a license issued by the governor in consultation with the island council. However, the island is unguarded and tourists as well as

V. Claudino-Sales, *Coastal World Heritage Sites*, Coastal Research Library 28,
https://doi.org/10.1007/978-94-024-1528-5_60

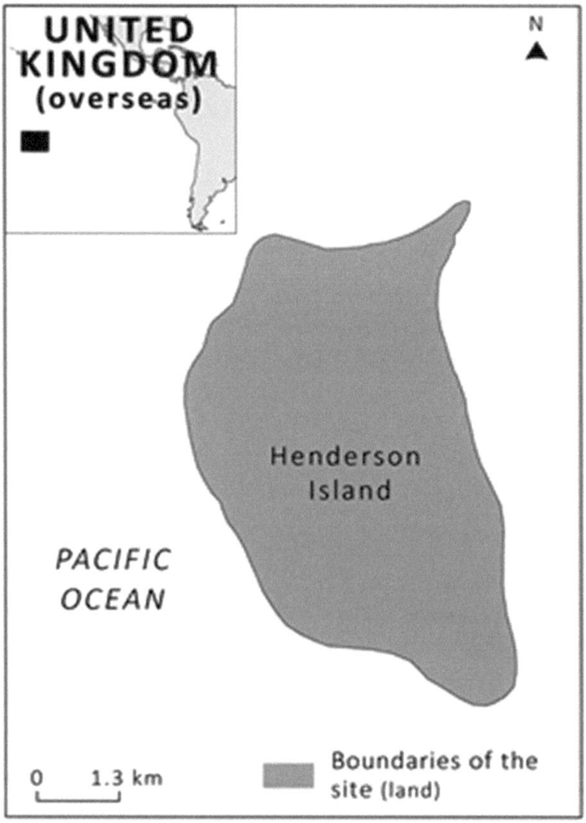

Fig. 60.1 Location of Henderson Island, part of the Pitcairn Archipelago (United Kingdom of Great Britain and Northern Ireland) in the southeastern Pacific Ocean

fishermen regularly land there, sometimes not even knowing that access is by permit only (UNESCO WHC 2017). It lacks tourist attractions but has a key importance for natural sciences (UNEP/WCMC 2011).

60.2 Natural Settings

60.2.1 Climatic Conditions

Henderson Island has a moderate subtropical climate. The mean annual rainfall is 1500 mm. There is no discernible seasonal pattern to the rainfall. The average temperature is around 26 °C. The maximum temperature is registered in June (29.6 °C) and the minimum (15.7 °C) in February (Spencer 1995).

The island lies in the path of the southeast trade winds and is subjected to the action of El Niño, which usually results in cooler and wetter weather. The humidity is around 80% all year (Brooke et al. 2004).

60.2.2 Geology and Geomorphology

Henderson Island is an uplifted coralline limestone platform 9.6 km long by 5.1 km wide (Brooke et al. 2004). It represents a raised fossil atoll (Spenser and Paulay 1989). Quaternary volcanic activity in the nearby Pitcairn Island resulted in systematic uplift of Henderson due to volcanic loading and bulge rebound process related to lithospheric flexure (McNutt and Menard 1978; Ducan et al. 1974).

The island is thus a reef-capped seamount of the makatea type tectonically raised from a depth of some 3.5 km as a result of the weight of the Pitcairn volcano formed over the hotspots under the archipelago some 13 million years ago. The limestones are of late Tertiary age and were exposed about 380,000 years ago (Blake 1995; Spenser and Paulay 1989).

The fossil records show that coral reefs grew prolifically in Henderson Island during marine isotopic stages 15 (ca 600,000 years) and 9 (ca 320,000 years) and, to a lesser extent, also in marine isotopic stage 7 (ca 240,000 years). It means that these paleo-reefs correlated with warm periods of stable sea level, allowing sufficient time for a well-developed reef structure to form (Andersen et al. 2010). The uplift history of Henderson caused the reefs to be raised above sea level to form the fossil-raised reef and a dry, fossil lagoon. This process, still continuing at a rate of 0.1 mm/y, has caused this coral atoll to be elevated to a maximum of 33.5 m (Blake 1995).

The central plateau of the island forms a slight, dish-shaped depression, which corresponds to the fossil lagoon floor, dated of more than 500 ka (Andersen et al. 2010). The island is characterized by eroded karst surfaces and is extremely rugged with areas of steep-sided pits, frequently 2–3 m deep and reported to reach 3–7 m deep. In other words, the surface of the island is largely a plateau of reef rubble interspersed with areas of sharp coral pinnacles and with dissected karstic limestone around the edges (Brooke et al. 2004).

Henderson is surrounded by vertical cliffs rising up to 30 m (Figs. 60.2 and 60.3) which have protected the island from hurricanes and storms. It is surrounded by a fringing reef averaging 50–100 m wide and extending to 200 m offshore, backed by narrow sand beaches (St. John and Philipson 1962). Marine erosion at the exposed southern end of Henderson has contributed to the formation of arches in the seacliffs (Brooke et al. 2004).

The process of uplift has not been continuous and is reflected in several reef terraces preserved in the eastern, northern, and northwestern embayments. During periods of little uplift, wave action has eroded caves into the base of the seacliffs, and, when uplift resumed, these caves have also been elevated (Spenser and Paulay 1989).

Henderson's geology, combining indicators of sea level change, and paleo-coral growth register the history of climate change in the second half of the Pleistocene and therefore testifies drivers of climatic cycles such as Milankovitch cycle (Andersen et al. 2010; Stirling et al. 2001).

Fig. 60.2 Raised, weathered, and eroded limestones shaped as vertical cliff on Henderson Island. Blocks and boulders produced by marine erosion rest in the sandy beach. (Picture reproduced by permission of Dr. Morten B. Anderson, University of Bristol)

60.2.3 Biodiversity

The terrestrial vegetation of the island is almost undisturbed. Due to its relative freedom from man and storms, the flora and fauna are unique relicts of the original condition. There are 71 species of vascular plants, 7 of which are endemic to the island and 10 introduced. The island has high endemicity for its size: only 15% of the flora is introduced (Brooke et al. 2004).

There are six vegetation communities: beachfront, beach-back woodlands, open scrub on cliff slopes, cliff and ledge communities, exposed cliff top, and plateau forests. The most diverse forests are on the top of the plateau. Most of the surface of the island is a dense mass of tangled vines, scrub forest, and semi-recumbent trees 4–7 m tall. The center of the depression is more sparsely covered, and the southern tip is exposed to salt spray and wind (Flenley et al. 1987).

The island is very important for its large breeding population of seabirds, estimated at 50,000–80,000 pairs, at least 9 species of which breed there. There is no native species of land mammal. Only the rat was introduced. All of the island's 16 species of land snail and about 30% of the 180 insect species may be endemic (Brooke et al. 2004).

Fig. 60.3 Steep vegetated cliff shaped on late Pleistocene raised coral reefs delimiting a steep sandy beach in Henderson Island. (Picture reproduced by permission of Dr. Morten B. Andersen, University of Bristol)

The marine fauna is also rich. Whales and turtles are present. A hundred ninety species of fish are recorded. There is a diverse echinoderm fauna. The coral is growing at the limit of reef growth, and they are sparse spatially and in diversity (e.g., Andersen et al. 2010).

60.3 Conservation Challenges

One large threat to the conservation of Henderson Island is the reproduction of the introduced Polynesian rat, which is imposing a decrease of the avifauna. In 2014, an ambitious program was initiated to eradicate the rat and also to ensure that no new introductions of alien invasive species occur through regulated tourism (UNESCO WHC 2017; IUCN 2013).

Climate change and global warming are also threatening the environment on the island, considering that it would impose limitations to the survival of corals and other species of the fauna and flora. It would also increase storms and coastal erosion, what would menace the integrity of the raised limestone and fossil coral reefs.

The International Union for Conservation of Nature (IUCN) found that the state of conservation of Henderson Island presents "significant concerns," due mostly to these two situations (invasive species and climate changes) (IUCN 2014). In the last conservation outlook, published at the end of 2017, the IUCN attested that the situation was the same, showing stability of the values to the site. Nevertheless, the area appears as being very highly threatened by rat predation and competition (IUCN 2017; Osipova et al. 2017).

References

Andersen MB, Stirling CH, Potter EK, Halliday AN, Blake SG, McCulloch MT, Ayling BF, O'Leary MJ (2010) The timing of sea-level high-stands during marine isotope stages 7,5 and 9: constraints from the uranium-series dating of fossil corals from Henderson Island. Geochim Cosmochim Acta 74:3598–3520

Blake S (1995) Late quaternary history of Pitcairn Island, Pitcairn Group. Biol J Linn Soc 56:43–62

Brooke ML, Hepburn I, Trevelyan RJ (2004) Henderson World Heritage Site. Management Plan 2004–2009, conserving a unique site. Foreign & Commonwealth Office, London

Ducan RA, McDougal I, Carter RM, Coombs DS (1974) Pitcairn Island – another Pacific hot spot. Nature 251(5477):679–682

Flenley J, Parkes A, Johnson M (1987) Vegetation survey of Henderson Island. Report, Operation Raleigh, London, England

IUCN (International Union for Conservation of Nature) (2013) Herderson Island State of conservation. IUCN 2014 Conservation Outlook. Report, Gland, Switzerland

IUCN (International Union for Conservation of Nature) (2014) IUCN world heritage outlook. A conservation assessment of all natural World Heritage sites. Gland, Switzerland

IUCN (International Union for Conservation of Nature) (2017) Henderson Island Conservation Outlook. http://www.worldheritageoutlook.iucn.org/node/1020. Accessed 27 Dec 2017

McNutt M, Menard HW (1978) Lithosphere flexure and uplifted atoll. J Geophys Res 83:1206–1212

Osipova E, Shadie P, Zwahlen C, Osti M, Shi Y, Kormos C, Bertzky B, Murai M, Van Merm R, Badman T (2017) IUCN world heritage outlook 2: a conservation assessment of all natural world heritage sites. IUCN, Gland

Spencer T (1995) The Pitcairn Islands, South Pacific Ocean: plate tectonic and climatic contexts. Biol J Linn Soc 56:13–42

Spenser T, Paulay G (1989) Geology and geomorphology of Henderson Island. Atoll Res Bull 323:1–18

St. John H, Philipson W (1962) An account of the flora of Henderson Island, South Pacific Ocean. Trans Royal Soc N Z 1:175–194

Steadman D, Olson S (1985) Bird remains from an archaeological site on Henderson Island, South Pacific: man-caused extinctions on an "uninhabited" island. Proc Natl Acad Sci USA 82:6191–6195

Stirling CH, Esat TM, Lambeck K, McCulloch MT, Blake SG, Lee D-C, Halliday AN (2001) Orbital forcing of the marine isotope stage 9 interglacial. Science 291:290–293

UNEP/WCMC (United Nations Environmental Programme/World Conservation Monitoring Centre) (2011) Henderson Island, United Kingdom of Great Britain and Northern Ireland. Report, Cambridge, England

UNESCO WHC (United Nations Educational, Scientific and Cultural Organization World Heritage Centre) (2017) Henderson Island. http://whc.unesco.org/en/list/487. Accessed 23 July 2017

Weisler M (1995) Henderson Island prehistory: colonisation and extinction a remote Polynesian island. Biol J Linn Soc 56:377–404

Chapter 61
Hawaii Volcanoes National Park, USA

Abstract Hawaii Volcanoes National Park, located in the central north Pacific, on the south of Hawaii Island, contains two of the world's most active and accessible volcanoes, Kilauea and Mauna Loa, where ongoing geological processes are easily observed. It shows the most recent activity in the continuing process of changing the Hawaiian Archipelago. The park is an example of island building through volcanic processes. Volcanic eruptions create a constantly changing landscape in the park, and the lava flows reveal unexpected geological formations. Mauna Loa, measured from the ocean floor, is the greatest volcanic mass on Earth. Rare birds can also be found, and there are forests of giant ferns.

61.1 Introduction

The site is located in the south of Hawaii Island, the easternmost island of the Archipelago of Hawaii, on the central north Pacific. It covers the summit, southeast, and southwest slopes of Mauna Loa Volcano and the lower summit, southwest, south, and part of the southeast slopes of Kilauea Volcano, reaching to the sea. The park has an area of 879 km^2 (UNESCO WHC 2017) (Fig. 61.1).

The Hawaii National Park was created on the islands of Hawaii and Maui by the US Congress in 1916. The area of the park was extended in 1922, 1928, and 1938. In 1961, the national park was divided between Hawaii Volcanoes National Park on Hawaii Island and Haleakala National Park on Maui Island. Hawaii Volcanoes National Park World Heritage site was stablished in 1987 (UNEP/WCMC 2011).

The islands were discovered by the explorer James Cook in 1779. Christian influences started around 1823, with churches, schools, and the introduction of cattle, goats, and tree fern fiber harvesting (NPS 2017). There are many archaeological remains, particularly along the coast where there are native villages, temples, graves, paved trails, canoe landings, cave shelters, and house sites, some dating back 700 years, as well as a large concentration of petroglyphs (Smart 1965).

The islands are very important in the history of evolutionary biology and for the study of effusive volcanism since the first observations in 1823. The US Geological Survey's Hawaiian Volcano Observatory, founded on Kilauea in 1912, monitors

V. Claudino-Sales, *Coastal World Heritage Sites*, Coastal Research Library 28,
https://doi.org/10.1007/978-94-024-1528-5_61

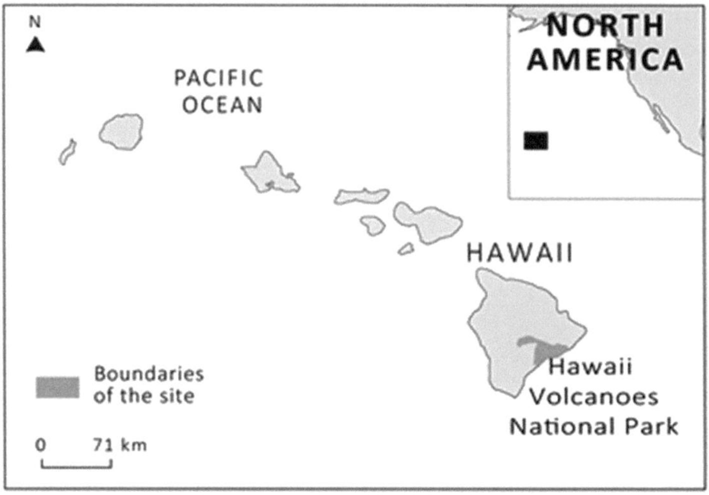

Fig. 61.1 Location of the Hawaii Volcanoes National Park in the north central Pacific, USA

Mauna Loa and Kilauea which are the most studied and best understood volcanoes in the world. The national park is managed by the National Park Service of the Department of the Interior (UNEP/WCMC 2011).

Mauna Loa was first climbed by visitors in 1794 and as a result of descriptions written in 1823, Kilauea had become a tourist attraction by 1840. The first volcano house was constructed in 1866 and successive structures were built later in the 19th Century for visitors. Almost two million people visited the park in 2015 (NPS 2017).

61.2 Natural Settings

61.2.1 Climatic Conditions

The climate in the Hawaiian Islands is tropical oceanic, dominated by northeast trade winds. Altitudinal climatic change is reinforced by an abrupt climatic gradient from east to west: windward mid-slopes receive a mean annual rainfall of 3810 mm, while leeward slopes receive around 400 mm. Summer storms and showery conditions are common, and violent hurricanes can occur between June and November (UNEP/WCMC 2011).

Annual average temperatures range from 22 °C at sea level to 7 °C at 3400 m, and the summits can carry snow in winter (Stone and Scott 1985). Thunderstorms, lightning, hail, floods, tornadoes, and droughts are not unknown. However, these phenomena are usually less frequent and less severe than their counterparts in continental regions. The eruptions of Kilauea can cause acid rainfall (NOAA 2016).

61.2.2 Geology and Geomorphology

The volcanism and seismicity at the Hawaiian Islands have been monitored for more than a century (Tilling 2012). It has been stated that they represent a straight and young chain of islands and submarine seamounts built on a large ridge which trend west-northwest across the eastern Pacific. The chain has inactive, active, and recent volcanoes progressively older toward the northwest, away from the East Pacific rise which is underlain by an ascending and spreading convection current (Wilson 1963).

The geographic trend of the islands is consistent with continued northwestern motion of the Pacific Plate over the past 80 million years, combined with a relatively stationary and stable source below leading to the observed volcanic activity (Tarduno et al. 2003). The volcanism is attributed to the presence of a hotspot, a narrow plume that rises from the core-mantle boundary below the Pacific Plate (Tosi and Yuen 2011; Wolfe et al. 2009).

The oldest of the Hawaiian Islands is at the western end of the chain, Kure Atoll, formed ca 25 million years. The islands are wholly volcanic, principally shield volcanoes, in a total of 129. The big island of Hawaii is the nearest existing island to the Hawaiian hotspot (Neall and Trewick 2008). It is also the largest and youngest island of the Hawaii Archipelago, rising to the two peaks of Mauna Kea (4205 m of altitude) and Mauna Loa (4169 m), with, on its southeast flank, Kilauea (1250 m) and the crater of Halemaumau. Kīlauea Volcano has been vigorously producing tholeiitic magmas for at least the last 200,000 years (Garcia et al. 2017).

Due to the load of the lavas, as well as the cooling of the ocean crust with time, the islands are suffering subsidence and are sinking (Parker 2016). With increasing distance from the source, the 129 volcanoes of the chain gradually subside with their accompanying oceanic plate to depths of 2 km (Neall and Trewick 2008).

The Hawaii Volcanoes National Park extends from the summit calderas of Mauna Loa and Kilauea to the southern coast (see Fig. 61.1) and is an example of island building through volcanic processes. Mauna Kea is a flat-domed shield volcano built up by lava from 6096 m below sea level, making it at 10,265 m, the highest mountain and volcano in the world (Hazlett 1993). Halemaumau fire pit was an active lava lake until the 1930s. Mauna Loa last erupted in 1984, but Kilauea remains active, expelling lava which extends far out under the sea (Holcomb 1987).

Kilauea is the Earth's most active volcano: since 1983, it has coated much of the southeast coast of Hawaii in fresh lava. Such flows often form unstable lava deltas and continuously contribute to the extension of the coastline (NASA 2017) (Fig. 61.2). Volcanic eruptions have created a constantly changing landscape, and the park includes volcanic features such as calderas, pit craters, cinder cones, spatter ramparts, fumaroles, lava tubes, tree molds, volcanic seacliffs, and black sand beaches (Hazlett 1993) (Fig. 61.3).

Fig. 61.2 Lava coming from Kilauea Volcano into seacliffs and flowing directly to the sea in Hawaii Volcanoes National Park. (Picture reproduced by permission of Judy McIlrath, University of South Florida, USA)

61.2.3 Biodiversity

The rate of endemism in the flora of Hawaii Islands is extremely high – 90% – because of the islands' geographic isolation, although it is impoverished, compared with continental sites. About 70% of the park is bare lava rock with very scattered vegetation. The vegetation is 15% scrub, 10% forest, and 5% grassland (Doty and Mueller-Dombois 1966).

The flora is characteristic of a mosaic of successional and climax stages distributed in seven zones: coastal lowlands, rainforest, grassland, mid-elevation woodland, upland forest and woodland, desert scrub, and subalpine to alpine scrub. Its greatest diversity is found in mesic and rainforests. Ferns form a noticeable proportion of the native flora, creating forests of giant exemplars (NPS 2017).

The islands' isolation resulted in relatively impoverished fauna, though it is rich in endemic species. Most island native species descend from animals that could fly in like birds, bats, and insects or were light enough to be carried by birds, such as snails, insects, and spiders, or were blown or washed ashore and survived by adaptive radiation (UNEP/WCMC 2011). The marine fauna is represented specially by the hawksbill turtle (see Fig. 61.3), by the endangered Hawaiian monk seal, and by the humpback whale (NPS 2017).

Fig. 61.3 Black coarse sand and pebbles originated by the erosion of volcanic rocks in Hawaii Volcanoes National Park. The beaches are used for the hatching of eggs by giant marine turtles. (Picture reproduced by permission of Judy McIlrath, University of South Florida, USA)

61.3 Conservation Challenges

Hawaii Volcanoes National Park faces an ecological crisis. Plants that have survived for millennia are facing tremendous threats from alien invasive plants and introduced animals (NPS 2017). Feral pigs, goats, mouflon sheep, mongoose, feral cats and rats, ants, wasps, and mosquitos and invasion by at least 25 exotic plants threaten native species (UNEP/WCMC 2011).

Climate changes already impose problems to the fauna and flora in the park, due to the frequency and intensity of wildfire (NPS 2017). In addition, coastal erosion tends to increase because of the rising of sea level associated with global warming, which increases erosion resulting from the subsidence that the islands are suffering.

The International Union for the Conservation of Nature (IUCN) found in its 2014 assessment related to the conservation conditions of the World Heritage sites (IUCN 2014) that the stage of conservation of the Hawaii Volcanoes National Park was good with some concerns, the concerns being related to the situations exposed above. Nevertheless, in the conservation outlook published at the end of 2017, the IUCN rated the site's level of conservation to be good, without concerns (Osipova et al. 2017). This change in the state of conservation in 3 years is linked to better management of the site (IUCN 2017).

References

Doty M, Mueller-Dombois D (1966) Atlas of bioecological studies at Hawaii Volcanoes National Park. Hawaii Botanical Science Paper No. 2. University of Hawaii, Honolulu

Garcia MO, Jicha BR, Marske JP, Pietruszka AJ (2017) How old is Kīlauea Volcano (Hawai'i)? Insights from $^{40}Ar/^{39}Ar$ dating of the 1.7-km-deep SOH-1 core. Geology 45:79–82

Hazlett R (1993) Geological Field Guide to Kilauea Volcano. Hawai'i Natural History Association, Honolulu

Holcomb R (1987) Eruptive history and long-term behavior of Kilauea Volcano. In: Decker R, Wright T, Stauffer P (eds) Volcanism in Hawaii. United States Geological Survey Professional Paper, vol 1350, pp 261–350

IUCN (International Union for Conservation of Nature) (2014) World heritage conservation outlook. Report, Gland, Switzerland

IUCN (International Union for Conservation of Nature) (2017) http://www.worldheritageoutlook.iucn.org/node/1031. http://www.worldheritageoutlook.iucn.org/node/1031. Accessed 27 Dec 2017

NASA (National Aeronautics and Space Administration) (2017) Earth observatory – lava flows on the Kilauea's Pali. https://earthobservatory.nasa.gov/NaturalHazards/view.php?id=78260. Accessed 28 July 2017

Neall VE, Trewick SA (2008) The age and origin of the Pacific Islands: a geological overview. Philos Trans R Soci London B Biol Sci 363:3293–3308

NOAA (National Oceanic and Atmospheric Administration) (2016) Honolulu Hi. http://www.prh.noaa.gov/hnl/pages/climate_summary.php. Accessed 26 July 2017

NPS (National Park Services) (2017) Hawaii Volcanoes National Park. https://www.nps.gov/havo/index.htm. Accessed 29 July 2017

Osipova E, Shadie P, Zwahlen C, Osti M, Shi Y, Kormos C, Bertzky B, Murai M, Van Merm R, Badman T (2017) IUCN world heritage outlook 2: a conservation assessment of all natural world heritage sites. IUCN, Gland

Parker A (2016) Rates of subsidence and relative sea level rise in Hawaii Islands. Nonlinear Eng 5:255–268

Smart CL (1965) The archaeological resources of Hawaii Volcanoes National Park: part I. An archaelogical survey of parts of Hawaii Volcanoes National Park. Report. Pacific Area Office, National Park Service, Honolulu, USA

Stone C, Scott J (eds) (1985) Hawaii's terrestrial ecosystems: preservation and management. University of Hawaii Cooperative National Park Resources Studies Unit, Honolulu

Tarduno JA, Duncan RA, Scholl DW, Cottrell RD, Steinberger B, Thordarson T, Kerr BC, Neal CR, Frey FA, Torii M, Carvallo C (2003) The emperor seamounts: southward motion of the Hawaiian hotspot plume in Earth's mantle. Science 301:1064–1069

Tilling RI (2012) Hawaiian volcano observatory: its first 100 years of advancing volcanology. American Geophysical Union Conference on Hawaiian Volcanoes Waikoloa Hawaii, August 2012

Tosi N, Yuen DA (2011) Bent-shaped plumes and horizontal channel flow beneath the 660 km discontinuity. Earth Planet Sci Lett 312:348–359

UNEP/WCMC (United Nations Environmental Programme/World Conservation Monitoring Centre) (2011) Hawaii Volcanoes National Park, USA. Report, Cambridge, England

UNESCO WHC (United Nations Educational, Scientific and Cultural Organization World Heritage Centre) (2017) Hawaii Volcanoes National Park. http://whc.unesco.org/en/list/409. Accessed 29 July 2017

Wilson JT (1963) A possible origin of the Hawaiian islands. Can J Phys 41:863–870

Wolfe CJ, Solomon SC, Laske G, Collins JA, Detrick RS, Orcutt JA, Bercovici D, Hauri EH (2009) Mantle shear-wave velocity structure beneath the Hawaiian hot spot. Science 326(2009):1388–1390

Chapter 62
Papahānaumokuākea, USA

Abstract Located in the north-central Pacific Ocean, Papahānaumokuākea, a part of the Hawaii Archipelago, is an isolated group of small islands and atolls 250 km to the northwest of Hawaii Island. Much of the site is made up of pelagic and deep-water habitats with features such as seamounts and submerged banks, extensive coral reefs and lagoons, and also emergent lands distributed between a number of eroded high islands, pinnacles, atoll islands, and cays. The area has deep cosmological and traditional significance for Hawaiian culture, as it is considered the place where life originates and to where the spirits return after death. It is the second largest marine protected area in the world and is a mixed cultural and natural World Heritage site.

62.1 Introduction

Located in the north-central Pacific Ocean, Papahānaumokuākea, a part of the Hawaii Archipelago know also known as Northwestern Hawaiian Islands, represents an isolated group of small islands and atolls 250 km to the northwest of Hawaii Big Island. The site comprises Papahānaumokuākea Marine National Monument, which extends for almost 2000 km with a linear distribution. It has an area of 362,075 km^2, with all boundaries located in the ocean. The marine national monument is the second largest marine protected area in the world (UNESCO WHC 2017) (Fig. 62.1).

The preservation of the area started early, mostly by action of the Federal Government of the USA. In 1909, the islets and reefs were placed within the Hawaiian Islands Reservation. In 1940, the reservation became the Hawaiian Islands Wildlife Refuge. In 2001, the area was established as the Northwestern Hawaiian Islands Coral Reef Ecosystem Reserve. In 2009, the Papahānaumokuākea Marine National Monument was created, and it was designated as a mixed cultural and natural World Heritage site in 2010 (UNEP/WCMC 2011).

Papahānaumokuākea is a sacred cultural landscape, a region of deep cosmological and traditional significance to the Hawaiian culture that contains a host of significant archaeological sites, as an expression of a culture that evolved cross-Pacific

V. Claudino-Sales, *Coastal World Heritage Sites*, Coastal Research Library 28,
https://doi.org/10.1007/978-94-024-1528-5_62

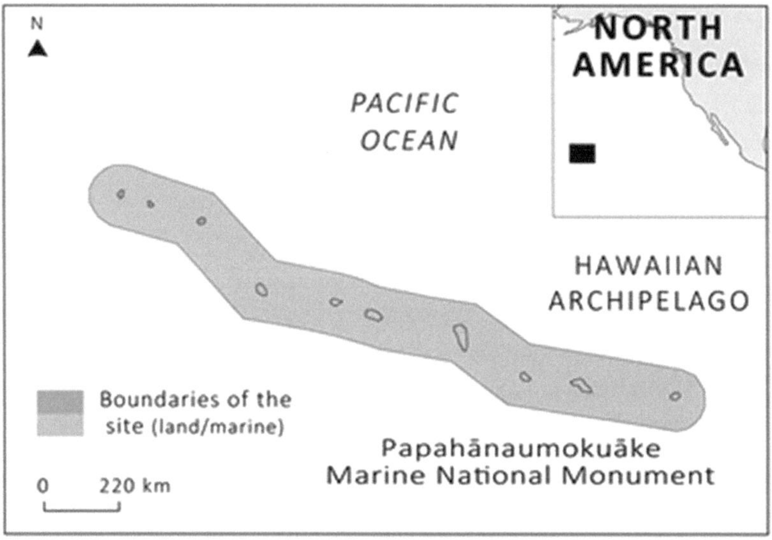

Fig. 62.1 Location of Papahānaumokuākea Marine National Monument and World Heritage Site, north-central Pacific Ocean, USA

Polynesian migration. The site represents the place where life originates and to where spirits return after death (Papahānaumokuākea Government 2017). The well-preserved shrines and their associated still-living traditions are distinctive to Hawaii and, positioned within a 3000-year-old Pacific/Polynesian cultural continuum, can be seen as an exceptional testimony to the strong cultural affiliation between Hawaii, Tahiti, and the Marquesas, resulting from long periods of migration (Papahānaumokuākea Government 2017).

There is no private ownership of the property, and the area is owned by the Governments of the USA and the State of Hawaii. The islands are uninhabited, except for the presence of less than 150 people working mostly as staff for the marine protected area. The islands are owned and controlled by the Government of the USA through the National Oceanic and Atmospheric Administration (NOAA) and the US Fish and Wildlife Service (FWS), with the Government of the State of Hawaii (UNEP/WCMC 2011).

Assessment, monitoring, and mapping of the flora and fauna in the Northwestern Hawaiian Islands began nearly a century ago as exploratory research voyages set sail primarily to collect data and specimens for cataloguing purposes. Since then, the property has been the object of extensive research (UNESCO WHC 2017).

Tourism increased after the marine reserve was designated as a World Heritage site, but the managers of the marine reserve allow recreational visitors only in one atoll (Midway Atoll). Numbers of overnight visitors are limited to no more than 50 at any one time. The number of larger day visits of 50–800 people to Midway is limited to no more than 3 per year, with no more than 400 people ashore at any one time (Papahānaumokuākea Government 2017).

62.2 Natural Settings

62.2.1 Climatic Conditions

The climate is mild year-round, with moderate humidity, persistent northeasterly trade winds, and infrequent storms. Annual rainfall averages 730 mm at French Frigate Shoals, the largest atoll in the Northwestern Hawaiian Islands (NOAA 2009a).

The area is influence by El Niño and La Niña ocean circulation patterns. In addition, annual extratropical storms create high waves during the winter, greatly affecting marine and terrestrial areas, as the elevation of a large portion of terrestrial habitat is less than the height of some of the waves that pass through (UNEP/WCMC 2011).

62.2.2 Geology and Geomorphology

Papahānaumokuākea constitute the northwest three-fourths of the vast chain of the Hawaiian Archipelago. Formed millions of years ago, these islands were created by a hotspot located south of the island of Hawaii, the Hawaii Hotspot. They are older than the main Hawaiian Islands and have been transported northwest from the main Hawaiian Archipelago by the movements of the Pacific Plate to their current locations (Dalrymple et al. 1974). The marine features comprise a portion of the world's longest and oldest volcanic chain (NOAA 2009b).

Effectively, the islands are part of a submerged volcanic chain that tracks the movement of the Pacific tectonic plate over the past 80 million years and has provided evidence upon which current theories of hotspot-mediated island formation and global plate tectonic movements have been based (Sharp and Clague 2006). The geological values of the property are directly connected to the values in Hawaii Volcanoes National Park, also a World Heritage site, and jointly present a very significant testimony of hotspot volcanism (NOAA 2009b).

Kure Atoll, with 29.9 million years, is the oldest and most northwestern island of the Hawaiian chain (Tarduno et al. 2003). It is located at a unique biogeographic position known as the Darwin Point, which represents the northern extent of coral reef development, beyond which cooler ocean temperatures prevent coral growth from keeping pace with the rate of geological subsidence (NOAA 2009a).

Due to the forces of subsidence and erosion, all that remains today of the former group of large islands are small patches of ancient land, shoals, and reefs that lie where significant mountains once loomed. Nowhere else in the world is this hotspot progression illustrated in such an unambiguous and linear fashion (NOAA 2009b).

The group of marine features is formed by 10 islands and atolls, none more than 5 km^2 in size, with heights varying from 4600 m below sea level to 275 m above sea level, including abyssal areas, shallow lagoons, sand and coarse beaches (Fig. 62.2), dunes, and a hypersaline lake. There are also more than 30 submerged banks and seamounts (UNESCO WHC 2017).

Fig. 62.2 Beach composed of coarse sediments surrounded by coral reefs in the shores of East Island at French Frigate Shoals. Dozens of the threatened green sea turtle that find habitat in Papahānaumokuākea are resting on the beach. (Picture reproduced by permission of Andy Collins and NOAA)

Six of the islands have an average altitude of less than 10 m. They consist of highly eroded islands with steep cliffs (Fig. 62.3), near-atolls with volcanic pinnacles jutting from surrounding lagoons, true ring-shaped atolls with roughly circular rims and central lagoons, and secondarily raised atolls (Papahānaumokuākea Government 2017; UNEP/WCMC 2011).

62.2.3 Biodiversity

The size of biogeographical processes is illustrated by its exceptional ecosystems and very high degree of marine and terrestrial endemism. Its terrestrial and marine habitats are crucial for the survival of many endangered or vulnerable species found only in the area (UNESCO WHC 2017).

The marine flora is characterized by the presence of 353 species of macroalgae and 2 seagrass species. Large numbers of Indo-Pacific algal species have been documented there that are not present in the main Hawaiian Islands. The reefs are largely free of alien algae, and high natural herbivory results in natural algal assemblages. In contrast to the marine systems, the terrestrial flora of the site is comparatively small but supports significant endemic biodiversity. At least six species of terrestrial plants found only in the region are listed as endangered (NOAA 2009a).

Fig. 62.3 Highly eroded and steep cliffs surrounded by wave-cut platform, where endangered Hawaiian monk seals rest on the Island of Mokumanamana, one of the group of the Northwestern Hawaiian Islands – Papahānaumokuākea. (Picture reproduced by permission of Hang Legario Eharis and NOAA)

For the marine fauna, the surveys show that 57 species of stony corals are known in the shallow waters (depths of less than 33 m) on the area, with an additional 28 species that are currently either undetermined or undescribed. There is also deepwater precious coral beds at depths of 365–406 m; these include corals whose growth rate is only a few centimeters every 100 years, with ages calculated as exceeding 2500 years. At depths below 500 m, a diverse community of octocorals and sponges flourishes (NOAA 2009b). The structure of reef fish assemblages as a whole showed a clear gradient from shallow to mesophotic depths. Fish assemblages at mesophotic depths had higher total densities than those in shallower waters (Fukunaga et al. 2016).

The area is a critical habitat for the Hawaiian monk seal (see Fig. 62.2), the world's second most endangered pinniped. Only 1200–1400 individuals exist, and the population probably will fall below 1000 individuals in the next years. The islands and atolls are also a crucial breeding, nesting, and basking habitat for the Hawaiian population of green turtles (see Fig. 62.3) (UNEP/WCMC 2011).

More than 14 million seabirds rely on the tiny islets in the chain, 5.5 million of which nest annually. This includes 99% of the world's Laysan albatrosses and 98% of the world's black-footed albatrosses. Within the terrestrial biota, four species of endemic birds have been identified, including some remarkably isolated and one of the world's rarest ducks (Laysan duck) (NOAA 2009b).

62.3 Conservation Challenges

Papahānaumokuākea conservation status is exceptional, largely due to its isolation as well as a combination of management and protection efforts, some dating back more than 100 years. The archaeological sites remain undisturbed by cultural factors. They are not under severe threat but should be better protected against damage from plants and wildlife (UNEP/WCMC 2011).

Threats to the natural values of the site are linked to the high presence of marine litter on the shores of the islands, coming mostly from the northern Pacific coast of North America (Agustin et al. 2015; Dameron et al. 2007). There are also problems related to the presence of hazardous cargo; possible future exploration and mining; illegal, unregulated, and unreported fishing; commercial fishing; anchor damage; vessel strikes; and invasive alien species (UNESCO WHC 2017).

Nevertheless, the most crucial of the threats to the site is, without doubt, climate changes, with the consequences of global warming. It includes rising of sea level, acidification of the oceans, increasing of ocean temperature, more frequent storms, and stronger coastal erosion, among others. These facts impose limits to the survival of coral, as well as to the integrity of the carbonate-based island atolls. Coastal erosion is also capable of destroying archaeological artifacts, considering that many of the islands are low-lying features and will have part of their land swept away by waves.

The International Union for Conservation of Nature (IUCN) published a report at the end of 2017 indicating the state of conservation of World Heritage sites (Osipova et al. 2017) in which Papahānaumokuākea appears well placed, having a good level of preservation.

References

Agustin AE, Merrifield MA, Potemra JT, Morishige C (2015) Temporal variability of marine debris deposition at Tern Island in the Northwestern Hawaiian Islands. Mar Pollut 101:200–207

Dalrymple GB, Lanphere MA, Jackson ED (1974) Contributions to the petrography and geochronology of volcanic rocks from the Leeward Hawaiian Islands. Bull Geol Soc Am 85:727–738

Dameron OJ, Parke M, Albins MA, Brainard R (2007) Marine debris accumulation in the Northwestern Hawaiian Islands: an examination of rates and processes. Mar Pollut Bull 54:423–433

Fukunaga A, Kosaki RK, Wagner D, Kane C (2016) Structure of mesophotic reef fish assemblages in the Northwestern Hawaiian Islands. PLoS One 11:e0157861. https://doi.org/10.1371/journal.pone.0157861

NOAA (National Oceanic and Atmospheric Administration) (2009a) The United States nomination of Papahānaumokuākea marine National Monument for inscription on the world heritage list. Report, Honolulu, USA

NOAA (National Oceanic and Atmospheric Administration) (2009b) Papahānaumokuākea marine national monument condition report 2009. Report, Honolulu, USA

Osipova E, Shadie P, Zwahlen C, Osti M, Shi Y, Kormos C, Bertzky B, Murai M, Van Merm R, Badman T (2017) IUCN world heritage outlook 2: a conservation assessment of all natural world heritage sites. IUCN, Gland

Papahānaumokuākea Government (2017) Papahānaumokuākea Marine National Monument. http://www.papahanaumokuakea.gov/. Accessed 2 Aug 2017

Sharp WD, Clague DA (2006) 50-Ma initiation of Hawaii-Emperor bend records major change in Pacific plate motion. Science 313:1281–1284

Tarduno JA, Duncan RA, Scholl DW, Cottrell RD, Steinberger B, Thordarson T, Kerr BC, Neal CR, Frey FA, Torii M, Carvallo C (2003) The emperor seamounts: southward motion of the Hawaiian hotspot plume in Earth's mantle. Science 301:1064–1069

UNEP/WCMC (United Nations Environmental Programme/World Coastal Monitoring Centre) (2011) Papahānaumokuākea United States of America. Report, Cambrigde, England

UNESCO WHC (United Nations Educational, Scientific and Cultural Organization World Heritage Centre) (2017) Papahānaumokuākea. http://whc.unesco.org/en/list/1326. Accessed 2 Aug 2017

Part XII
World Coastal Heritage in Islands, Gulfs and Reefs: Southern Ocean

Chapter 63
Heard and McDonald Islands, Australia

Abstract The volcanically active islands of Heard and McDonald are in the Southern Ocean, 1700 km from the Antarctic continent, as part of Australian territory. They are one of the most pristine ecosystems in the world, with complete absence of alien animal and vegetal species and with minimal impact from human activities. As the only volcanic active subantarctic islands, they provide good examples of ongoing tectonic activity, offering a view of plume volcanism, giving direct geological evidence of the action of the longest operational plume system known in the world. The islands also furnish crucial habitat for large populations of marine birds and mammals, and their endemic species demonstrate ongoing evolutionary processes.

63.1 Introduction

Heard and McDonald islands are remote subantartic volcanic features located in the Southern Ocean (Fig. 63.1), halfway between Australia and South Africa, just 1700 km from the Antarctic continent and 4000 km southwest of mainland Australia (Australian Government 2016).

Heard Island is 43 km long and 21 km wide. McDonald Islands are a group of uninhabited rocky islets, 40 km west of Heard Island (Encyclopaedia Britannica 2006). Heard Island has approximately 362.5 km^2 of area and the McDonald Island, 2.6 km^2. The site includes the adjacent offshore rocks and shoals and all territorial waters to a distance of 12 nautical miles (UNEP/WCMC 2008; see Fig. 64.1).

The islands were unknown to humanity until the 19th Century. The first discovery was by the American Captain Heard in 1853, followed in 1854 by a British sealer, Captain McDonald (Australian Government 2016).

In 1953, the islands were formalized as part of Australian external territory. In 1933, they were designated as a "Strict Nature Reserve" by the International Union for Conservation of Nature (IUCN). They became a World Heritage site in 1997, because of pristine subantarctic ecosystems and ongoing geological activity (UNEP/WCMC 2008). In 2002, the Australia government established a marine reserve, grouping the islands, which, with 64 km^2, is the world's largest marine reserve. The

V. Claudino-Sales, *Coastal World Heritage Sites*, Coastal Research Library 28,
https://doi.org/10.1007/978-94-024-1528-5_63

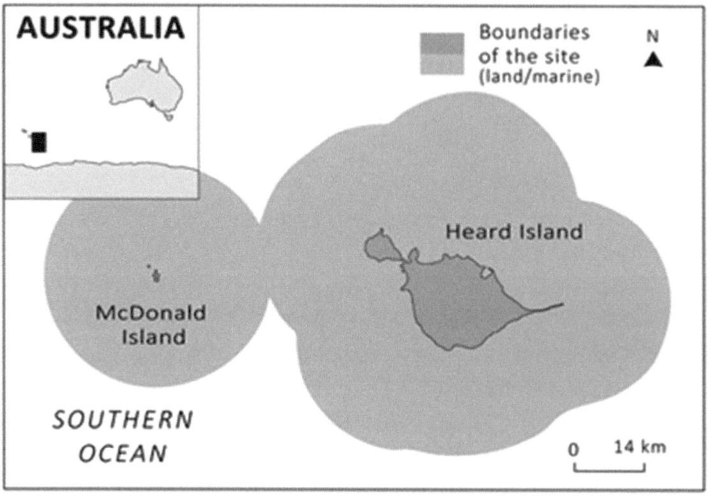

Fig. 63.1 The subantartic islands of Heard and McDonald, part of Australia territory

reserve not only protects many species of plant, bird, and mammal but allows researchers to monitor climate change and pollution (McCarthy 2003).

Land tenure is the Government of Australia. The area is managed by the Australian Government Heard Island and McDonald Islands Marine Reserve and World Heritage site management. The islands are inhabited (UNESCO WHC 2016). With the cessation of sealing on Heard Island in the early 1900s, extreme geographic isolation and severity of climate have deterred visitors (Keage 1981).

63.2 Natural Settings

63.2.1 Climatic Conditions

The islands are surrounded by stormy seas and cloudy skies, caused by temperate waters from the Indian Ocean meeting the icy Antarctic waters (Australian Government 2016).

The climate is a cool maritime, with strong prevailing westerly winds. The turbulence caused by the active volcanic massif of Big Ben creates extremely changeable weather. The mean annual temperature is 1 °C, with a summer mean of 3.2 °C and a winter mean of 0.1 °C. Annual precipitation is about 1400 mm. Snowfalls occur throughout the year, with maximum frequency in winter and spring (UNEP/WCMC 2008).

Fig. 63.2 Glaciers and snow in a cloudy sky in McDonald Islands. (Picture reproduced by permission of G. Budo and Department of Environment of Australia)

63.2.2 Geology and Geomorphology

Volcanic activity has occurred on Heard Island since the mid-1980s, and activity on McDonald Islands occurred during the 1990s. The activity on the McDonald Islands group resulted in the island doubling in size and increasing in elevation by about 100 m (Australian Government 2016).

Along with the French Kerguelen group of islands 440 km to the northwest, Heard and McDonald Islands form the only exposed parts of the submarine Kerguelen plume (Australian Government 2016). Their recent volcanism may correspond to the Tertiary hot spot track of the Kerguelen plume, which belongs to the large igneous province of the Southern Ocean (Weis et al. 2002).

The geomorphology of the islands is primarily the product of close interplay between volcanism, glaciation, and vigorous marine processes in a stormy environment. The dominant landform is the stratovolcano Big Ben (2745 m altitude), in Heard Island. The volcano is Pleistocene, and just a million years old (Kiernan and McConnel 1999).

A large variety of erosional and depositional glacial landforms is present, including major lateral moraines. The glaciers are steep and rapidly growing, and a number of them deposit sediment directly into the sea (Kiernan and McConnel 1999) (Figs. 63.2 and 63.3).

Fig. 63.3 Moraines in a glacial valley in the Spit Bay, Heard Island. (Picture reproduced by permission of Harry Burton)

The water circulation in the area is controlled by the Antarctic Circumpolar Current, an ocean current that flows clockwise from west to east around Antarctica. The net eastward transport to the south of the Heard/McDonald Islands is estimated as 560 m/s (Park et al. 2009).

Vigorous longshore drift and abundant sediment supply are present, forming bars from reworked glacigenic sediment that impound proglacial estuarine lagoons (Kiernan and McConnel 1999).

63.2.3 Biodiversity

The islands are a classic example of a subantarctic island group with low species diversity but with huge populations of certain species (Dasett 1990). They are one of the most pristine ecosystems in the world, with complete absence of alien animal and vegetal species and with minimal impact from human activities (UNESCO WHC 2016).

The extensive preserved marine area ensures the ecological integrity of the islands, since virtually all species living there depend on the ocean for their survival (UNEP/WCMC 2008). They furnish crucial habitat for large populations of marine birds and mammals, and endemic species there demonstrate ongoing evolutionary processes (UNESCO WHC 2016).

The seabird and seal community on the islands comprises an estimated total biomass of 27,893 tons, of which 15 breeding species of seabirds make up 70% (Woehler et al. 1982). Twenty-three species of marine mammals are recorded. Population levels of most species of seabirds and marine mammals breeding in the islands are reported to be stable (UNEP/WCMC 2008).

The distribution and occurrence of vegetation are strongly influenced by salt deposition, exposure to wind, substrate stability, and biotic influences, particularly the very large assemblages of seabirds (Jenkin 1997).

Five species of vascular plant occur on McDonald Islands and 11 on Heard Island. The isolation of the plant communities, combined with the increasing growth of habitat resulting from glacial retreat, makes Heard Island of significance for the study of plant colonization (UNEP/WCMC 2008).

63.3 Conservation Challenges

Any activity that may interfere with fauna or flora is prohibited on the islands, as well as introducing any animal, plant, or parasite. Also prohibited are motorized vehicles, fires, permanent structures, and the carrying or use of firearms (UNEP/WCMC 2008). The islands are subject to low anthropogenic pressures except for the largely unknown impact of commercial fisheries on the marine ecosystem, outside the property (UNESCO WHC 2016).

While isolation has been a key in conserving the islands, it also makes it difficult to manage a site so far from the mainland (UNEP/WCMC 2008). The main management requirements are the maintenance of strict visitation to keep natural conditions and ecologic integrity and preventing the introduction of nonnative species (UNESCO WHC 2016). Global warming is a threat. It is already producing the retreat of glaciers (UNEP/WCMC 2008) and changing the ecology of the islands. These changes are expected to be larger in the immediate future.

The International Union for Conservation of Nature (IUCN) released a report at the end of 2017 about the state of conservation of the World Heritage sites, in which it was found that the values of the site remain well preserved, largely as a result of the isolation, and challenging logistic needs to access the islands (IUCN 2017). The level of conservation is hence considered good (Osipova et al. 2017).

References

Australian Government (2016) World Heritage Places – Heard and McDonald Islands. https://www.environment.gov.au/heritage/places/world/heard-mcdonald. Accessed 16 Sept 2016

Department of Arts, Sports, the Environment, Tourism and Territories of Australia – DASETT (1990) Nomination of subantarctic heard island and McDonald Islands by the government of Australia for inclusion in the world heritage list. Report, Australia

Encyclopaedia Britannica (2006) Heard Island and McDonald Islands. https://www.britannica.com/place/Heard-and-McDonald-Islands. Accessed 14 Sept 2016

IUCN (International Union for Conservation of Nature) (2017) Heard and McDonalds islands conservation outlook. http://www.worldheritageoutlook.iucn.org/node/1077. Accessed 27 Dec 2017

Jenkin JF (1997) Vegetation of the McDonald Islands, sub-Antarctic. Polar Biol 18(4):260–272

Keage PL (1981) The conservation status of heard island and the McDonald Islands. Thesis of Master in Environment, University of Tasmania

Kiernan K, McConnel A (1999) Geomorphology of the sub-Antartic Australian territory of heard island – McDonald Islands. Aust Geogr 30(2):159–195

McCarthy R (2003) World's largest marine reserve (Australia). Geographical 17:10–14

Osipova E, Shadie P, Zwahlen C, Osti M, Shi Y, Kormos C, Bertzky B, Murai M, Van Merm R, Badman T (2017) IUCN world heritage outlook 2: a conservation assessment of all natural world heritage sites. IUCN, Gland

Park YH, Vivier F, Roquet F, Kestenare E (2009) Direct observations of the ACC transport across the Kerguelen Plateau. Geophys Res Lett 36(18):1–5

UNEP/WHCM (United Nations Environmental Programme/World Conservation Monitoring Centre) (2008) Heard and McDonald Islands, Australia. Report, Cambridge, England

UNESCO WHC (United Nations Educational, Scientific and cultural organization, World Heritage Centre) (2016) Heard and McDonald Islands. http://whc.unesco.org/en/list/577. Accessed 18 Sept 2016

Weis D, Frey F, Schlich R, Schaming M, Montigny R, Damasceno D, Mattielli N, Nicolaysen K, Scoates J (2002) Trace of the Kerguelen mantel plume: evidence form seamounts between the Kerguelen archipelago and heard island, Indian Ocean. Geochem Geophys Geosyst 3(6):1–27

Woehler E, Green K, Woehler E (1982) Consumption of marine resources by seabirds and seals at Heard Island and the McDonald Islands. Polar Biol 12(6–7):659–665

Chapter 64
Macquarie Island, Australia

Abstract Macquarie Island lies 1500 km southeast of Tasmania, halfway between Australia and the Antarctic continent. It is the only place on Earth where rocks from the Earth's mantle are being actively exposed above sea level. Its unique exposures include active faults, examples of pillow basalts and other extrusive rocks, all geological evidence for seafloor spreading, and continental drift. It provides a unique opportunity to study, in detail, geological features and processes of oceanic crust formation and plate boundary dynamics. In addition, its remote and windswept landscape of steep escarpments, lakes, and dramatic changes in vegetation provides an outstanding spectacle of wild, natural beauty complemented by vast congregations of wildlife.

64.1 Introduction

Macquarie Island lies in the Southern Ocean, about halfway between Australia and Antarctica, 1500 km southeast of Tasmania. The site includes Macquarie Island, Judge, Clerk and Bishop Islets, rocks, reefs, and the surrounding waters to a distance of 22 km from the shore. The site has 540 km^2, 127 km^2 of which is land (UNESCO WHC 2017) (Fig. 64.1).

In 1933, the island was declared a wildlife sanctuary. In 1971, it was designated as a Tasmanian State conservation area. In 1977, the island was entered on the register of the Australian National Estate and declared a UNESCO Biosphere Reserve. In 1997, the Macquarie Island Marine Park was created. It became a World Heritage site in the same year (UNEP/WCMC 2013).

Discovery of the island is attributed to Captain Hasselborough in 1810, who named it after the Governor of New South Wales. He may have been preceded by Polynesians and other earlier visitors. Sailors almost exterminated the endemic fur seals for their pelts and greatly reduced elephant seal and penguin populations, rendering them down for oil. This tradition ended in 1919 with the establishment of the first scientific station on the island (Australian Government 2017a).

The islands are not inhabited, except for the presence of around 20 personnel of the Australian National Antarctic Research Expedition (ANARE) station. The State

V. Claudino-Sales, *Coastal World Heritage Sites*, Coastal Research Library 28,
https://doi.org/10.1007/978-94-024-1528-5_64

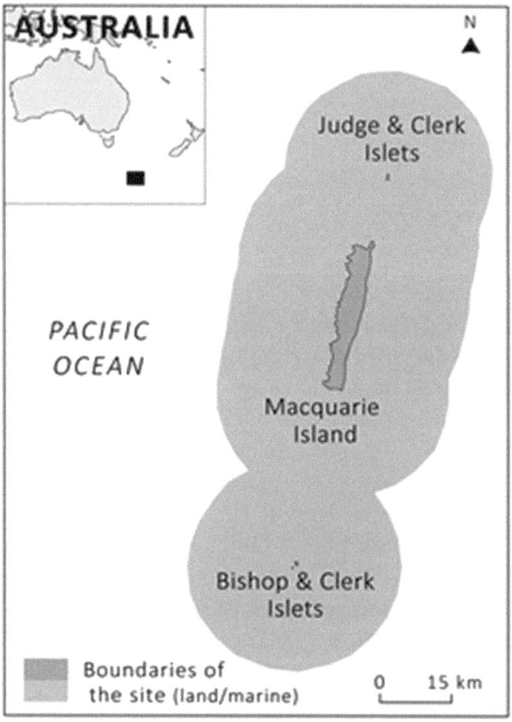

Fig. 64.1 Location of Macquarie Island in the Southern Ocean, southeast of Tasmania, Australia

of Tasmania owns the island, islets, rocks, reefs, and surrounding sea to 5.5 km from the low-water mark. The Commonwealth of Australia has jurisdiction over the marine area from there to 22 km off shore. It is managed by the Tasmanian Parks and Wildlife Service (PWS). Tourists can visit the island by applying for a permit to the park authority (UNEP/WCMC 2013).

64.2 Natural Settings

64.2.1 Climatic Conditions

Macquarie Island has an extreme oceanic cold temperate climate with an average rainfall of approximately 900 mm a year. Average temperatures are above freezing, ranging between 3 °C and 7 °C, and snow is frequent. During the peak of summer, there are only about 4 h of darkness. Cloud cover averages 85% of the time in all months (Australian Government 2017a).

The wind is almost constant, averaging around 25 km/h. Some 70% of winds are westerly to northwesterly. Low pressure systems passing the island produce wind speeds of over 170 km/h (UNEP/WCMC 2013). The island lies in latitudes known as the "Furious Fifties" because of the very strong wind and stormy seas.

64.2.2 Geology and Geomorphology

Macquarie Island is the sole complete subaerial exposure of a complete oceanic crust found in the ocean basin in which it formed, offering a unique window into the tectonic, magmatic, and hydrothermal processes associated with seafloor spreading and transform faulting (Coggon et al. 2016; Varne et al. 2000). The crust formed over 10 million years at a spreading ridge segment along the Indo-Australian-Pacific plates' boundary (Ducan and Varne 1988).

The spreading started in Eocene until the extension was nearly parallel to the present-day plate boundary and created the transform plate boundary observed today (Cande et al. 2000; Massell et al. 2000). Transpression along the plate boundary deforms adjacent oceanic crust to produce the Macquarie Ridge Complex with Macquarie Island as its apex. The Finch-Langdon fault, the most significant spreading-related structure on the island, juxtaposes upper-crust rocks against lower-crust and upper-mantle rocks. This is the only place in the world where rocks from the earth's mantle (6 km below the ocean floor) are being actively exposed above sea level (Wertz et al. 2003). Actually, Macquarie Island is the only island in the world composed entirely of oceanic crust and rocks from the mantle (Australian Government 2017b).

The island probably appeared only 600,000 years ago, and its geological evolution continues today with earthquakes and rapid rate of uplift, which are related to the active geological processes in action along the boundary between the two tectonic plates (Marmion 2008).

The north part of the island is mainly intrusive rocks, apparently derived from deeper crustal levels than the southern section; the north section is composed mainly of serpentinized peridotite and gabbro masses, although there are small areas of extrusive volcanic rock (Varne and Rubenach 1972). Dolerite dyke swarms are extensive in the south. These rocks, which are not deformed, as well as active faults, are the geological evidence of seafloor spreading and continental drift (Griffin and Varne 1980).

Macquarie Island is 34 km long and 5 km wide. It is made by a narrow undulating plateau 300 m above sea level surrounded by steep cliffs from the foot of which a coastal platform of beach slopes and terraces extends to the coast. Glacial deposits up to 20 m thick cover much of the plateau (UNEP/WCMC 2013). Vast waterlogged areas on the coastal platform are heavily vegetated, forming mire based on deep peat beds. Old sea stacks testify to the continual uplifting of the island, some of them now being several hundred meters from the existing coastline (UNESCO WHC 2017). Behind the coastal terrace, steep escarpments rise more than 200 m to the undulating central plateau which has three peaks over 400 m, the highest being Mount Hamilton at 433 m (Australian Government 2017b).

The ophiolite exposure includes good examples of pillow basalts, which have been sculptured by the strong winds and waves that characterize the area. Along the plateau and the raised platform, there are large lakes and numerous small lakes, tarns, and pools. Fluctuations in sea level and marine erosion have cut away the original escarpments, leaving some lakes perched on the edge of the plateau, while

Fig. 64.2 Rocky coast made by extrusive and intrusive rocks on the shores of Macquarie Island. The slopes on the cliffs are colonized by tussock grassland. In the horizon there are the buildings of the Australian National Antarctic Research Expedition Station. (Picture reproduced by permission of Melanie Lancaster)

others have been partially or totally drained (Australian Government 2017b). The coastline is generally rocky with a several offshore islets and stacks. Visually, the island can be grand but forbidding (UNESCO WHC 2017) (Fig. 64.2).

64.2.3 Biodiversity

Among the most aesthetically appealing features of the island are the vast congregations of wildlife, which provides an outstanding spectacle of natural beauty (UNESCO WHC 2017). Its isolation and the short geological time since its emergence exemplify the colonization of a site by long-distance dispersal and the evolution of endemic species (Marmion 2008).

There are five main vegetation formations: tall tussock grassland, short tussock grassland (herbfield), fen, bog, and feldmark. Changes in topography result in dramatic changes in the vegetation cover which can vary from lush grassland to sparse feldmark with the space of a few meters (Selkirk et al. 1990). The vascular flora contains at least 45 species, plus 80 moss species, 50 liverworts, and some 141 species of lichen. There are 20 species of freshwater algae and at least 90 diatoms and 110 species of marine and littoral algae (Ricker 1981).

With some 100,000 seals and 3 to 4 million penguins, the island teems with life, though the number of species is relatively low (UNEP/WCMC 2013) (Fig. 64.3).

Fig. 64.3 Colony of penguins in a pebble beach on Macquarie Island. Rocks from the ocean floor form the steep cliffs around the cove. (Picture reproduced by permission of Melanie Lancaster)

There are no native terrestrial mammals, and the indigenous mammals are all marine. They include southern right whale, killer whale, southern elephant seal, New Zealand fur seal, subantarctic fur seal, and Antarctic fur seal, all having recovered from the destruction of the 19th Century sealing industry. But the indigenous fur seal, of unknown species, was exterminated long ago (Cumpston 1968). Four species of introduced mammal still survive, the most frequent being rabbits and rats (Australian Government 2017a, b).

Sixty-seven species of birds are recorded at Macquarie Island, representing one of the greatest concentrations of sea birds in the world. No native passerines have been found (UNEP/WCMC 2018). Twelve benthic and 21 pelagic species of fish have been registered around the island (Williams 1988). Of some 27 species of marine mollusk, 64% are endemic. The island's fauna has probably less than 300 species of terrestrial invertebrates, approximately 10% being endemic (Cumpston 1968).

64.3 Conservation Challenges

A large threat to the integrity of the site is the spread of pathogens and introduced species, such as rabbits, black rats, and mice. A project to eradicate the pest species is on course and expected to be complete soon. The site also suffers the effects of climate change (UNESCO WHC 2017).

Global warming tends to increase sea erosion and speed winds, as well as the frequency of storms, threatening landscape, vegetation, and wildlife. This caused the International Union for Conservation of Nature (IUCN) to declare the state of conservation of the site to be good but presenting some concerns (IUCN 2014). In 2017, the IUCN released a new report about the state of conservation of the World Heritage sites, and the evaluation did not change, meaning that the management and level of conservation of the island is going through a period of stability (IUCN 2017; Osipova et al. 2017).

References

Australian Government (2017a) Macquarie Island Station. http://www.antarctica.gov.au/living-and-working/stations/macquarie-island. Accessed 3 Aug 2017

Australian Government (2017b) World Heritage Places – Macquarie Island. http://www.environment.gov.au/heritage/places/world/macquarie-island. Accessed 6 Aug 2017

Cande SC, Stock JM, Muller RD, Ishihara T (2000) Cenozoic motion between East and West Antarctica. Nature 404:145–150

Coggon RM, Teagle DAH, Harris M, Davidson GJ, Alt JC, Brewer TS (2016) Hydrotermal contributions to global biogeochemical cycles: insights from the Macquarie Island ophiolite. Lithos 264:329:348

Cumpston J (1968) Macquarie Island. ANARE Scientific Reports Series A (1): 93, Perth, Australia

Ducan RA, Varne R (1988) The age and distribution of the igneous rock of Macquarie Island. In: Banks MR, Smith SJ (eds) Proceedings of the Symposium on Macquarie Island. Royal Society of Tasmania, Papers and Proceedings 122:45–50

Griffin B, Varne R (1980) The Macquarie Island ophiolite complex: mid-tertiary oceanic lithosphere from a major ocean basin. Chem Geol 30:285–308

IUCN (International Union for Conservation of Nature) (2014) IUCN world heritage outlook 2014. Report, Gland, Switzerland

IUCN (International Union for Conservation of Nature) (2017) Macquarie Island Conservation Outlook. http://www.worldheritageoutlook.iucn.org/node/1078. Accessed 27 Dec 2017

Marmion I (2008) UNESCO-MAB biosphere reserve information Australia. Macquarie Island. Parks and Wilde Service, Tasmania, Australia

Massell C, Coffin MF, Mann P, Mosher S, Frohlic C, Schuur CL, Karner GD, Ramsay D, Lebrun LF (2000) Neotectonic of the Macquarie ridge complex, Australia-Pacific plate boundary. J Geophys Res 105:13457–13480

Osipova E, Shadie P, Zwahlen C, Osti M, Shi Y, Kormos C, Bertzky B, Murai M, Van Merm R, Badman T (2017) IUCN world heritage outlook 2: a conservation assessment of all natural world heritage sites. IUCN, Gland

Ricker R (1981) Macquarie Island: a blend of cold temperate, subantarctic and Antarctic seaweeds. 8th International Biological Congress. Saint Louis, Missouri, USA

Selkirk P, Seppelt R, Selkirk D (1990) Subantarctic Macquarie Island: environment and biology. Cambridge University Press, Cambridge

UNEP/WCMC (United Nations Environmental Programme/World Conservation Monitoring Centre) (2013) Macquarie Island Australia. Report, Cambridge, England

UNESCO WHC (United Nations Educations, Scientific and Cultural Organization World Heritage Centre) (2017) Macquarie Island. http://whc.unesco.org/en/list/629. Accessed 6 Aug 2017

Varne R, Rubenach M (1972) Geology of Macquarie Island in relation to tectonic environment. In: Coleman P (ed) The western Pacific islands: islands areas, marginal seas, Geochimistry. University of Western Australia Press, Perth

Varne R, Brown AV, Faltoon T (2000) Macquarie Island: its geology, and the timming and tectonic settings of its N-MORB to E-MORB magmatism. In: Dilek Y (ed) Ophiolites and oceanic crust: New insights from field studies and the Ocean Drilling Geological Society of America Special Paper 349:301–320

Wertz KL, Mosher S, Daczko NR, Coffin MF (2003) Macquarie Island finch-Langdon fault: a ridge-transform inside-corner structure. Geology 31:661–664

Williams R (1988) The nearshore fishes of Macquarie Island. Proceedings Royal Society of Tasmania 122:233–245

Chapter 65
New Zealand Subantarctic Islands, New Zealand

Abstract The New Zealand Subantarctic Islands consist of five island groups (Snares, Bounty, Antipodes, Auckland, and Campbell) in the Southern Ocean southeast of New Zealand. The Snares and Bounty Island groups are formed of basement granite and metamorphic rocks. The three more southerly groups are of basic volcanic rocks. They have been extensively glaciated. The islands display a pattern of immigration of species, diversification, and emergent endemism and are remarkable for their high level of biodiversity, population densities, and endemism in birds, plants, and invertebrates, which are unique to the islands. The group of islands is the most remote protected area of New Zealand and includes some of the world's least-modified islands.

65.1 Introduction

New Zealand Subantarctic Islands comprise five island groups (Snares, Bounty, Antipodes, Auckland, and Campbell) in the Southern Ocean 220 to 650 km south and southeast of New Zealand's South Island. The site has 764 km^2 of area (UNESCO WHC 2017) (Fig. 65.1).

Between 1910 and 1978, the islands have been declared reserves for the preservation of fauna and flora. In 1993, the Auckland Islands Marine Mammal Sanctuary was proclaimed. The islands became a serial World Heritage site in 1998 (UNEP/WCMC 2011).

The land in the islands is state-owned, and the properties are managed by the Department of Conservation of New Zealand (UNESCO WHC 2017). Land claims from two Maori groups have been filed with the Waitangi Tribunal (UNEP/WCMC 2011).

Traces of the thirteenth-century Polynesian settlements have been found in the islands. They were discovered by Europeans in the eighteenth century, followed by a scramble of sealing expeditions in the early nineteenth century. Rabbits, goats, and pigs were released on several islands by sealers during the early nineteenth century. An era of farming started in the 1870s with sheep farming at Port Ross, followed on Campbell Island in 1894. Sheep numbers at the latter peaked at 8500 in 1910, but the

V. Claudino-Sales, *Coastal World Heritage Sites*, Coastal Research Library 28,
https://doi.org/10.1007/978-94-024-1528-5_65

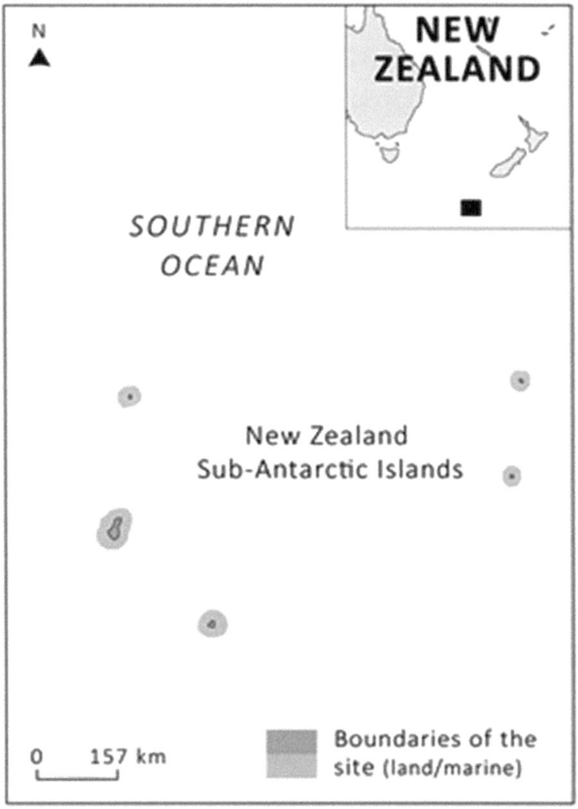

Fig. 65.1 Location of the New Zealand Subantarctic Islands in the Southern Ocean

station was abandoned in 1931. The last sheep were eliminated in 1992 and the last rats by 2003. Several coast watcher stations were built on Auckland and Campbell Islands and manned during World War II (Government of New Zealand 1997).

Today, the islands are uninhabited. Although a strict management plan restricts the number of people allowed ashore each year, some companies operate tours to the islands at various times throughout the year. The group of islands is the remotest protected area of New Zealand and includes some of the world's least-modified islands (Government of New Zealand 2017).

65.2 Natural Settings

65.2.1 Climatic Conditions

The islands have a cool-temperate climate with a fairly low seasonal range: the mean annual temperature ranges from 11 °C in the Snares to 6 °C on Campbell Island. The annual rainfall is 1200–1500 mm on most of the islands, and there is frequent cloud cover (UNEP/WCMC 2011).

The average wind is 40 km/s, but it can be much stronger in winter, considering that the climate on the islands is tempestuous (Dall 1911). The winds come mostly from the west-northwest. The average humidity is 84% on most of the islands. The seas and the weather surrounding them are strongly influenced by relatively warm ocean currents (MeteoBlue 2017).

65.2.2 Geology and Geomorphology

The New Zealand Subantarctic Islands lie close to the triple junction of the Indo-Australian, Pacific, and Antarctic plates, close to the eastern margin of an ancient continental shelf. The Campbell Plateau and the Bounty Platform are submerged platforms to the south of mainland New Zealand and constitute the southern extension of the New Zealand micro-continent, which was connected to the West Antarctic in the Gondwanaland until its separation 80 million years ago (Michaux and Leschen 2005).

The Snares, Auckland, and Campbell Islands lie on the Campbell Plateau and the Bounty and Antipodes in the Bounty Platform. The Snares and Bounty Islands are made by Cretaceous granitic basements, but the other three islands are volcanic. The volcanism is part of a long-lived ($\sim$100 million years), low-volume, diffuse alkaline magmatic province that includes deposits on the North and South Islands of New Zealand as well as portions of West Antarctica and SE Australia (Panter et al. 2006).

The volcanism took place 25 million to 10 million years ago (Auckland), 11 to 6 million years (Campbell) and from 5 million to less than 1 million years in the case of Antipodes. Auckland and Campbell are impressive basalt shield volcanoes, extensive and high enough to have carried large valley glaciers in the Pleistocene glacial times. Glacial features are visible in the form of U-shaped valleys, cirques, moraines, dammed lakes, and fjords (Government of New Zealand 1997).

Steep seacliffs characterize most of the islands. Sea erosion has produced deep, narrow caverns, sink holes, and gulches, and there are columnar joints in the cliffs. There are shallow shore platforms and dome-shaped masses protruding through the peat in the volcanic islands. Basalt boulders mantle the beaches (Government of New Zealand 1997) (Figs. 65.2 and 65.3).

The soils are made by blanket peat, several meters thick in some places, except in Bounty Islands. This peat reflects the steady accumulation of decomposing vegetation in a cool, moist climate spanning several thousand years (McGlone et al. 2000).

65.2.3 Biodiversity

The fauna and flora of these isolated islands have a very special interest, taking into account the modification that plants and animals have undergone in the area (Dall 1911). Isolation, climatic factors, and 7 degrees of latitudinal spread have combined to significantly influence the biota on the islands, which constitute a true laboratory

Fig. 65.2 The Snares Islands, the closest to New Zealand mainland, made by Cretaceous granites. The soil is colonized mostly by herbaceous and shrub vegetation. On the shallow adjacent sea, sandy bars are present. (Picture reproduced by permission of Department of Conservation of the New Zealand Government)

for the study of genetic variation, speciation, and adaptation, particularly in the Insular Antarctic biogeographic province (UNESCO WHC 2017).

Except for the Bounty Islands which have no higher plants, the islands have rich flora and are a center for plant diversity. The terrestrial flora has approximately 250 taxa, of which 35 are endemic to the region. They have the southernmost tree ferns and forests of the region (UNEP/WCMC 2011).

The Snares, Antipodes, and Auckland Islands are also remarkable because their vegetation is essentially unmodified by humans or alien animals. There are no comparable forests on the Campbell and Antipodes Islands, where only woody shrubs and tussock grasses grow. The size and density of the seaweed beds are vast (UNESCO WHC 2017).

The islands lack both native herbivores and terrestrial carnivores and are therefore vulnerable to invasions of alien substitutes. The highly productive marine ecosystem, however, supports a huge abundance and diversity of marine mammal and seabirds (Hiscok and Chilvers 2016). Ten of the world's 24 species of albatross breed in the islands. The 120 recorded species of birds include 40 seabirds, of which 5 breed nowhere else (UNEP/WCMC 2011).

The islands are home to the rare New Zealand sea lion with an estimated population of 10,000 of which some 95% breed in the Auckland Islands (Collins et al. 2016). They are also home to the New Zealand fur seal of which the largest breeding sites are in the Bounty Islands. Southern right whales breed in the waters off the

Fig. 65.3 Steep seacliffs shaped on the Miocene/Pliocene volcanic rocks (basalts) of Campbell Island. (Picture reproduced by permission of Dale Lorna Jacobsen/Shutterstock Stock Photos)

Campbell and Auckland Islands. There is one species of freshwater fish found in streams in the Campbell and Auckland Islands with 19 endemic species of freshwater invertebrates. The insect fauna has high endemism (UNEP/WCMC 2011).

65.3 Conservation Challenges

The impacts of alien mammal species, currently restricted to pigs, cats, and mice, on Auckland Island and mice on Antipodes Island, along with a range of alien plant and invertebrate species, are real threats to the site. Increased tourism demand has resulted in a significant increase in tourist numbers and activity within the property, which is also challenging the management of the islands (UNESCO WHC 2017).

Climate change and global warming are factors of stress within the integrity of the environment, as is the case for most of the small islands in the world's oceans. The increase of sea erosion and impacts on the weather and biota are the most expected effects of climate changes.

In general, the state of conservation of the site, as stated by the International Union for Conservation of Nature (IUCN) in its report about the conditions of World Heritage sites (IUCN 2014), is good without concerns. The same evaluation appeared in the state of conservation released by the IUCN at the end of 2017 (Osipova et al. 2017). The sites' internationally important ecological values are safe under current management policies which have been progressively improved through the creation of additional adjoining marine reserves, ambitious invasive species eradication goals, strengthened biosecurity protocols, and a robust research program (IUCN 2017).

References

Collins CJ, Chilvers BL, Taylor M, Robertson BC (2016) Historical population size of the threatened New Zealand Sea lion Photocarctos hookeri. J Mammal 97:436–444

Dall WH (1911) The subantarctic islands of New Zealand. Science 34:157–158

Government of New Zealand (1997) Nomination of the New Zealand Subantartic Islands by the government of New Zealand for inclusion on the world heritage list. Report, Wellington, New Zealand

Government of New Zealand (2017) NZ Subantarctic Islands. https://www.beehive.govt.nz/release/world-heritage-status-nz-subantarctic-islands. Accessed 7 Aug 2017

Hiscok JA, Chilvers BL (2016) Snares crested penguins Eudyptes robustus population estimates 2000-2013. N Z J Ecol 40:1–7

IUCN (International Union for Conservation of Nature) (2014) IUCN world heritage outlook 2014. Report, Gland, Switzerland

IUCN (International Union for Conservation of Nature) (2017) New Zealand Subantarctic Islands. http://www.worldheritageoutlook.iucn.org/node/1085. Accessed 27 Dec 2017

McGlone MS, Wilmshust JM, Wiser SK (2000) Late glacial and Holocene vegetation and climatic change on Auckland Island, Subantarctic New Zealand. The Holocene 10:719–729

MeteoBlue (2017) Climate Auckland and Campbell Islands. https://www.meteoblue.com/en/weather/forecast/modelclimate/auckland-islands_new-zealand_2193727. Accessed 7 Aug 2017

Michaux B, Leschen RAB (2005) East meets west: biogeology of the Campbell Plateau. Biol J 86:95–115

Osipova E, Shadie P, Zwahlen C, Osti M, Shi Y, Kormos C, Bertzky B, Murai M, Van Merm R, Badman T (2017) IUCN world heritage outlook 2: a conservation assessment of all natural world heritage sites. IUCN, Gland

Panter KS, Blusztajn J, Hart SR, Kyle PR, Esser R, McIntosh WC (2006) The origin of HIMU in the SW Pacific: evidence from intraplate volcanism in southern New Zealand and Subantarctic Islands. J Petrol 47:1673–1705

UNEP/WCMC (United Nations Environmental Programme/World Conservation Monitoring Centre) (2011) New Zealand subantarcitc islands, New Zealand. Report, Cambridge, UK

UNESCO WHC (United Nations Educational, Scientific and Cultural Organization World Heritage Centre) (2017) New Zealand Subantarctic Islands. http://whc.unesco.org/en/list/877. Accessed 8 Aug 2017

Part XIII
World Coastal Heritage in Islands, Gulfs and Reefs: Caribbean Sea

Chapter 66
Belize Barrier Reef System, Belize

Abstract The Belize shelf, in Central America, is a submarine extension of the Yucatan Platform, forming a drowned low-relief karst surface composed of drowned river channels and sinkholes. The system of reefs that developed in this area, the world's second-largest reef and the longest in both the Northern and Western hemispheres, provides a classic example of the evolutionary history of reefs, containing superlative natural phenomena and areas of exceptional natural beauty. The system has inshore, mid-shelf, and shelf-edge coral reefs. There are also offshore atolls, coastal lagoons, large seagrass meadows, mangrove forests, and sand cays. The site illustrates the whole range of coral reef development and provides the habitat for a very large number of fish species, as well as for threatened animals such as marine turtles, manatees, and the American marine crocodile.

66.1 Introduction

The Belize Barrier Reef System lies off the Caribbean coast of Belize, in Central America. It extends for 235 km between the Mexican and Guatemalan borders, 0.5–80 km from the mainland (Fig. 66.1). The Belize Barrier Reef System is the largest reef system in the Atlantic Ocean (Gischler and Hudson 2004), the world's second-largest reef and the longest in both the Northern and Western Hemispheres (UNESCO WHC 2016).

The World Heritage property established in the reef system is composed of seven different areas, such as marine reserves, national parks, and natural monuments. The seven properties are located along the length of the barrier reef, the shelf lagoon, offshore atolls, and a segment of coastal zone with estuaries (UNESCO WHC 2014). The entire site totals 963 km^2 (UNESCO WHC 2016).

The Belize Barrier Reef System is an ecosystem of great biological diversity and beauty and of great scientific value and a major habitat for threatened species (UNESCO WHC 2016). It was referred to "as the most remarkable reef in the West Indies" by Charles Darwin (UNESCO WHC 2014).

The reef system was the object of preservation for the first time in 1928, when a small part of the property was declared a bird sanctuary by the Belize Government.

V. Claudino-Sales, *Coastal World Heritage Sites*, Coastal Research Library 28,
https://doi.org/10.1007/978-94-024-1528-5_66

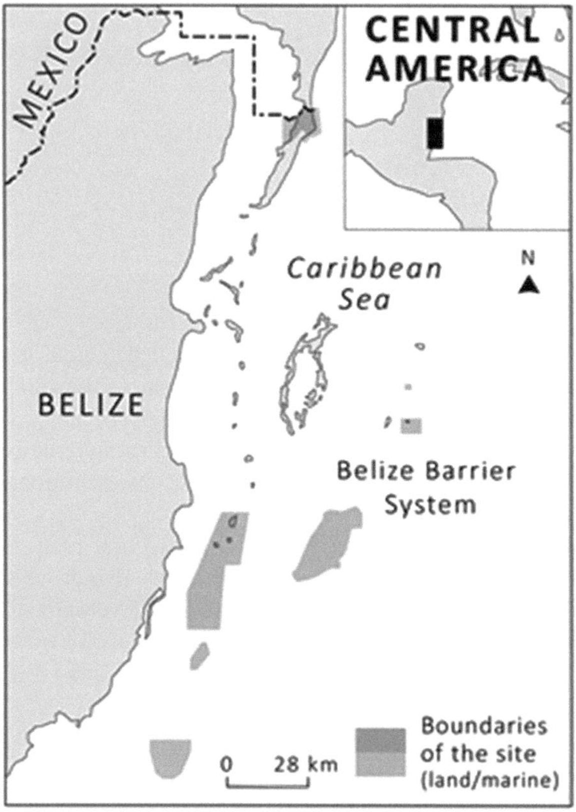

Fig. 66.1 Location of the Belize Barrier Reef System World Heritage site, formed by seven protected areas

In 1983, all the marine areas were designated protected sites. In 1990, the Government of Belize created the Coastal Zone Management Unit, and a Coastal Zone Management Plan was approved (UNEP/WCMC 2008). In 1996, the sites were transformed into marine reserves or into national parks and natural monuments. They also became a World Heritage site in 1996 (UNEP/WCMC 2008).

The seven sites are property of the Belize Government, except for the cays, that are privately owned. Marine areas are managed by the Fisheries Department of the Ministry of Agriculture and Fisheries and terrestrial areas by the Forest Department of the Ministry of Natural Resources (UNESCO WHC 2016).

Shell middens at Mayan sites along the coast show that the reefs were used for fishing some 2500 years ago. With the decline of the Mayan civilization, the resources were probably little used for centuries. Later, Spanish explorers used the area to repair boats and collect freshwater. By the seventeenth century, the coastal waters of Belize had become a refuge for pirates (UNEP/WCMC 2008).

Fishing records show that commercial fishing began to develop in the mid-nineteenth century. In the twentieth century, the economic importance of the reef increased with the growth of the coastal population. The fishery evolved during the 1960s to the local cooperatives which dominate the market today (UNEP/WCMC 2008).

Belize is a popular international tourism destination, and one of its main attractions is the Belize Barrier Reef System. The reef system spans the entire coastal area, earning the country the reputation of one of the premier sites for marine ecotourism in the world (Diedrich 2007).

The reef contributes to approximately 30% of the nation's gross domestic product through the tourism it attracts, the provision of commercial fisheries products, and private sector investments in aquaculture and coastal development (Cho 2005).

66.2 Natural Settings

66.2.1 Climatic Conditions

The Belize coast lies in the subtropics of the Northern Hemisphere. Mean annual temperatures range from 23–26 °C in winter to 28–31 °C in summer near the mainland. Mean annual precipitation ranges between 1500 mm in the north to 4000 mm in the south (Dakar 2008).

A dry season with strong easterly winds runs from January to June. Strong winter storms blow from October to February (UNEP/WCMC 2008). A hurricane season goes from August to November (Adomat and Gischler 2016).

66.2.2 Geology and Geomorphology

The Belize shelf, where the Belize Barrier Reef System is found, is a submarine extension of the Yucatán Platform. The Yucatán Platform is attached to a large, mature pull-apart basin set within a wide paleo-transform zone (Rosencrantz 1990). The zone defines the northwestern portion of the Caribbean-North American convergent plate boundary. It preserves a record of the Late Cretaceous to Paleogene Caribbean-North American convergent history and is largely unaffected by Neogene strike-slip tectonics of the current plate boundary (Rosencrantz 1990).

The coast and the platform shelf felt the effects of the rise and fall of sea level over the millennia. Coupled with natural karst topography and clear waters, it resulted in a diverse submarine seascape of patch reefs, fringing reefs, pinnacle reefs and barrier reefs, outer barrier reef platform, and oceanic atolls (Perkins and Carr 1985) (Fig. 66.2). The area also presents off-shelf atolls, rare deep water coral reefs, and some unique geological features, such as the contact between barrier reefs and the shoreline (UNESCO WHC 2014).

The barrier system also contains an intact terrestrial ecosystem, formed by littoral segments, wetland areas, marshes, mangrove swamps, beaches, and coastal lagoons (UNEP/WCMC 2008) (Fig. 66.3). Holocene reefs were established from 8260 to 6680 years BP on Pleistocene reef limestones presumably of oxygen isotope stage 5 age (ca. 125,000–130,000 years: Gischler and Hudson 2004). From

Fig. 66.2 Reefs in the clear waters of Belize littoral, near a coastal area with mangrove, beaches, and palm vegetation cover. (Picture reproduced by permission of Luigi Tura, webpage www.wnhs-experience.org)

Fig. 66.3 Coastal landscapes in the coast of Belize Barrier Reef System, with beaches and coastal vegetation formed by mangrove and palms. (Picture reproduced by permission of Luigi Tura, webpage www.wnhs-experience.org)

north to south, Pleistocene elevation decreases along these structural trends, probably reflecting differential subsidence and variations in karst topography (Gischler and Hudson 2004).

Differential subsidence and sequential drowning took place along a series of continental margin fault blocks (Gischler et al. 2000). The subsidence is presumably related to extension and left-lateral movements along the adjacent active boundary between the North American and Caribbean plates (Gischler et al. 2000).

The sea level rise linked to the Holocene transgression produced varied sediment deposits (Adomat and Gischler 2015). Mud is present in lagoon sediments. Peats and peat-rich sequences were deposited in mangrove, whereas sandy facies occur in the shoreface, beach, barriers, bars, barrier spits, and overwash deposits (Adomat and Gischler 2015). The anatomy of the Belize reef system is characterized by the presence of three facies: (1) corals, (2) unconsolidated sand and rubble, and (3) well-cemented coral grainstones-rudstones (Gischler and Hudson 2004).

66.2.3 Biodiversity

The ecosystem gradient ranging from the terrestrial to the deep ocean provides for a full complement of life cycle needs, supporting critical spawning, nesting, foraging, and nursery ecosystem functions (UNESCO WHC 2014). This structural diversity benefits a very complex and productive trophic chain based on abundant micro- and macroinvertebrate life and three thriving fisheries: commercial, artisanal, and recreation (UNEP/WCMC 2008).

More than 500 species of fish are recorded. Thousands of birds visit the lagoons during migrations. The reef system is luxuriant and diverse: 126 corals (90% of the Caribbean total), 350 mollusks, and a great diversity of algae, crustaceans, echinoderms, sponges, and marine worms are recorded from the sites (UNEP/WCMC 2008). The reef system also provides the habitat for a very large number of fish species, as well as for threatened animals such as marine turtles, manatees, and the American marine crocodile (UNESCO WHC 2016).

66.3 Conservation Challenges

Marine recreational tourism imposes a large threat to the Belize Barrier Reef System, although it appears to exist as a positive correlation between tourism development, coral reef conservation awareness, and support in the coastal communities (Diedrich 2007). Management challenges and threats that impact the integrity of the property include overharvesting of marine resources, coastal development, tourism, industrial development, and proposed oil and gas exploration and exploitation (UNESCO WHC 2014).

These threats justified the inclusion of this World Heritage site in the list of sites in danger in 2009, and it continues with this status (UNESCO WHC 2018). Despite this situation, the geographic spread and diversity enhance the resilience of the Belize Barrier Reef System, what is an essential factor today, also considering the dangers of climate change with its risks of coral bleaching, stronger and more frequent hurricanes, and sea level rise (UNESCO WHC 2014).

Significant progress has recently been achieved in developing an overarching framework for addressing the problems, particularly through the development of the Integrated Coastal Zone Management Plan (IUCN 2017). For this reason, the state of conservation of the site in the last report of the International Union for Conservancy of Nature (IUCN) was not considered critical, but only presenting significant concerns (Osipova et al. 2017).

References

Adomat F, Gischler E (2015) Sedimentary patterns and evolution of coastal environments during the Holocene in Central Belize, Central America. J Coast Res 31(4):802–826

Adomat F, Gischler E (2016) Assessing the suitability of Holocene environments along the Central Belize coast, Central America, for the reconstruction of hurricane records. Int J Earth Sci 106:283. https://doi.org/10.1007/s00531-016-1319-y

Cho L (2005) Marine protected areas: a tool for integrated coastal management in Belize. Ocean Coast Manage 48(11):932–947

Dakar E (2008) Coastal Belize: a paradoxal paradise. Weatherscapes 61(2):12–13

Diedrich A (2007) The impacts of tourism on coral reef conservation awareness and support in coastal communities in Belize. Coral Reefs 26(4):985–996

Gischler E, Hudson JH (2004) Holocene development of the Belize Barrier Reef. Sediment Geol 164(3):223–236

Gischler E, Lomando AJ, Hudson JH, Holmes CW (2000) Last interglacial reef growth beneath Belize Barrier and isolated platform reefs. Geology 28(5):387–390

IUCN (International Union for Conservation of Nature) (2017) Belize Barrier Reef System Conservation Outlook. http://www.worldheritageoutlook.iucn.org/node/1070. Accessed 27 Dec 2017

Osipova E, Shadie P, Zwahlen C, Osti M, Shi Y, Kormos C, Bertzky B, Murai M, Van Merm R, Badman T (2017) IUCN world heritage outlook 2: a conservation assessment of all natural world heritage sites. IUCN, Gland

Perkins JS, Carr A (1985) The Belize Barrier Reef: Status and prospects for conservation management. Biol Conserv 31(4):291–301

Rosencrantz E (1990) Structure and tectonics of the Yucatan Basin, Caribbean Sea, as determined from seismic reflection studies. Tectonics 9(5):1037–1059

UNEP/WCMC (United Nations Environmental Programme/World Conservation Monitoring Centre) (2008) Belize Barrier Reserve System, Belize. Report, Cambridge, England

UNESCO WHC (United Nations Educational, Scientific and Cultural Organization World Heritage Centre) (2014) Thirty-eighth session. Report, Doha, Qatar

UNESCO WHC (United Nations Educational, Scientific and Cultural Organization World Heritage Centre) (2016) Belize Barrier Reef Reserve. http://whc.unesco.org/en/list/764. Accessed 19 Oct 2016

UNESCO WHC (United Nations Educational, Scientific and Cultural Organization World Heritage Centre) (2018) List of World Heritage in Danger. http://whc.unesco.org/en/danger/. Accessed 15 Jan 2018

Chapter 67
Alejandro de Humboldt National Park, Cuba

Abstract Alejandro de Humboldt National Park is in northeastern Cuba, in the provinces of Guantanamo and Holguin. The site includes a complex system of mountains, coastal plains, bays, and coral reefs. The coastal plains represent the highest and elongated system of watersheds in the Caribbean region. Complex geology and varied topography create a high diversity of ecosystems: it represents one of the most biologically diverse tropical sites on Earth and is the largest and best-conserved remnant of forested mountain ecosystem in the Caribbean area. It is also a representative example of the combination of neotectonic evolution under conditions of sea shelf formation on ancient structures resulting from island arcs. Historically, it is lightly affected by humans.

67.1 Introduction

Alejandro de Humboldt National Park is located in the Nipe-Sagua-Baracoa Mountains, on the north coast of eastern Cuba. The site is a total of 710 km^2, with 688 km^2 of terrestrial area and 22 km^2 of marine area (UNESCO WHC 2016) (Fig. 67.1).

The first measures to ensure the preservation of the area date from the 1960s. In 1963, the Jaguani and Cupeyal del Norte Natural Reserves were declared, and in 1991, the Ojito de Agua Fauna Refuge was established. These three areas were unified in 1996 as the Alejandro de Humboldt National Park, which was legally declared in 2001. It became a World Heritage site in 2001 (UNEP/WCMC 2008).

Alejandro de Humboldt National Park is the most important strictly protected area in Cuba and the most extensive park in the national system of protected areas (Cuba Naturaleza 2016). The park contains the most important and significant natural habitats for in situ conservation of terrestrial biological diversity in the entire insular Caribbean and is one of the most important sites for conservation of endemic flora in the entire Western Hemisphere (IUCN 2014).

The area has historically been poorly inhabited and used. During the eighteenth and nineteenth centuries, runaway slaves took advantage of the remoteness of the region and lived there. In the twentieth century, some valleys were converted to

V. Claudino-Sales, *Coastal World Heritage Sites*, Coastal Research Library 28,
https://doi.org/10.1007/978-94-024-1528-5_67

Fig. 67.1 Location of Alejandro Humboldt National Park, in the northeast coast of the Island of Cuba, in the Caribbean Sea

farmland. More recently, mining and logging took place, until the area was declared protected in the 1980s (UNESCO WHC 2016).

The land of the park is now state owned and is held and managed by the Ministry of Science, Technology and Environment (República de Cuba 1998). The park is entirely within the Cuchillas del Toa Biosphere Reserve (UNESCO WHC 2016). It also lies within an international conservation hotspot, in World Wildlife Fund for Nature (WWF) Global 200 (e.g., the list of ecoregions identified by WWF as priorities for conservation) as both Marine and Freshwater Eco-regions, and is an Endemic Bird Area (UNEP/WCMC 2008).

67.2 Natural Settings

67.2.1 Climatic Conditions

The park has a hot, subtropical climate year-round. Most rain falls between May and October. It is subject to hurricanes between August and October. Humidity is high, varying between 75% and 95%. The cooler months are January to April. The maximum temperature is 32 °C, and the minimum is 15 °C (UNEP/WCMC 2008).

The area is bathed by the trade winds. The climatic conditions are different from the rest of the country, since the precipitation is above 2200 mm a year, higher than the average in the island. Within the park, there are mountainous areas with microclimates with annual average precipitation of 3600 mm, the highest in the country (Baracoa Organization 2017).

67.2.2 Geology and Geomorphology

Alejandro de Humboldt National Park's landscape is characterized by the presence of mountains, valleys, plains, tablelands, marine terraces, keys, bays, and coral reefs (Scarpaci and Portela 2009). The altitude goes from 220 m below sea level to 1168 m (UNEP/WCMC 2008).

The coastal plain resulted from neotectonic activity creating a sea shelf on island arc type of structures dating from the Oligocene, with a foundation made by ophiolites (República de Cuba 1998). The Nipe-Sagua-Baracoa mountain range on the northern edge of the park is sustained by older rocks, dating from the Jurassic period (Scarpaci and Portela 2009).

The coastal plains are the highest most elongated system of planation surfaces in the Caribbean. Large weathering crusts formed during the course of the Quaternary tectonic uplift and of the changes in sea level related to Quaternary climatic fluctuations. The coastal plains also represent the highest system of watersheds in the Caribe region (República de Cuba 1998). In contact with the sea, there are also vegetated seacliffs (Fig. 67.2).

Within the park, there is an important world example of the development of karstic forms, as well as of "pseudokarst," this developed on non-carbonated lithologies. The soils are largely toxic, lateritic, and acidic, except in the karstic area which has many flooded dolines, swallow holes, closed basins, and caves. Several clear water rivers flow from the park through pools and over cascades (UNEP/WCMC 2008).

Fig. 67.2 Coastal plain with vegetated seacliffs in Alejandro de Humboldt National Park. (Picture reproduced by permission of Oro Verde - Forest Tropical Foundation, website www.oroverde.de)

67.2.3 Biodiversity

Alejandro Humboldt National Park has the highest levels of biodiversity and endemism in the West Indies and one of the highest levels of endemism in the world. The region was not disturbed by the climate changes that took place during the Quaternary glaciations and therefore is a biogeographic bridge and a Miocene-Pleistocene refuge site of the Caribbean and the American biota (República de Cuba 1998). The antiquity and relative stability of the area together with geological complexity have determined a great variety of habitats (Scarpaci and Portela 2009).

The site contains one of the most complete neotropical wet island forests. Its terrestrial vegetation contains 16 of the 28 plant formations of the island, ranging from lowland rainforest between 200 m and 400 m, submontane rainforest to 600 m, montane rainforest above 600 m to cloud forest above 1000 m (República de Cuba 1998). These species are associated with Cuban pine forest, evergreen forest, semi-deciduous forest, river rainforest, xeromorphic shrub (80% endemic species), coastal xerophytic scrub on both sands and rocks, and mangrove forest in the coastal area (UNEP/WCMC 2008) (Fig. 67.3).

The toxic soil of the region poses particular challenges to plants and plays an important role in the evolution of the outstanding ecological features of the site, including its high degree of endemism (UNESCO WHC 2016).

Plant endemism in the region is over 70%, the highest in Cuba and one of the highest in the world. There are 1302 spermatophytes (69% of Cuba's endemic species) and 145 pteridophytes. In total, 905 of the area's species are endemic (29% of

Fig. 67.3 Estuary of Miel River, in Baracoa, Alejandro de Humboldt National Park, showing mangroves, seacliffs on sedimentary rocks and a wood bridge across the area. (Picture reproduced by permission of Tupungato/Shutterstock Stock Pictures)

all the country's endemics), of which 343 grow exclusively in the park (UNEP/WCMC 2008).

The fauna is very rich and also presents a high level of endemism: 10 species of mammals (30% endemic), 95 birds (21% endemic), 45 reptiles (83.3% endemic), 21 amphibians (95.8% endemic), 59 arachnids, and 191 insects (27.7% endemic) are recorded. The region's forests are also an important refuge for many resident and migratory birds (UNEP/WCMC 2008). Due to the exposure to trade winds and the mountainous topography, the area is Cuba's rainiest and coolest region, so important rivers rise in the forested mountains, boasting remarkable freshwater biodiversity (UNESCO WHC 2016).

The ongoing evolutionary processes in a largely intact and conserved area provide extraordinary insights for scientists and conservation practitioners. There is a consensus that many species remain to be discovered in the property (UNESCO WHC 2016).

67.3 Conservation Challenges

While historically little affected by humans and currently in a relative good state of conservation (IUCN 2014), important mineral deposits within the property represent a potential threat to the values of the park (UNESCO WHC 2016).

Natural threats are related to the action of hurricanes and fires, with several examples in recent years. Disaster preparedness and restorations related to these natural events are among the management challenges (UNESCO WHC 2010).

Introduction of new species and the arrival of invasive species are also threatening the endemic fauna in the site. Such a situation is exemplified by the presence and action of wild dogs, wild cats, and black rats in the natural niches of the park (Borroto-Páez 2009). Global warming may change the precipitation rates in the area, imposing new situations to the environment, especially to the forest cover. The coastal areas within the park will also suffer with the action of wave erosion.

The International Union for Conservation of Nature (IUCN) released a report about the state of conservation of the World Heritage sites in the end of 2017, in which the Alejandro de Humboldt National Park conservation level was found to be good, with some concerns (IUCN 2017; Osipova et al. 2017). The concerns are related to the situations indicated above.

References

Barracoa Organization (2017) Alexander Von Humboldt National Park. http://www.baracoa.org/geography/alexander-von-humboldt.html. Accessed 15 Nov 2016

Borroto-Páez R (2009) Invasive mammals in Cuba: an overview. Biol Invasions 11(10):2279–2290

Cuba Naturaleza (2016) Alejandro de Humboldt National Park, the most important protected area in Cuba. http://www.cubanaturaleza.org/en/nature-activities/-birdwatching/42-cuba-naturaleza-org-alejandro-de-humboldt-national-park-the-most-important-protected-area-in-cuba.html. Accessed 15 Nov 2016

IUCN (International Union for Conservation of Nature) (2014) World heritage outlook Alejandro de Humboldt National Park. Report, Gland, Switzerland

IUCN (International Union for Conservation of Nature) (2017) Alejandro de Humboldt National Park Conservation Outlook. http://www.worldheritageoutlook.iucn.org/es/explore-sites/wdpaid/900628. Accessed 27 Dec 2017

Osipova E, Shadie P, Zwahlen C, Osti M, Shi Y, Kormos C, Bertzky B., Murai M, Van Merm R, Badman T (2017) IUCN world heritage outlook 2: a conservation assessment of all natural world heritage sites. IUCN, Gland

República de Cuba (1998) Lista del Patrimonio Mundial Parque Nacional Alejandro de Humboldt. Havana, Cuba

Scarpaci JL, Portela AH (2009) Cuban landscapes, heritage, memory and place. The Guilford Press, London

UNEP/WCMC (United Nations Environmental Programme/World Conservation Monitoring Centre (2008) Alejandro de Humboldt Cuba. Report, Cambridge, UK

WHC (United Nations Educational, Scientific and Environmental Organization World Heritage Centre) (2010) State of Conservation Alejandro Humboldt National Park Cuba. http://whc.unesco.org/en/list/839/documents/. Accessed 24 Nov 2016

UNESCO WHC (United Nations Educational, Scientific and Environmental Organization World Heritage Centre) (2016) Alejandro de Humboldt National Park. http://whc.unesco.org/en/list/839. Accessed 24 Nov 2016

Chapter 68
Desembarco del Granma National Park, Cuba

Abstract Desembarco del Granma is located in and around Cabo Cruz, in the southeast coast of Cuba, at the western end of the Sierra Maestra Mountains, lying within the tectonically active zone between the Caribbean and the North American plates. The site has spectacular marine terraces, the world's largest and best preserved coastal limestone terrace system, as well as some of the most pristine and impressive coastal cliffs bordering the western Atlantic. Thea area is also representative of ongoing development of karst topography. The little disturbed landscape offers a wide spectrum of karst phenomena, such as giant sinkholes, cliffs, canyons, and caves. It is a globally significant illustration of geomorphologic and physiographic features and of ongoing geological processes.

68.1 Introduction

Desembarco del Granma National Park is on the coast of southeastern Cuba. It extends for 47 km, 160 km west of Santiago de Cuba. The site has 325.76 km^2, with 261 km^2 terrestrial and 64 km^2 marine (UNEP/WCMC 2008) (Fig. 68.1).

Desembarco del Granma is the first national park of Cuba and was created in 1986. It is state owned and is managed by the "Empresa Nacional para la Protección de la Flora y la Fauna," of the Ministry of Science, Technology and Environment (UNEP/WCMC 2008). It was declared a World Heritage site in 1999 (UNESCO WHC 2016).

The territory of the park was historically known as "Marine Terraces of Cabo Cruz and Maisí." During the process of converting the region to a protected area, it came to be known as Desembarco del Granma National Park (Republic of Cuba 1998). It is named after the landing ("desembarco") of Fidel Castro and rebels coming from Mexico in 1956. The ship they had was called "Granma," a name that was later also given to the province where the park is located (Scarpaci and Portela 2009).

There are noteworthy archaeological sites within the park, including ceremonial caves of the original indigenous people who inhabited the area. Petroglyphs, pictographs, pottery, and pre-agrarian and Aboriginal Caribbean agrarian artifacts are spread across the property (UNESCO WHC 2016; Scarpaci 2012).

V. Claudino-Sales, *Coastal World Heritage Sites*, Coastal Research Library 28,
https://doi.org/10.1007/978-94-024-1528-5_68

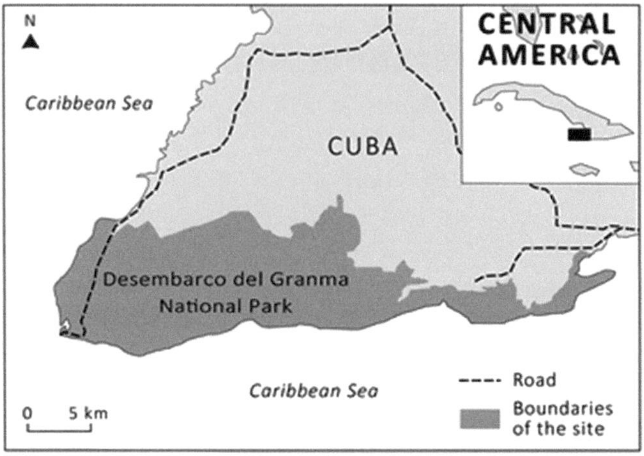

Fig. 68.1 Location of Desembarco del Granma National Park, in the southeastern coast of Cuba

Tourism is not an important activity in Desembarco del Granma National Park, but this situation is likely to change in the future. The park is included as one of the 15 zones defined as priority areas for the development of nature-based tourism by the Caribbean Zone of Sustainable Tourism agreement made by the Association of Caribbean States (Pérez et al. 2013).

68.2 Natural Settings

68.2.1 Climatic Conditions

The weather in Desembarco del Granma National Park is mostly dry, because the area is located in the rain shadow of the Sierra Maestra. The rainfall has totals from 300 to 1200 mm a year. With this volume of rain, the area ranks with the driest in Cuba (Republic of Cuba 1998).

The average temperature is 26 °C, but it reaches a maximum of 28 °C in summer. The area has one of the highest temperature averages in the country. The coast is subject to increasingly frequent hurricanes (UNEP/WCMC 2008).

68.2.2 Geology and Geomorphology

The park is located within the western part of the mountainous massifs of Sierra Maestra. It has the world's largest and best preserved coastal limestone terraces – the so-called "terrace system of Cabo Cruz." They form a singular coastal

Fig. 68.2 Marine terraces at the horizon, covered with a coastal vegetation in Desembarco del Granma National Park. (Picture reproduced by permission of Archivo Fotográfico del Sistema Nacional de Áreas Protegidas – SNAP, Cuba)

landscape, with pristine coastal cliffs of extraordinary beauty. Jointly with the diverse, mostly native vegetation, the cliffs form an extraordinary visual ensemble of forms, contours, color and texture within a spectacular coastal setting (IUCN 2015) (Figs. 68.2 and 68.3).

The marine terraces are modeled in calcareous rocks (limestones and dolomites of Jurassic and Cretaceous age) and comprise a series of seven to eight 30 m steps extending from 460 m above sea level to 180 m below the sea level, with vertical cliffs up to 100 m high (UNEP/WCMC 2008). They were formed by tectonic uplift, global climate change and changes in the sea level, which took place during the end of the Tertiary and beginning of Quaternary (Plio-Pleistocene periods) (Republic of Cuba 1998).

The karst contains caves, canyons, seacliffs, and giant sinkholes. Geologically they are in a pseudo-pericline formation – elongated domes developed during the deposition of Miocene to Quaternary sediments, now exposed on the southern slopes of the Sierra Maestra Mountains (UNEP/WCMC 2008). There are also coral reefs, where the clear currents sustain vigorous marine life. There are also seagrass beds, mangroves, and dunes near Cabo Cruz, in the southeast portion of the park (Republic of Cuba 1998).

The area is still a tectonically active zone, because it is located where the Caribbean plate is being subducted below the North American plates: 30 km offshore the floor of the Cayman Trough is 6000 m deep (UNEP/WCMC 2008).

Fig. 68.3 Coastline at Desembarco del Granma National Park, with mountains and vegetated uplifted marine terrace. (Picture reproduced by permission of PHB.cz (Richard Semik)/ Shutterstock Stock Photos)

68.2.3 Biodiversity

While recognized on the World Heritage List primarily for its landscape beauty and geology, Desembarco del Granma National Park also hosts noteworthy biodiversity values (UNESCO WHC 2016).

The flora is characterized by its great diversity, endemism, and high degree of conservation (Escalona et al. 2014). The main formations are semi-deciduous forest over limestone (see Fig. 70.2), coastal and subcoastal scrub, and forest of mangrove. The park has approximately 93 botanical families, 341 genres, and 619 species, 246 of them endemic (39.7% of total) (Palacio et al. 2012).

As to fauna, there are 13 mammals (23% of endemism), 110 birds (22.7% of endemism), 44 reptiles (90.9% of endemism), and 7 amphibians (85.7% of endemism) (Republic of Cuba 1998). There are 97 species of fish in the corals (Navarro-Martínez and Angulo-Valdéz 2015).

68.3 Conservation Challenges

The park encompasses intact limestone terrace systems on land and sea. It also provides the conservation of valuable plant and animal species, both terrestrial and marine, some of which are restricted to the site. The legislative framework assures a prominent position for national parks in Cuba and a high degree of protection (UNESCO WHC 2016).

Due to the hardness of rocks and the dry weather, relief forms are well preserved. The ecosystems developed there are also very well preserved because of the inaccessibility of these areas and the poor development of soils in their surfaces. They are affected by human activity in only some scarce patches (Republic of Cuba 1998). In the marine area, preservation is also good. Nevertheless, in spite of the good conditions of the corals, the richness of species of fish is lower than in other areas of Cuba. This condition is attributed to fishing for subsistence in the area (Navarro-Martínez and Angulo-Valdéz 2015).

Invasive species are the most serious current threat to the site's terrestrial biodiversity values. The number of introductions and resulting established populations of amphibians and reptiles in Cuba is alarming (Borroto-Páez et al. 2015).

The most serious threat to the marine environment is climate change, producing acidification of the sea, modifications on the climatic regime, and sea level changes (IUCN 2015). The coastal area is already suffering flooding by penetration of the seawater due to the increase of the sea level resulting from climate change (Arenal 2004). Finally, it is also necessary to consider that the increase in tourism could potentially have an impact on a park not prepared to receive visitors, especially in the coastal zone (IUCN 2015).

Generally, the state of conservation of the site is good, with some concerns. The values related to geologic features are in excellent condition, while scenic values have been degraded to some degree by climate change and introduced species (IUCN 2014). The same consideration appears in the last conservation outlook published by the International Union for Conservation of Nature (IUCN) at the end of 2017 (IUCN 2017; Osipova et al. 2017).

References

Arenal IM (2004) Las Inundaciones Costeras en Cuba y su Repercusión social. Bioética, Septiembre-Deciembre, Habana, Cuba

Borroto-Páez R, Bosch RA, Fabres BA, Garcia OA (2015) Introduced amphibians and reptiles in the Cuban Archipelago. Herpetol Conserv Biol 10(3):985–1012

Escalona RY, Riveron FB, Garcia A, Hernandez Y, Palacio E (2014) Parque Nacional Desembarco del Granma, maravillas de su flora. Flora e Fauna 11:25–31

IUCN (International Union for Conservation of Nature (2014) World Heritage Conservation Outlook. Report, Gland, Switzerland

IUCN (International Union for Conservation of Nature) (2015) IUCN technical evaluation of System of Terraces of Cabo Cruz, Cuba. Report, Gland, Switzerland

IUCN (International Union for Conservation of Nature) (2017) Desembarco del Granma National Park Conservation Outlook. http://www.worldheritageoutlook.iucn.org/node/1094. Accessed 28 Dec 2017

Navarro-Martínez ZM, Angulo-Valdéz JA (2015) Estado de Conservación de la Ictiofauna arrecifal en parques nacionales Cubanos: una revision. Revista Investigaciones Marinas35(1):82–99

Osipova E, Shadie P, Zwahlen C, Osti M, Shi Y, Kormos C, Bertzky B, Murai M, Van Merm R, Badman T (2017) IUCN world heritage outlook 2: a conservation assessment of all natural world heritage sites. IUCN, Gland

Palacio E, Escalona R, Cala YR, Calaña S, Ocano CA, Alayón L, Pérez J, Cisneros G, Sariego O, Ramón A (2012) Plan de Manejo Parque Nacional Desembarco del Granma (2012–2016). Empresa Nacional para la Protección de la Flora y la Fauna, MINAGRI, Cuba

Pérez V, Guerrero F, Gonzalez M, Pérez F, Caballero R (2013) Composite indicator for the assessment of sustainability: the case of Cuban nature-based tourism destinations. Ecol Indic 29:316–324

Republic of Cuba (1998) Lista del Patrimonio Mundial, Sistema de Terrazas Marinas de Cabo Cruz Y Maisi. Parque Nacional Desembarco del Granma, Reserva Ecologica Maisi, Elemento Natural Destacado Calleta. UNESCO World Heritage Centre Documentation Unit, Paris

Scarpaci JL (2012) Urban heritage, representation and planning: comparative approaches in Habana Vieja and Trinidad, Cuba. In: Ruggles DF (ed) On location. Heritage Cities and sites. Springer, New York

Scarpaci JL, Portela AH (2009) Cuban landscapes, heritage, memory and place. The Guilforf Press, London

UNEP/WCMC (United Nations Environmental Programme/World Conservation Monitoring Centre) (2008) Desembarco del Granma National Park Cuba. Report, Cambridge, UK

UNESCO WHC (United Nations Educational, Scientific and Cultural Organization World Heritage Centre) (2016) Desembarco del Granma National Park. http://whc.unesco.org/en/list/889. Accessed 5 Dec 2016

Chapter 69
Pitons Management Area, Saint Lucia

Abstract Located in the Lesser Antilles, the volcanic island of Saint Lucia is surrounded by the Eastern Caribbean Sea on one side and by the Atlantic Ocean in the other side. The Pitons Management Area in the southwest of Saint Lucia is two towering volcanic spires, Gros Piton and Petit Piton, which rise side by side from the sea to more than 700 m of altitude. The area encompasses a wide range of diverse geological features, such as a site of geothermal activity with fumaroles and hot springs. The Pitons are an area of iconic, spectacular, and unique coastal volcanic cones with a large fringing coral reef extremely well preserved. The volcanic features are clothed in tropical moist and subtropical wet forest and harbor endemic species.

69.1 Introduction

Belonging to the Lesser Antilles, the island of Saint Lucia is located in the Caribbean Sea and surrounded by the Atlantic Ocean. The Pitons Management Area, made by two towering volcanic spires (Gros Piton and Petit Piton), an accessible caldera-like formation known as the Qualibou Depression and a site of geothermal activity with fumaroles and hot springs known as the Sulphur Springs, is situated in the southwest of Saint Lucia, near the city of Soufriere. The site represents a management area with 29 km^2, −11 km^2 of land and 0.8 km^2 of sea (UNESCO WHC 2017) (Fig. 69.1).

In 1986, most of the reef system fringing Gros Piton was declared a marine reserve. In 1994, the Soufriere Marine Management Area was established. In 2002, the Pitons Management Area was created. The land tenure in the management area is mixed public (53%) and private ownership (47%). Terrestrial Conservation Areas are administered by the Ministry of Physical Development and Housing with the Forestry Department, and Marine Areas are administered by the Soufriere Marine Management Area under the Department of Fisheries (UNEP/WCMC 2011).

V. Claudino-Sales, *Coastal World Heritage Sites*, Coastal Research Library 28,
https://doi.org/10.1007/978-94-024-1528-5_69

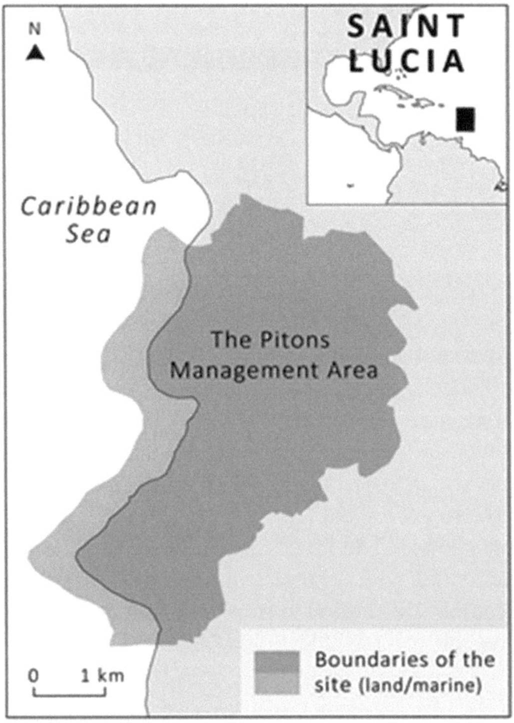

Fig. 69.1 Location of Pitons Management Area, in the Lesser Antilles, Saint Lucia

Archaeological sites – megalith, petroglyphs, and middens – prove the existence of an Amerindian Carib population which was displaced by Africans brought by the French who settled in the mid-seventeenth century. In 1814, the island was ceded to Britain from which independence was gained in 1979. The Soufriere area was a base camp for runaway slaves and their settlements. This, aided by the rugged topography, nurtured a local tradition of independence. There are also old plantation estates and the ruins of sugar mills (UNESCO WHC 2004).

No one lives permanently in the Pitons Management Area. The rough topography and local traditions have favored the presence of diversified medium-sized estates. Around 1500 people live within the terrestrial multiple use zone of the Management Area. There is a thriving artisanal fishery based on Soufriere (UNEP/WCMC 2011).

Tourism is an important economic activity to the island, the only one showing growth in 30 years. The "Sulphur Springs Park" is an important tourist attraction, touted as the "world's only drive-through volcano," with some of the hot pools being developed into recreational pools. Some 200,000 people visit the park each year (Joseph et al. 2013).

69.2 Natural Settings

69.2.1 Climatic Conditions

The island has a moist tropical climate with a rainfall of 2000–2500 mm a year. It is dry between January and May and wet in November and December. The temperature is constant at around 26.7 °C all over the year (UNESCO WHC 2004).

Because of its tropical climate and maritime dimension, Saint Lucia stays humid all-year-round, with humidity around 80%. Typical wind speeds fluctuate between 6 m/s and 11 m/s, rarely exceeding 14 m/s (Weather and Climate 2017). The area is under the influence of hurricanes (UNEP/WCMC 2011).

69.2.2 Geology and Geomorphology

The Lesser Antilles, home to the Pitons Management Area of Saint Lucia, are a volcanic arc resulting from the west-northwest subduction of the Atlantic lithosphere, part of the South America Plate, beneath the Caribbean Plate in the Lesser Antilles subduction zone. Arc volcanism has been active in the area for 40 million years (Bouysse et al. 1990). The process is ongoing and is responsible for many of the islands and for volcanic and seismic activity in the region (Defant et al. 2001; Wadge 1986).

The Soufriere Volcanic Center is the only active volcano of the island of Saint Lucia, which last erupted in 1766 (Tomblin 1965). The main part of the volcanic edifice is occupied by the large depression opened to the Caribbean Sea known as "the Qualibou Depression," a ~10-km-diameter-wide sector flank-collapse structure (Lindsay et al. 2013). The flank-collapse related to the Qualibou structure probably took place 100,000 years ago (Boudon et al. 2007).

In this composite volcano structure, acidic magmas erupted immediately after the flank-collapses. They are voluminous lava domes, forming the so-called Gros Piton and Petit Piton (Boudon et al. 2013). The domes are dated at 95,000 and 77,000 years, respectively (Samper et al. 2008) (Fig. 69.2).

The Pitons are two steep forested cone-shaped mountains which rise side by side from the sea with spectacular abruptness, having more than 700 m of altitude. Gros Piton is 3 km wide at the base; Petit Piton is 1 km wide and is linked to it by the high Piton Mitan ridge. In the center of the depression are the Sulphur Springs, the geothermal field with sulfurous fumaroles and hot springs surrounded by a variety of other volcanic features, such as extensive pyroclastic flow deposits, lava flows, and dome collapse block-and-ash-flow deposits (Boudon et al. 2013).

Fig. 69.2 Gros Piton (right) and Petit Piton (left), two vegetated lava domes in the Soufriere Volcanic Center, Saint Lucia Island, in the Caribbean Sea. (Picture reproduced by permission of Tab Hauser, webpage www.tabhauser.com)

The Marine Management Area is a shoreline 11 km long by 1 km wide, made of boulders and sandy plains (Fig. 69.3) and a steeply sloping continental shelf down with fringing and patch reefs. Fifty-seven of the marine segments of the World Heritage site are covered by healthy coral (UNEP/WCMC 2011).

The volcanic spires – e.g., the Pitons – predominate over the Saint Lucia landscape, being visible from virtually every part of the island and providing a distinctive landmark for seafarers. They are not necks or plugs, which are relatively common, but remnants of cumulo-domes, what make them geologically unique (UNEP/WCMC 2011). The combination of the varied topography and the lush natural tropical vegetation gives the site a spectacular natural beauty (UNESCO WHC 2017).

69.2.3 *Biodiversity*

Despite the small length of the site, there is a high diversity of terrestrial habitats, flora and fauna. The dominant vegetation is tropical moist forest grading to subtropical wet forest with small areas of dry forest near the coast and on steep slopes. On the Pitons, undisturbed natural forests remain, preserved by the steepness of the

Fig. 69.3 Gros Piton viewed from the sandy shoreline colonized with tropical flora at the Marine Management Area that integrates the Pitons Management Area World Heritage site. (Picture reproduced by permission of Tab Hauser, webpage www.tabhauser.com)

land. At least 148 species of plants have been recorded on Gros Piton and 97 on Petit Piton (Anthony 1997).

Twenty-seven bird species have been recorded on the site, including five endemic. There are the indigenous black-eared opossum, three species of bat, three rodents, eight reptiles, three amphibians, and many butterflies among the numerous species which have not yet been completely surveyed (Cox 1999; UNEP/WCMC 2011).

The coral reefs are healthy and diverse, comprised of fringing and patch reefs. A survey to a depth of 20 m revealed 168 species of finfish, 60 species of cnidaria, 8 mollusks, 14 sponges, 11 echinoderms, 15 arthropods, and 8 annelid worms (UNEP/WCMC 2011).

69.3 Conservation Challenges

A large part of the site is in private hands and risks being approved for luxury development. Despite the Integrated Development Plan and specific guidelines, the designation seems to have accelerated the sale of plots for resorts and multiple homes on the almost pristine and visually sensitive land between the Pitons (UNESCO WHC 2017).

Solid wastes and sediments from construction and mining are threats to coral and marine life. Hurricanes, dry-season fires, deforestation for timber, fuel and agriculture, livestock grazing, and noise from tourist helicopters are also a risk to the integrity of the site (UNEP/WCMC 2011). Global warming and climate changes can kill corals, raise the incidence of hurricanes, increase coastal erosion, and increased the occurrence of slides in the areas of rough topography, among other things, putting fauna, flora, and landscape in danger.

The International Union for Conservation of Nature (IUCN) published a report at the end of 2017 announcing the state of conservation of the World Heritage sites, in which the conservation level of the Pitons appears as presenting significant concerns (Osipova et al. 2017). The pressure from tourism and housing development coincided with the absence of a structured and enforced development control system and has been aggravated by limited financial and technical management capacity (IUCN 2017), which threatens the values of the World Heritage site.

References

Anthony D (1997) Inventory of flora and fauna of gros Piton, Saint Lucia. Environmental and Coastal Resources Project, Castries, Saint Lucia

Boudon G, Le Friant A, Komorowski J-C, Deplus C, Semet MP (2007) Volcano flank instability in the lesser Antilles arc: diversity of scale, processes, and temporal recurrence. J Geophys Res 112. https://doi.org/10.1029/2006JB004674

Boudon G, Villemant B, Le Friant A, Paterne M, Cortijo E, Pavel E, Gordeev E, Eichelberger J, West ME (2013) Role of large flank-collapse events on magma evolution of volcanoes; insights from the Lesser Antilles arc. J Volcanol Geotherm Res 263:224–237

Bouysse P, Westercamp D, Andreieff P (1990) The Lesser Antilles island arc. Proc Ocean Drill Program Part B Sci Results 110:29–44

Cox C (1999) A rapid inventory of the Flora and Fauna of petit piton and the ridge between the Pitons. Forestry Department, Ministry of Agriculture, Castries

Defant MJ, Sherman S, Maury RC, Bellon H, de Boer J, Davidson J, Kepezhinskas P (2001) The geology, petrology and petrogenesis of Saba Island, Lesser Antilles. J Volcanol Geotherm Res 107:87–111

IUCN (International Union for Conservation of Nature) (2017) Pitons Management Area Conservation Outlook. http://www.worldheritageoutlook.iucn.org/node/1132. Accessed 28 Dec 2017

Joseph EP, Fournier N, Lindsay JM, Robertson R, Beckles D (2013) Chemical and isotopic characteristics of geothermal fluids from Sulphur Springs, Saint Lucia. J Volcanol 254:23–36

Lindsay JM, Trumbell RB, Schmidt AK, Stockli D, Shane PA, Howe TC (2013) Volcanic stratigraphy and geochemistry of the Soufriere Volcanic Centre, Saint Lucia, with implications for volcanic hazards. J Volcanol Geotherm Res 258:126–142

Osipova E, Shadie P, Zwahlen C, Osti M, Shi Y, Kormos C, Bertzky B, Murai M, Van Merm R, Badman T (2017) IUCN world heritage outlook 2: a conservation assessment of all natural world heritage sites. IUCN, Gland

Samper A, Quidelleur X, Boudon G, Le Friant A, Komorowski JC (2008) Radiometric dating of three large volume flank collapses in the Lesser Antilles. Arc J Volcanol Geoth Res 176(4):485–492

Tomblin JF (1965) The geology of the Soufrière Volcanic Centre, St Lucia. Fourth Caribbean Geological Conference, Trinidad Tobago

UNEP/WCMC (United Nations Environmental Programme/World Coastal Monitoring Centre) (2011) Pitons management area. Report, Cambridge, UK

UNESCO WHC (Unesco Nations Educational, Scientific and Cultural Organization World Heritage Centre) (2017) Pitons management area. http://whc.unesco.org/en/list/1161. Accessed 12 July 2017

UNESCO WHC (United Nations Educational, Scientific and Cultural Organization World Heritage Centre) (2004) Nomination to inclusion in the world heritage list the Pitons management area, Saint Lucia. Report, Paris, France

Wadge G (1986) The dykes and structural settings of the volcanic front in the Lesser Antilles islands arc. Bull Volcanol 48:349–372

Weather and Climate (2017) Climate Soufriere. https://weather-and-climate.com/average-monthly-Rainfall-Temperature-Sunshine,soufriere,Saint-Lucia. Accessed 9 July 2017

Part XIV
World Coastal Heritage in Islands, Gulfs and Reefs: China Sea

Chapter 70
Ha Long Bay, Vietnam

Abstract Ha Long Bay, in Vietnam, is known for its emerald waters and towering limestone islands topped by rainforests. It includes 1600 islands and islets, which creates a majestic natural scenery of pillars with a variety of coastal erosional features such as arches and caves. The multitude of limestone waterscapes rise directly from the sea in a variety of sizes and shapes, presenting picturesque, unspoiled nature. The exceptional scenic beauty is complemented by its biological interest. It contains a diverse ecological system that includes tropical forests, mangrove forests, seagrass beds, and productive coral reefs. Because of the steep geomorphology, most of the islands are unaffected by human presence, though many archaeological sites have been found dated between 25,000 and 3000 years ago. It was voted one of the New Seven Wonders of the World in 2011.

70.1 Introduction

Ha Long Bay is in the Gulf of Tonkin, in northeast Vietnam, 165 km from the capital, Hanoi. Running along approximately 120 km of coastline, it covers an area of 434 km^2, including over 1600 islands and islets (Baker 2007) (Fig. 70.1).

The area was established as a historical and cultural relict and national scenic spot by the Vietnamese government in 1962. It was designated a World Heritage site in 1994 for its scenic values. Later, the Government of Vietnam presented a case to UNESCO for its universal geological value, and the site was reinscribed according to these criteria in 2000. The area is managed by the Ha Long Bay Management Department of Ha Long City, a huge urban agglomeration that is a great contrast with the natural majesty of the heritage site (UNESCO WHC 2016).

Although not listed as a cultural site, many archaeological locations have been found in the caves and grottos that characterize the site, dated between 25,000 and 3000 years BP (UNEP/WCMC 2008). Archaeological sites have yielded so many artifacts that they have been grouped as the Ha Long Culture, typical of the northeastern coast of Neolithic Vietnam: Ha Long was a significant port, located on the trade routes between China, Japan, and other Southeast Asian countries. Its waters

V. Claudino-Sales, *Coastal World Heritage Sites*, Coastal Research Library 28,
https://doi.org/10.1007/978-94-024-1528-5_70

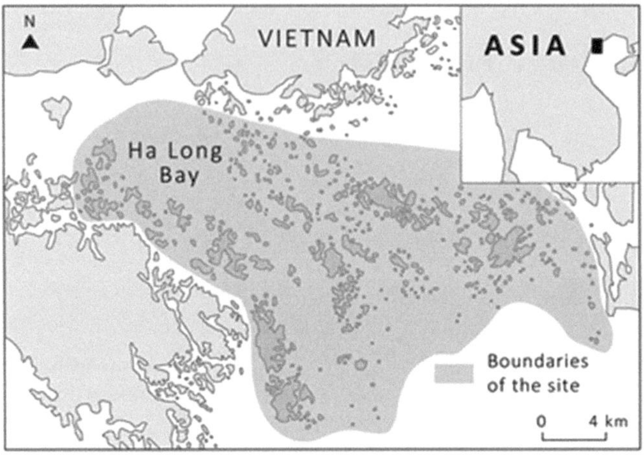

Fig. 70.1 Ha Long Bay, in the northern Vietnam, in the Gulf of Tonkin

have been the location of historical naval battles with Chinese and Mongol invaders (UNEP/WCMC 2008).

Today, the bay is a major center for fishing, agriculture, maritime transport, and tourism, and major shipping routes run through the archipelago. About 300 people live within the site in floating villages of houseboats and bamboo rafts, but most of the islands are uninhabited and unaffected by humans (UNESCO WHC 2016). The site was voted as one of the new Seven Wonders of the World in 2011.

The area has been heavily visited both by international and national tourists. Junk boat tours and cave tours led by multilingual guides, kayaking, camping, and swimming are all very popular. Junk boats and cruise ships run regularly through the bay, offering overnight stays and visits to the caves, grottos, and sandy beaches (Fig. 70.2). Tourism facilities exist on some of the islands, and the caves are well presented for visitors (UNEP/WCMC 2008).

Hang Dau Go is the most famous cavern, an underground feature with many stalagmites and stalactites. Ba Mun is the only one of the islands with a primitive forest, and in 1997, the government made it a national park to ensure its protection (Tran et al. 2013).

70.2 Natural Settings

70.2.1 Climatic Conditions

The weather conditions in Ha Long Bay are characterized by an equable warm climate with four distinct seasons, subject to typhoons and tropical storms. The storms come from the southeast during the spring and summer, from May to July (UNEP/WCMC 2008).

Fig. 70.2 Carbonate-rich sandy pocket beach surrounded by karst islands in Ha Long Bay. Tourist oriental boats (junks) float attached to wharves. (Photo reproduced by permission of Perfect Lazybones/Shutterstock Stock photos)

Following the monsoons, the precipitation is characterized by high rainfall from April to September, and little rainfall from October to March. The average annual rainfall is between 1600 mm and 1800 mm. Only 150–400 mm fall during the rest of the year (Kim et al. 2014).

The average annual temperature ranges between 19 °C and 25 °C. January–February are the coldest and driest months in the year. The average relative humidity is 84%. The many islands buffer the waves and the wind so that the seas are usually calm, except during the storms (Nguyen et al. 2014). The caves' microclimate is dependent on the outside atmosphere and on human activities (Duc and Garcia-Guinea 2014).

70.2.2 *Geology and Geomorphology*

Ha Long Bay is a large bay near the mouth of the Bach Dang River, hosting a diversity of ecosystems such as reefs, swamp forests, mangroves, and sandy beaches. Nevertheless, it is particularly known by its mature karst landscape of a multitude of sparsely vegetated limestone pinnacles, which rise directly from the sea (Fig. 70.3). A multitude of caves present well-developed stalactites and stalagmites.

Fig. 70.3 Limestone islets forming karst pinnacles rising directly from the sea in Ha Long Bay, Vietnam. (Picture reproduced by permission of Els Slots, webpage www.worldheritagesite.org)

The bay also has islands of schist in the southeast portion, plus a few more formed from decayed lateritic hills (UNESCO WHC 2016).

The surrounding sea is only 6–10 m deep except along old river channels, the result of a marine transgression caused by the sinking of the underlying limestone plateau (UNEP/WCMC 2008).

The carbonate rocks were deposited during three periods: Late Cambrian, Early Devonian to Early Carboniferous, and Carboniferous to Earliest Triassic. These sedimentary units have undergone multiple deformational events, leading to significant thickening of the original carbonate units (Tran et al. 2013). The combination of the primary structures with those formed during the deformational history is an important controlling factor in geomorphological and hydrogeological systems in the region (Tran et al. 2013).

The limestone karst of the bay developed since the Miocene. In the Quaternary, the rocks underwent five marine and continental intercalations. The present Ha Long Bay appeared after the transgressions in the Middle Holocene (UNEP/WCMC 2008). The limestone cliff bases typically have wave-cut notches reflecting erosion at the modern and two former higher Holocene sea levels (Boyd 2004).

Characteristically, the caves have enormous entrances, roomy interiors, but short length, typical of karst caves in tropical climates. Large islands, rising to 100–200 m, are found in the south, interspersed with smaller islets only 5–10 m high. On the east side of the bay, the middle-sized islands have almost vertical sides (Duc and Garcia-Guinea 2014).

The pinnacles are separate 50–100 m towers with a height-to-width ratio of 6:1 of a classic "fenglin" landscape form (isolated tower features) where the plain from which they rise has been flooded by the sea. There are also rows of cone-shaped islands characteristic of a "fengcong" landscape (clusters of conical peaks). Numerous lakes occupy drowned dolines within the fengcong islands, and springs appear on the shores of the bay (UNEP/WCMC 2008). Saltwater lakes disconnected from the sea are also found, developed in the caves carved out of the limestone (Cerrano et al. 2006).

70.2.3 Biodiversity

The sites' exceptional scenic beauty is complemented by its biological interest (UNESCO WHC 2016). The mostly warm weather means that many of the karst islands are thickly covered with jungle vegetation. There are 499 limestone-adapted plants including seven new to science (UNEP/WCMC 2008). Moreover, Ha Long Bay has more than 1000 species of flora (Baker 2007).

Fauna in the shores and in the bay include 14 mammals, 40 birds, 4 amphibians, and 8 reptiles (UNEP/WCMC 2008). There are antelopes, bantams, and iguanas. There are also deer, pheasants, wild boars, tortoises, flying squirrels, bats, and silver-headed parrots (Baker 2007).

The surrounding, shallow waters contain 1000 different species of fish and 450 types of mollusks. When people were less common in the area, elephants, rhinoceroses, tigers, panthers, and bears could be found. Today, a wildlife sanctuary exists on Cat Ba Island (Baker 2007).

70.3 Conservation Challenges

Located within an area of high tourism, marine transport, fisheries, and the daily activities of people living and conducting their business on Ha Long Bay City, management of the heritage site applies strict regulation and control of activities in an attempt to minimize impacts (UNESCO WHC 2016).

Generally, the natural scenic features, geomorphology, landform values, and cultural heritage and key features such as islands, caves, and grottoes remain intact, and the property retains a high level of naturalness despite the long history of human use in the area (UNEP/WCMC 2008).

Nevertheless, the limestone cave ecosystems are being degraded by tourism. The mouths of some caves have been widened to allow access, and the increase in light has led to imbalance in the delicate links between the flora and the fauna, to decreasing humidity inside the caves and increasing carbon dioxide which accelerates the growth of algae (Duc and Garcia-Guinea 2014).

With a booming tourist trade, mangroves and seagrass beds have been dredged and jetties and wharves built for tourist boats (see Fig. 70.2). Heavy metals, fuel, and oil have polluted the islands and their waters which have become turbid (UNEP/WCMC 2008). Human waste from portable toilets for tourists seeps into the soil and waters around the islands, altering the ecosystems through nutrient enrichment. Game fishing, often near coral reefs, threatens many species. Coral is sold as trinkets by the locals, but despoiling it expels the fish that they depend upon for their survival (UNEP/WCMC 2008).

Besides tourism, there are risks of degradation from the natural resources from the growth of the urban population and industries such as coal mining. Efforts to preserve the property have been made by the Ministry of Culture, Information and Sports of Vietnam, which is responsible for the overall management of the bay, and improvements are expected (UNESCO WHC 2016).

The International Union for Conservation of Nature (IUCN), in its last report about the state of conservation of the World Heritage sites, concluded that the values of the site remain largely intact due to their very nature and the property's natural defenses but that management capacity has recently improved to keep better pace with the rapid growth in tourism numbers (IUCN 2017). The level of conservation was considered "good with some concerns" (Osipova et al. 2017).

References

Baker S (2007) The 'dragon' islands of Ha long Bay, geographic overview. World and I 22(6):16–24

Boyd WE (2004) Holocene elevated sea levels on the north coast of Vietnam. Aust Geogr Stud 42(1):77–88

Cerrano C, Azzini F, Bavestrello G, Calcinai B, Pansini M, Sarti M, Thung D (2006) Marine Lakes of karst islands in Ha Long Bay (Vietnam). Chem Ecol 22(6):489–500

Duc TA, Garcia-Guinea J (2014) Vulnerability, pressures, and protection of karst caves and their speleothems in Ha Long Bay, Vietnam. Environ Earth Sci 71(11):4899–4913. https://doi.org/10.1007/s12665-013-2884-z

IUCN (International Union for Conservation of Nature) (2017) Ha Long Bay Conservation Outlook. http://www.worldheritageoutlook.iucn.org/node/1128. Accessed 28 Dec 2017

Kim D, Jin CS, Ho CH, Kim J, Kim JH (2014) Climatological features of WFR-simulated tropical cyclones over the western North Pacific. Clim Dyn 44:3223–3235

Nguyen K, Kafzfey J, McGregor J (2014) Downscaling over Vietnam using the stretched-grid CCAM: verification of the mean and internannual variability of rainfall. Clim Dyn 43(3):861–879

Osipova E, Shadie P, Zwahlen C, Osti M, Shi Y, Kormos C, Bertzky B, Murai M, Van Merm R, Badman T (2017) IUCN world heritage outlook 2: a conservation assessment of all natural world heritage sites. IUCN, Gland

Tran VT, Tran ST, Waltham, T, Li SA, LaI HA (2004) The Ha Long Bay world heritage: outstanding geological values. National Committee for ICCP, Vietnam

Tran BV, Hoang QD, Nguyen QM (2013) Structural controls on the occurrence and morphology of karstified assemblages in northeastern Vietnam: a regional perspective. Environ Earth Sci 70:511–520

UNEP/WCMC (United Nations Environmental Programme/World Conservation Monitoring Centre) (2008) Ha Long Bay, Vietnam. Report, Cambridge, England

UNESCO WHC (United Nations Educational, Scientific and Cultural Organization World Heritage Centre) (2016) Ha Long Bay Vietnam. http://whc.unesco.org/en/list/672. Accessed 29 Dec 2015

Chapter 71
Yakushima, Japan

Abstract Located in the interior of Yaku Island, at the meeting-point of the Palearctic and oriental biotic regions, Yakushima exhibits a primeval temperate rainforest and also contains a remnant of a warm-temperate ancient forest that is unique in this region. The site has a complex, sinuous boundary, which is less than 1 km wide in some places. It lies in the center of the island, with arms stretching west, south, and east, the western extending down to the coast. Despite being a small island, it boasts several key features, including impressive mountains which rise to nearly 2000 m and an outstanding gradient from the high peaks of the central core down to the seacoast. The site is home to a number of extremely large-diameter cedar trees that are thousands of years old.

71.1 Introduction

Yakushima, or Yaku Island, an almost round-shaped island, is located in the southern segment of the Japan Archipelago. Lying in the center of the island, the World Heritage site has a boundary with arms stretching west, south, and east. The western arm extends down to the coast at the East China Sea. It has 107 km^2 of area, which equals 20% of the island (UNESCO WHC 2017) (Fig. 71.1).

The history of preservation of the island by the Japanese Government started many decades ago. In 1964, the Yaku National Park was created. A wilderness area was designated in 1975. A forest ecosystem reserve was established in 1991. In addition to the government's actions, other preservation movements took place: the center of Yakushima Island and parts of the coastal lowlands were recognized as a Biosphere Reserve under UNESCO's Man and the Biosphere Program in 1980. Yakushima was inscribed on the World Heritage List in 1993, representing the first Japanese World Heritage site (UNEP/WCMC 1993).

The island's mountains, together with the cedar trees, are considered to be sacred and to have a spiritual value. Beyond the spirituality, the forests have been the subject of detailed ecological studies since the area was selected as a Biosphere Reserve. Today, the site is managed by the Yakushima World Heritage Area Liaison Committee (UNESCO WHC 2017).

V. Claudino-Sales, *Coastal World Heritage Sites*, Coastal Research Library 28,
https://doi.org/10.1007/978-94-024-1528-5_71

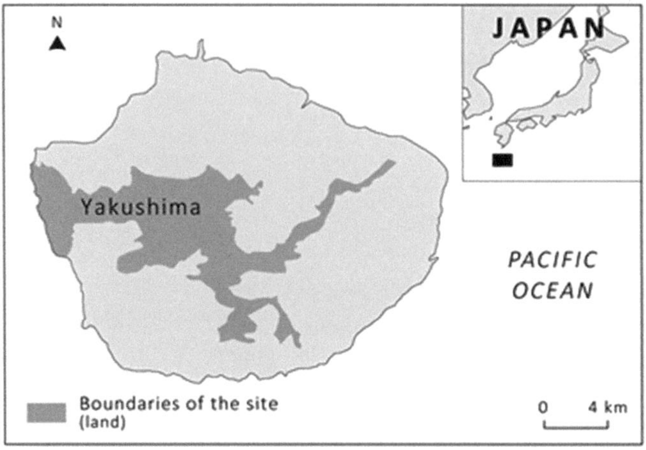

Fig. 71.1 Yakushima Island World Heritage site, between the Pacific Ocean and the East China Sea, on the southern segment of Japan

The land is 96% property of the state and 4% privately owned. The access to the island is by hydrofoil ferry, by slow car ferry, or by air to Yakushima Airport. A road loops around the entire island. No one lives inside the World Heritage site (UNEP/WCMC 1993).

71.2 Natural Settings

71.2.1 Climatic Conditions

Climate conditions are characterized by annual mean temperatures of 15.4 ° C; the monthly mean temperature ranges from 5.4 °C in January to 24.8 °C in July. The annual precipitation is 7296 mm, the lowest monthly precipitation is 389 mm in March, and the highest is 1054 mm in June. The average number of days with rainfall per year is 224 days. The maximum snow depth is 185 cm (Suzuki and Tsukahara 1987).

Yakushima is the rainiest area of Japan and one of the rainiest in the world. The climate is also tough in to other ways: strong typhoons frequently hit or pass Yakushima Island almost every year, with wind velocities exceeding 55 m/s (Takashima et al. 2007).

The influence of the climate and air masses on the environment is very complex, and the main feature of this area is that temperature and seasonal changes are the largest on Earth (Japan National Tourism Organization 2017).

71.2.2 Geology and Geomorphology

The site spans an area from the western coastline to the 2000 summit of the island, which is the highest mountain in southern Japan. Several peaks are over 1800 m with mountain ridges over 1000 m surrounding these central high peaks. Topography from coastline to the mountainous summits is extremely steep. The mountainous landscape explains the island's nickname of "alps of the ocean" (Government of Japan 1995).

The predominant bedrock on the island is granite, but small areas of sandstone and shale occur at the foot of the central mountain (Numata 1986). The granite is Paleogene (15.5 million years). It was intruded 10 km below Yakushima and forced the surface upward due to the Pacific Plate tectonic movements (Koji 2012).

In other words, the magma chamber that formed under bottom of the ocean solidified into granite. The granite thrust upward faster than erosion by wind and rain and became the mountainous Yakushima Island. The granite rose at a steady 1 meter per thousand years and continues to rise at a slower rate, creating the dramatic mountain landscape that exists today on the island (Government of Japan 2017).

The Yakushima granite is soft, due to the presence of large potassium feldspar crystals. Because of the high humidity, it is strongly weathered, producing large boulders along the hills. Eventually, the boulders slide to the foot of the mountains and into the coastline, creating a unique landscape (Wakita 2013; UNEP/WCMC 1993) (Figs. 71.2 and 71.3).

The extremely wet climate also means that crystal-clear rivers tumble down from the high peaks, resulting in spectacular waterfalls, which carve their way through the granite bedrock. Geothermal hot springs are everywhere, and some of the island's best are right on the beach, on the rocky shoreline (The Guardian 2015).

71.2.3 Biodiversity

The island is located at the interface of the Palearctic and oriental biotic regions, with an outstanding gradient form the high peaks of the central core down to the seacoast (UNESCO WHC 2017).

The flora is very diverse for such a small island, with more than 1900 species and subspecies. A variety of vegetation from the subtropical zone to the cool temperate zone, depending on elevation, can be seen on the island. The area up to 800 m is the laurel forest zone. Subtropical trees are also found near the coastline. The area between 800 and 1600 m is covered with mixed broadleaf forests. The area above the tree line at 1600 m becomes a shrub zone under the effect of strong winds. The area above 1800 m is grassland of dwarf bamboo (Government of Japan 2017).

Fig. 71.2 Stepped and mountainous landscape, from summits to the western coastal area in Yakushima World Heritage site. (Picture reproduced by permission of Slyellow/Shutterstock Stock Photos)

Of great significance to the area is the presence of indigenous Japanese cedar, which can reach more than a 1000 years on stable sites under the climate of the island; older specimens may reach 3000 years, found between 600 and 1800 m (Numata 1986).

The fauna of the island is diverse, with 16 mammal species. Four mammal subspecies, including Japanese macaque and deer, are endemic to the island. Among the 150 bird species present, 4 have been designated as natural monuments (UNEP/WCMC 1993).

There are also 15 species of reptile, 8 species of amphibians, and about 1900 species of insects. Yakushima Island is also a major spawning ground for loggerhead turtles. The annual number of spawning loggerhead turtles on the island account for nearly half the total number in Japan (Government of Japan 2017).

71.3 Conservation Challenges

The site seems to be in a good state of conservation, although there are some concerns, especially related to the impact of tourism on the soil and vegetation around the hiking trails, as well as associated with water pollution (IUCN 2014). Areas

Fig. 71.3 Rocky coast of Yakushima, with granites and sedimentary rocks forming seacliffs, which are colonized by forest-type vegetation. Pocket beaches with outcrops of granite are also present. (Picture reproduced by permission of Sara Winter/Shutterstock Stock Photos)

within the site also show negative impacts on vegetation from grazing by deer as a result of overpopulation.

The consequences of climate changes are not yet analyzed for this site, and it is not clear if the changes go in the direction of an increase of precipitation and winds or the opposite. Nevertheless, the fauna and flora have to adapt to new climate situations in the future.

The last conservation outlook about the World Heritage sites published by the International Union for Conservation of Nature (IUCN) at the end of 2017 found that the site has inappropriate boundaries. The boundaries are considered to be so inextricably linked, both physically and ecologically, with the surrounding forested lands that management will fail if confined within the rather arbitrary limits (IUCN 2017).

The site is also very vulnerable to external threats from activities in surrounding lands. Threats of an ecological nature are already appearing and can be expected to increase over time, especially from invasive species and overpopulation of native species (IUCN 2017). The conservation outlook confirmed that the level of conservation of the site is "good with some concerns" (Osipova et al. 2017).

References

Government of Japan (1995) Yakushima world heritage area management plan. Rapport, Tokyo

Government of Japan (2017) Yakushima (Island) National Park. https://www.env.go.jp/en/nature/nps/park/yakushima/. Accessed 21 June 2017

IUCN (International Union Conservation of Nature) (2014) IUCN World Heritage Outlook 2014. A conservation assessment of all natural World Heritage sites. Rapport, Gland, Switzerland

IUCN (International Union for Conservation of Nature) (2017) Yakushima Conservation Outlook. http://www.worldheritageoutlook.iucn.org/node/1061. Accessed 28 Dec 2017

Japan National Tourism Organization (2017) World Heritage Sites Yakushima. http://www.jnto.go.jp/eng/indepth/scenic/worldheritage/c_12_yakushima.html. Accessed 21 June 2017

Koji O (2012) Paleoseismology of fault displacement. Quat Int 279–280:280–358

Numata M (1986) The natural characteristics of Yaku Island. Memoirs Shukutoku University 20:15–20

Osipova E, Shadie P, Zwahlen C, Osti M, Shi Y, Kormos C, Bertzky B, Murai M, Van Merm R, Badman T (2017) IUCN world heritage outlook 2: a conservation assessment of all natural world heritage sites. IUCN, Gland

Suzuki E, Tsukahara J (1987) Age structure and regeneration of old-growth Cryptomeria japonica forests on Yakushima Island. Bot Mag 100:223–224

Takashima A, Kume A, Yoshida S, Murakami T, Kajisa T, Mizoue N (2007) Discontinuous DBH–height relationship of Cryptomeria japonica on Yakushima Island: effect of frequent typhoons on the maximum height. Ecol Res 24:1003–1011

The Guardian (2015) Yakushima Japan, an island of natural wonders. https://www.theguardian.com/travel/2015/oct/14/yakushima-island-japan-guide-sights-hotels-fooders. Accessed 20 June 2017

UNEP/WCMC (United Nations Environmental Programme/World Conservation Monitoring Centre) (1993) Yakushima (Yaku-Island). http://www.unep-wcmc.org/sites/wh/yaku.html. Accessed 21 June 2017

UNESCO WHC (United Nations Environmental, Scientific and Cultural Organization World Heritage Centre) (2017) Yakushima. http://whc.unesco.org/en/list/662. Accessed 21 June 2017

Wakita K (2013) Geology and tectonics of Japanese islands: a review – The key to understanding the geology of Asia. J Asian Earth Sci 72:75–88

Chapter 72
Jeju Volcanic Island and Lava Tube, Korea

Abstract Jeju Island, in the East China Sea, is the low crest of a dormant shield volcano. The island is characterized by the presence of lava tube made by lava flows that formed over a period of 1.2 million years until about 25,000 years ago. The lava flows are made of fine-grained basalt in the form of great tubes of rock, creating an exceptional system of lava tubes. In addition to the lava tubes, the World Heritage site includes a segment of volcanic tuffs and a segment of volcanic peak, the highest in Korea. The site bears testimony of the history of the planet, its features and processes. The biocharacteristics of Jeju Island, extremely diverse for such a small area, are also important, showing influences from East Asiatic, East Siberian, and Indo-Malaysian domains.

72.1 Introduction

Jeju Island is an oval some 73 km long by 32 km wide. It is 80 km south of the Korean mainland between the East China Sea and the Korea Strait. The World Heritage site located on the island is made up of three segments of land and sea, having an area of 93.7 km^2 (UNESCO WHC 2017) (Fig. 72.1).

The beginning of the preservation of the volcanic features dates from 1962, when two areas of lava tubes were declared a national monument by the Korean government. Between 1962 and 2006, other segments were preserved, and some national parks were created. In 2002, the central island was designated a UNESCO Biosphere Reserve, and in 2005, three segments were declared a World Heritage site (UNEP/WCMC 2007).

The area has an ancient history. Many Paleolithic artifacts, such as stoneware, bronzeware, and ironware, have been discovered, and in one place, there are hominid fossil footprints belonging to late Neolithic humanoids who lived in the mid- to late Holocene (Sohn et al. 2012). There are also evidences of the occupation of the peninsula in the first centuries before Christ. A thirteenth-century Mongol invasion left it with a tradition of raising horses. Seaweed harvesting by its women is another distinctive tradition (Cultural Heritage Administration 2006).

V. Claudino-Sales, *Coastal World Heritage Sites*, Coastal Research Library 28,
https://doi.org/10.1007/978-94-024-1528-5_72

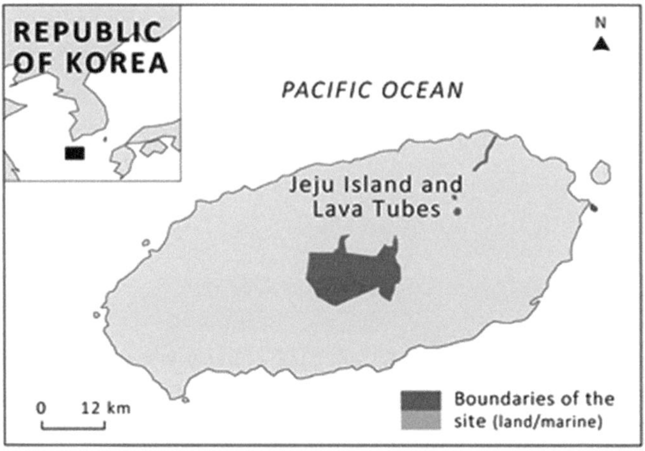

Fig. 72.1 Location of Jeju Island and Lava Tube, Korea

The segments of the island defined as World Heritage site are the Geomunoreum Lava Tube, the Seongsan Tuff Cone, and the volcanic peak Mount Halla. No one lives in the core zones. The main occupations are cattle and horse-raising, fishing, and tourism – the island is a very popular tourist destination for Koreans. They come for the volcanic features, for the local culture, and for the biodiversity (UNEP/WCMC 2007).

The sites are property of the state and of private owners. They are under the supervision of the Cultural Heritage Administration through the Jeju Provincial Government and the mayors of two cities. They are managed by the Jeju World Natural Heritage Management Committee, but each site is separately managed by the board of the parks, preserves, and national monuments (UNESCO WHC 2017).

72.2 Natural Settings

72.2.1 Climatic Conditions

The island has a subtropical oceanic climate, moderated by a cold current coming from the south. The climate grades with altitude through warm temperate to subarctic. There are four distinct seasons (UNEP/WCMC 2007).

The average annual temperature of the site is 11.7 °C. Winters are short and relatively warm, with only 17 days of frost. Rainfall averages 2044 mm, rising with height. Snow falls between November and May. Winds are strong, especially from the northeast in winter. Typhoons are common on the island, arriving in a frequency of more than two a year, and mostly during the late summer monsoon season (Cultural Heritage Administration 2006).

72.2.2 Geology and Geomorphology

Jeju Island is an intraplate basaltic volcano built on the 100-m-deep East China Sea continental shelf off the southern coast of the Korean Peninsula (Brenna et al. 2012). The island was produced by widespread and prolonged hydrovolcanic activity between 1.8 and 0.5 million years, and later, lava effusion after the hydrovolcanic field emerged above the fluctuating Quaternary sea level (Sohn and Park 2004). The former activity produced a subsurface hydrovolcanic succession, whereas the latter produced the lava shield dotted with hundreds of volcanic cones. Several tuff rings and tuff cones were produced on top of the lava shield in the Late Pleistocene and Holocene (Cheong et al. 2007).

A mantle plume was proposed to explain Jeju volcanism (Tatsumi et al. 2005), but the spatial-temporal, geochemical, and volumetric evolution of the magmatic system makes a traditional hot plume model an unlikely driver of Jeju volcanic activity. It seems, in reality, that the activation of shear zones close to the area (those associated with the subduction zones around Japan) allowed upward movement of some mantle pockets, which subsequently partially melted and caused the volcanism on Jeju (Brenna et al. 2015).

Even if the eruptive flux was low, the successive eruptions on the flank of the volcano have formed 120 large lava tubes and 360 parasitic cinder and scoria cones all over the island. The tubes contain a variety of both lava and carbonate speleothems, some formed, unusually, by secondary carbonates leached from overlying sand dunes (Go et al. 2017; Cultural Heritage Administration 2006). The eruptions also produced a large cliff-like cone of tuff offshore, called Songaksan Ilchulbong. Songaksan Ilchulbong is the youngest volcanic center on Jeju Island and was produced by an eruption 3700 in a coastal setting (Sohn et al. 2002) (Fig. 72.2).

The area of lava tubes included in the World Heritage site, Geomunoreum Lava Tube System, was formed 300,000–100,000 years ago, comprising a scoria cone and eight large lava tubes along one segmented line of flow to the sea. It represents the world's largest number of cones to be derived from a single volcano. The tubes contain a great variety of lava and carbonate speleothems, the latter normally found only in limestone caverns. It is regarded as the finest such cave in the world (UNEP/WCMC 2007).

The other feature included in the site is the volcanic peak Mount Halla, the highest point in Korea. It represents the main dome in the middle of the island, surrounded by a plateau and lava plains spattered with 368 parasitic cinder and scoria cones formed by fountains of lava (Cultural Heritage Administration 2006). Mount Halla has multiple waterfalls, multi-shaped rock formations, a lake-filled crater and displays columnar-jointed cliffs (UNEP/WCMC 2007).

The volcanic features in the three components of the World Heritage site and their great accessibility create a distinctive and important contribution to the understanding of global volcanism and of the Surtseyan type of volcanic eruptions in particular and also to the comprehension of the planet's dynamics (UNESCO WHC 2017).

Fig. 72.2 The fortress-like Seongsan Ilchulbong tuff cone rising out of the East China Sea, offshore of Jeju Island. (Picture reproduced by permission of Dave Warburton, website http://freelanceflaneur.blogspot.com)

72.2.3 *Biodiversity*

The fauna and flora of Jeju Island, extremely diverse for such a small area, are important elements of the World Heritage site, showing influences from East Asiatic, East Siberian, and Indo-Malaysian domains, besides being influenced by the characteristics of the island (Oh 2002).

Jeju Island as a whole has 1453 vascular plant species with 2 subspecies, 275 varieties, and 65 forms, totaling 1795 taxa in 214 genera. These include 344 angiosperms, 12 being endemic, and 6 gymnosperms. Ninety-six vascular species and 2 genera are endemic; 59 mostly alpine or arctic plants are found on the mountain (Kang 1992).

Mount Halla rises from subtropical through temperate to subalpine zones. Above 1700 m, there are alpine relicts which have speciated by adapting to the harsh climate of very strong winds and the dry soils derived from volcanic ash. On the tuff cone, there are endemic flowers, and more than 300 marine plants are recorded on the coast a (Lee 1985) (Fig. 72.3).

Continental, Japanese, and southern characteristics coexist in the island's fauna. There are 19 species of mammals, mostly rodents and bats. On the Ilchulbong tuff cone, 220 species of land animals have been recorded. There are 236 species of birds (Cultural Heritage Administration 2006).

As with the flora, most animal species endemic to Jeju live on the higher levels of Mount Halla. However, guano and organic sediments, especially near breaks in

Fig. 72.3 Volcanic tuff outcrops in Seongsan Ilchulbong tuff cone, with an iconic symbol of Jeju Island, the yellow flower rapeseed (*Brassica napus*). (Picture reproduced by permission of Maxim Tupikov/Shutterstock Stock Photos)

the tube caves, support troglodytic life. For instance, the Yongcheon cave lake contains copepod crustaceans which have not yet been closely studied (UNEP/WCMC 2007).

72.3 Conservation Challenges

The sites are well managed and resourced. Key management issues include avoiding potential agricultural impact on the underground and managing the high number of visitors (UNESCO WHC 2017). Despite these potential threats, the state of conservation of the composite site is good (IUCN 2014). Climate change and global warming, nevertheless, impose some concerns, related to the number of typhoons annually, what can damage the lava tubes, and coastal erosion.

The last report published by the International Union for Conservation of Nature (IUCN) about the state of conservation of the World Heritage sites published at the end of 2017 confirmed that the level of conservation of Jeju Volcanic and Lava Tube site is good without concerns (Osipova et al. 2017). The outstanding universal value of the site is under no immediate threat and generally effectively managed and protected. The inscription of the site in the World Heritage List has brought considerable benefits to the local community through employment and through promotion of tourism (IUCN 2017).

References

Brenna M, Cronin SJ, Smith IEM, Sohn YK, Maas R (2012) Spatio-temporal evolution of a dispersed magmatic system and its implications for volcano growth, Jeju Island volcanic field, Korea. Lithosphere 148:337–352

Brenna M, Cronin SJ, Kereszturi G, Sohn YK, Smith IEM, Wijbrans J (2015) Intraplate volcanism influenced by distal subduction tectonics at Jeju Island, Republic of Korea. Bull Volcanol 77:7

Cheong CS, Choi JH, Sohn YK, Kim JC, Jeong GY (2007) Optical dating of hydromagmatic volcanoes on the southwestern coast of Jeju Island, Korea. Quat Geochronol 2:266–271

Cultural Heritage Administration (2006) Jeju Volcanic Island and lava tubes. A candidate for world heritage inscription. Report, Seo-gu, Daejeon, Korea

Go SY, Kim GB, Jeong JO, Sohn YK (2017) Diatreme evolution during the phreatomagmatic eruption of the Songaksan tuff ring, Jeju Island, Korea. Bull Volcanol 79:1–26

IUCN (International Union for Conservation of Nature (2014) IUCN World Heritage Outlook 2014. Report, Gland, Switzerland

IUCN (International Union for Conservation of Nature) (2017) Jeju Volcanic Island and Lava Tube Conservation Outlook. http://www.worldheritageoutlook.iucn.org/node/1146. Accessed 28 Dec 2017

Kang S (1992) Natural environment of Jeju Island. Report, Seoul, Korea

Lee C (1985) Native and rare plants of Mt. Halla. Scientific Survey Report, Cheju Province, Korea

Oh H (2002) Report of survey of natural resources in Hallasan National Park. National Parks Authority of Korea report, Seoul, Korea

Osipova E, Shadie P, Zwahlen C, Osti M, Shi Y, Kormos C, Bertzky B, Murai M, Van Merm R, Badman T (2017) IUCN world heritage outlook 2: a conservation assessment of all natural world heritage sites. IUCN, Gland

Sohn YK, Park KH (2004) Early-stage volcanism and sedimentation of Jeju Island revealed by the Sagye Borehole, SW Jeju Island, Korea. Geosci J 8:73–84

Sohn YK, Park JB, Khim BK, Park KH, Koh GW (2002) Stratigraphy, petrochemistry and quaternary depositional record of the Songaksan tuff ring, Jeju Island, Korea. J Volcanol Geotherm Res 119:1–20

Sohn YK, Cronin SJ, Brenna M, Smith IEM, Nemeth K, White JDL, Murtagh RM, Jeon YM, Kwon CW (2012) Ilchulbong tuff cone, Jeju Island, Korea, revisted: a compound monogenetic volcano involving multiple magma pulses, shifting vents, and discrete erutpive phases. Geol Soc Am Bull 124(3–4):259–274

Tatsumi Y, Shukuno H, Yoshikawa M, Chang Q, Sato K, Lee MW (2005) The petrology and geochemistry of volcanic rocks on Jeju Island: plume magmatism along the Asian continental margin. J Petrol 46:523–553

UNEP/WCMC (United Nations Environmental Programme/World Conservation Monitoring Centre) (2007) Jeju Volcanic Island and lava tubes, South Korea. Report, Cambridge, England

UNESCO WHC (United Nations Educational, Scientific and Cultural Organization World Heritage Centre) (2017) Jeju Volcanic Island and Lava Tubes. http://whc.unesco.org/en/list/1264. Accessed 27 June 2017

Part XV
World Coastal Heritage in Islands, Gulfs and Reefs: Mediterranean Sea

Chapter 73
Gulf of Porto: Calanche of Piana, Gulf of Girolata, Scandola Reserve, France

Abstract This nature reserve on the west coast of Corsica, Mediterranean Sea, is centered on the gulfs of Porto and Girolata between the craggy Scandola Peninsula to the north and the turrets of the Calanche (a type of "fjord" of the Mediterranean Sea) of Piana to the south. The reserve is part of the Regional Natural Park of Corsica and is made by an impressive porphyritic rock mass. The coastline is known for its orange-red cliffs and rugged headlands up to 900 m high, sheltered sand beaches, and the coves and serrated red ochre pinnacles of the Calanche of the Piana coast – the livid, sheer, and jagged cliffs of the coast contain many grottos and are flanked by numerous stacks, islets, and coves. The vegetation is also outstanding and gives examples of Mediterranean maquis, a densely growing evergreen shrub. In addition, the clear waters, their islets, and inaccessible caves harbor a rich marine life.

73.1 Introduction

The site lies along the northwestern coast of Corsica, Mediterranean Sea, including the Scandola (Elbo) Peninsula and the Gulfs of Girolata and Porto. The site has 118 km^2 of area, with 72 km^2 terrestrial and 42 km^2 marine. The altitude of the site goes from −100 m to 900 m (UNESCO WHC 2017) (Fig. 73.1).

From 1943 to 1961, the two gulfs were declared a "classified site" by the French government. In 1972, the "Corsica Regional Natural Park" was created, including the two gulfs and the peninsula. In 1979, the "Scandola Nature Reserve" was created. The area related to the gulfs and peninsula was declared a World Heritage site in 1983 (IUCN 1983).

In Corsica, there are sites of prehistoric pastoralists. The coastal and near coastal zone has been the focus for habitation since the Early Neolithic, but the interior was not extensively exploited until the recent past. There are also evidences of occupa-

The original version of this chapter was revised. A correction to this chapter is available at https://doi.org/10.1007/978-94-024-1528-5_85

V. Claudino-Sales, *Coastal World Heritage Sites*, Coastal Research Library 28,
https://doi.org/10.1007/978-94-024-1528-5_73

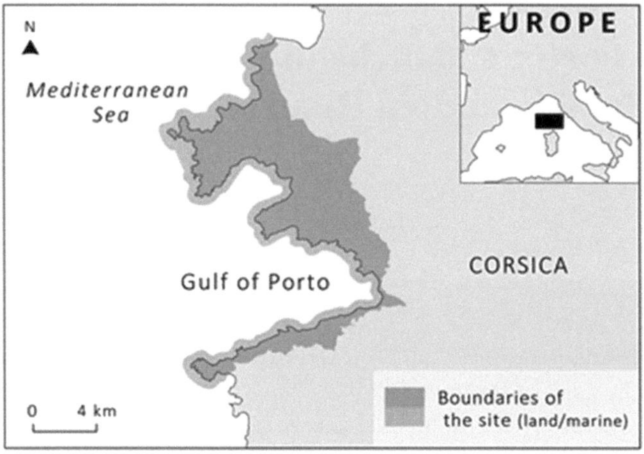

Fig. 73.1 Location of Gulf of Porto World Heritage site, in Corsica, France

tion during the Roman period, in addition to coastal forts built in the sixteenth and seventeenth centuries (Llobera et al. 2010).

The island has been French since 1798, but preserves Corsican character. The area conserves traditional agriculture and grazing, which used to include transhumance from the mountain pastures in winter. There are several small villages within the site. Since the turn of the twenty-first century, the area has been depopulated, except for the villages along the main road, and many of the terraces around the villages have been abandoned (UNEP/WCMC 2012).

The ownership of the land is mainly communal, although some sections of Scandola Reserve are in state and private ownership. The park and reserves are administered by the “Parcs Naturels” under the Ministry of the Environment of France. The site lies within a Conservation International Hotspot, a WWF Global 200 Ecoregion and a WWF/IUCN Center of Plant (UNEP/WCMC 2012).

73.2 Natural Settings

73.2.1 Climatic Conditions

Corsica is surrounded by water and soaks in the warmth of the Mediterranean climate. The littoral is characterized by a microclimate controlled by the presence of the sea, with precipitation around 650 mm a year and mild temperatures all year (Regions of France 2017).

The average maximum temperature is 28 °C, between July and August. The average minimum temperatures are 3 °C in winter. The winds are not too strong, with average around 3 m/s all year (Weather and Climate 2017).

73.2.2 Geology and Geomorphology

Politically, the Island of Corsica belongs to France, but geologically it forms the northern part of the Corsardinian Massif, a microplate which was separated from the European continent during the middle Tertiary (Gibbons 1985).

The island is located in the center of the western Mediterranean Sea, a marginal basin generated by the N-NW subduction of the African Plate beneath the European Plate, in the so-called Alpine Orogeny. Oceanic accretion occurred between 21 and 18 million years ago, following an Oligocene rifting phase (Rehault et al. 1984). This process produced two back-arc basins in the area, the Island of Corsica being between them (Platt 2007). It thus occupies a unique position in the western Mediterranean, since it has recorded both the Cenozoic Alpine orogenic history of the area as well as subsequent extensional collapse and oceanic basin formation (Zarki-Jakni et al. 2004).

Corsica is broadly separable into two unequal geological units (Durand-Delga 1978): to the west and south is "Hercynian Corsica" (e.g., segment resulting from the orogeny which formed the continent Pangea), dominated by Carboniferous granitic rocks topped with a Permian-Eocene sedimentary cover (Malasoma and Marroni 2007; Rossi et al. 2003), whereas to the northeast are polydeformed schists belonging to "Alpine Corsica," related to the Alpine Orogeny, made up of a nappe stack largely composed of ophiolitic rocks with associated metasedimentary cover and slivers of continental basement (Maggi et al. 2012). The contact between these two major units is tectonic, being marked by a major shear zone produced as the Alpine schists were obducted on to the Hercynian basement (Mattauer et al. 1981).

The Eocene sediments that unconformably overlapped the Hercynian crystalline basement also indicate the presence of relict elevated planation surfaces. Protected by a sedimentary cover, the planation surface was preserved. In the latest Oligocene to Miocene times, the surface was reexposed and offset by reactivated faults, with individual basement blocks differentially uplifted in several phases to elevations of, in some cases, >2 km. Currently, the planation surface remnants occur at different altitudes and with variable tilt (Danišík et al. 2012).

Three great ridges of the mountainous landscape divide the World Heritage site into three distinct segments: (1) the coast between Punta Stollo and Punta Muchillina, including the Peninsula of Scandola; (2) the Gulf of Girolata from Punta Scandola to Capo d'Osani, on Cape Senino; and (3) the Gulf of Porto from Capo Senino to Punta Bianca (Fig. 73.2) plus its southern coast to Capo Rosso including the calanche (drowned ravines, resulting from Quaternary sea level changes; they are a type of fjords or rias of the Mediterranean Sea) below Piana (Desjeux and Desjeux 1984).

The coastline is characterized by orange-red cliffs and craggy headlands up to 900 m high and also by the presence of sheltered sand beaches and the serrated red ochre pinnacles of the Calanche of the Piana coast, all much eroded by storms (Figs. 73.2 and 73.3). The sheer and jagged seacliffs contain many grottos and are flanked by numerous stacks, islets, and coves almost inaccessible except from the sea (Féderation des Parcs Naturels de France 1982).

Fig. 73.2 Gulf of Porto, sustained by metamorphic rocks of Late Permian, with rugged and partially vegetated high seacliffs and sheltered dark sandy beach. (Picture reproduced by permission of Luigi Tura, webpage www.wnhs-experience.org)

Fig. 73.3 Capo Rosso rocks with an arch produced by wave erosion. The rocks make part of the Calanques de Piana. (Picture reproduced by permission of Sasha64f/Shutterstock Stock Photos)

73.2.3 *Biodiversity*

The natural vegetation is Mediterranean dry forest. The coastal plains are part of the Tyrrhenian-Adriatic sclerophyllous and mixed forest ecoregion, of forests and woodlands of evergreen oaks and pines. Typical Mediterranean maquis vegetation grows on the more gentle slopes. The coastal area around Scandola has open cliff-top grass swards and tall maquis (a densely growing Mediterranean evergreen shrub) degraded by past fires or ancient pasturage (Gamizans and Marzocchi 1996).

The littoral zone is dominated by sea lavender. All the 450 species of marine algae known in this part of the Mediterranean are found, including *Halimeda* and several species such as red algae unrecorded anywhere else in France. Because of the highly transparent waters, meadows of the sublittoral are found as deep as 35 m; also calcareous alga are present where the substrate is rocky, and spectacular coralligenous concretions can be found over large areas (Aymonin 1975).

The Scandola Nature Reserve contains a rich resident and migrant fauna, and the avifauna is very important. There are numerous species of snakes, lizards, and geckos. The marine fauna contains a wide range of littoral and sublittoral invertebrates and the precious Mediterranean red coral. The bottlenose dolphin is seen in the surrounding seas. The Mediterranean monk seal once had a colony near Galeria, but it has disappeared. The strong currents offshore attract a large number of pelagic fish (UNEP/WCMC 2012).

73.3 Conservation Challenges

The International Union for Conservation of Nature, in its last assessments about the conservation of all natural World Heritage sites (IUCN 2014, 2017), found the state of conservation of this site as "good with some concerns." The concerns are linked especially to the rapid expansion of tourism, which poses a threat to both cultural and natural elements. Coastal industries are also polluting the Scandola coastline. UNESCO WHC (2018) considers that the Government of France is working to reverse these threats. Nevertheless, there are also risks of exploitation of liquid or gaseous hydrocarbons in the vicinity of the site (UNESCO WHC 2015).

The pluviometry in Corsica is not very high, with values under 700 mm a year, which characterizes a semiarid region. With the global warming and climatic changes coming, it is very possible that the region will become drier, which would be a threat to fauna, flora, and society. The rising of sea level linked to these conditions would also increase coastal erosion, and part of the features that characterizes the island morphology (calanches, grottos, pinnacles, stacks, coves) possibly will disappear.

From a general point of view, nevertheless, the site values are good. Protection is effective, but requires planning and reinforcement throughout the site, in particular to manage the marine part (IUCN 2017; Osipova et al. 2017).

References

Aymonin GG (1975) La nature Corse: menaces et espoirs. Bulletin de la Societe Botanique de France 121:5–8

Danišík M, Kuhlemann J, Dunkl I, Evans NJ, Székely B, Frisch W (2012) Survival of ancient landforms in a collisional setting as revealled by combined fission track and (U-Th)/He thermochronometry: a case study for Corsica. J Geol 120(2):87–98

Desjeux C, Desjeux B (1984) Les Parcs Naturels Régionaux de France. Ed. Créer, Nonette

Durand-Delga M (1978) Corse, Guides Geologiques Regionaux. Masson, Paris

Fédération des Parcs Naturels de France (1982) Les Parcs Naturels de Régioneaux et la Protection de la Nature. Ministère de l'Environnement, Report, Paris, France

Gamisans J, Marzocchi J (1996) La Flore Endémique de la Corse. EDISUD, Aix-en-Provence, France

Gibbons W (1985) Corsican geology: an excursion guide based on G.A. Trip to Corsica June 25th – July 2n 1983. Proc Geol Assoc 96:171–182

International Union for Conservation of Nature (1983) World Heritage Nomination IUCN Technical Review Capes Girolata and Porto and Scandola Nature Reserve. Report, Gland, Switzerland

IUCN (International Union for Conservation of Nature) (2014) IUCN World Heritage Outlook 2014. Rappord, Gland, Switzerland

IUCN (International Union for Conservation of Nature) (2017) Gulf of Porto, Calanche of Piana, Gulf of Girolata, Scandola Conservation Outlook. http://www.worldheritageoutlook.iucn.org/node/999. Accessed 28 Dec 2017

Llobera M, Wilkinson KN, Weiss MC, Flaming RJ, Marini NAF, Mazet S (2010) Into the Maquis Methodological and Interpretational Challenges in Surveying La Balagne, Northwest Corsica. J Mediterr Archeol 23:169–197

Maggi M, Rossetti F, Corfu F, Theye T, Torgeir BA, Faccenn C (2012) Clinopyroxene–rutile phyllonites from the East Tenda Shear Zone (Alpine Corsica, France): pressure –temperature–time constraints to the Alpine reworking of Variscan Corsic. J Geol Soc 169:723–732

Malasoma A, Marroni M (2007) HP/LT metamorphism in the Volparone Breccia (Northern Corsica, France): evidence for involvement of the Europe/Corsica continental margin in the alpine subduction zone. J Metamorph Geol 25:529–545

Mattauer M, Faure M, Malavielle J (1981) Transverse lineation and large-scale structures related to Alpine obduction in Corsica. J Struct Geol 3:401–409

Osipova E, Shadie P, Zwahlen C, Osti M, Shi Y, Kormos C, Bertzky B, Murai M, Van Merm R, Badman T (2017) IUCN world heritage outlook 2: a conservation assessment of all natural world heritage sites. IUCN, Gland

Platt JP (2007) From orogenic hinterlands to Mediterranean-style back-arc basins: a comparative analysis. J Geol Soc 164:297–311

Regions de France (2017) Corsica weather and climate. http://www.regions-of-france.com/regions/corsica/weather/. Accessed 29 Aug 2017

Rehault JP, Boillot G, Maufret A (1984) The western Mediterranean Basin geological evolution. Mar Geol 55:447–477

Rossi PH, Durand-Delga M, Lahondere JC, Lahandore D (2003) Carte géologique de la France à 1/50.000, feuille Santo Pietro di Tenda. BRGM, Orleans

UNEP/WMCM (United Nations Environmental Programme/World Conservation Monitoring Centre) (2012) Gulf of Porto: Calanche of Piana, gulf of Girolata, Scandola reserve, France. Report, Cambridge, England

UNESCO WHC (United Nations Educational, Scientific and Cultural Organization World Heritage Centre) (2017) Gulf of Porto: Calanche of Piana, Gulf of Girolata, Scandola Reserve. http://whc.unesco.org/en/list/258. Accessed 4 Sept 2017

UNESCO WHC (United Nations Educational, Scientific and Cultural Organization World Heritage Center) (2015) Golfe de Porto: calanche de Piana, golfe de Girolata, réserve de Scandola (France) Sommaire de l'État de conservation 2015 – Décision 37 COM 7B.19. Report, Paris, France

UNESCO WHC (United Nations Educational, Scientific and Cultural Organization World Heritage Centre) (2018) Gulfo of Porto: Calanche of Piana, Gulf of Girolata, Scandola Reserve. http://whc.unesco.org/en/list/258/documents/. Accessed 17 Jan 2018

Weather and Climate (2017) Climate in Corsica. http://www.regions-of-france.com/regions/corsica/weather/. Accessed 29 Aug 2017

Zarki-Jakni B, van der Beek P, Poupeau G, Sosson M, Labrin E, Rossi P, Ferrandini J (2004) Cenozoic denudation of Corsica in response to Ligurian and Thyrrhenian extension: results from apatite fission track thermochronology. Tectonics 23:123–135

Chapter 74
The Aeolian Islands, Italy

Abstract The Aeolian Archipelago lies in the Tyrrhenian Sea between Italy and Sicily. It consists of seven main islands, all steep-sided volcanoes either active or dormant, with six offshore islets. The volcanic landforms represent classic features of two of the six types of volcanic eruptions, the "Vulcanian" and the "Strombolian," with a history of volcanic island building and destruction and ongoing volcanic phenomena. The easily accessible occurrence within a small area of so many volcanic processes and landforms provides a natural museum and study site. The islands have been well studied, documented, and monitored since the eighteenth century and continue to provide a rich field for volcanological studies and related geomorphology.

74.1 Introduction

The Aeolian Islands – also known as "the Lipari Islands" – lie in the Tyrrhenian Sea, 20 km off the north coast of Sicily and 55 km from the Italian mainland, in the Mediterranean Sea. The group consists of seven main islands and six islets. The site has 12.6 km^2 of area (UNESCO WHC 2017) (Fig. 74.1).

In 1966, the "Lipari Reserve" was created, including the islands of Lipari, Alicudi, Filicudi, Panarea, Vulcano, and Stromboli. In 1979, a "Landscape Protection" decree was announced to Salina Island. In 1991, the islands of Alicudi, Filicudi, Panarea and Stromboli were designated natural reserves by the Sicilian Regional Authority for Territories and Environment. In 1998, a heritage management consortium was founded by the municipalities of Lipari and Salina. The archipelago was added to the World Heritage site list in 2000 (UNEP/WCMC 2011).

The islands lie in the Sicilian province of Messina. Most land is in private ownership, with some regional and state ownership of protected lands. The site is administered by the "National Park of the Aeolian Islands" (UNESCO WHC 2010).

The archipelago has noteworthy archaeological and ethno-anthropological features. Archaeological remains reveal a Neolithic and Bronze Age occupation and

V. Claudino-Sales, *Coastal World Heritage Sites*, Coastal Research Library 28,
https://doi.org/10.1007/978-94-024-1528-5_74

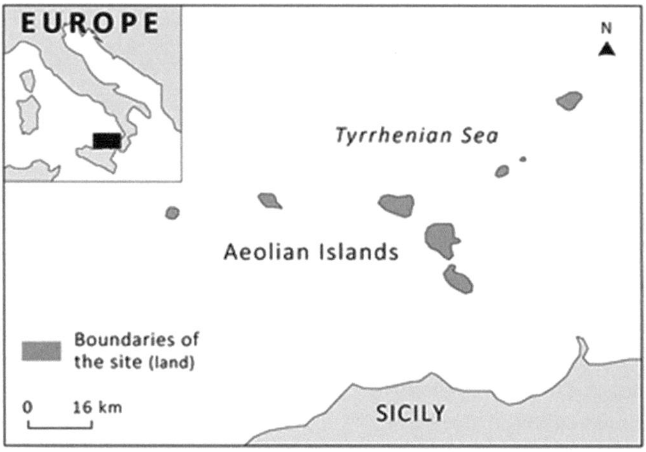

Fig. 74.1 The Aeolian Islands, located in the Mediterranean Sea (Thyrrhenian Sea, Italy)

subsequently the presence of Greeks, Phoenicians, Romans, Arabs, Normans, and Spanish. The islands were terraced for the cultivation of wheat and olives but were abandoned in the late nineteenth and early twentieth centuries. Today, there are several settlements but few inhabitants, due to the steepness of the slopes and risk of volcanic activity (UNEP/WCMC 2011).

The islands have been well studied, documented, and monitored since the eighteenth century and are the classic examples for two of the six types of volcanism, the "Vulcanian" and "Strombolian." The easily accessible occurrence within a small area of so many volcanic processes, landforms, and ongoing volcanic phenomena provides a natural volcanological museum and study site (UNESCO WHC 2017). Tourism is also an important activity in the archipelago, attracting more than 300,000 people a year (Italian Tourism Government Tourism Board 2017).

74.2 Natural Settings

74.2.1 Climatic Conditions

The climate is Mediterranean, moderated by the sea. Winter is mild and relatively rainy, while summer is hot and sunny, tempered by the breeze. In spring, morning fog can form. Although temperatures are generally mild, the wind, which blows especially in the cold half of the year, can increase the feeling of cold (Meteoblue 2017).

The average annual rainfall varies from 600 to 700 mm, and the average annual temperature varies between 14 and -18 °C. The average wind speed is 5.5 m/s, and the winds come mostly from the northwest (UNEP/WCMC 2011).

74.2.2 Geology and Geomorphology

The seven volcanic islands of the Aeolian Archipelago – Lipari, Vulcano, Salina, Stromboli, Filicudi, Alicudi, and Panarea (Figs. 74.2 and 74.3) – are the emerged portion of a half-ring-shaped volcanic complex that also includes seven seamounts (Albert et al. 2017). Magmatism in the region stared at around 1.3 million years ago (Beccaluva et al. 1985). The geodynamic setting responsible for this volcanic activity is still debated, but it seems possible that the magmatism is the result of active subduction of the African Plate beneath the Arc of Calabria (Mandarano et al. 2016; Francalanci et al. 2013).

In reality, the magmatism on the archipelago reflects a combination of tectonic related influences, including back-arc extension and subduction with a remnant plate present beneath the islands (Chiarabba et al. 2008). Consequently, the effusive and explosive eruptions on the islands have produced a large variety of products with differing magmatic associations (Pecerillo et al. 2013).

The reduced dimensions of the volcanic arc in relation to the great Pacific arcs and the reduced energy of the phenomena give the possibility of using the Aeolian Islands as an anatomical and physiological model for interpretation of magmatic arcs. The two volcanoes still active, Vulcano and Stromboli, identify in international terminology two typical volcanic activities, called "Vulcanian" and "Strombolian." Stromboli has continuous regular spurts of lapilli, bombs, and blazing gas, and

Fig. 74.2 Feature in arc produced by marine erosion in the volcanic rocks of Alicudi Island, in the Aeolian Archipelago. (Picture reproduced by permission of Luigi Tura, webpage www.wnhs-experience.org)

Fig. 74.3 Volcanic island of Lipari, with the municipality of Lipari at the foot of the dormant volcano, in the Aeolian Archipelago. (Picture reproduced by permission of Luigi Tura, web page www.wnhs-experience.org)

Vulcano presents explosions of greater violence at longer intervals, projecting a high cloud of ash (UNEP/WCMC 2011).

Both volcanoes remain a risk: although Vulcano last erupted in 1888–1990, Stromboli is in a permanent state of low-level eruption and in 2002, 2004, and 2007 erupted with greater force, causing local tsunamis. Seismic activity also occurs in the north of Salina Island at Pollara and Malfa, on Alicudi and Filicudi islands, and at Vulcanello on Vulcano. A geothermal field underlies the group, and there are fumaroles and hot springs, particularly on Vulcano island. The volcanic soils are fertile but often on steeply sloping ground (Sopritendenza di Messina 1999).

From a lithological point of view, the islands are a complete museum of all types of volcanic rocks: there are effusive types as well as the lightest pyroclastic rocks of subaerial deposit or flux produced by the explosive hydromagmatic activity. In terms of geomorphology, the islands include archipelago, island, isthmus, headland, promontory, gulf, bay, mountain, hills, volcano, crater, and volcanic scores. With the action of exogenous agents, this ensemble was shaped as a remarkable landscape (Luchi et al. 2013).

74.2.3 Biodiversity

The terrain provides a variety of microclimates that host many species found on mainland Italy and Sicily. Most areas are dominated by a man-modified landscape. The many abandoned terraces are reverting to Mediterranean maquis (a type of Mediterranean shrub) and being dominated by plants typical of this vegetation. Nine hundred species are recorded, including 33 endemics (UNEP/WCMC 2011).

The most interesting characteristic of the Aeolian fauna is the presence of continental European species. Seven-four species endemic or subendemic to Italy are listed, 31 local to the Aeolian Islands and several confined to a single island. Mammals include one endemic subspecies of garden dormouse and bats. Seven reptile species are present, and over 40 birds are recorded. The invertebrate fauna is well known, and the assemblage of marine fish is remarkable. The wealth of fauna of the islands is due mainly to the presence of seagrass that is still widespread along the coasts (Lo Cascio 1998).

74.3 Conservation Challenges

Most of Aeolian Islands' important habitats are threatened by activities connected to high tourism. Tourism has changed the originality of the islands and caused the abandonment of farming, and now the islands' food supplies depend totally on imports. This has caused a loss of the rural landscape and also has brought an increase in maritime traffic, disturbing the environment (Aeolian Preservation Funds Organization 2017).

Global warming and climate changes can produce drier weather, threatening fauna and flora of an area that already has low annual pluviometry. Coastal erosion is also predicted to be higher, threating coastal morphological features and settlements.

The International Union for Conservation of Nature (IUCN) released a report at the end of 2017 about the state of conservation of the natural World Heritage sites, in which The Aeolian Islands appear as "good with some concerns" (Osipova et al. 2017). The concerns are all the more noteworthy in light of the major threat of mass tourism, which has been compromising the conservation of the site's natural heritage and threatening the islands' notable endemic species (IUCN 2017).

References

Aeolian Preservation Funds Organization (2017) The Aeolian Sea: treasures and threats. https://www.aeolianpreservationfund.org. Accessed 28 Dec 2017

Albert PG, Tomlinson EL, Smith VC, Di Traglia F, Pistolesi M, Morris A, Donato P, De Rosa R, Sulpizio R, Keller J, Rosi M, Menzies M (2017) Glass geochemistry of pyroclastic deposits from the Aeolian Islands in the last 50 ka: a proximal database for tephrochronology. J Volcanol Geotherm Res 336:81–107

Beccaluva L, Gabbianelli G, Lucchini F, Rossi PL, Savelli C (1985) Petrology and K/Ar ages of volcanics dredged from the Aeolian seamounts: implications for geodynamics evolution of the southern Tyrrhenian basin. Earth Planet Sci Lett 74:187–208

Chiarabba C, De Gori P, Speranza F (2008) The southern Tyrrhenian subduction zone: deep geometry, magmatism and Plio-Pleistocene evolution. Earth Planet Sci Lett 268:408–423

Francalanci L, Lucchi F, Keller J, De Astics G, Tranne CA (2013) Eruptive, volcano-tectonic and magmatic history of Stromboli Volcano (northeastern Aeolian archipelago). In: Lucchi F, Peccerillo A, Keller J, Tramme CA, Rossi PL (eds) The Aeolian Islands. Geological Society of London Memoirs, London, pp 397–471

Italian Tourism Government Board (2017) The Aeolian Islands, Sicily. http://www.italiantourism.com/island6a.html. Accessed 5 Sept 2017

IUCN (International Union for Conservation of Nature) (2017) The Aeolian Islands (Isole Eolie) Conservation Outlook. http://www.worldheritageoutlook.iucn.org/node/1115. Accessed 28 Dec 2017

Lo Cascio P (1998) Aspects Fauniques et Zoogéographiques de l'Archipel Eolien. Report, Messina, Italy

Luchi F, Peccerillo A, Keller J, Tranne CA, Rossi PL (2013) The Aeolian Islands. Memoirs of the Geological Society of London 37

Mandarano M, Paonita A, Martelli M, Viccaro M, Nicotra E, Millar IL (2016) Revealing magma degassing below closed-conduit active volcanoes: geochemical features of volcanic rocks versus fumarolic fluids at Vulcano (Aeolian Islands, Italy). Lithos 249-251:272–287

Meteoblue (2017) Climate Aeolian Islands. https://www.meteoblue.com/en/weather/forecast/modelclimate/stromboli_italy_6691846. Accessed 5 September 2017

Osipova E, Shadie P, Zwahlen C, Osti M, Shi Y, Kormos C, Bertzky B, Murai M, Van Merm R, Badman T (2017) IUCN world heritage outlook 2: a conservation assessment of all natural world heritage sites. IUCN, Gland

Pecerillo A, De Astis G, Faraone D, Forni F, Frezzotti ML (2013) Compositional variation of magmas in the Aeolian arc: implications for petrogenesis and geodynamics. In: Luchi F, Peccerillo A, Keller J, Tranne CA, Rossi PL (eds) The Aeolian Islands. Memoirs of the Geological Society of London, London, p 37

Sopritendenza di Messina (1999) Iles Eoliennes ou Lipari. Patrimoine Naturel et Culturel. Report, Messina, Italy

UNEP/WCMC (United Nations Environmental Programme/World Conservation Monitoring Centre) (2011) Isole Eolie (Aeolian Islands), Italy. Report, Cambridge, England

UNESCO WHC (United Nations Educational, Scientific and Cultural Organization World Heritage Centre) (2010) Thirty-fourth session of the Committee. Report, Brasilia, Brazil

UNESCO WHC (United Nations Educational, Scientific and Cultural Organization World Heritage Centre) (2017) Isole Eolie (Aeolian Islands). http://whc.unesco.org/en/list/908. Accessed 6 Sept 2017

Chapter 75
Ibiza, Spain

Abstract Ibiza Island is a mixed natural and cultural site. Its natural importance is from the evolution of the island's shoreline, controlled by the influence of *Posidonia oceanica* on the interaction of coastal and marine ecosystems. *Posidonia* is a seagrass found only in the Mediterranean Basin and is threatened in most Mediterranean locations. The local well-preserved *Posidonia*, together with the flanking salt pan lagoons, contains and supports a diversity of marine life. The coastal area also shows impressive seacliffs and pebbles beaches. From the cultural perspective, Ibiza's archaeological sites show the important role played by the island during the Phoenician-Carthaginian period. A fortified town is also an outstanding example of Renaissance military architecture, very influential on the fortification of ports in the Spanish Americas.

75.1 Introduction

The site is located in the western Balearic Islands of Ibiza and Formentera in the western Mediterranean, 90 km east of the Spanish mainland. It has 85.6 km^2 of area, with approximately 5.5% as land, 4.5% as wetland, and 90% as marine (Fig. 75.1).

The area has been the object of preservation for many decades. In 1989, the Salinas of Ibiza and Formentera were designated a natural reserve by the Spanish government and declared a Special Protection Area under the European Community Birds Directive in 1992. In 1995, the Salinas of Ibiza and Formentera Terrestrial and Marine Natural Reserve was created. In 2000, the natural reserve became a natural park. In 2008, its largest lagoon was declared a national biosphere reserve. The park became a mixed World Heritage site by UNESCO in 1999 (IUCN 2017).

The importance of the culture is shown by the Phoenician ruins of Sa Caleta and the Phoenician-Punic cemetery of Puig des Molins that are exceptional evidence of urbanization and social life in the Phoenician colonies of the western Mediterranean. The Upper Town of Ibiza is an excellent example of a fortified acropolis which preserves in an exceptional way in its walls and in its urban fabric successive imprints of the earliest Phoenicians settlements and the Arab and Catalan periods through to the Renaissance bastions (UNESCO WHC 2017).

V. Claudino-Sales, *Coastal World Heritage Sites*, Coastal Research Library 28,
https://doi.org/10.1007/978-94-024-1528-5_75

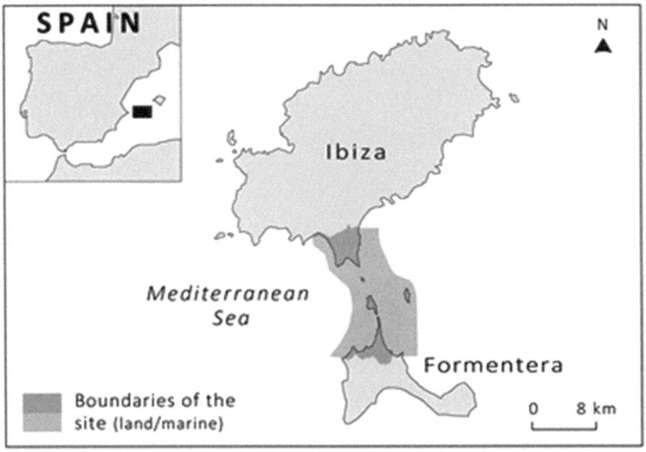

Fig. 75.1 Location of the mixed World Heritage site Ibiza, called Ibiza biodiversity and culture by UNESCO

The island also has more than ten underwater archaeological sites of the Late Bronze Age which have revealed the extent of early trade in the western Mediterranean. Most of them have not yet been adequately studied. The intact sixteenth century fortifications of Ibiza bear unique witness to the military architecture and engineering and the aesthetics of the Renaissance. This Italian-Spanish model was very influential, especially in the construction and fortification of towns in the New World (UNEP/WHMC 2011).

In the World Heritage site, the land tenure is publicly owned except for 16 km^2 of an area where there is a wetland reserve, which is private property. The management authority is the General Directorate for Nature Conservation under the Balearic Government Council for the Environmental Management of Lands and Shorelines and the National Ministry of Public Works and the Economy (UNEP/WHCM 2011).

The island is subject to intense tourism. In the last 50 years, holiday tourism has become the mainstay of the Ibizan economy and the island receives hundreds of thousands of visitors annually (Roig-Munar et al. 2013).

75.2 Natural Settings

75.2.1 Climatic Conditions

The climate of the island is semiarid with moderate average temperatures. The mean annual temperature is 19 °C, but it can freeze from December to March. Rainfall, usually torrential, is irregular, around 380 mm, falling mainly between August and December (UNESCO WHC 2010).

Annual evaporation is high year-round, with the water deficit spread over 9 months. The dominant winds are from the west and southwest in winter and from the east in summer. The area is characterized by a high level of insolation (2800 h per year) (UNEP/WCMC 2011).

75.2.2 Geology and Geomorphology

Ibiza is in the Mediterranean Basin, on the Iberian margin, in the Balearic Promontory. The Balearic Promontory consists of three thrust faulted and folded blocks with a complex geological evolution (Acosta et al. 2001). The Ibiza block is characterized by the presence of an ~20 km thick crustal horst formed of Paleozoic rocks deformed during the Variscan Orogeny, which is a geologic event caused by Late Paleozoic continental collision between Euramerica and Gondwana to form the supercontinent of Pangaea. The horst is unconformably covered by about 1–3 km thick Mesozoic-Cenozoic cover (Etheve et al. 2016).

The area underwent an important Cenozoic tectonic evolution, linked to the back-arc opening of the northwestern Mediterranean due to the collision with blocks located between Africa and Europe. This Oligo-Miocene structural heritage gave rise to a continuous deformation with compression through the Quaternary in the Balearic area, responsible for the existence of offshore earthquakes (Maillard et al. 2013).

Most of the sea floor between the islands of Ibiza and Formentera is a wide submarine platform with a sandy substrate averaging 15 m deep, showing indications of structural control (Acosta et al. 2002). The islands' shorelines are mostly flat plain formed by Quaternary calcareous deposits, with two enclosing wooded hills, with a maximum altitude of 144 m (IUCN/ICOMOS 1999). It includes coastal plains, lagoons, marshes, salt flats, fixed and mobile dunes, seacliffs (Fig. 75.2), sand and pebble beaches, and underwater caves. There are also many lagoons, which have long been used commercially as salt pans (UNEP/WCMC 2011). In Formentera, there are coastal reefs 4 meters high, the highest known formation of this kind (San Félix 1997).

Seagrass prairies of *Posidonia* (Fig. 75.3), which exist exclusively on the Mediterranean Basin, exemplify the interaction of coastal and marine ecosystems: the ongoing accumulation and decomposition of their litter have led to development of the sandy beaches of the site (UNESCO WHC 2017). The reef aggradation of its litter, besides stabilizing and building beaches, buffers the coast and coral reefs from storms (IUCN 2017).

75.2.3 Biodiversity

The flora is strongly influenced by the proximity of the sea, the high humidity, salinity, and the mobility of the sand dunes, which leads to the formation of microclimates. Underwater, the vast, dense prairies of *Posidonia oceanica* seagrass,

Fig. 75.2 Seacliffs shaped on Cenozoic sandstones retreating by the combined action of sea and continental processes. On the foot, there is a wave-cut platform with pebbles. (Picture reproduced with permission of Dr. Miquel Mir-Gual, University of Balearic Islands)

threatened throughout the Mediterranean, are very well preserved. With the coral reefs, it supports a great diversity of marine life: 1 hectare of *Posidonia* produces 21 tons per year of biomass, similar to the productivity of a tropical forest of 22 ton/year/ha (UNEP/WCMC 2011).

The grass also purifies coastal waters by retaining sediments and oxygenating the environment, and the meadows are a nursery for many varieties of fish, helping to maintain a sustainable fishery. The main terrestrial ecosystems are associated with the salt pans and grow in bands parallel to the coast, graded by their tolerance of the salinity of the soil (UNESCO WHC 2010).

The seas around the site contain a population of the Mediterranean monk seal, one of the 12 most endangered mammals in the world. The loggerhead turtle and bottlenose dolphin also occur. A highly diverse marine community exists, dominated by Mediterranean endemic conditions. There are also several species of sponge which have disappeared from other parts of the Mediterranean. Thirty-six species of fish are recorded, belonging to 15 families (Government Balear 1988).

The salt marshes of Ibiza and Formentera are important above all for waterfowl. There is a highly diverse avifauna with 210 recorded species, 171 of which are migratory. This richness results from the diversity of environments as breeding and

Fig. 75.3 Accumulations of incrusted *Posidonia* overlocking a pebble beach on Western Ibiza. In the horizon, the prolonging of the thrust faulted and folded block of Ibiza. (Picture reproduced by permission of Dr. Miquel Mir-Gual, University of Balearic Islands)

wintering habitats: marine channels, shorelines, salt marshes, islets, woodland, and even low cliffs on Ibiza. Most are concentrated on the coastal lagoons and salt pans (IUCN/ICOMOS 1999).

75.3 Conservation Challenges

The natural values of the site are under high threat from uncontrolled tourism. High existing and potential threats have been growing during the last years and several cumulative impacts seem to be approaching a critical point (UNESCO WHC 2017). Some decline in the site's values has already been observed, and its conservation outlook shows significant concern (Osipova et al. 2017).

The principal problems are physical damage and sewage pollution from thousands of tourist boats and yachts, as well as from untreated effluents from nearby urban areas. Additionally, threats come from the visitation pressure on its terrestrial parts and disposal of dredged material nearby. Invasive alien species are also a great problem in the island (IUCN 2017). At the same time, the management has so far been insufficient to properly address these pressures on the site's values. Some decline in the site's values has already been observed (IUCN 2017).

Climate change with global warming can be an important danger for the next years in the area, considering that the seagrass *Posidonia oceanica* may be sensitive to the increasing of the temperature of the ocean water and associated acidification, which could threaten the rich local marine ecosystem.

References

Acosta J, Muñoz A, Herranz P, Palomo C, Ballesteros M, Vaquero M, Uchupi E (2001) Geodynamics of the Emile Baudot Escarpment and the Balearic Promontory, western Mediterranean. Mar Pet Geol 18(3):349–369

Acosta J, Canals M, Lopez-Martinez J, Munoz A, Herranz P, Urgeles R, Palomo C, Casamor JL (2002) The Balearic Promontory geomorphology (western Mediterranean): morphostructure and active processes. Geomorphology 49:177–204

Etheve N, Frizon de Lamotte D, Mohn G, Martos R, Roca E, Blanpied C (2016) Extensional vs contractional Cenozoic deformation in Ibiza (Balearic Promontory, Spain); integration in the West Mediterranean back-arc setting. Tectonophysics 682:35–55

Government Balear (1988) Demande d'inscription sur la liste du patrimoine de l'humanite des biens naturels and culturels de l'Ile d'Ibiza. Report, Ibiza, Spain

IUCN (International Union for Conservation of Nature) (2017) Ibiza: biodiversity and culture conservation outlook. http://www.worldheritageoutlook.iucn.org/node/1100. Accessed 28 Dec 2017

IUCN/ICOMOS (1999) World heritage nomination IUCN technical evaluation. Ibiza, biodiversity and culture (Spain). Report, Gland, Switzerland

Maillard A, Mauffret AG, Gaulier V, Basile C, Roure F, Scheck-Wenderoth M (2013) Structure and present day compression in the offshore area between Alicante and Ibiza Island (eastern Iberian margin). Tectonophysics 591:116–130

Osipova E, Shadie P, Zwahlen C, Osti M, Shi Y, Kormos C, Bertzky B, Murai M, Van Merm R, Badman T (2017) IUCN world heritage outlook 2: a conservation assessment of all natural world heritage sites. Gland, Switzerland

Roig-Munar FX, Mir-Gual M, Rodriguez-Perea A, Pons GX, Martin-Prieto JA, Gelabert B, Blázquez-Salom M (2013) Beaches of Ibiza and Formentera (Balearic Islands): a classification based on their environmental features, tourism and management. J Coast Res Spec Issue 65:1851–1855

San Félix M (1997) Guía Submarina de Ibiza y Formentera. Ayuntamiento de Ibiza, Ibiza

UNEP/WCMC (United Nations Environmental Program/World Conservation Monitoring Centre) (2011) Ibiza: biodiversity and culture. Report, Cambridge, England

UNESCO WHC (United Nations Educational, Scientific and Cultural Organization World Heritage Centre) (2010) Report on the 34th session of the committee. Paris

UNESCO WHC (United Nations Educational, Scientific and Cultural Organization World Heritage Centre) (2017) Ibiza, biodiversity and culture. http://whc.unesco.org/en/list/417. Accessed 1/12/2017

Part XVI
World Coastal Heritage in Islands, Gulfs and Reefs: North Sea

Chapter 76
The Wadden Sea: Denmark, The Netherlands, and Germany

Abstract The Wadden Sea is the largest unbroken system of intertidal sand and mud flats in the world. It is distinctive in being almost entirely a tidal flat and barrier system with only minor river influences, and an outstanding example of the large-scale development of an intricate and complex temperate-climate sandy barrier coast under conditions of rising sea level. It is a large, temperate, relatively flat coastal wetland environment, formed by a multitude of transitional habitats with tidal channels, sandy shoals, seagrass meadows, mussel beds, sandbars, mudflats, salt marshes, estuaries, beaches, and dunes. It is considered one of the most important areas for migratory birds in the world: up to 6.1 million birds can be present at the same time, and an average of 10–12 million pass through it each year.

76.1 Introduction

The Wadden Sea is a tripartite site in Europe, located at the southeastern part of the North Sea. It covers the Dutch Wadden Sea Conservation Area, the German Wadden Sea National Parks of Lower Saxony and Schleswig-Holstein, and most of the Danish Wadden Sea Maritime Conservation Area. It has a total length of some 500 km and a total area of about 10,000 km^2. It became a World Heritage Site in 2009 (UNESCO WHC 2016) (Fig. 76.1).

Human occupation was small throughout most of the region's history, but since the eleventh century dikes have been built to protect settlements and reclaim land. In the fifteenth and sixteenth centuries, the population increased, followed by agricultural production (Alberts 2015).

Over 10 million people visit the Wadden Sea region every year, and tourism is the major source of income for many areas along the coast. A large variety of tours are offered: boat tours to observe the seals on the sandbanks, guided tours over the mudflats to explore the flora, or specialized tours for birders who want to see the large numbers of migratory birds (Alberts 2015).

V. Claudino-Sales, *Coastal World Heritage Sites*, Coastal Research Library 28,
https://doi.org/10.1007/978-94-024-1528-5_76

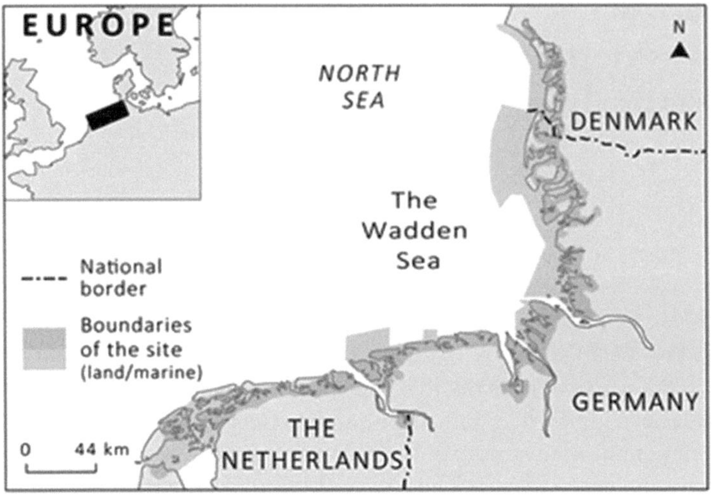

Fig. 76.1 Location of the tripartite site the Wadden Sea, along the coast of the Netherlands, Germany, and Denmark

The Trilateral Wadden Sea Cooperation provides the framework and structure for integrated conservation and management of the property and coordination between all three countries. Global measures of protection are in place within each country (UNESCO WHC 2016).

76.2 Natural Settings

76.2.1 Climatic Conditions

The climatic conditions dominant in the Wadden Sea area characterized by a permanent atmospheric low-pressure system near Iceland and a high-pressure system over the Azores (Kabat et al. 2012).

The temperature is variable, and the westerly winds prevail. The mean annual air temperature is 8.5°C, and precipitation is moderate, between 700 and 800 mm/year (Philippart and Epping 2010).

76.2.2 Geology and Geomorphology

The Wadden Sea is part of the North Sea, which sits on the continental shelf and is rather shallow. It is a good example of a depositional coastline which is largely modeled by tidal influences (Alberts 2015). It is a coastal wetland of exceptional

Fig. 76.2 Tidal flat creeks in the Wadden Sea. (Picture reproduced by permission of Dr. Martin Stock, Landesbetrieb für Küstenschutz Nationalpark und Meeresschutz Schleswig-Holstein)

size, great beauty and richness in unique natural assets. It contains the largest coherent tidal flat area in the world (Wang et al. 2012) (Figs. 76.2 and 76.3).

The Wadden Sea is made up of three areas: (1) the sublittoral, constantly covered by water; (2) the supralittoral, usually above sea level but covered with water during severe floods; and (3) the eulittoral, covered by water only during high tides. The eulittoral is commonly understood to be the Wadden Sea itself (Alberts 2015). Some also considers the eulittoral as a larger lagoon (e.g., Philippart and Epping 2010; Kennish and Paerl 2010).

The Wadden Sea is characterized by extensive intertidal flats along the coast, separated from the open sea by barrier islands (Burchard and Badewien 2015). Estuarine circulation represents a generic process, independent of the amount of river runoff (Flser et al. 2013). In geological terms, the Wadden Sea Region is ephemeral. Only around 6000 years BC, after a period of sea level rise since the latest Ice Age, this coastline reached its present location (van Leeuwe and Haartsen 2012).

The sea is actually a large and relatively flat coastal wetland, formed by the interactions between physical and biological factors that result in a multitude of transitional habitats such as tidal channels, sandy shoals, seagrass meadows, mussel beds, sandbars, mudflats, salt marshes, estuaries, beaches, and dunes (UNESCO WHC 2016). The barrier island system exhibits highly dynamic behavior, considering that

Fig. 76.3 Tidal flats with shelves in the Wadden Sea. (Picture reproduced by permission of Dr. Martin Stock, Landesbetrieb für Küstenschutz Nationalpark und Meeresschutz Schleswig-Holstein)

processes in action result in an ever-changing morphology (topography/bathymetry) of the environment. The multitude of islands is marked by dunes and wide sandy beaches (Wang et al. 2012).

76.2.3 Biodiversity

The vast stretches of intertidal sandflats in the Wadden Sea harbor about 150 macrozoobenthic species, mainly mollusks, polychaetes, and crustaceans (Thieltges et al. 2013). But the most remarkable of the elements of the fauna are the birds (UNESCO WHC 2016).

Effectively, water birds are an important aspect of the Wadden Sea ecosystem. By their migratory behavior, they connect the Wadden Sea with other sites, ranging from the Arctic to the western seaboards of Europe and Africa, forming the East Atlantic Flyway (van Roomen et al. 2012). In this sense, the Wadden Sea is one of the most productive biosystems in the world: over 6 million birds are present at the same time, with about twice passing through each year attracted by the availability of food (Alberts 2015).

The area is also home to rich fauna and flora. About 2300 species of plants and animals are present in the saltmarshes, and 2700 species can be found in brackish and saltwater habitats (Alberts 2015). There are also marine mammals such as the harbor seal, grey seal, and harbor porpoise (UNESCO WHC 2016).

76.3 Conservation Challenges

Changes in the environment of the Wadden Sea over the past millennium resulted from anthropogenic influences such as coastal protection, land claim from the sea, and drainage of wetland for agriculture, besides fishing, extraction of groundwater, gas and oil, and tourism. These activities are a source of threats to the conservation of the site (Kabat et al. 2012).

Besides these uses, the Wadden Sea receives contaminants from various sources and via various transport routes. The contaminants are various metals (Cd, Cu, Hg, Pb, and Zn) and of various organic origins. The main sources of contaminants are the rivers Rhine (via Dutch coastal zone), Elbe, and Weser (Laane et al. 2013). The pollution can be a threat to the local biodiversity. The fauna and flora are also under permanent stress from another source: the invasions of exotic species, what may transform the ecosystem (Kabat et al. 2012).

Nevertheless, the biggest risk for the Wadden Sea is global warming and climate change: climate change and sea level rise may seriously impact the biodiversity of the ecosystem, as well as the safety of the inhabitants in the region (Wang et al. 2013). Climate change also may lead to changes in hydrographic patterns and species distribution.

Nevertheless, the International Union for Conservation of Nature (IUCN 2017) released a report about the state of conservation of the natural World Heritage sites at the end of 2017, in which the level of conservation of the Wadden Sea appears as good (Osipova et al. 2017). The site is an example of an area that has undergone serious conservation damage in the past and continues to be faced with significant challenges, but conservation measures being undertaken are demonstrating positive results (IUCN 2017).

References

Alberts HC (2015) The Wadden Sea: an intertidal ecosystem of global importance. Focus Geogr 58:27–35

Burchard H, Badewien T (2015) Thermohaline residual circulation of the Wadden Sea. Ocean Dyn 65(12):1717–1730

Flser G, Riethmller R, Nauw J, Burchard H (2013) Observational evidence for the general presence of estuarine circulation in the Wadden Sea. J Coast Res 65(sp2):1527–1532

IUCN (International Union for Conservation of Nature) (2017) The Wadden Sea conservation outlook. http://www.worldheritageoutlook.iucn.org/node/1110. Accessed 28 Dec 2017

Kabat P, Bazelmans J, van Dijk J, Herman P, van Oijen T, Pejrup M, Reise K, Speelman H, Wolff W (2012) The Wadden Sea region: towards a science for sustainable development. Ocean Coast Manag 68:4–17

Kennish MJ, Paerl HW (eds) (2010) Coastal lagoons, critical habitats of environmental changes. CRC Press, Boca Raton

Laane RWPM, Vethaak AD, Gandrass J, Vorkamp K, Köhler A, Larsen MM, Strand J (2013) Chemical contaminants in the Wadden Sea: sources, transport, fate and effects. J Sea Res 82:10–53

Osipova E, Shadie P, Zwahlen C, Osti M, Shi Y, Kormos C, Bertzky B, Murai M, Van Merm R, Badman T (2017) IUCN world heritage outlook 2: a conservation assessment of all natural world heritage sites. Gland, Switzerland

Philippart KJM, Epping EG (2010) The Wadden Sea, A coastal ecosystem under continuous change. In: Kennish MJ, Paerl HW (eds) Coastal lagoons, critical habitats of environmental changes. CRC Press, Boca Raton

Thieltges DW, Engelsma MY, Wendling CC, Wegner KM (2013) Parasites in the Wadden Sea food web. J Sea Res 82:122–133

UNESCO WHC (United Nations Educational, Scientific and Cultural Organization World Heritage Centre) (2016) Wadden Sea. http://whc.unesco.org/en/list/1314. Accessed 29 Mar 2016

van Leeuwe M, Haartsen T (2012) Dynamics as key value of the Wadden Sea region. Ocean Coast Manag 68:1–3

van Roomen M, Laursen K, van Turnhout C, van Winden E, Blew J, Eskildsen K, Gunther K, Halterlein B, Kleefstra R, Potel P, Schrader S, Luerssen G, Ens BJ (2012) Signals from the Wadden Sea: population declines dominant among waterbirds depending on intertidal mudflats. Ocean Coast Manag 68:79–88

Wang ZB, Hoeskstra P, Burchard H, Ridderinkhof H, De Swart HE, Stive MJF (2012) Morphodynamics of the Wadden Sea and its barrier island system. Ocean Coast Manag 58:39–57

Chapter 77
Dorset and East Devon Coast, the United Kingdom of Great Britain and Northern Ireland

Abstract The Dorset and East Devon represents a stunning coast along the English Channel. The site has an outstanding combination of internationally renowned geological features. The property includes an exposure of approximately 185 million years of the Earth's history along a near-continuous stretch of coastline, with a number of internationally important fossil localities along its length. The property also contains a range of spectacular examples of coastal geomorphological features, landforms, and processes. The site includes a range of globally significant fossil localities – vertebrate and invertebrate, marine and terrestrial – which have produced well-preserved and diverse evidence of life during Mesozoic times.

77.1 Introduction

Located on the south coast of Britain, along the English Channel, the site, also known as Jurassic Coast, has eight sections along 155 km. The total area is 25 km^2, 80% of which is cliffed coastline (UNESCO WHC 2016) (Fig. 77.1).

The site lies almost entirely within two areas that are legally designated as "Areas of Outstanding Natural Beauty" (87%) and "Site of Special Scientific Interest," following a national law of 1981 (UNEP/WCMC 2011). It became a World Heritage site in 2001 (UNESCO WHC 2016).

The surrounding area is heavily populated, but the site itself is relatively unoccupied. The majority of the property is under private ownership (UNEP/WCMC 2011). The responsibility for management of the Dorset and East Devon Coast World Heritage Site lies with a nonexecutive committee known as the Steering Group. The group is led by Dorset and Devon County Councils (Dorset and East Devon Coast World Heritage Site Management Plan 2009).

Evidence of Bronze Age inhabitants is illustrated in the form of ancient "barrows," located a little inland from the boundary of the site. This is further supported by the discovery of Bronze Age artifacts. Iron Age hill forts were also built along the coast, many of which are still in relief. There is also evidence of Roman and Medieval settlements (UNEP/WCMC 2011).

V. Claudino-Sales, *Coastal World Heritage Sites*, Coastal Research Library 28,
https://doi.org/10.1007/978-94-024-1528-5_77

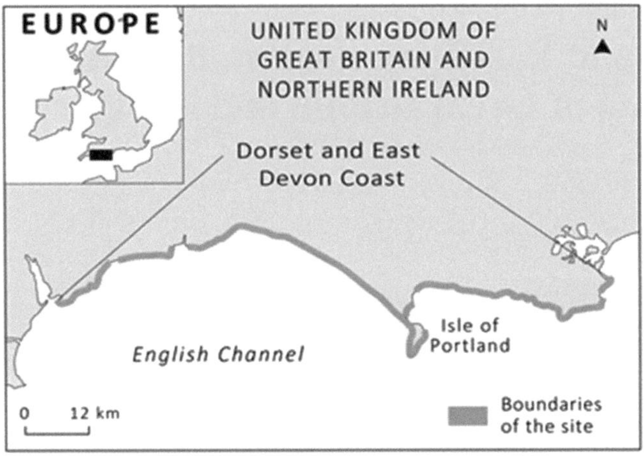

Fig. 77.1 Location of the Dorset and East Devon Coast, the United Kingdom of Great Britain and Northern Ireland, in the North Sea

The Jurassic Coast has a high number of visitors, and tourism is a very important activity (D'Amore and Kalifungwa 2013). The site is well served by a number of well-developed and professionally managed information centers, museums, accommodation, roads, and public transport. Although there are no car parks within the nominated sites, adequate facilities are located close to the property within gateway towns and villages (UNEP/WCMC 2011).

77.2 Natural Settings

77.2.1 Climatic Conditions

The area experiences a temperate Atlantic climate. It is a part of the UK closest to continental Europe and as such can be subject to continental weather influences that bring cold spells in winter and hot, humid weather in summer (UNEP/WCMC 2011). It is also furthest from the paths of most Atlantic depressions, with their associated cloud, wind, and rain, so the climate is relatively mild. The mean annual temperature is 10.5°C (Met Office 2016).

77.2.2 Geology and Geomorphology

The 155 km long Dorset and East Devon Coast World Heritage Site is recognized for its 185,000 years of Earth history (UNESCO WHC 2016).

Fig. 77.2 Arch created by wave and wind erosion in Dorset ("the Durdle Door"). The beach around it is made by pebbles. (Picture reproduced by permission of David Evison/Shutterstock Stock Photos)

Fig. 77.3 Segment of Jurassic rocks exposed as a seacliff at Dorset. (Picture reproduced by permission of Bbofdon/Shutterstock Stock Photos)

Effectively, the site holds one of the most spectacular geological sequences in the world (D'Amore and Kalifungwa 2013). The high rate of erosion and mass movement in the area creates a very dynamic coastline, which maintains both rock exposures and geomorphological features (Brunsden and Goudie 1997) (Fig. 77.2). It is a naturally active coast, ranging from vertical hard cliffs to very active landslides (Vincent 2015).

The property includes a near-continuous sequence of Triassic, Jurassic, and Cretaceous rock exposures, representing almost the entire Mesozoic Era (between 251 and 66 million years ago) (IUCN 2001) (Fig. 77.3).

The Triassic succession is a virtually continuous exposure of ca 1100 m of sediments representing most of the Triassic Period in continental, terrestrial, and shallow marine facies (Dorset County Council 2000). The Jurassic rocks are one of the finest marine sequences of this age anywhere in the world (Callomon and Cope 1995). Every stage of the Jurassic is represented (Cope et al. 1980). The succession includes rocks of all stages of the Cretaceous Period, with the exception of the uppermost stage (Dorset County Council 2000).

The site includes a range of globally significant fossil localities – both vertebrate and invertebrate and marine and terrestrial – which have produced well-preserved and diverse evidence of life during Mesozoic times (UNESCO WHC 2016).

The Dorset coast was abandoned by the sea during the last glacial period. Major mud sliding occurred, and at the foot of the old cliffs, an apron of debris accumulated. Since ca. 5500 years BP, the rising sea has initiated the removal of the deposits and destabilized many of the ancient slides (Brunsden and Moore 1999).

There are many small beaches within the site. The largest is Chesil Beach, one of the most famous coastal landforms on the British coast (Dorset County Council 2000). It is formed by gravels, which extend for more than 18 km and is separated for much of its length from land by a tidal lagoon known as The Fleet, one of the most important saline lagoons in Europe (Bennett et al. 2009).

The altitude ranges from 203 m above sea level to below sea level, at Fleet lagoon. The rock strata dip gently to the east, the oldest rocks being found in the west of the site, with progressively younger strata outcropping to the east (UNEP/WCMC 2011).

77.2.3 Biodiversity

A variety of important coastal habitats occur in the property, which supports several rare species, and the coastline is rich in wildlife (IUCN 2001).

There is a variety of important coastal habitats, especially the landslipped cliffs and cliff-top grasslands of west Dorset that support several rare plants. The Exe Estuary Special Protected Area supports over 20,000 wildfowl in the winter months, including the internationally important populations of avocet, dark-bellied goose and the Slavonian grebe (UNEP/WCMC 2011).

Short-beaked and bottle-nosed dolphins occur offshore. Lyme Bay Sensitive Marine Area contains a number of reefs that form one of the most easterly English locations for a number of Mediterranean and Atlantic species, and it is one of the few sites in Great Britain where the sunset star coral has been found (UNEP/WCMC 2011).

77.3 Conservation Challenges

The site has strong legal protection, a clear management framework, and a strong involvement of all stakeholders with responsibilities for the property and its setting. The tourism is well managed (UNESCO WHC 2016). Nevertheless, the reefs are being devastated by uncontrolled scallop dredging (UNEP/WCMC 2011).

However, the greatest threat to the site is the rising of sea level due to global warming and climate change. The results may tend to destabilize many of the ancient slide aprons so that coastal erosion may diffuse inland.

But the International Union for Conservation of Nature (IUCN) in its last report about the state of conservation of the natural World Heritage sites published at the end of 2017 considered that Dorset and East Devon Coast's level of conservation is good without concerns (Osipova et al. 2017). Geological and paleontological values of the site have remained well preserved and are very robust against most current and potential threats. The protection and management of the site is mostly effective to highly effective overall, with several examples of international best practice in World Heritage management (IUCN 2017).

References

Bennett MR, Cassidy NJ, Pile J (2009) Internal structure of a barrier beach as revealed by ground penetrating radar (GPR): Chesil Beach, UK. Geomorphology 104(3):218–229

Brunsden D, Goudie A (1997) Classic landforms of the West Dorset Coast. The Geographical Association, Sheffield

Brunsden D, Moore R (1999) Engineering geomorphology on the coast: lessons from west Dorset. Geomorphology 31(1):391–409

Callomon JH, Cope JCW (1995) The Jurassic geology of Dorset. In: Taylor PD (ed) Field geology of the British Jurassic. Geological Society, London

Cope JCW, Getty TA, Howarth MK, Morton N, TORRENS HS (1980) A correlation of Jurassic rocks in the British Isles. Introduction and Lower Jurassic. Geological Society Special Reports 14. London, England

D'Amore L, Kalifungwa P (2013) Meeting the challenges of climate change to tourism. Cambridge Scholars Publishing, Cambridge

Dorset and East Devon Coast World Heritage Site Management Plan (2009) Dorset County council. file:///C:/Users/User/Downloads/CD.ENV6_Dorset_and_East_Devon_Coast_World_Heritage_Site_Management_Plan.pdf. Accessed 1 May 2016

Dorset County Council (2000) Nomination of the Dorset and East Devon Coast for inclusion in the world heritage list. Report, Dorset, England

IUCN (International Union for Conservation of Nature) (2017) Dorset and East Devon Coast conservation outlook. http://www.worldheritageoutlook.iucn.org/node/1121. Accessed 28 Dec 2017

IUCN (International Union of Conservation of Nature) (2001) World heritage nomination – IUCN technical evaluation Dorset and East Devon Coast (United Kingdom). Report, Gland, Switzerland

Met Office (2016) http://www.metoffice.gov.uk/climate/uk/regional-climates/so. Accessed 1 May 2016

Osipova E, Shadie P, Zwahlen C, Osti M, Shi Y, Kormos C, Bertzky B, Murai M, Van Merm R, Badman T (2017) IUCN world heritage outlook 2: a conservation assessment of all natural World Heritage Sites. Gland, Switzerland

UNEP/WCMC (United Nations Environmental Programme/World Conservation Monitoring Centre) (2011) Dorset and East Devon Coast, United Kingdom. Report, Cambridge, England

UNESCO WHC (United Nations Educational, Scientific and Cultural Organization World Heritage Centre) (2016) Dorset and East Devon Coast. http://whc.unesco.org/en/list/1029. Accessed 1 May 2016

Vincent M (2015) Coastal cliff conservation and management: the Dorset and East Devon Coast World Heritage Site. J Coast Conserv 19(6):821–829

Chapter 78
Giant's Causeway and Causeway Coast, the United Kingdom of Great Britain and Northern Ireland

Abstract The Giant's Causeway and Causeway Coast lies at the foot of high cliffs along the North Sea in the coast of Northern Ireland. It is composed of some 40,000 massive columns of basalt sticking out of the sea. The Giant's Causeway itself is a sea-level promontory outcrop and pavement running out to sea. Exposure of the columns in a pavement of such perfect horizontal sections is considered unique and is also scenically remarkable. The Causeway Coast is an unparalleled displays of geological formations, remnants of volcanic activity some 50–60 million years ago, forming the largest remaining lava plateau in Europe. Geological studies of this lava flow over the last 300 years have greatly contributed to the analysis of Tertiary volcanic events in the North Atlantic.

78.1 Introduction

Giant's Causeway and the Causeway Coast are located in the Plateau of Antrim, on the northeast coast of Northern Ireland, in the North Sea. The World Heritage site has 0.71 km^2 and occupies a thin, approximately 5-km-long strip of mainly coastal cliff (UNESCHO WHC 2016) (Fig. 78.1).

In 1961, part of the site was bequeathed by its owner, Sir P. McNaghten, to the National Trust (a charitable Non-Governmental Organization). The National Trust, in turn, bought other parts from private owners – approximately 5% of the property still remains in private ownership – and now owns the terrestrial part of the property (UNESCO WHC 2016). The Crown Estate is considered the legal owner of all lands between high- and low-water mark and has rights over the seabed within territorial water (UNESCO WHC 2016). In 1986, the area was declared a World Heritage site by UNESCO. In 1987, the British government declared the segment of coast a National Nature Reserve (UNEP/WCMC 2005).

In the nineteenth century, and up to the acquisition of the site in 1961 by the National Trust, many temporary commercial stalls and huts were either owned or managed by local people for tourists and visitors. The site is uninhabited now, but several small settlements exist immediately inland of the Causeway cliffs. The sublittoral area is a protected nautical archaeological site (UNEP/WCMC 2005).

V. Claudino-Sales, *Coastal World Heritage Sites*, Coastal Research Library 28,
https://doi.org/10.1007/978-94-024-1528-5_78

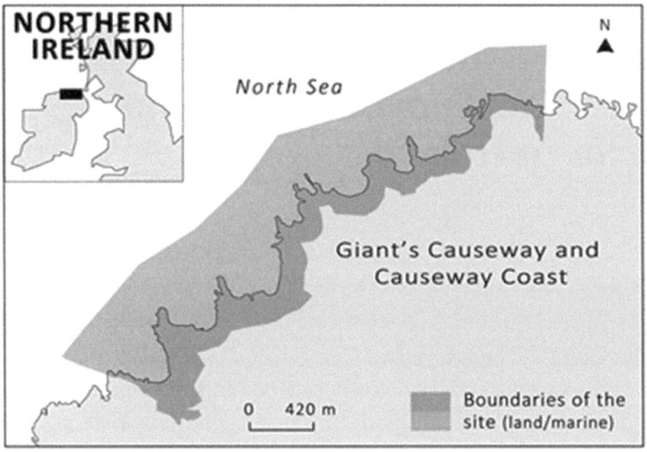

Fig. 78.1 Location of Giant's Causeway and Causeway Coast on Northern Ireland

The spectacular area has inspired legends of giants striding over the sea to Scotland, and the name of the site derives from the legend. The landscape has been celebrated in the arts and in science and has been a visitor attraction for at least 300 years. It is regarded as a symbol of Northern Ireland and is its most visited tourist attraction (UNESCO WHC 2016).

The site makes a classic locality for the study of basaltic volcanism. The features of the Giant's Causeway and Causeway Coast site and in particular the strata exposed in the cliff faces have been key to shaping understanding of the sequences of activity in the Earth's geological history (Crawford and Black 2012).

The National Trust manages the cliffs, and it is monitored by the Environment and Heritage Service of the Department of the Environment for Northern Ireland (UNEP/WCMC 2005).

78.2 Natural Settings

78.2.1 Climatic Conditions

The climate is cold oceanic. The maximum temperature is in July, and it reaches 18 °C. The minimum takes place in January, and it averages 2 °C. The climate is dominated by mild, damp westerly winds, which come mostly from the southeast (Meteoblue 2016). Winds speeds average 12 km/h, but it sometimes reaches speeds of 61 km/h. The annual average rainfall is 1016 mm (UNEP/WCMC 2011).

78.2.2 Geology and Geomorphology

The Giant's Causeway forms a sea-level promontory consisting of basalt columns which were formed in the Tertiary (60 million years ago), during a series of intense volcanic eruptions (Crawford and Black 2012). The volcanism is linked to the opening of what is now the North Atlantic Ocean. The lavas cover some 3800 km^2, forming the largest remaining lava plateau in Europe (Smith 2005).

In the site, there are approximately 40,000 almost entirely regular polygonal columns averaging 45 cm in diameter, formed by rapid cooling of extruded lavas (Smith 2005) (Fig. 78.2). The occurrence of a natural polygonal network is attributed to contraction during cooling and desiccation of the lavas, with further development of the crack patterns into columnar fabrics (Budkewitsch and Robin 1994).

The coastline is a series of 100 m headland cliffs on the regular columns of basalt, with intervening bays. The cliffs are stepped due to a succession of five or six differentially eroded lava flows (Wilson and Manning 1978) (Fig. 78.3). The coastline is also cut through by olivine and tholeiitic dikes. Geomorphological features include coastal erosional landforms contained within the embayed coastline and dynamic slope instability features (Smith et al. 2011).

The intertidal area is dominated by bedrock with wave-cut platforms often reaching substantial widths (National Trust 2005). The coast is also characterized by the

Fig. 78.2 Regular polygonal basaltic columns known as Giant's Causeway. The columns' pathway leads to the sea. (Picture reproduced by permission of Nahlik/Shuterstock Stock Photos)

Fig. 78.3 Basaltic columns in the first plan and seacliffs developed in the lava flow at Giant's Causeway coast. The lava formed an extensive plateau. (Picture reproduced by permission of Els Lots, web site www.worldheritagesite.org)

presence of very large boulders (mega-clasts), probably resulting from exhumation in situ as a result of storm waves acting on heavily weathered, jointed basalts (Mckenna et al. 2011).

Exposure of the columns in a pavement of such perfect horizontal sections is considered unique and is scenically remarkable (UNESCO WHC 2016). Other major features are the Giant's Organ of about 60 12 m-high regular columns, Chimney Tops, and columns separated from the cliffs by erosion (UNEP/WCMC 2011).

Much of the local coastal scenery dates from the end of the last glaciation, when ice would have covered and surrounded the cliffs, with the North Channel filled with pack ice (Carter 1991). Ice sculptured forms were later modified by periglacial processes. These processes continue to some degree today in conjunction with active marine erosion and the ongoing collapse and movement of the cliffs (National Trust 2005).

78.2.3 *Biodiversity*

Along the cliff tops, there is a valuable area of coastal heath supporting rich and diverse vegetation dominated by gorse. The basalt bedrock generates base-rich but acidic soils, resulting in interesting mixtures of calcicole and calcifuge plant species (National Trust 2005).

The cliffs and cliff tops include patches of heath, flushes, slips and screes, rock exposure with lichen cover, and ledge/crevice flora. Scree is widespread, varying from species-poor block fields, to more diverse types comprising smaller grade rock. The large extent of unimproved grassland is particularly notable and includes a good range of maritime and non-maritime species (National Trust 2005). It is considered that more than 200 plant species representing a range of habitats covering seashore, cliff, scree, grassland, scrub, heathland, and marsh are present (UNEP/WCMC 2011).

The site is regarded by the Royal Society for the Protection of Birds as one of regional importance within the United Kingdom, based on the number of breeding species. More than 50 resident and 30 migrant species have been recorded (UNEP/WCMC 2011). The area is also important for wintering birds. At the same time, it supports a rich and diverse invertebrate assemblage with a number of notable species (National Trust 2005).

78.3 Conservation Challenges

The International Union for Conservation of Nature (IUCN 2014, 2017) considers that Giant's Causeway and Causeway Coast World Heritage site is under significant concern in relation to its degree of conservation. This means that its values are threatened by a number of current and potential problems and that additional measures are required to preserve these values over the medium to long term. The threats requiring effective protection and management include direct damage to natural features through human impact due to tourism (IUCN 2014, 2017).

The continuing effects of natural erosion, which gradually alter the cliff exposures, are also a threat. Changes in sea level and increase in frequency of storms due to global warming and climatic changes may also affect the degree to which the causeway is accessible or visible, and path routes, as well as the site boundaries, may need to be changed in the future to accommodate the effects of these processes (UNESCO WHC 2016).

The IUCN also considers that because of its inherent robustness against anthropogenic impact and the effective management of the property itself, the conservation outlook for the site's geological values is good. At the same time, the conservation outlook for the exceptional natural beauty of the site, which depends on the integrity of the surrounding landscape, is of significant concern, mainly because of the planned and approved tourism development project in the immediate vicinity of the site and potential additional development projects in the future (IUCN 2017; Osipova et al. 2017).

References

Budkewitsch P, Robin PY (1994) Modelling the evolution of columnar joints. J Volcanol Geotherm Res 59:219–239

Carter RWG (1991) Shifting sands: a study of the coast of Northern Ireland from Magilligan to Larne. Countryside & Wildlife Research Series 2. Belfast, Ireland

Crawford KR, Black R (2012) Visitor understanding of the geodiversity and the geoconservation value of the Giant's Causeway World Heritage Site, Northern Ireland. Geoheritage 4:115–126

IUCN (International Union for Conservation of Nature) (2014) World heritage conservation outlook. Report, Gland, Switzerland

IUCN (International Union for Conservation of Nature) (2017) Giants Causeway and Causeway Coast Conservation Outlook. http://www.worldheritageoutlook.iucn.org/node/1018. Accessed 28 Dec 2017

Mckenna J, Jackson DWT, Cooper JAG (2011) In situ exhumation from bedrock of large rounded boulders at Giant's Causeway, Northern Ireland: an alternative genesis for large shore boulders (mega-clasts). Mar Geol 283(1):25–35

Meteoblue (2016) Climate Giants Causeway. https://www.meteoblue.com/en/weather/forecast/modelclimate/giants-causeway_united-kingdom_2648688. Accessed 11 Dec 2016

National Trust (2005) Management plan, final draft, Giant's Causeway and Causeway Coast World Heritage Site. Report, Belfast, Northern Ireland

Osipova E, Shadie P, Zwahlen C, Osti M, Shi Y, Kormos C, Bertzky B, Murai M, Van Merm R, Badman T (2017) IUCN world heritage outlook 2: a conservation assessment of all natural World Heritage Sites. Gland, Switzerland

Smith BJ (2005) Management challenges at a complex geosite: the Giant's Causeway World Heritage Site, Northern Ireland. Géomorphol Relief Processus Environ 11(3):219–226

Smith BJ, Pellitero Ondicol R, Alexander G (2011) Mapping slope instability at the Giant's Causeway and Causeway Coast World Heritage Site: implications for site management. Geoheritage 3:253–266

UNEP/WCMC (United Nations Environment Program/World Conservation Monitoring Center) (2005) Giant's Causeway and Causeway Coast, United Kingdom and Nothern Ireland. Report, Cambridge, UK

UNEP/WCMC (United Nations Educational Programme/World Conservation Monitoring Centre) (2011) Giants Causeway and Causeway Coast Northern Ireland, United Kingdom. Report, Cambridge, England

UNESCO WHC (United Nations Educational, Scientific and Cultural Organization World Heritage Centre) (2016) Giant's Causeway and Causeway Coast. http://whc.unesco.org/en/list/369. Accessed 11 Dec 2016

Wilson H, Manning P (1978) Geology of the Causeway Coast. Geological Survey of Northern Ireland. Belfast, Northern Ireland

Part XVII
World Coastal Heritage in Islands, Gulfs and Reefs: Red Sea

Chapter 79
Sanganeb Marine National Park and Dungonab Bay: Mukkawar Island Marine National Park, Sudan

Abstract The site is a serial one, made by Sanganeb, an isolated coral reef structure and the only atoll in the Red Sea, 25 km off the shoreline of Sudan, and by the Dungonab Bay and Mukkawar Island, 125 km from Port Sudan. The site includes a high diversity of coastal and marine systems, such as coral reefs, mangroves, seagrass beds, beaches, and islets. These natural systems provide habitat for large populations of marine and coastal animals, including mammals, fish, turtles, and birds. Dungonab Bay has a globally significant population of dugongs (sea cow). The clarity of the water, coral diversity, marine species, and pristine habitats create a striking landscape and seascape.

79.1 Introduction

The serial site is in the central Red Sea, off the coast of Sudan. It is formed by three areas, which are Sanganeb Marine National Park and Dungonab Bay and Mukkawar Island Marine National Park, located, respectively, 25 km and 125 km from mainland Sudan. The site has 2607 km^2 of area, mostly marine. The areas became a World Heritage site in 2016, the first in the Red Sea (UNESCO WHC 2017) (Fig. 79.1).

Sanganeb is the first national park to be designated in Sudan, in 1990 (IUCN 2016). A bonus for tourists is the lighthouse built by the British between 1958 and 1964 on the coast. The lighthouse is 50 m tall, has 268 steps, and is a highly photographed landmark. It is also the best place for a tourist to view all the beauty of the Sanganeb (World Atlas 2017) (Fig. 79.2).

The site is public land and is managed by the Wildlife Conservation General Administration (WCGA) of the Ministry of Interior of the Sudanese Government (IUCN 2016). Together with non-governmental organizations, the administration promotes a series of activities to protect and improve the site.

The serial site attracts tourists from far and wide, which is a source of revenue to the Sudanese government. The corals are particularly famous, being one of the most unspoiled in the world. The Sanganeb Marine National Park is especially popular

V. Claudino-Sales, *Coastal World Heritage Sites*, Coastal Research Library 28,
https://doi.org/10.1007/978-94-024-1528-5_79

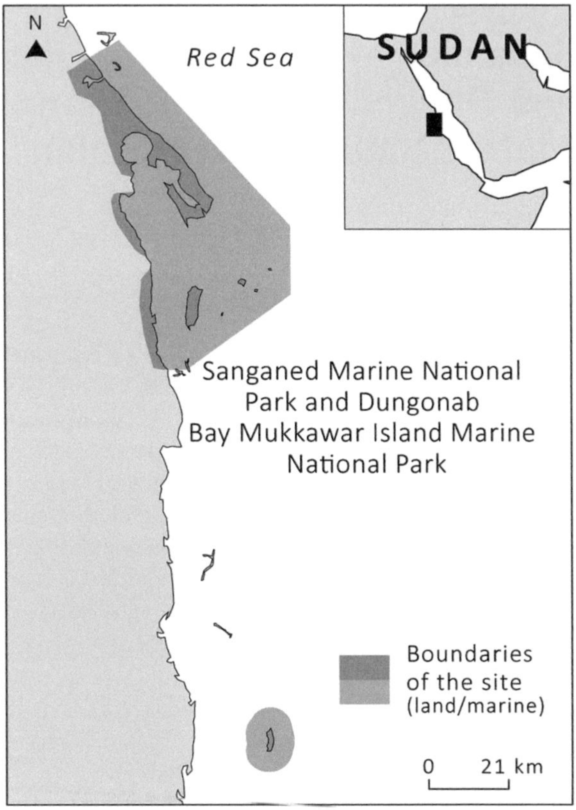

Fig. 79.1 Location of Sanganeb Marine National Park and Dungonab Bay – Mukkawar Island Marine National Park, Sudan, in the Red Sea

during international diving holidays, because its largely pristine marine ecosystem provides some of the most impressive dive sites on Earth (WCGA 2016).

A lack of terrestrial fresh water runoff, combined with relatively low productivity, gives rise to low levels of turbidity in the Sudanese coastal waters. Sea water clarity is particularly high in deeper water areas and around offshore reefs, where visibility typically ranges between 20 and 30 m, but may reach 70 m (PERSGA 2001).

79.2 Natural Settings

79.2.1 Climatic Conditions

The Red Sea region receives very little precipitation in any form. In general, the temperatures vary between 8°C and 28 °C. Summer temperatures, however, are much higher, up to 40 °C, and the air humidity is high (Encyclopaedia Britannica 2017). In the coast of Sudan, the winds are variable, but from June through August,

Fig. 79.2 Sanganeb Atoll, the only one in the Red Sea, with the lagoon seen in contact with the pier of the lighthouse built on its coastal border. (Picture reproduced by permission of Bram de Bruin, website htpp://members.chello.nl/a.bruim32)

strong northwest winds move down from the north. The rest of the year south to southwest winds prevail. The speeds vary from a 5 km/h average in summer to a 12 km/h average in winter (Meteoblue 2017).

79.2.2 Geology and Geomorphology

The Red Sea occupies part of a large rift valley in the continental crust of Africa and Arabia and has resulted from the gradual increasing separation between these two tectonic plaques. This break in the crust is part of a complex rift system that includes the East African Rift System (Sheppard et al. 1992).

It is considered a relatively new sea, formed in at least two complex phases of land motion. The movement of Africa away from Arabia began about 55 million years ago, and the northern part of the Red Sea opened about 20 million years ago. The second phase began about 3 to 4 million years ago, creating the trough in the southern half of the Red Sea valley. This motion, estimated as amounting to 15.0 to 15.7 mm per year, is still proceeding (Ligi et al. 2011).

The rifting has created a stepped profile to the continental margins on either side of the Red Sea. Its maximum depth is around 3000 m, but the shallowest of the rift valley steps lies less than 5 m below sea level in the northern Sudanese coast (WCGA 2016).

The Sudanese Red Sea coast is about 750 km long, including bays and inlets. Three distinct depth zones are recognized: shallow reef shelves less than 50 m deep, deep shelves 500 to 1000 m deep, and a central trench more than 1000 m deep, reaching a maximum of 3000 m off the city of Port Sudan. The serial site is an expanse of coastline including coral reef, mangroves, offshore islands, soft-bottom mud flats, sand beaches, and hard-bottom rocky shores in addition to salt marshes, sabkhas (Fig. 79.3), and khore basins (formed by tidal creeks) (Gass 1960).

Sanganeb is an isolated, atoll-shaped coral reef structure, representing a largely pristine marine ecosystem (see Fig. 79.2). Surrounded by 800 m deep water, the atoll coral reef system is part of the northernmost coral reef system in the world (IUCN 2016).

Dungonab Bay is approximately 13 km across at its southern end, and extends 31 km from north to south, containing several islands and islets. The bay has three notable features: a large southern basin separated from the deeper water outside the bay to the south by a wide shallow sill, an almost circular northern basin partially isolated from the southern basin by islands and another shallow sill, and a long narrow khore basin (WCGA 2016).

Mukkawar Island is composed of uplifted sedimentary rocks and fossil reef with a maximum 100 m of altitude. The mainland shore is backed by a gently sloping coastal plain varying in width from 5 km to over 30 km (Farah 1982). The coastal plain is composed of sand and gravel deposits, in many places overlying fossil reefs, and is backed by the Red Sea Hills that rise to over 1500 m. In some areas, low raised areas of fossil reef up to 10 or 15 m high extend almost to the waters' edge (WCGA 2016).

79.2.3 Biodiversity

The Red Sea is an embryonic ocean partially isolated from the rest of the global ocean, resulting in a place with significant marine ecosystem. It is renowned for its unique marine and species richness, which includes many found nowhere else on Earth (WCGA 2016).

The Red Sea is one of the most biologically diverse tropical seas in the world (Spalding et al. 2001). The site represents an outstanding marine ecosystem that sustains intact ecological functions. It covers shallow habitats as well as deep-sea areas (PERSGA 2001).

There are more than 200 species of corals, of which about 6% are endemic. The Sudanese coast is characterized by the most diverse reefs of the Red Sea. Most of the coast is bordered by fringing reefs 1–3 km wide separated by deep channels from a barrier reef 1–14 km wide (UNESCO WHC 2017). The rich biodiversity is also attested by the presence of 170 species of echinoderms, of which 5.3% are endemic (Sheppard et al. 1992). About 500 species of benthic algae have been

Fig. 79.3 A sebkha in Dungonab Bay coastline, showing salinization of the strand and lithification of sands. These processes attest to the aridity of the environment. (Picture reproduced by permission of Maria Cristina Pulliero, website www.blusudan.com)

recorded (Chiffings 2005) and over 1000 species of fish. The level of endemism among Red Sea fish is about 17% (Goren and Dor 1994).

Dungonab Bay and Mukkawar Marine National Park support a globally significant dugong (sea cow) population, one of the last remaining healthy populations of the species in the Indian Ocean. The whale and manta ray seasonal aggregation in the area is unique to the entire Western Indian Ocean, and the marine park is recognized as an important bird area (UNESCO WHC 2017).

For the flora, the arid environmental conditions are a factor of stress: Vegetation is sparse, consisting of a low cover of halophytes and grasses in scattered locations (see Fig. 79.3). In many areas, there are wide expanses of gravel plain (largely eroding fossil reef) completely devoid of vegetation (PERSGA 2001).

79.3 Conservation Challenges

The ecosystems and landscapes in the serial site remain relatively weakly affected by human activities. However, growing regional coastal development in the last years can be a source of degradation, reinforcing the need to protect these landward and marine areas. The increase of activities from local residents and tourism could amplify pollution and direct damage on ecosystems such as anchor, boat, and diving damage (UNESCO WHC 2017).

The level of conservation of the site is considered to be good with some concerns (Osipova et al. 2017). Besides the problems connected to development, the biggest threat is climate change and global warming (IUCN 2017), which is affecting coral communities and exposing them to risk of coral bleaching.

The International Union for Conservation of Nature (IUCN) in its last report related to the state of conservation of the natural World Heritage site published at the end of 2017 found that the site is a relatively undisturbed area that serves as a standard to assess the health of the central Red Sea's regional ecosystems. Its marine habitats are well preserved and remain largely untouched and unspoiled, largely due to isolation and low visitation. However, some degradation of the site's biodiversity values through exploitation by the local community is a potential threat (IUCN 2017).

References

Chiffings AW (2005) Arabian seas. A global representative system of marine protected areas. In: Kelleher G, Bleakley C, Well S (eds) Central Indian Ocean, Arabian Seas, East Africa and East Asian Seas. IUCN, Gland

Encyclopaedia Britannica (2017) Red Sea. https://www.britannica.com/place/Red-Sea. Accessed 3 Dec 2017

Farah OM (1982) The Bathymetry, oceanography and bottom sediments of Dungonab Bay (Red Sea) Sudan. Ph.D thesis, University of Delaware, Newark

Gass IG (1960) The geology of Sudan Coastal Province. Report, General Exploration Company of California, San Diego, USA

Goren M, Dor M (1994) Updated check list of the fishes of the Red Sea. Israel Academy of Science and Humanities. Jerusalem, Israel

IUCN (International Union for Conservation of Nature) (2016) World heritage nomination – IUCN technical evaluation, Sanganeb Marine National Park and Dungonab Bay/Mukkawar Island Marine National Park (Sudan). Report, Gland, Switzerland

IUCN (International Union for Conservation of Nature) (2017) Sanganeb Marine National Park and Dungonab Bay – Mukkawar Island conservation outlook. http://www.worldheritageoutlook.iucn.org/node/1192. Accessed 28 Dec 2017

Ligi M, Bonatti E, Tontini FC, Cipriani A, Cocchi L, Schettino A, Bortoluzz G, Ferrante V, Khalil S, Mitchell N, Rasul N (2011) Initial burst of oceanic crust accretion in the Red Sea due to edge-driven mantle convection. Geology 39(11):1019–1022

Meteoblue (2017) Climate Red Sea Hills. https://www.meteoblue.com/en/weather/forecast/modelclimate/red-sea-hills_egypt_350154. Accessed 3 Dec 2017

Osipova E, Shadie P, Zwahlen C, Osti M, Shi Y, Kormos C, Bertzky B, Murai M, Van Merm R, Badman T (2017) IUCN world heritage outlook 2: a conservation assessment of all natural World Heritage Sites. Gland, Switzerland

PERSGA (Regional Organization for the Conservation of the Environment of the Red Sea and Gulf of Aden) (2001) Marine environmental protection in the Red Sea and Gulf of Aden – a new initiative. Report, Jeddah, Saudi Arabia

Sheppard CRC, Price A, Roberts C (1992) Marine ecology of the Arabian region – patterns and processes in extreme tropical environments. Academic, London

Spalding MA, Ravilious C, Green EP (2001) World atlas of coral reefs. University of California Press, Berkeley

UNESCO WHC (United Nations Educational, Scientific and Cultural Organization World Heritage Centre) (2017) Sanganeb Marine National Park and Dungonab Bay -Mukkawar Island Marine National Park. http://whc.unesco.org/en/list/262. Accessed 3 Dec 2017

WCGA (Wildlife Conservation General Administration) (2016) Nomination of Sanganeb Marine National Park and Dungonab Bay/Mukkawar Island Marine National Park (Sudan – Red Sea) for inscription on the World Heritage List. Rapport, Cartum

World Atlas (2017) Sanganeb Marine National Park, Sudan. http://www.worldatlas.com/articles/sanganeb-marine-national-park-sudan.html. Accessed 2 Dec 2017

Part XVIII
World Coastal Heritage in Islands, Gulfs and Reefs: East Indian Archipelago

Chapter 80
Komodo National Park, Indonesia

Abstract Komodo is a volcanic island inhabited by a population of giant lizards that exist nowhere else in the world, whose appearance and aggressive behavior have led to them being called "Komodo dragons." The topography of the island is steep and rugged, dominated by a range of rounded north-south oriented hills. The rugged hillsides and pockets of thorny green vegetation contrast with the white sandy beaches and the blue waters colonized by corals. The seas around the island have been among the most productive in the world due to upwelling and a high degree of oxygenation resulting from the strong tidal currents. The marine site has sea mounts, semi-enclosed bays, and seagrass beds.

80.1 Introduction

Komodo National Park is in south-central Indonesia, between the islands of Sumbawa and Flores, and is composed of three major islands and numerous smaller ones. The World Heritage site includes Komodo National Park and other coastal segments around it, making an area of 2193 km^2 (UNESCO WHC 2016) (Fig. 80.1).

In 1915, the Komodo dragon was formally protected, 4 years after its discovery to science. In 1938, two islands of the present-day park were established as nature reserves. In 1965, Komodo Island Nature Reserve was established. In 1980, the islands of Komodo and other three were declared a national park. The national park and surroundings became a World Heritage site in 1991 (UNEP/WCMC 2011).

The state is the owner of the land. The area is administered by the Komodo National Park Authority under the Directorate-General of Forest Protection and Nature Conservation (PHKA) of the Ministry of Forestry of Indonesia (UNESCO WHC 2016).

Annual visitor numbers increased rapidly in the last decades in Komodo National Park. Visitors are mostly foreign, mainly attracted to dragons. They can see the animals at the viewing stations at the visitor center where they are baited twice a week. Komodo is also a natural laboratory for studying evolutionary history (IUCN 2017).

V. Claudino-Sales, *Coastal World Heritage Sites*, Coastal Research Library 28,
https://doi.org/10.1007/978-94-024-1528-5_80

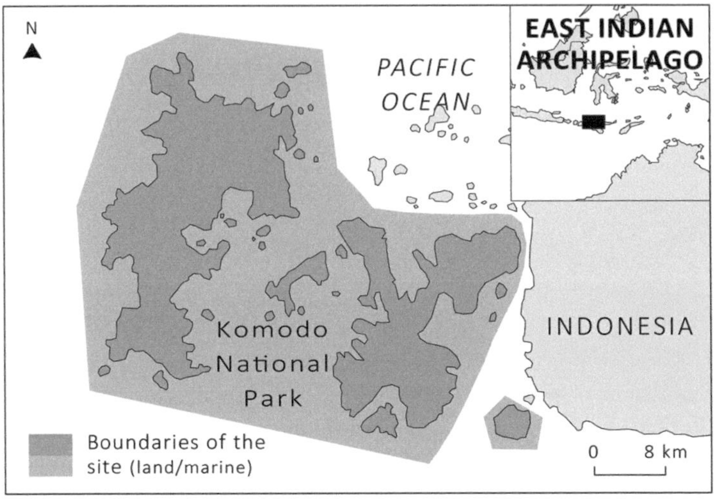

Fig. 80.1 Location of Komodo World Heritage site, in the Indonesian Archipelago

80.2 Natural Settings

80.2.1 Climatic Conditions

The park is in one of the driest regions of Indonesia, with an annual rainfall of less than 1000 mm. The heaviest rainfall, highest humidity, strongest winds, and lowest temperatures are recorded during the monsoon between November and March. This pattern is reversed during the long dry season from April to October, when mean daily temperatures can reach 40 °C (UNEP/WCMC 2011).

80.2.2 Geology and Geomorphology

Located among three major plates, namely, the Indian-Australian, the Pacific, and the Eurasian, Indonesia archipelago is one of the most seismically active regions in the world. Between 1994 and 2014, 528 earthquakes occurred in Indonesia, about 26 earthquakes per year (Muhamad et al. 2016).

Komodo, in Central Indonesia, is one of the long arcs of islands along the line of a tectonic subduction between the Sunda (or Java) shelf, a crustal extension of Southeast Asia, and the Sahul shelf, which extends northwest from Australia (Letouzey et al. 1990). This volcanic arc developed linearly along the leading edge of the Sunda platform as the result of subduction of the Indian Ocean plate beneath the Southeast Asian plate (Hamilton 1979). The subduction of the Indian oceanic

Fig. 80.2 Rugged volcanic rocky coast with vertical seacliffs framing a cove with dark sandy beach in Komodo National Park, Indonesia. (Picture reproduced by permission of Achmad Ariefiandy)

crust was continuous during the entire Cenozoic period. The collision with the Australian continent took place in the Late Miocene-Pliocene period (from 10 million years to 3 million years) and affected the eastern part of the arc, in the Sumbawa area (Daly et al. 1987).

The topography is generally steep and rugged, due to its volcanic origin, and is dominated by a range of rounded north-south oriented hills between 500 and 600 m high, crowned by deep, rocky dry gullies. The coastline is irregular, with little flat ground, numerous bays, beaches, and inlets separated by headlands, often with sheer cliffs falling vertically into the sea (Hayes et al. 2012) (Fig. 80.2).

The mainland sections of the park lie on the rugged coast of northwestern Flores Island, where surface fresh water is more abundant than on the other segments. The islands were formed by volcanism in the Eocene, although the west side preserves Jurassic rocks. Deposits are generally volcanic ash, conglomerates, and coral formations raised by tectonic movement. There are frequent tremors, but no active volcanoes. The soils are rocky and shallow (Tsang et al. 2015; Sumardja 1981).

The site has seamounts, semi-enclosed bays, and seagrass (Kvalvagnaes and Halim 1979). Fringing and patch coral reefs are extensive and best developed in west- and north-facing areas. However, recent disturbance by man has reduced this diverse reef to isolated patches (UNEP/WCMC 2011).

80.2.3 *Biodiversity*

Komodo lies between the Oriental and Australian zoogeographic provinces and on the western edge of a biotic transition zone extending to the New Guinean and Australian coasts. The dry climate has triggered specific evolutionary adaptation within the terrestrial flora. The predominant vegetation, covering some 70% of the Park, is dry open grass-woodland savanna, mainly of anthropogenic origin. Most of its species are xerophytic with water-retaining adaptions and many are fire-adapted. There are also patches of tropical rainforest, deciduous monsoon forest, and mangrove (Sumardja 1981).

A quasi cloud forest occurs above 500 m on pinnacles and ridges. Although covering only small areas on Komodo Island, it harbors a relict flora of many endemic species. It is characterized by moss-covered rocks, rattan, bamboo groves, and many trees generally absent at lower elevations (Auffenburg 1980).

The transitional nature of the biota has resulted in the presence of many mammals from Asia and several reptiles and birds from the Australasian bioregion. Compared with the rich marine life, the terrestrial fauna is small. The park is best known for the Komodo monitor, the world's largest lizard and venomous animal. The lizard can grow to 3 m long, an example of island gigantism. It is a relict Australasian species, isolated by postglacial sea level rise and only discovered to science in 1910 (Auffenburg 1981).

The giant lizard is endemic to five islands in Eastern Indonesia, existing nowhere else in the world. The volcanic islands are inhabited by a population of around 5700 lizards, whose appearance and aggressive behavior have led to them being called "Komodo dragon" (Purwandana et al. 2014) (Fig. 80.3).

80.3 Conservation Challenges

The park receives strong support from the government of Indonesia. As a tourism location known worldwide, the Indonesian government has a specific program for ecotourism management to promote the park at the international level and to ensure sustainability of tourism activities. Nevertheless, illegal fishing and poaching pose important threats to the values of the World Heritage site (UNESCO WHC 2016).

Climate change and global warming threaten several features of this site. Increased CO_2 concentrations and sea surface temperature threaten coral reefs. Sea level rise and changes in air temperature will affect sea turtle nesting beaches and hatching of turtle eggs. Furthermore, sea level rise will threaten the conservation of mangrove forests (UNESCO WHC 2009).

The IUCN (International Union for Conservation of Nature), in its last update about the state of conservation of the World Heritage sites (IUCN 2017) published at the end of 2017, found that Komodo National Park's state of conservation presents significant concerns (Osipova et al. 2017). They found that the exceptional

Fig. 80.3 A Komodo dragon facing the ocean in a sandy and pebble beach in Komodo National Park, Indonesia. (Picture reproduced by permission of Adhi Rachdian/Shutterstock Stock Photos)

land- and seascape beauty of the site remains well-preserved and largely intact but that the marine area of the site is being impacted by a number of threats, including fishing, boat anchoring, and pollution (IUCN 2017).

References

Auffenburg W (1980) The herpetofauna of Komodo, with notes on adjacent areas. Bull Fla State Mus 25:136–137

Auffenburg W (1981) The Behavioural ecology of the Komodo monitor. University of Florida, Gainsville

Daly MC, Hooper BGD, Smith DG (1987) Tertiary plate tectonics and basin evolution in Indonesia. Proceedings of the Indonesian Petroleum Association Annual Convention, Jakarta

Hamilton W (1979) Tectonics of the Indonesian region. US geological survey professional papers 1078, San Diego, USA

Hayes GP, Wald DJ, Johnson RL (2012) Slab1.0: a three-dimensional model of global subduction zone geometries. J Geophys Res 117:B01302. https://doi.org/10.1029/2011JB008524

IUCN (International Union for Conservation of Nature) (2017) Komodo National Park conservation outlook. http://www.worldheritageoutlook.iucn.org/node/1051. Accessed 27 Dec 2017

Kvalvagnaes K, Halim M (1979) Report on a survey of marine areas of the proposed Komodo National Park. UNDP/FAO nature conservation and wildlife management project FO/INS/78/061. Field report 4, Bogor

Letouzey J, Werner P, Marty A (1990) Fault reactivation and structural inversion. Backarc and intraplate compressive deformations. Example of the eastern Sunda shelf (Indonesia). Tectonophysics 183:341–362

Muhammad A, Katsuichiro G, Nicholas A (2016) Tsunami hazard analysis of future Megathrust Sumatra earthquakes in Padang, Indonesia using stochastic Tsunami simulation. Front Built Dev 2(33):1–19

Osipova E, Shadie P, Zwahlen C, Osti M, Shi Y, Kormos C, Bertzky B, Murai M, Van Merm R, Badman T (2017) IUCN world heritage outlook 2: a conservation assessment of all natural world heritage sites. IUCN, Gland

Purwandana D, Ariefiandy A, Jerilmanyah M, Agantoseno HR, Damien CC, Fortham TA, Jessop TS (2014) Demographic status of Komodo Dragons populations in Komodo National Park. Biol Conserv 171:29–35

Sumardja E (1981) First five national parks in Indonesia: Komodo National Park. Parks 6:4–8

Tsang LLH, Meltzner AJ, Philibosian B, Hill EM, Freymueller JT, Sieh KN (2015) A 15 year slow-slip event on the Sunda megathrust offshore Sumatra. Geophys Res Lett 42(16):6630–6638

UNEP/WCMC (United Nations Environmental Program/World Conservation Monitoring Centre) (2011) Komodo National Park, Indonesia. Report, Cambridge, UK

UNESCO WHC (United Nations Environmental, Scientific and Cultural Organization World Heritage Centre) (2009) Case studies on climate change and world heritage. UNESCO WHC, Paris

UNESCO WHC (United Nations Environmental, Scientific and Cultural Organization World Heritage Centre) (2016) Komodo National Park. http://whc.unesco.org/en/list/609. Accessed 15 Dec 2017

Chapter 81
Lorentz National Park, Indonesia

Abstract Lorentz National Park, located in Indonesian Papua, is one of the world's last great wildernesses, comprising two main biomes, lowland and highland. It is the only protected area to incorporate a continuous transect from snowcap to tropical marine environment, including extensive wetlands in the Arafura Sea between New Guinea and Australia. The mountains result from the collision of two tectonic plates and have a complex geology with glacially sculpted peaks. The lowland, on the other hand, is continually being extended by shoreline accretion. The area also contains fossil sites which provide evidence of the evolution of life in Papua, a high level of endemism and the highest level of biodiversity of the region.

81.1 Introduction

Lorentz National Park, located in Indonesian Papua, is the only protected area to incorporate a continuous transect from snowcap to a tropical sea. It has 23,500 km^2 of area, the largest protected segment of territory in Southeast Asia. The park extends 200 km along the central Pegunungan Maoke mountain range and south across some 120–170 km of wet lowland plain to the Arafura Sea (UNESCO WHC 2017) (Fig. 81.1).

The first preservation act of the area took place in 1919 by the Dutch Colonial Government, which created the Lorentz Nature Monument. In 1956, the protected status was abolished due to conflicts with local people over unresolved land claims. In 1978, the protection was reestablished, with the creation of a strict nature reserve by the Indonesian Government. In 1997, the area was declared a national park by the Ministry of Forestry, and the use of its resources by the local people was restored. The park became a World Heritage site in 1999 (UNEP/WCMC 2011).

Papua New Guinea was first discovered by the West in the seventeenth century. The west half was annexed by the Netherlands in 1848 and became part of Indonesia in 1969 as Irian Barat, then Irian Jaya. In 2002, it became Papua and West Papua (Government of the Republic of Indonesia 1999).

Some of the distinctive tribal cultures existing in the area have lived in isolation for more than 20,000 years, and it is quite possible that there are still communities

V. Claudino-Sales, *Coastal World Heritage Sites*, Coastal Research Library 28,
https://doi.org/10.1007/978-94-024-1528-5_81

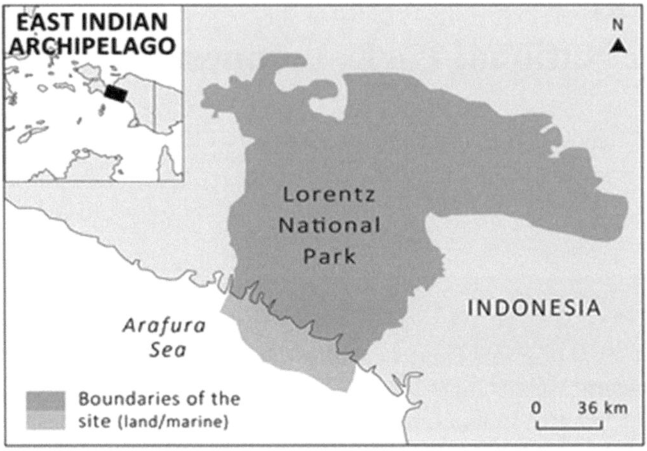

Fig. 81.1 Location of Lorentz National Park, in the Western New Guinea, in Indonesian Papua

which have never had contact with the modern world. They have to a great extent maintained their traditional lifestyles. They use the park's resources for hunting and fishing and gathering, with little impact. They practice rotational agriculture of root crops, mainly taro and sweet potatoes. Pigs play an important role in rituals (Handelsman 2009).

The management of Lorentz National Park World Heritage site is under the authority of the Directorate General of Forest Protection and Nature Conservation of the Ministry of Forestry. The land is public and also owned by the traditional and indigenous people (UNEP/WCMC 2011). The managers are acknowledging the importance of involving indigenous communities in the effective protection of the park and communication channels area being forged with local indigenous organizations. Tourism is not yet an important activity in the site, which is under the pressure of development of infrastructure and disputes between the local communities (UNESCO WHC 2017).

81.2 Natural Settings

81.2.1 Climatic Conditions

Papua lies in the humid tropics with an annual rainfall between 3160 and 4100 mm in the lowlands and between 3500 and 5000 mm in the mountains. West winds prevail between October and March, and from December to March, there are usually high waves on the coast. East winds prevail from April to September (Government of the Republic of Indonesia 1999).

Daytime temperatures range from 29–32 °C in the lowlands and to below freezing above 4800 m. Early morning snow on the summits of Mount Trikora and Mount Jaya occurs regularly down to 3800 m, but permanent snow and ice are found only around the latter. In the mountains, the weather conditions are strongly influenced by the immediate topography (UNEP/WCMC 2011).

81.2.2 Geology and Geomorphology

Eastern Indonesia is one of the most active tectonic regions in the world. Three major crustal plates converge in this area – the Pacific, Eurasia, and Australian Plates. It has caused a complex pattern of subduction zones, volcanic arcs, mountain ranges, and sedimentary basins (Simandjuntak and Barber 1996). The Western Papua and the Arafura Sea are a continental shelf area situated on the northern part of the Australian Plate. To the north, it is bordered by the Tertiary collision zone between the Australian craton and the northern Papua island arc, while to the south, it adjoins the stable Australian craton (Harahap 2012).

The formation of the Papuan Fold Belt was initiated during the collision of the Australian continent and Pacific oceanic plates 24 million years ago (Late Oligocene) (Dow 1997). Most of the deformation is a result of an oblique convergent motion that led to an arc-continent collision between the Australian and Pacific plates (Cloos et al. 2005). Most of the area is made by non-volcanic islands containing fragments of continental crust correlated with the Australian northwestern shelf (Pigram and Panggabean 1981). The resemblance to the Australian margin forms the basis of the generally accepted idea that many islands of eastern Indonesia are the detached and displaced fragments of that margin (Sapiie 2016).

Eastern Indonesia has the most complete stratigraphic succession and contains the oldest rock (Cambrian) in Indonesia (Sapiie 2016). But geologically, most of the area consists of Jurassic-Cretaceous sedimentary and metamorphic rocks, Tertiary limestones and Quaternary alluvium (Harahap 2012) (Fig. 81.2).

The park comprises two main biomes, lowland and highland. The wide, swampy lowlands with mangrove of the southern half border the Arafura Sea between New Guinea and Australia. The highlands in the north are 210 km of the glaciated snow-capped mountain spine of Pegunungan Maoke or Snow Mountains, the highest mountains in Southeast Asia (Petocz 1989).

The park covers various main types of landform: beaches, tidal swamps, meander belts, peat swamps, alluvial valleys, alluvial fans, dissected terraces, hills, mountain slopes, and alpine summits. The descending rivers cut deep valleys through the slopes and foothills. Across extensive alluvial plains, a large number of rivers carry massive amounts of mountain sediments, which build up the shoreline where 5 m tides blend land and sea imperceptibly over a wide area (Petocz 1989).

The Snow Mountains divide geologically and by vegetation type into east and west sectors near Kwiyawagi village, separating the high peaks of the Sudirman Range from Gunung Daam (4922 m) and Trikora Peak (4730 m) in the Jayawijaya

Fig. 81.2 Hill made up of Tertiary limestones in the coastal area of Lorentz National Park. There is a strait between the limestone and the Quaternary sandy island of the left, which is colonized by tropical coastal vegetation. (Picture reproduced by permission of By Igor Tichonow/Shutterstock Stock Photos)

Range. Rising and falling sea levels during the glacial and interglacial periods of the Pleistocene and continuous tectonic activity created the present great topographic variety of mountain ridges, valleys, outwash slopes, plateau, escarpment, moraines, glaciated valleys, glacial lakes, and outwash fans. Sandstone and limestone karst topography is represented by features such as pinnacle, sinkholes, hills, dolines, and caves, with rivers diving underground (Government of the Republic of Indonesia 1999).

81.2.3 Biodiversity

Fluctuating sea levels and tectonic activity created the conditions for the great floristic biodiversity of the area. The Snow Mountains also form a nearly impassable barrier which has accelerated speciation and produced high levels of endemism. The park has five main vegetation zones, lowland, montane, subalpine, alpine, and nival, and within these at least nine subzones (UNEP/WCMC 2011).

The fauna contains 64 mammal species (Government of the Republic of Indonesia 1999). There are elements of Gondwanan fauna in its 16 marsupial species. At least four species of lemur-like marsupial opossums and five species of tree kangaroo are founded in the site. Twenty-five species of rats and mice and 19 species of bat have also been identified recently, many being new records for Papua (Petocz 1989).

Fig. 81.3 School of fish in the marine area of Lorentz National Park, Indonesian Papua. (Picture reproduced by permission of Mike Veitch/Shutterstock Stock Photos)

The park has 650 bird species, 72% of the total number of Papuan birds. Ninety species of amphibians were recorded. Two species of crocodiles also occur, and it is estimated that more than 100 species of freshwater fish species live in the park's waters. Green and hawksbill turtles feed in the seagrass beds offshore, and the Arafura Sea harbors an important biodiversity (Fig. 81.3). Besides the present-day flora and fauna, extensive fossils of Ice Age plants and animals found in mountain caves provide evidence of the evolution of life on the island (Government of the Republic of Indonesia 1999).

81.3 Conservation Challenges

The extensive size of the site is one of the guarantees ensuring the integrity of the habitats it hosts. Nevertheless, several threats need to be addressed to ensure its protection, such as development pressures, including plans for road construction, boundary demarcation, mining activity, petroleum exploration, illegal logging, impacts from human residents, forest dieback caused by natural disease, and limited management capacities and resources. To assist in addressing and overcoming these problems, the "Friends of Lorentz" initiative has been endorsed by a broad range of national and international partners (UNESCO WHC 2017).

Climate change and global warming are also important threats, especially to the tropical forests and to the snowcapped mountains. Ice has been retreating in the last decades. The area was still covered by 13 km^2 of ice caps in 1936 which melted to just 6.9 km^2 by 1972 and had further shrunk to 3 km^2 by 1999, a trend that will almost certainly continue (UNEP/WCMC 2011).

The International Union for Conservation of Nature (IUCN), in its last conservation outlook about natural World Heritage sites, found that management is ill-prepared to deal with the threats and its ineffectiveness augurs poorly for protection of the park and its values (IUCN 2017). The level of conservation of the park is considered to present "significant concerns" (Osipova et al. 2017).

References

Cloos M, Sapiie B, van Ufford AQ, Weiland RJ, Warren PQ, McMahon TP (2005) Collisional delamination in New Guinea: the geotectonics of subducting slab breakoff. Geol Soc Am Spec Pap:400

Dow DB (1997) A geological synthesis of Papua New Guinea. Bureau of Mineral Resources Bulletin, 201. Jakarta, Indonesia

Government of the Republic of Indonesia (1999) Submission for nomination of the Lorentz National Park to the World Heritage List. Report, Jakarta, Indonesia

Handelsman S (2009) Realities and perceptions of human rights in the mining industry – a case study. PhD thesis, University of British Columbia, Canada

Harahap BH (2012) Tectnostratigraphy of the Southern part of Papua and Arafura Sea, Eastern Indonesia. Indones J Geosci 7:167–187

IUCN (International Union for Conservation of Nature) (2017) Lorentz National Park conservation outlook. http://www.worldheritageoutlook.iucn.org/node/1095. Accessed 29 Dec 2017

Osipova E, Shadie P, Zwahlen C, Osti M, Shi Y, Kormos C, Bertzky B, Murai M, Van Merm R, Badman T (2017) IUCN world heritage outlook 2: a conservation assessment of all natural World Heritage Sites. Gland, Switzerland

Petocz R (1989) Conservation and development in Irian Jaya: a strategy for rational resource utilization. EJ Brill Editions, Leiden

Pigram CJ, Panggabean H (1981) Pre-tertiary geology of western Papua and Misool Island: implication for the tectonic development of eastern Indonesia. In: Proceedings of 1st annual convention of Indonesian Petroleum Association, Jakarta, pp 385–399

Sapiie B (2016) Kinematic analysis of fault-slip data in the Central Range of Papua, Indonesia. Indones J Geosci 3:1–16

Simandjuntak TO, Barber AJ (1996) Tonic styles in the Neogene orogenic belts of Indonesia. In: Hall R, Blundell D (eds) Tectonic evolution of Southeast Asia, Geological Society Special Publication, vol 106. Geological Society, London, pp 185–201

UNEP/WCMC (United Nations Environmental Program/World Conservation Monitoring Centre) (2011) Lorentz National Park, Indonesia. Report, Cambridge, England

UNESCO WHC (United Nations Educational, Scientific and Cultural Organization World Heritage Centre) (2017) Lorentz National Park. http://whc.unesco.org/en/list/955. Accessed 29 Dec 2017

Chapter 82
Tropical Rainforest Heritage of Sumatra, Indonesia

Abstract The site is made up of three national parks that contain some of the richest and most varied of the world's remaining rainforests. They are in forested volcanic mountains, ranging from coastal lowlands to highlands, having a wide range of soils, hydrological conditions, and habitats from marine to subalpine volcanoes, supporting a high diversity of flora and fauna. They retain their importance as critical refugia for future evolutionary processes. The site holds the greatest potential for long-term conservation of the distinctive and diverse biota of Sumatra. The protected area is home to an estimated 10,000 plant species and more than 200 mammal species. Of the mammal species, 22 are Asian, not found elsewhere in the archipelago.

82.1 Introduction

The serial site comprises three national parks (Gunung Leuser National Park, Keinci Seblat National Park, and Bukit Barisan Selatan National Park) widely separated along the prominent main spine of the Barisan Mountains, known as the "Andes of Sumatra." The range runs down the western side of the Island of Sumatra, straddling the Equator from north to south. The area of the site is 25,000 km^2, ranging from coastal lowlands to highlands (UNESCO WHC 2017) (Fig. 82.1).

The preservation of the areas that form the site started in 1914. During the twentieth century, many acts of protection were signed by the Indonesian Government. In the beginning of the 1980s, all three areas were already designated as national parks. They became a World Heritage site in 2004. In 2011, all three parks entered the list of World Heritage sites in danger (UNEP/WCMC 2011), where they still remain (UNESCO WHC 2017).

Sumatra has a great variety of ancient and sophisticated ethnic tribal groups and subgroups. There are many different tribes in the national parks, with their own languages and cultures. Most farm various crops, living in longhouses with traditional clothing and dances, but are being challenged by outside influences. Ten sites with tools of stone and wood 10,000 years old have been discovered, proving long settlement of the island. The parks are inhabited only by these tribal groups, but the population is expanding very fast (Government of the Republic of Indonesia 2003).

V. Claudino-Sales, *Coastal World Heritage Sites*, Coastal Research Library 28,
https://doi.org/10.1007/978-94-024-1528-5_82

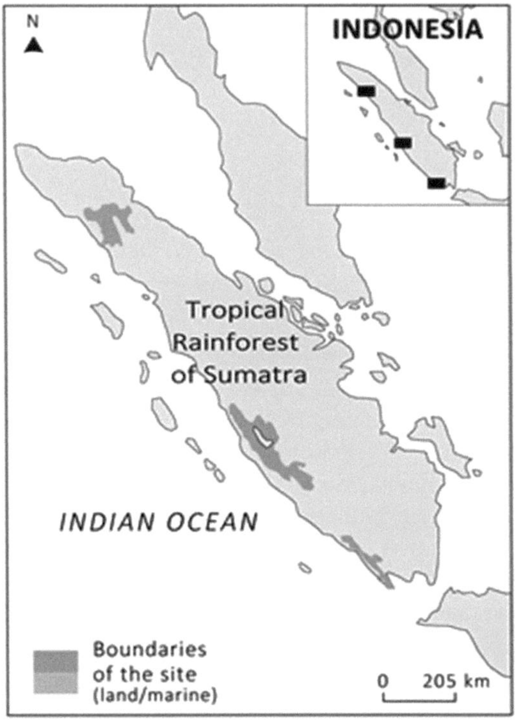

Fig. 82.1 Location of the serial site "Tropical Forest Heritage of Sumatra," in Indonesia

The land is state owned. Each park follows its own policies and is administered by a separate authority under the national Directorate General of Forest Protection and Forest Conservation in the Ministry of Forestry. All three parks have outstanding tourism potential. However, a variety of factors impede its development: inadequate strategic planning, totally inadequate and derelict infrastructure, and poor roads in some places outside the protected areas (UNEP/WCMC 2011).

82.2 Natural Settings

82.2.1 Climatic Conditions

The areas at low altitude belong to the wet tropical climate, while the higher areas belong to the montane or subalpine climate (Government of the Republic of Indonesia 2003). Year-round, the mountains have little varying weather. Temperatures average between 21 °C and 28 ° C and the humidity is always above 60%. The rainfall is an average 3000 mm in the north and 4650 mm in the lowland south. The rainy season is November to May during the wet west monsoon, and dry season is June to August as a result of east monsoon wind (UNEP/WCMC 2011).

The winds are not strong, averaging 5 m/s almost all year long. However, during the monsoons, from December to April, it can double speed. The dominant directions are north and south. The area is subjected to tsunamis, due to the tectonic patterns (Meteoblue 2017).

82.2.2 Geology and Geomorphology

Sumatra, forming the southwestern margin of the Southeast Asian promontory (Sundaland), is considered to be composed of fragments of continental plates and magmatic arcs which were derived from Gondwana during the late Paleozoic and Mesozoic (Barber and Crow 2003).

The area is characterized by the presence of an oblique subduction between the Australian Plate and the Sunda Plate (Asian Plate). The Sunda trench and the Sumatran Fault System are prominent examples of a particular tectonic setting that supports lithospheric-scale strike-slip fault, responsible for the area being highly prone to large, hazardous earthquakes (Fernández-Blanco et al. 2016). The strike-slip Sumatran Fault System accommodates the high-angle oblique subduction of the Australian Plate below the Sunda Plate (Natawidjaja et al. 2006).

The topographic and prominent geological and geomorphological feature response of this subduction in the serial site area is the presence of the Barisan Mountains. This mountain range is composed largely of sedimentary rocks uplifted at the same time as the Himalayas about 70 million years ago, being characterized by the presence of limestone, shale, schist, sandstone, quartzite, and breccias. Three million years ago, a period of faulting created the central rift valley (Semangko Zone) which parallels the Barisan on the east, almost dividing the island in two segments (Berglar et al. 2017). Relatively recent magmatic activity associated with tectonic movement created volcanoes still active today, including the volcano Gunung Kerinci at 3805 m, the highest peak in Sumatra and highest volcano in Indonesia (Pesicek et al. 2008).

Many soil types in the humid climate parks result from these rocks, especially in Gunung Leuser National Park, where brown and red-yellow podzols with lithosols and latosols predominate. In Keinci Seblat National Park, the dominant soils, podzols, are erodable and infertile. Latosols in the lowlands are somewhat more fertile, and in the volcanic uplands, moderately to highly fertile andosols are common. Most of Bukit Barisan Selatan National Park has easily eroded rather infertile podzols (UNEP/WCMC 2011).

In the geomorphological point of view, with its chains of hills and mountains, the area is marked by steep slopes in more than 70% of their area. It has the magnificence of the volcanoes, coastal and glacial lakes in natural forested settings, fumaroles belching smoke from forested mountains, numerous waterfalls, and cave systems in lush rainforest setting, combined with the spectacular beautiful Lake Gunung Tujuh, the highest lake in Southeast Asia. These features emphasize the outstanding beauty of the serial site (UNECO WHC 2017).

Fig. 82.2 Shoreline with corals, pebbles, rainforest, and coastal vegetation in the coastal area of Bukit Barasan Salasan National Park, part of the Tropical Rainforest Heritage of Sumatra World Heritage site. (Picture reproduced by permission of George Martinus/Shutterstock Stock Photos)

Two parks have segments of coast: Gunung Leaser National Park (very small coastal segment) and Bukit Barasan Selatan National Park, in the southeast promontory. The rainforest goes until the coastal plains of thc two parks, which have fringing corals and are made up of coarse grains and pebbles transported by the top of the high mountains by catastrophic slides and rivers during the rain period of monsoons (Government of the Republic of Indonesia 2003) (Fig. 82.2).

82.2.3 Biodiversity

This once vast island of tropical rainforest, in the space of only 50 years, has been reduced to isolated remnants including those centered on the three nominated sites. They contain some of the richest and most varied of the world's remaining rainforests. All three parks would undoubtedly have been important climatic refugia for species over evolutionary time and have now become critically important refugia for future evolutionary processes (UNESCO WHC 2017).

In general, the characteristics of the vegetation and floristic composition of Sumatra show similarities to those of the rest of West Malaysia. All three protected areas exhibit altitudinal zonation of vegetation, from lowland rainforest to montane forest, extending to subalpine low forest, scrub, and shrub thickets. They cover a large diversity of ecosystems (Government of the Republic of Indonesia 2003).

Fig. 82.3 Remnant segment of rainforest in the Tropical Rainforest Heritage Sumatra World Heritage site, with tall, large, and well-developed specimens of trees. (Picture reproduced by permission of feathercollector/Shutterstock Stock Photos)

The parks that comprise the Tropical Rainforest Heritage of Sumatra are areas of very diverse habitat and exceptional biodiversity. Collectively, the three sites include more than 50% of the total plant diversity of Sumatra. There are an estimated 10,000 species of plants, including 92 endemic genera. The heritage site also contains populations of both the world's largest flower and the tallest flower (*Rafflesia arnoldii* and *Amorphophallus titanum*, respectively) (UNEP/WCMC 2011).

The relict lowland forests in the sites are very important for conservation of the biodiversity, as well as the montane forests. They retain their importance as critically refugia for future evolutionary processes. The site holds the greatest potential for long-term conservation of the distinctive and diverse biota of Sumatra (UNESCO WHC 2017) (Fig. 82.3).

Animal diversity in Sumatra is impressive, with 201 mammal species in the World Heritage site. There are some 580 bird species, 465 being resident, and 21 endemic (Hitchcock and Meyers 2006). The mammalian fauna of Sumatra, which includes some mainland Asian species, is comparable in diversity with those of much larger Borneo and New Guinea. Twenty-two are Asian species not found elsewhere in the Indonesian archipelago, and 15 are confined to the region, including the endemic Sumatran orangutan (IUCN 2008). There are large animals in the rainforest, including the orangutan, the Sumatran tiger, the Sumatran elephant, and the Sumatran rhinoceros (Government of the Republic of Indonesia 2003).

82.3 Conservation Challenges

The unique shape and size of the serial site provide significant habitat for in situ conservation. Biological and ecological processes are preserved because the area contains a sufficient number of ecosystems, forest types, ranges of altitude, and topographies. However, a number of situations are threatening the parks, such as agricultural encroachment and road development. The roads provide access to the protected areas, which facilitates illegal logging, encroachment, and poaching (UNESCO WHC 2017). These situations justify the permanence of the site in the list of World Heritage sites in danger for the last 6 years.

The International Union for Conservation of Nature (IUCN) published a conservation outlook of the natural World Heritage sites at the end of 2017 in which the Tropical Rainforest Heritage of Sumatra site appears as being in a critical stage of conservation (Osipova et al. 2017). Recent official reports indicate that environmental crimes within the property continue and may even be increasing. The damage occurring to lowland tropical rainforests and the habitat of species such as orangutans, rhinoceros, tigers, and elephants is therefore continuing at a high level. There is no authoritative evidence that populations of key species have stabilized. Management and law enforcement are totally inadequate (IUCN 2017).

Climate changes and global warming probably will also degrade the environment. It has been considered that tropical wet areas will probably get more rain in the context of climate change, and it may interfere in the fauna and flora of the parks. The corals and beaches of the small segment of coast of the serial site may also suffer with bleaching and increasing erosion, respectively.

References

Barber AJ, Crow MJ (2003) An evaluation of plate tectonics models for the development of Sumatra. Gondwana Res 6:1–28

Fernández-Blanco D, Philippon M, von Hagke C (2016) Structure and kinematics of the Sumatran fault system in North Sumatra (Indonesia). Tectonophysics 693:453–464

Government of the Republic of Indonesia (2003) Submission for nomination of the tropical rainforest heritage of Sumatra to the World Heritage list. Report, Jakarta, Indonesia

Hitchcock P, Meyers K (2006) Report on the IUCN-UNESCO world heritage monitoring mission to the tropical rainforest heritage of Sumatra, Indonesia. IUCN/UNESCO, Gland, Switzerland

IUCN (International Union for Conservation of Nature) (2008) State of conservation report tropical rainforest heritage of Sumatra (Indonesia). Report, Gland, Switzerland

IUCN (International Union for Conservation of Nature) (2017) Tropical rainforest heritage of Sumatra conservation outlook. http://www.worldheritageoutlook.iucn.org/node/1129. Accessed 30 Dec 2017

Meteoblue (2017) Climate of Sumatra. https://www.meteoblue.com/en/weather/forecast/modelclimate/sumatra_united-states-of-america_4174274. Accessed 30 Dec 2017

Natawidjaja D, Sieh K, Chlieh M, Galetzka J, Suwargadi B, Cheng H, Edwards R, Avouac JP, Ward S (2006) Source parameters of the great Sumatran megathrust earthquakes of 1797 and 1833 inferred from coral microatolls. J Geophys Res 111:B06403

Osipova E, Shadie P, Zwahlen C, Osti M, Shi Y, Kormos C, Bertzky B, Murai M, Van Merm R, Badman T (2017) IUCN world heritage outlook 2: a conservation assessment of all natural World Heritage Sites. Gland, Switzerland

Pesicek J, Thurber C, Widiyantoro S, Engdahl E, DeShon H (2008) Complex slab subduction beneath northern Sumatra. Geophys Res Lett 35:L20303

UNEP/WCMC (United Nations Environmental Programme/World Conservation Monitoring Centre) (2011) Tropical rainforest heritage of Sumatra, Indonesia. Report, Cambridge, England

UNESCO WHC (United Nations Educational, Scientific and Cultural Organization World Heritage Centre) (2017) Tropical rainforest heritage of Sumatra. http://whc.unesco.org/en/list/1167. Accessed 30 Dec 2017

Chapter 83
Ujung Kulon National Park, Indonesia

Abstract Ujung Kulon National Park is in the extreme southwestern tip of Java, on the Sunda shelf. It includes the Ujung Kulon Peninsula and several offshore islands, besides the nature reserve of Krakatoa Volcano. The site is of great interest for the study of volcanoes and also contains the largest remaining area of lowland rainforests in the Java plain. The site possesses a very diverse flora and fauna, demonstrating ongoing evolution processes since Krakatoa erupted in 1883, which became the most well known and studied of all modern volcanic eruptions. The physical feature of Krakatoa Island combined with the surrounding sea, succession of vegetation, and volcanic activity combine to form a landscape of exceptional beauty. Several species of endangered plants and animals can be found there, the Javan rhinoceros being the mostly seriously threatened.

83.1 Introduction

The Ujung Kulon National Park is a peninsula in the Sunda Strait between Java and Sumatra at the extreme southwestern tip of Java. The western boundary encloses the Ujung Kulon Peninsula, offshore islands, and the Krakatoa Nature Reserve some 60 km north across the Sunda Straits. The eastern boundary runs along the eastern foothills of the Gunung Honje mountain range. The site is 785 km^2 (UNESCO WHC 2018) (Fig. 83.1).

In 1921, the Krakatoa area was declared a nature reserve. In 1937, two more natural reserves were established in the surroundings. In 1958, the Ujun Kulon nature area was created. In 1992, Ujung Kulon Peninsula, Panaitan Island, the Krakatoa Islands' nature reserves, and the other reserves already existing were merged into a national park. The national park became a World Heritage site in 1991 (UNEP/WCMC 2011).

The land is a property of the Indonesian Government, and the site is administered by the Ujung Kulon National Park Authority, under the Directorate of Nature Conservation of the Ministry of Forestry. The Krakatoa Nature Reserve is managed separately by Nature Resource Conservation, Sumatra (UNESCO WHC 2018).

V. Claudino-Sales, *Coastal World Heritage Sites*, Coastal Research Library 28,
https://doi.org/10.1007/978-94-024-1528-5_83

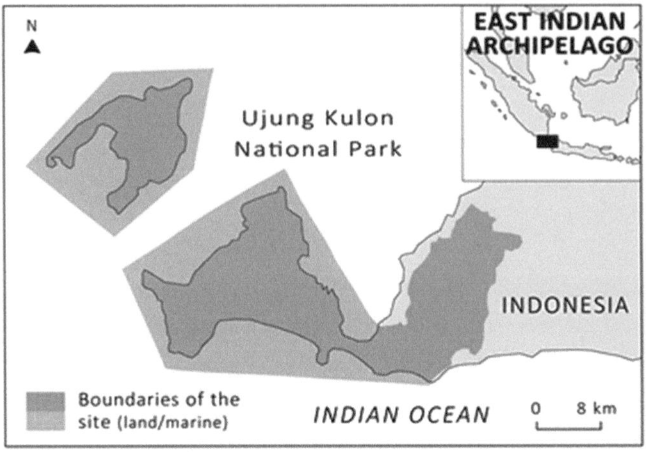

Fig. 83.1 Location of the Ujung Kulon National Park, in the Sunda Strait, southwestern tip of Java Island, Indonesia

The occupation of the area is ancient, considering that archaeological relic from the first century has been found in one of the islands that form the national park. Natives traditionally used the area for hunting, fishing and gathering, now all illegal within the park (Government of the Republic of Indonesia 1990). There are many villages surrounding the site, but the island of Krakatoa is uninhabited. Tourism is increasing in the site, which is especially visited by foreigners (UNESCO WHC 2018).

83.2 Natural Settings

83.2.1 Climatic Conditions

The climate is tropical maritime, with a mean annual rainfall of approximately 3250 mm. The heaviest rainfall is between October and April during the northwest monsoon. A noticeably drier period occurs between May and September during the southeast monsoon. Mean temperatures range between 25 °C and 30 °C, and the relative humidity ranges between 65% and 100% (Blower and van der Zon 1977; Hommel 1987).

The climate in Java is strongly influenced by monsoon trade winds and the waves of La Niña and El Niño. During the rainy season, the weather is dominated by west winds, particularly in the north coast of the island (Java Indonesia Organization 2017). Typhoons can hit the islands between September and December, causing rainstorms and heavy winds (Weather Online 2017).

83.2.2 Geology and Geomorphology

The Island of Java is dominated by a subduction-induced volcano-plutonic back-arc with the prominent presence of a fold and thrust belt resulting from the local subduction between the Indian-Australian Plate and the Asian plate (Sunda Shelf) (Queisser et al. 2017). The Ujung Kulon National Park is characterized by volcanos and by a Tertiary mountain system linked to this subduction (Bohm et al. 2013).

Ujung Kulon National Park terrains may have formed the southernmost extension of the Sumatran Bukit Barisan Mountains during the Pleistocene, having become separated from Sumatra on the collapse of the Sunda Strait dome. Central and eastern Ujung Kulon comprises raised Miocene limestone formations overlain in the north by alluvial deposits and in the south by sandstones. To the west, the Gunung Payung Massif is of Miocene sedimentary origin, and Pulau Panaitan Island is similar although volcanic material and breccias form outcrops in the northwest (Hommel 1987).

Krakatoa lies on the edge of the tectonically active Sunda Shelf. It is the remnant of a single andesitic volcano which exploded and collapsed some 1500 years ago, leaving three remnant cones. These eventually coalesced into an island, Krakatoa, which erupted in 1883, expelling some 18–21 km^3 of material and killing more than 36,000 people in the immediate area (Government of the Republic of Indonesia 1990). Volcanic activity lasted 40 years, resulting in 1930 in the emergence of Anak Krakatoa ("child of Krakatoa"), which is now 181 m high and 2 km across (Gardner et al. 2013).

The area continues to experience volcanic activity, with significant eruptions taking place in 1952, 1972, 1992, and 1994. The Krakatoa Volcano is the best well known and studied of all modern volcanic eruptions, due primarily to the devastating effects registered throughout the northern hemisphere (UNESCO WHC 2018) (Fig. 83.2).

83.2.3 Biodiversity

The site includes the Ujung Kulon Peninsula and several offshore islands of natural beauty and demonstrating evolution of geological processes since the Krakatoa eruption of 1883, creating different environments. The combination of natural vegetation of the lowlands, tropical rainforest, grasslands, beach forest, mangrove forest, and coral reefs within the site is of exceptional magnificence (UNEP/WCMC 2011) (Fig. 83.3).

Ujung Kulon National Park has the best and most extensive lowland forest on the Island of Java, a habitat that has virtually disappeared elsewhere in the area and is under severe pressure in Indonesia and Southeast Asia. The Peninsula of Ujung Kulon provides invaluable habitat critical for the survival of a number of threatened plants and animals (UNESCO WHC 2018).

Fig. 83.2 Krakatoa Volcano, in the Island of Krakatoa, part of the Ujung Kulon National Park. The image was shot in the second half of 1993, just before the eruption of 1994. (Picture reproduced by permission of Juan Carlos Munoz/AGE Fotostock/Visualphotos)

Fig. 83.3 Peucang Island, with quartz sand forming dissipative beaches. Tertiary limestone colonized by lowland tropical rainforest and coastal forest forms the rest of the scenery. In Ujung Kulon National Park, Indonesia. (Picture reproduced by permission of MikeMarcs/Shutterstock Stock Photos)

The vegetation is semi-evergreen rainforest, which has suffered a number of anthropogenic and natural modifications. It is mainly secondary growth, following the destructive eruption of Krakatoa and the associated tsunami of 1883. It is dominated by palms and wild sugarcane. The palms which grow on thick ash may be dominant as a result of long-past volcanic disturbance. For this reason, the natural vegetation cover, primary lowland rainforest, now occupies only 50% of the total area. Other vegetation types are beach forest, mangrove, and swamp. At least 50 species of rare plants are present (UNEP/WCMC 2011).

The vegetation of the Krakatoa group is characterized by a number of different stages of succession. The volcanically active island of Anak Krakatoa is characterized by vegetation in the early stages of succession following effective sterilization by the eruption of 1952 (Thornton et al. 1990). The eruptions between 1992 and 1994 again severely retarded the colonization of Anak Krakatoa (Whittaker et al. 1989).

Ujung Kulon probably has the most complete surviving fauna on Java (Whitten et al. 1996) and is the last remaining viable natural refuge in the world for the Javan rhinoceros, for which the most recent estimate indicates a small, fairly stable total population of 60 individuals (UNESCO WHC 2018).

Primates include various endemic monkeys. Carnivores include Asiatic wild dog, Eurasian otter, Asian small-clawed otter, Javan mongoose, Javan leopard, and fishing cat. Several ungulates range within the park, with a population of around 700 on Ujung Kulon Peninsula. Others include wild boar, barking deer, lesser mouse deer, and rusa deer (Whitetaker et al. 1989).

A rich avifauna of more than 270 species is recorded. Terrestrial reptiles and amphibians include two species of python, estuarine crocodile, and many frogs. Various kinds of turtles are also present (UNEP/WMCM 2011). The marine areas of Ujung Kulon National Park also support some of the richest fish fauna in the archipelago, with both deep water and reef species well represented (Halim and Kvalvagnaes (1980).

83.3 Conservation Challenges

The long history of conservation of the site has helped to protect the place, and it contains all the necessary habitat for the in situ conservation of its unique biological diversity (UNEP/WCMC 2011). Implementation of the management plan has helped to control the problems of illegal encroachment, logging, and commercial fishing within the boundaries of the site, although it continues to pose a threat to the conservation of the park. Impacts from tourism are a new problem, but poaching of the Javan rhino has historically been the main management issue within the property (UNESCO WHC 2018). The current situation of the Javan rhinoceros as a species is considered to be critical, and Ujung Kulon National Park remains a stronghold and the only hope for the survival of this species (IUCN 2017).

The state of conservation of the park is considered to be "good with some concerns" (Osipova et al. 2017). The concerns are related to the situations pointed out above. Climatic change and global warming are also a problem, because it can change the environment and threaten the fauna and flora. Corals and the coastal area would suffer the most impact with these changes.

References

Blower J, van der Zon A (1977) Ujung Kulon National Park Management Plan 1977–1981. Field report. UNDP/FAO Nature Conservation & Wildlife Management Project INS/73/013. FAO, Bogor, Indonesia

Bohm M, Haberland C, Asch G (2013) Imaging fluid related subduction processes beneath Central Java (Indonesia) using seismic attenuation tomography. Tectonophysics 590:175–188

Gardner MF, Troll VR, Gamble JA, Gertisser R, Hart GL, Ellam RM, Harris C, Wolff JA (2013) Crustal differentiation processes of Krakatoa Volcano, Indonesia. J Petrol 54(1):149–182

Government of the Republic of Indonesia (1990) Nomination of Ujun Kulon National Park to the list of World Heritage Sites. Report, Jakarta, Indonesia

Halim M, Kvalvagnaes K (1980) Marine resources of the proposed Ujung Kulon National Park. Field report. UNDP/FAO National Park Development Project INS/78/061. FAO, Bogor, Indonesia

Hommel P (1987) Landscape-ecology of Ujung Kulon (West Java, Indonesia). Soil Survey Institute, Wageningen

IUCN (International Union for Conservation of Nature) (2017) Ujung Kulon National Park conservation outlook. http://www.worldheritageoutlook.iucn.org/node/1052. Accessed 2 Jan 2018

Java Indonesia Organization (2017) Climate, weather and temperature in Java Indonesia. http://www.javaindonesia.org/general/climate-weather-temperature-java-indonesia/. Accessed 2 Jan 2018

Osipova E, Shadie P, Zwahlen C, Osti M, Shi Y, Kormos C, Bertzky B, Murai M, Van Merm R, Badman T (2017) IUCN world heritage outlook 2: a conservation assessment of all natural World Heritage Sites. Gland, Switzerland

Queisser M, Burton MR, Arzilli F, Chiarugi A, Marliyani GI, Anggara F, Harijoko A (2017) CO_2 flux from Javanese mud volcanism. J Geophys Res Solid Earth 122(6):4191–4207

Thornton I, Zann R, Stephenson D (1990) Colonization of the Krakatau Islands by land birds, and the approach to an equilibrium number of species. Philos Trans R Soc Lond B 328:55–94

UNEP/WCMC (United Nations Environmental Programme/World Conservation Monitoring Centre) (2011) Ujung Kulon National Park. Report, Cambridge, England

UNESCO WHC (United Nations Educational, Scientific and Cultural Organization World Heritage Centre) (2018) Ujung Kulon National Park. http://whc.unesco.org/en/list/608. Accessed 2 Jan 2018

Weather Online (2017) Indonesia. https://www.weatheronline.co.uk/reports/climate/Indonesia.htm. Accessed 2 Jan 2018

Whittaker R, Bush M, Richards K (1989) Plant recolonization and vegetation succession on the Krakatau Islands, Indonesia. Ecol Monogr 59:59–123

Chapter 84
Tubbataha Reefs Natural Park, the Philippines

Abstract Tubbataha Reefs Natural Park is in the center of the Coral Triangle between Indonesia, Malaysia, and the Philippines. It is the only coral atoll in the Philippines and consists of North and South Atoll, 8 km apart, and Jessie Beazley Reef, 20 km north. The site is the locus of the world's greatest coral diversity and is one of the few remaining examples of a highly diverse near-pristine coral reef, with a spectacular 100 m perpendicular wall, an extensive lagoon, and two coral islands. It is also a unique example of an atoll reef with a very high density of marine species. Its huge assemblages of fish, marine mammals, and seabirds are invaluable to science as a wildlife refuge and field laboratory and a dive site of exceptional quality and beauty.

84.1 Introduction

Tubbataha Reefs Natural Park is in the center of the Coral Triangle between Indonesia, Malaysia, and the Philippines, in the Sulu Sea, East Indian Archipelago. It is the only coral atoll in the Philippines and consists of North and South Atoll, 8 km apart, and Jessie Beazley Reef, 20 km north. The Tubbataha Reef Marine Park covers an area of 1300 km^2 (UNESCO WHC 2018) (Fig. 84.1).

The Tubbataha Reef Marine National Park was established by the Filipino government in 1988. It was designated a Ramsar Wetland site under the Ramsar Wetland Convention in 2000. It became a World Heritage site in 1993, with an extension of the initial area by UNESCO in 2009 (UNEP/WCMC 2011).

The government owns the land. The site is managed by the Tubbataha Management Office for the Tubbataha Protected Area Management Board. Tubbataha is under the political jurisdiction of the island municipality of Cagayancillo (Government of the Philippines 2008).

There are no permanent inhabitants on the Tubbataha Reefs, although during the fishing season, fishermen from Cagayancillo used to visit for a month, gathering turtle and birds' eggs, seabirds, and fish. Patrolling park rangers, navy sailors, and coastguard staff use the ranger station (Government of the Philippines 2008).

V. Claudino-Sales, *Coastal World Heritage Sites*, Coastal Research Library 28,
https://doi.org/10.1007/978-94-024-1528-5_84

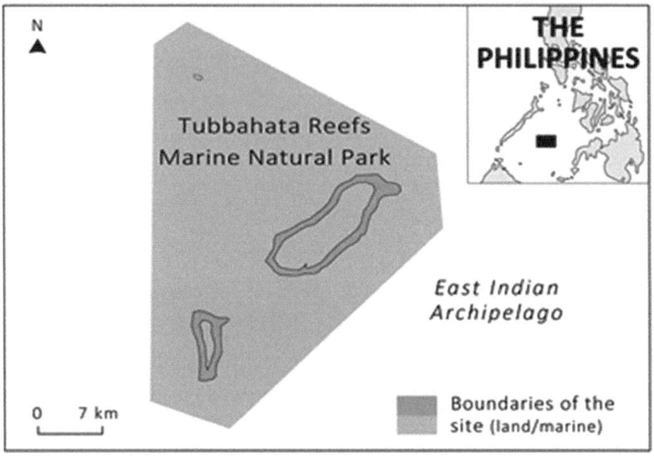

Fig. 84.1 Location of Tubbataha Reefs Natural Park, the Philippines, in the Sulu Sea

A wide range of fisheries is practiced in the Sulu Sea, including traditional line fishing, trawling, spear fishing, offshore long-line fishing, and reef gleaning. The Cagayagon annual fishing expedition to the reefs is also an initiatory experience for young boys (UNEP/WCMC 2011).

The reefs can only be reached by live-aboard boats and only from the end of March to June. Visitors may land only at the ranger station. All other islands are off-limits. Currents are strong and changeable, and strong winds and high waves make mooring very difficult as there is no shelter. However, the site is considered one of the top 10 places to dive in the world, and the number of visitors is rapidly increasing (UNESCO WHC 2018; Jackson 2003).

84.2 Natural Settings

84.2.1 Climatic Conditions

Tubbataha is exposed to two monsoons a year. The seas are rough from June to October during the southwest monsoon and again from November to March during the northeast monsoon. Breaks between monsoons bring a week of calmness, usually before monsoonal shifts. Moderate winds from the northeast are common between April and June (UNEP/WCMC 2011).

Temperatures can reach 35 °C in summer time. Thunderstorms are common in the season. Annual rainfall is 1475 mm on average. The area is rarely visited by typhoons (Palawan Weather 2018).

84.2.2 Geology and Geomorphology

Tubbataha Reefs Natural Park is located in the Sulu Sea, which is around 750 m deep. The Sulu Sea experienced a long history of Mesozoic paleo-Pacific subduction and Cenozoic subduction with interactions with the South China Sea (Li et al. 2012; Li and Li 2007). Different types of terrains (oceanic crust, continent and ocean transition zone, continental block, island arc) amalgamated here, but their boundaries are not well defined, and the origin of some tectonic units remains uncertain, because of complicated rifting and collision processes that have occurred (Franke et al. 2008, 2011).

Two models are presented for the evolution of the Sulu Sea, either as a back-arc basin triggered by the subduction of the proto-South China Sea under the Asian Plate (Rangin and Silver 1991) or a marginal basin analogous to the South China Sea (Rangin and Silver 1991). Moreover, the formation timing of the Sulu Sea, either the Early Miocene (Rangin and Silver 1991) or the Oligocene (Roeser 1991) is also debated (Liu et al. 2014).

Later, in the Miocene (15 million years ago), an ocean ridge of the Sulu Sea collided with the Philippine arc, which represents a complex portion of the tectonic boundary between the Eurasian Plate and the Philippine Sea Plate (Aurelio and Pena 2010). A line of volcanos where formed during this collision (Briais et al. 1993). Tubbataha Reefs developed over these now extinct volcanos (Government of the Philippines 2008).

The reefs consist of two main very different habitats: outer reef slope and lagoon. The North Atoll is an oblong platform 16 km long by 4.5 km wide. The South Atoll is a small triangular-shaped reef about 5 km long by 3 km wide. Jessie Beazley Reef is 5 km long by 3 km wide and becomes a cay at low tide. These are classic reef formations with extensive reef flats, which to seaward fall 100 m below sea level with often perpendicular 40–50 m walls of coral-hung crevices, overhangs, ledges, and caverns (UNEP/WCMC 2011) (Fig. 84.2).

The outer reef slopes have very clear water, strong wave action and currents, and a very wide depth range from about 1 m to over 40 m. They have much higher coral diversity and greater beauty than the lagoons. The lagoons have turbid water, little wave action, few currents, and a depth range from under 1 m to 25 m (Government of the Philippines 2008).

84.2.3 Biodiversity

The reefs, in the Indo-west Pacific faunal province, are in the world center of coral diversity where their remoteness preserved them almost untouched until the 1980s. Despite destructive exploitation since then, the park contains 396 species of corals, which corresponds to almost 90% of all species found in the Philippines, and to 80% of all coral species found in the Sulu Sea. Actually, Tubbataha Reefs have

Fig. 84.2 Different coral species exposed during low tide in the outer edge of the North Atoll, Tubbataha Reefs Natural Park. (Picture reproduced by permission of Dr. Roger G. Dolorosa, Western Philippines University)

higher species diversity per square meter than the Great Barrier Reef (Alinio and Licuanan 2006).

The marine life is very rich. Sharks, turtles, and rays are seen in the lagoons; 11 species of cetaceans and 11 sharks are recorded. Big schools of pelagic fish are common in the outer waters. The area also supports the world's highest densities of whitetip reef sharks. The presence of top predator species such as tiger shark and hammerhead shark validates the ecological balance of the reef. There are 479 species of fish (UNEP/WCMC 2011).

The park is one of the few remaining refuges for a diversity of seabirds in Southeast Asia. Ninety-nine species of birds, resident and migrant, have been recorded on the cays. Seven species of seabirds breed there and are found only in the park. The endangered Christmas Island frigate bird is a regular visitor (UNESCO WHC 2018) (Fig. 84.3).

84.3 Conservation Challenges

The site has an adequate size to ensure the preservation of the features and processes of the reefs. A key aspect of its integrity is the low level of fishing pressure, due to the no-take policies in the area. Nevertheless, the maintenance of these values requires measures in relation to some migratory species and the buffering of the property from threats to the marine environment that could occur in the wider area,

Fig. 84.3 Colonies of seabirds in Bird Islet, North Atoll, Tubbataha Reefs Natural Park. (Picture reproduced by permission of Dr. Roger G. Dolorosa, Western Philippines University)

such as shipping, marine litter, marine pollution, and oil exploration (UNESCO WHC 2018). Besides these threats, this offshore, no-fishing park remained attractive to illegal fishermen in spite of an active park enforcement system (Dolorosa et al. 2013).

The International Union for Conservation of Nature (IUCN) in its last conservation outlook of the natural World Heritage sites published at the end of 2017 found that the state of conservation of the park is good, with some concerns (Osipova et al. 2017). The concerns are related to the threats mentioned above.

The biggest threat to the site cannot be solved at a local level, because it is linked to climate change and global warming. The rising of the sea level associated with global warming puts the corals and the lagoons in danger of completely disappearing, and the increasing of the temperature and acidification of the sea water risks killing the coral reefs. Changes in rain and temperatures are also a source of problem to the fauna. These threats are considered to be very high (IUCN 2017).

References

Alinio P, Licuanan W (2006) Completing the connectivity cycle for adaptive management: coral reef ecosystem based MPA network management chain. Report, Manila, Philippines

Aurelio MA, Pena RE (eds) (2010) Geology of the Philippines, tectonics and stratigraphy, vol 1. Mines and Geosciences Bureau, Department of Environment and Natural Resources, Quezon City, Philippines

Briais A, Patriat P, Tapponnier P (1993) Updated interpretation of magnetic data anomalies and seafloor spreading stages in the South China Sea: implications for the tertiary tectonics of Southeast Asia. J Geophys Res 98:6299–6328

Dolorosa RG, Conales SF, Bundal NA (2013) Status of horned helmet cassis cornuta in Tubbataha Reefs Natural Park, and its trade in Puerto Princesa city, Philippines. Atoll Res Bull 595:1–17

Franke D, Barckhausen U, Heyde I, Tingay M, Ramli N (2008) Seismic images of a collision zone offshore NW Sabah/Borneo. Mar Petrol Geol 25:606–624

Franke D, Barckhausen U, Baristeas N, Engels M, Ladage S, Lutz R, Montano J (2011) The continent-ocean transition at the southeastern margin of the South China Sea. Mar Pet Geol 28:1187–1204

Government of the Philippines (2008) Re-nomination of Tubbataha Reefs Natural Park to the World Heritage List. Report, Cagayancillo, The Philippines

IUCN (International Union for Conservation of Nature) (2017) Tubbataha Reefs Natural Park conservation outlook. http://www.worldheritageoutlook.iucn.org/node/1062. Accessed 3 Jan 2018

Jackson J (2003) Top dive sites of the world. New Holland Publishers, London

Li Z-X, Li X-H (2007) Formation of the 1300-km-wide intracontinental orogen and post-orogenic magmatic province in Mesozoic South China: a flat-slab subduction model. Geology 35:179–182

Li X-H, Li Z-X, He B, Li W-X, Li Q-L, Gao Y, Wang X-C (2012) The early a collision zone offshore NW Sabah/Borneo: in situ U-Pb, Lu-Hf and O isotope analyses of detrital zircons. Chem Geol 328:195–207

Liu W-N, Li C-F, Li J, Fairhead D, Zhou Z (2014) Deep structures of the Palawan and Sulu Sea and their implications for the opening of the South China Sea. Mar Pet Geol 58:721–735

Osipova E, Shadie P, Zwahlen C, Osti M, Shi Y, Kormos C, Bertzky B, Murai M, Van Merm R, Badman T (2017) IUCN world heritage outlook 2: a conservation assessment of all natural World Heritage Sites. Gland, Switzerland

Palawan Weather (2018) Palawan Weather. https://www.accuweather.com/en/ph/plw/palawan-weather. Accessed 03 Jan 2018

Rangin C, Silver EA (1991) Neogene tectonic evolution of the Celebes-Sulu basins: new insights from Leg 124 drilling. In: Proceedings ocean drilling program scientific results, vol 124. Texas A&M University, College Station, pp 51–63

Roeser H (1991) Age of the crust of the southeast Sulu Sea basin based on magnetic anomalies and age determined at site 768. In: Proceeding ocean drilling program scientific results, vol 124. Texas A&M University, College Station, pp 343–339

UNEP/WCMC (United Nations Environmental Programme/World Conservation Monitoring Centre) (2011) Tubbataha Reefs Natural Park, Philippines. Report, Cambridge, England

UNESCO WHC (United Nations Educational, Scientific and Cultural Organization World Heritage Centre) (2018) Tubbataha Reefs Natural Park. http://whc.unesco.org/en/list/653. Accessed 3 Jan 2018

Correction to: Gulf of Porto: Calanche of Piana, Gulf of Girolata, Scandola Reserve, France

Correction to:
Chapter 73 in: V. Claudino-Sales, Coastal World Heritage Sites, Coastal Research Library 28, https://doi.org/10.1007/978-94-024-1528-5_73

The names of two authors in the reference section of this chapter, namely Martin D and Balazs S, were published incorrectly. The authors' names have been corrected to read as Danišík M and Székely B.

The updated online version of the chapter can be found at
https://doi.org/10.1007/978-94-024-1528-5_73

V. Claudino-Sales, *Coastal World Heritage Sites*, Coastal Research Library 28,
https://doi.org/10.1007/978-94-024-1528-5_85

Index

V. Claudino-Sales, *Coastal World Heritage Sites*, Coastal Research Library 28,
https://doi.org/10.1007/978-94-024-1528-5

B

F

G

J

K

P

MIX
Papier aus verantwortungsvollen Quellen
Paper from responsible sources
FSC® C105338

If you have any concerns about our products,
you can contact us on
ProductSafety@springernature.com

In case Publisher is established outside the EU,
the EU authorized representative is:
Springer Nature Customer Service Center GmbH
Europaplatz 3, 69115 Heidelberg, Germany

Printed by Libri Plureos GmbH
in Hamburg, Germany